CONSTRUCTING QUANTUM MECHANICS

Constructing Quantum Mechanics
Volume One

The Scaffold: 1900–1923

Anthony Duncan and Michel Janssen

OXFORD
UNIVERSITY PRESS

OXFORD

UNIVERSITY PRESS

Great Clarendon Street, Oxford, OX2 6DP,
United Kingdom

Oxford University Press is a department of the University of Oxford.
It furthers the University's objective of excellence in research, scholarship,
and education by publishing worldwide. Oxford is a registered trade mark of
Oxford University Press in the UK and in certain other countries

© Anthony Duncan and Michel Janssen 2019

The moral rights of the authors have been asserted

First Edition published in 2019

Published in the United States of America by Oxford University Press
198 Madison Avenue, New York, NY 10016, United States of America

British Library Cataloguing in Publication Data
Data available

Library of Congress Control Number: 2019909012

ISBN 978–0–19–884547–8

DOI: 10.1093/oso/9780198845478.001.0001

Printed and bound by
CPI Group (UK) Ltd, Croydon, CR0 4YY

Preface

How did physicists in the first three decades of the last century come to recognize that they needed a new framework to account for a growing list of phenomena that they could not account for with what we now call classical physics? That is the central question of this book. To answer it, we analyze what we have identified as the main strands in the development of this new framework. This first volume takes us from the late 1890s to the early 1920s. The second volume will take us from the early to the late 1920s. Our starting point can be compared to that of someone first encountering an arch. This person will wonder how it was built and come to realize that it was done with the help of a scaffold. If we think of quantum mechanics as an arch, we can likewise say that it was put together with the help of a scaffold, provided in this case by elements of the theory it replaced. By using the arch-and-scaffold metaphor in their subtitles, we acknowledge the backwards-looking perspective we are adopting in these volumes.[1] Decisions about which strands in the development of quantum mechanics we cover in detail and which ones we treat more cursorily (if at all) were informed by our assessment, in hindsight, of their relative importance for the scaffold on which the arch was erected. However, our analysis of the strands selected for detailed discussion is given entirely in terms of insights and concepts available to the historical actors at the time. To find meaningful historically sensitive answers to our central question, we scrupulously adhered to this methodological principle.

The authors of two previous histories of quantum physics, Max Jammer (1966) and Jagdish Mehra and Helmut Rechenberg (1982a,b,c,d, 1987, 2000–2001), addressed questions similar to ours and faced similar historiographical challenges. Our book, however, is not only more explicit about the approach taken but differs from these earlier efforts in several other respects. A reader familiar with these older histories will notice the most important differences just leafing through the pages of this book and glancing at its table of contents.

First of all, our choice of topics is more selective than either Jammer's or Mehra and Rechenberg's. Without being selective, we would have had no hope of covering the developments we cover at the level of detail we cover them in just two volumes. It took Mehra and Rechenberg, depending on how one counts, six or nine volumes to cover roughly the same period and Jammer only managed to do so in one volume by omitting the derivation of most results.

---

[1] The arch-and-scaffold metaphor for the development of scientific theories is presented and the historiographical issues it raises are discussed in Janssen (2019). The metaphor was inspired by Alexander Graham Cairns-Smith (1985, pp. 58–60), who used it to illustrate his suggestion that the complex nucleotides of RNA and DNA were first assembled on minute clay crystals.

This brings us to a second important difference. We include derivations of all key results at a level that a reader with a command of physics and mathematics comparable to that of an undergraduate in physics should be able to follow without having to take out pencil and paper. We frequently use modern notation and streamline derivations but conceptually we always remain faithful to the historical sources. As we will explain in more detail below, this feature makes our book suitable for classroom use. But first we want to emphasize another way in which our book goes beyond the books by Jammer and Mehra and Rechenberg.

Since the publication of these books, more research on the early history of quantum physics has been published and additional primary source material has become available. In the latter category, we mention the collected papers of Niels Bohr (1972–2008) and Albert Einstein (1987–2018) as well as editions of the scientific correspondence of Arnold Sommerfeld (2000, 2004) and Hendrik Antoon Lorentz (2008, 2018). We have taken full advantage of these primary sources and the annotation provided by the editors. We should also note that it made our work considerably easier that so much material is available on-line these days. This is true not only for many of the primary and secondary sources we used but also for biographical information for all but the most obscure characters in our story. Library support nonetheless remained crucial for our work and we thank the dedicated staff of the libraries of the University of Minnesota, the University of Pittsburgh, and, especially, the Max Planck Institute for History of Science in Berlin.

In terms of secondary sources, Thomas S. Kuhn's (1978) book on black-body radiation has led to a reevaluation of Planck's role in the early history of quantum theory, as illustrated, for instance, by the first part of Olivier Darrigol's (1992) book on the role of classical analogies in the development of quantum theory (see note 8 in Section 2.1). This literature has been taken into account in our discussion of both Planck's work and Einstein's reaction to it. Darrigol's book also deepened our understanding of Bohr's correspondence principle, which informed our discussion of this important principle. More recent books on topics covered in this volume include Helge Kragh's (2012) history of the Bohr model of the atom from its inception in 1913 to its demise in 1925 and Michael Eckert's (2013b) biography of Arnold Sommerfeld, who played a leading role in the elaboration of Bohr's model and its confrontation with empirical data. We also drew on Eckert's (2013a) introduction to an edition of Sommerfeld's (1915a, 1915b) first papers extending the Bohr model.[2] Books by younger scholars (e.g., Staley 2008, Seth 2010, Navarro 2012, Badino 2015) have also been useful to us even though their approach is sometimes markedly different from ours. Many older sources remain invaluable. In addition to Jammer (1966) and Mehra and Rechenberg (1982a), we single out the classic paper by John L. Heilbron and Kuhn (1969) on the genesis of the Bohr atom and Martin J. Klein's (1970a) magisterial partial biography of Paul Ehrenfest, covering its subject's most important contributions to the developments discussed in this volume. We should note, however, that much more has been written on the early

---

[2] For English translations and commentary, see Sommerfeld (2014a, 2014b) and Eckert (2014).

work covered in Part One of this volume than on the developments covered in Part Two, where, in some places, we worked almost exclusively from primary sources.

Research for this book was done in the context of a large international project in the history and foundations of quantum physics that ran from 2006 to 2013. This project was a joint initiative of Jürgen Renn's department at the Max Planck Institute for the History of Science and the theory department of the Fritz Haber Institute, both located in Berlin. The project led to the establishment of a series of conferences on the history of quantum physics. After a small workshop in Berlin in the Summer of 2006 (HQ0), four such conferences have been held so far: HQ1 (Berlin, July 2–6, 2007), HQ2 (Utrecht, July 14–17, 2008), HQ3 (Berlin, June 28–July 2, 2010), and HQ4 (San Sebastián, July 15–18, 2015). We were involved with this Berlin quantum project from the beginning and presented our work both at the original workshop and at all four subsequent conferences. These presentations resulted in four papers, all on topics that will be covered in the second volume of our book (Duncan and Janssen 2007, 2008, 2009, 2013).

A pair of papers on a topic covered in this volume grew out of the presentation we gave at a conference to commemorate the centenary of the Bohr atom, held June 11–14, 2013 in Copenhagen and attended by many members of the Berlin quantum project. These two papers are on the explanation of the Stark effect, the splitting of spectral lines in electric fields, both in the old and in the new quantum theory (Duncan and Janssen 2014, 2015). One of us co-authored a paper on Ehrenfest's adiabatic principle with another member of this large international collaboration, the Barcelona historian of physics Enric Pérez (Duncan and Pérez 2016). These papers formed the basis for parts of Sections 5.2 and 6.3 in this volume. We are grateful to Elsevier, publisher of *Studies in History and Philosophy of Modern Physics*, and the Royal Danish Academy of Sciences and Letters, publisher of the proceedings of the Bohr centenary conference, for permission to use this material.

As indicated above, we wrote this book with the idea that it could be used for courses in the history of modern physics aimed at undergraduate majors in physics or beginning graduate students in the history and philosophy of science. We both have experience teaching such courses. The past couple of years, one of us (MJ) has been using drafts of chapters of this book to teach the history of quantum physics to an audience of predominantly physics juniors and seniors at the University of Minnesota. We are grateful to the students in these classes for their feedback.

Several sections in Part One of this volume were written with the express purpose in mind of having students read them in tandem with some of the classic papers by Planck (1900d, 1901a), Einstein (1905a, 1907, 1909b, 1916b) (all available in English translation), and Bohr (1913b, 1913c, 1913d). Part Two likewise prepares the student for reading many of the original papers of the old quantum theory (though many of these have not been translated) as well as the four editions of the "bible of the old quantum theory," *Atombau und Spektrallinien* (Sommerfeld 1919, 1921, 1922a, 1924). An English translation (*Atomic Structure and Spectral Lines*) of the third edition appeared shortly after the German version was published (Sommerfeld 1923).

To encourage and support faculty in physics departments interested in offering a class on the history of quantum mechanics based on the latest scholarship in the field, we partnered with the Center for History of Physics and the Emilio Segrè Visual Archives of the Niels Bohr Library & Archives, both at the American Institute of Physics in College Park, MD, to set up a website that will make supplementary materials for a course based on our book available to students and instructors. This will include electronic versions of the primary sources mentioned above, materials from the course offered at the University of Minnesota, and selected images from the Emilio Segrè Visual Archives. Oxford University Press has set up a companion website featuring a series of "web resources" with more detailed treatments of various topics covered in the book. The URL for this companion website is: www.oup.com/companion/constructingquantum.

The present volume has two appendices with crucial background material on classical mechanics and spectroscopy. The additional information provided in our "web resources" is not critical for understanding the main text of this volume. To give two examples: one of these web resources is on classical dispersion theory and includes the derivation of a key result Planck needed to derive a formula for the spectral distribution of black-body radiation (see Eq. (2.7) in Section 2.3); another is on Woldemar Voigt's classical theory of the Zeeman effect, which played an important role in attempts to account for the effect in the old quantum theory (see Section 7.2.2).

We are grateful to Gregory Good, Director of the Center for History of Physics, for agreeing to partner with us in developing and maintaining a website in support of our book, and to Sonke Adlung, our editor at Oxford University Press, for agreeing to have the American Institute of Physics host a website connected to a book in their catalog. The materials from the course on history of modern physics at the University of Minnesota were developed as part of the project "Digital Essays in the History of Quantum Mechanics" led by Robert Rynasiewicz of Johns Hopkins University and supported by the National Science Foundation under NSF Grant No. SES-1027018.

It remains for us to thank a number of individuals and organizations that made our work possible. Our main debt of gratitude is to the Berlin quantum project. We wrote parts of this volume as guests in Jürgen Renn's department at the Max Planck Institute for History of Science and we thank the institute for its support. One of us (MJ) has been a regular visitor at the institute since its inception in the early 1990s and focused on this book project in the Spring of 2016 during the second half of a sabbatical from the University of Minnesota. This work in Berlin was supported by a Research Award from the Alexander von Humboldt Foundation. The recipient (MJ) would like to express his gratitude to the foundation for its support. For arranging our stays in Berlin, we thank Shadiye Leather-Barrow.

Most of the book was written in Pittsburgh where we have been getting together regularly to work on the history of quantum physics. We started the actual writing of this book in the Fall of 2015 when one of us (MJ) spent the first half of the aforementioned sabbatical at the Center for Philosophy of Science at the University of Pittsburgh. We thank the staff at the Center, especially its director John Norton, for its hospitality and support. The company of Edith Cohen, Donna Naples, Ted Newman, Max Niedermaier, and Merrilee Salmon has made our get-togethers in Pittsburgh all the more enjoyable.

We greatly benefited from discussions with Massimiliano Badino, Michael Eckert, Allan Franklin, Clayton Gearhart, Ted Jacobson, Christian Joas, Anne Kox, Joe Martin, Charles Midwinter, John Norton, Enric Pérez, Sergio Pernice, Jürgen Renn, Serge Rudaz, Robert Rynasiewicz, Robert Schulmann, Andrey Sharapov, Roger Stuewer, Jeroen van Dongen, and Brian Woodcock. We are especially grateful to Olivier Darrigol for his detailed comments on a complete draft of this volume. Although they are too numerous to mention individually, we also learned much from other historians, philosophers, and physicists involved in the Berlin quantum project and attending the HQ conferences. We thank Christian Joas for providing us with the LaTeX template that we used for the book and for solving various LaTeX problems we ran into along the way.

We thank Sonke Adlung, Ania Wronski, Cheryl Brant, and Alannah Santra at Oxford University Press and its partner production company SPi Global for shepherding our book through the production process. We thank Julian Thomas for his meticulous copy-editing of our manuscript.

Finally, we express our gratitude to the Emilio Segrè Visual Archives, the Deutsches Museum in Munich, the Niels Bohr Archive in Copenhagen, the Royal Society in London, and the Stichting "Pieter Zeeman-Fonds" in Amsterdam for granting us permission to include various photographs in their collection in this volume. These photographs, mostly portraits of physicists playing an important role in our story, can be found in the plates inserted between Part One and Part Two.

Contents

List of plates

1 Wilhelm Wien. Photo Gen. Stab. Lit. Anst. (*Generalstabens Litografiska Anstalt*). American Institute of Physics (AIP), Emilio Segrè Visual Archives, Weber Collection, E. Scott Barr Collection.
2 Friedrich Paschen. AIP Emilio Segrè Visual Archives, Landé Collection, and Gift of Jost Lemmerich.
3 Heinrich Rubens. AIP Emilio Segrè Visual Archives, Gift of Ludwig Genzel.
4 Max Planck. *Deutsches Museum Archiv*, BN04391.
5 Hendrik Antoon Lorentz. Stichting "Pieter Zeeman-Fonds," Amsterdam.
6 Tombstone of Ludwig Boltzmann at the *Zentralfriedhof* in Vienna with $S = k \log W$. Emilio Segrè Visual Archives.
7 Tombstone of Max Planck at the *Stadtfriedhof* in Göttingen with $h = 6.62 \cdot 10^{-34} W \cdot s^2$. Photograph by Karl-Henning Rehren, Göttingen.
8 Pieter Zeeman, Albert Einstein, and Paul Ehrenfest in Zeeman's Laboratory in Amsterdam. University of Michigan IST Willow Run Center Photographic Laboratory. AIP Emilio Segrè Visual Archives, W. F. Meggers Collection, Fermi Film Collection.
9 The first Solvay conference, Brussels, October 29–November 4, 1911. Photograph by Benjamin Couprie. Courtesy of the Solvay Institutes, Brussels.
10 Johannes Stark. Photograph by A. B. Lagrelius & Westphal. AIP Emilio Segrè Visual Archives, W. F. Meggers Gallery of Nobel Laureates Collection.
11 Peter Debye. AIP Emilio Segrè Visual Archives, Segrè Collection, Fermi Film Collection.
12 Hans Geiger and Ernest Rutherford in Schuster Laboratory in Manchester. AIP Emilio Segrè Visual Archives, *Physics Today* Collection.
13 George A. Schott. Copyright: The Royal Society, London.
14 John Williams Nicholson. Copyright: The Royal Society, London.
15 Arnold Sommerfeld and Niels Bohr in Lund, 1919. AIP Emilio Segrè Visual Archives, Margrethe Bohr Collection.
16 Johannes Rydberg. AIP Emilio Segrè Visual Archives, W. F. Meggers Collection.
17 Johann Jakob Balmer. AIP Emilio Segrè Visual Archives, W. F. Meggers Collection, *Physics Today* Collection.
18 Walter Ritz. AIP Emilio Segrè Visual Archives, W. F. Meggers Collection.
19 Woldemar Voigt. AIP Emilio Segrè Visual Archives, Landé Collection.
20 Max Laue (far left), Paul Sophus Epstein (to his left), Paul Ewald (far right), and other members of Sommerfeld's group in Munich in 1912. *Bildagentur Preußischer Kulturbesitz* (Nr. 10024059), Berlin.

1
Introduction to Volume One

1.1 Overview

This is the first of two projected volumes on the genesis of quantum mechanics, covering developments from Max Planck's work on black-body radiation around the turn of the twentieth century to John von Neumann's introduction of the Hilbert-space formalism of modern quantum mechanics in 1927. This first volume deals with developments up to the early 1920s. It is divided into two parts, each consisting of three chapters.

In the chapters comprising Part One, we take a broadly biographical approach, tracing the earliest stages of the development of quantum theory through the contributions of three main protagonists, Max Planck (Chapter 2), Albert Einstein (Chapter 3), and Niels Bohr (Chapter 4).[1] The focus in this early period was on black-body radiation and—starting around 1910 when Walther Nernst provided experimental support for a theory proposed by Einstein in 1907—the specific heat of solids at low temperatures. These were the topics dominating the discussion at the first Solvay Conference in Brussels in 1911 (see Plate 9), which was devoted to the fledgling quantum theory.

Part One ends with Bohr's derivation of the Balmer formula for the spectrum of hydrogen from a model of hydrogen and hydrogen-like atoms in the first part of his famous 1913 trilogy. This model is a precarious combination of elements from classical mechanics and electrodynamics and elements from quantum theory. Bohr's success with hydrogen led to a shift in focus in work on quantum theory to atomic structure and spectroscopy. His model became the germ for what has come to be known as the old quantum theory, which a little over a decade later was transmogrified into modern quantum mechanics. This transition will be covered in Volume Two of our book.

Part Two of this volume deals with the old quantum theory. As in Bohr's model, atoms are treated as miniature solar systems in this theory, which made the analysis of atomic structure amenable, as the astronomer Karl Schwarzschild (see Plate 21) first realized in 1916, to sophisticated techniques imported from celestial mechanics. The undisputed leader during this period, besides Bohr, was Arnold Sommerfeld

---

[1] See Plates 4, 8, and 15 in the section between Parts One and Two for photographs of Planck, Einstein, and Bohr, respectively.

Constructing Quantum Mechanics. Anthony Duncan and Michel Janssen, Oxford University Press (2019).
© Anthony Duncan and Michel Janssen. DOI: 10.1093/oso/9780198845478.001.0001

(see Plate 15), the author of *Atombau und Spektrallinien* (Atomic Structure and Spectral lines), which became known as the bible of the old quantum theory and went through four ever-expanding editions between 1919 and 1924 (Eckert 2013c; Seth 2010). Despite Sommerfeld's central role, however, the biographical approach of Part One is ill-suited to deal with the complexities of the old quantum theory, its various refinements, and the confrontation with a wealth of new data, in optical as well as in X-ray spectroscopy, and including the complicated patterns observed in the splitting of spectral lines by electric and magnetic fields, both internal and external to the atoms.

The old quantum theory itself will therefore take center stage in the three chapters comprising Part Two. In Chapter 5, we cover the principles the theory was based on, the quantization conditions in their various guises as well as Bohr's correspondence principle and the adiabatic principle of Paul Ehrenfest (see Plate 8). In Chapters 6 and 7, we cover the most important examples of the interplay between theory and experiment. Chapter 6 is devoted to what were seen at the time as important successes of the theory, even though several practitioners realized that some of these successes were incomplete. In hindsight, moreover, other successes turned out to be largely accidental. With these qualifications, the most important of these successes were the theory's agreement with the measurements by Friedrich Paschen (see Plate 2) of the relativistic fine structure in the spectra of hydrogen and helium, with the systematic examination by Henry Moseley (see Plate 22) of the X-ray spectra of the elements in the periodic table, and with the data of Johannes Stark (see Plate 10) on the effect named after him, the splitting of spectral lines by external electric fields, in hydrogen and helium (Duncan and Janssen 2014, 2015). In Chapter 7, we turn to what at the time were acknowledged failures of the theory, notably that it could not explain the Zeeman effect, the magnetic counterpart of the Stark effect, or the intricate multiplets in spectra of atoms more complex than hydrogen and that it gave the wrong value for the binding energy of helium. Like the successes, not all failures were complete failures. In the case of the Zeeman effect, for instance, Sommerfeld and others found empirical rules accounting for the line splittings observed under various conditions in impressive quantitative detail. That the Zeeman effect was nonetheless seen as a failure of the old quantum theory was because no consistent mechanical model could be found that captured these empirical regularities.

In the balance of this introduction we give an overview, as nontechnical as possible, of the contents of Volume One, identifying key players and their main contributions and highlighting points we think are new or not given the emphasis they deserve in the existing literature on the history of quantum theory.

1.2   Early developments: Planck, Einstein, and Bohr

1.2.1   Planck, the second law, and black-body radiation

In 1860, Gustav Kirchhoff showed that the spectral distribution of the heat radiation emitted by a perfect black body, i.e., a body absorbing all radiation that falls on it, is a universal function of frequency and temperature. In the late 1890s, Planck rose to the challenge of finding this function. Initially, however, his primary objective was not to find the black-body radiation law but to provide an electromagnetic foundation for the second law of thermodynamics. Although Ludwig Boltzmann and James Clerk Maxwell had argued decades earlier that the second law only has statistical validity, Planck still believed it to hold strictly, admitting no exceptions, no matter how improbable.

Planck thought he had identified a conservative time-asymmetric microscopic process in electromagnetism from which he could derive macroscopic irreversibility: a plane wave that is absorbed by a charged harmonic oscillator (or resonator as he called it) and then reemitted as a concentric wave. The inverse of this process is never observed in nature. As Boltzmann pointed out to him, however, this is just because the initial conditions for the inverse process are never realized. Maxwell's equations, Boltzmann reminded Planck, are time-symmetric, just as Newton's laws, and hence cannot serve as the basis for a derivation of the second law. Planck thereupon introduced the assumption of what he called "natural radiation" to rule out initial conditions that would lead to violations of the second law. Using this assumption, he could show that the total entropy of the system, the sum of the entropy of the resonators and the entropy of the radiation, never decreases. In this system, however, the entropy increases without changing the spectral distribution of the energy. As Ehrenfest, who first drew attention to this problem, put it: "radiation may over time become arbitrarily disordered—it certainly will not get any blacker" (1906, p. 529; quoted in Klein 1970a, p. 238). The problem was that the radiation's interaction with the oscillators is such that it cannot transfer energy from one frequency to another. To reach the equilibrium distribution, the system must be in contact with some other body, a "speck of soot" [*Kohlenstäubchen*] as Planck (1906a, p. 66) suggested later or the walls of the cavity, which will act as a heat bath kept at some fixed temperature. This aspect of the second law, the approach to equilibrium normally associated with the increase in entropy, could thus not be derived in Planck's approach but simply had to be assumed.

Planck's approach, however, did provide a way to find the black-body radiation law. The key result he needed for this was a relation between the average energy of an oscillator with a certain resonance frequency and the energy density of the radiation at that same frequency. With this relation and some basic thermodynamics, the choice of an expression for the entropy of one of Planck's oscillators as a function of its average energy translates directly into a specific black-body radiation law. In this way, Planck (1900a) recovered a law gleaned from experimental data by Paschen (1896–1897) and proposed independently by Wilhelm Wien (1896) (see Plate 1) on the basis of a superficial analogy with the Maxwell–Boltzmann distribution. Planck convinced himself that the expression for the oscillator entropy leading to the Wien law (or Paschen–Wien law as it was sometimes called in those days) was the only one compatible with the second law of thermodynamics.

Right around that time, however, Otto Lummer, Ernst Pringsheim, Heinrich Rubens (see Plate 3), and Ferdinand Kurlbaum, experimentalists at the *Physikalisch-Technische Reichsanstalt*, the German bureau of standards, began to find deviations from the Wien law (Kangro 1976). Reexamining his derivation of the Wien law, Planck realized that the second law does not uniquely determine the expression for the oscillator entropy after all. He used the leeway the second law still afforded him to change the expression to one giving a black-body radiation law that merges with the Wien law at high frequencies and merges with a law proposed by Lord Rayleigh (1900b) at low frequencies. Planck's experimental colleagues quickly established that this new law fit the data remarkably well at all frequencies. Planck had found Kirchhoff's universal function, the Planck law of black-body radiation.

1.2.2   Planck's first tenuous steps toward energy quantization

Planck's new law of black-body radiation initially was little more than an inspired guess. As soon as he found it, therefore, Planck set out to find a derivation of it. Even though he still did not accept Boltzmann's statistical interpretation of the second law (he would stick to his guns until 1914; Needell 1980), he borrowed some of Boltzmann's techniques to derive the expression for the oscillator entropy leading to his new law. This is where Planck's constant h made its first appearance.

Planck used Boltzmann's (1877) relation $S = k \log W$ (first written down in this form by Planck) between the entropy S of some macro-state and the number of micro-states W realizing that macro-state (with k being Boltzmann's constant) to compute the oscillator entropy. Following Boltzmann's lead, Planck discretized the energy of the oscillator to get a finite count. In Boltzmann's case, the discrete energy unit drops out of the final equation. In Planck's case, it does not. To get the Planck law, ε, the discrete energy unit, has to be proportional to the resonance frequency ν of the oscillator. The proportionality constant is Planck's constant h.

In the papers in which Planck (1900d, 1901a) first presented his derivation, he identified the relation $\varepsilon = h\nu$ as the crucial step but did not commit himself to a definite interpretation of it. However, his treatment of heat radiation in lectures published five years later (Planck 1906a) and his resistance for another two years to Einstein's (1906b) interpretation of $\varepsilon = h\nu$ as expressing the quantization of the oscillator energy strongly suggest that Planck at this point thought that an oscillator could take on the full continuum of values but that all values between $nh\nu$ and $(n + 1)h\nu$ should be lumped together in counting the number of micro-states realizing the same macro-state. This underscores a more general point about Planck's initial attitude toward his own results. Nothing in his papers of 1900–1 suggests that Planck realized that his new law of black-body radiation required a complete overhaul of what we now call classical physics.

1.2.3   Einstein, equipartition, and light quanta

Much more radical steps were taken in Einstein's (1905a) light-quantum paper, the only paper of his *annus mirabilis* that he himself, in a letter to a friend, called "very revolutionary" (Einstein 1987–2018, Vol. 5, Doc. 27). Einstein's starting point was a tension in the accepted picture of matter and fields at the time: matter was thought of as discrete, with a large but finite number of degrees of freedom; electric and magnetic fields as continuous, with an infinite number of degrees of freedom. This tension can be brought out with the help of the equipartition theorem, which is a general consequence of the kinetic theory of heat (see Section 3.2) and says (loosely speaking) that every degree of freedom of a system at temperature T contributes $\frac{1}{2}kT$ to the system's average energy. Since radiation has infinitely many more degrees of freedom than matter, all energy would end up in the radiation, and thermal equilibrium between matter and radiation would be impossible. Making radiation discrete, as Einstein was proposing, is one way to avoid this problem. Independently of Einstein and without committing himself to it, Ehrenfest (1905, 1906) made the same suggestion.

A much simpler way to avoid the problem was to deny that equipartition applies to radiation. This escape, moreover, was perfectly respectable. Equipartition was seen as highly problematic, though more so in England than in German-speaking countries (Klein 1970a, pp. 108–110). According to the equipartition theorem, every degree of freedom of a system should contribute $k/2$ to its specific heat. As Maxwell had pointed out a quarter of a century before, this was a serious problem:

> [T]he spectroscope tells us that some molecules can execute a great many different kinds of vibrations. They must therefore be systems of very considerable degree of complexity, having far more than six variables [i.e., degrees of freedom]...Every additional variable...increases the specific heat...I have now put before you what I consider to be the greatest difficulty yet encountered by the molecular theory (Maxwell 1875, p. 433).

The difficulty persisted unabated at the beginning of the twentieth century. "What would appear to be wanted," Lord Rayleigh reflected in 1900, "is some escape from the destructive simplicity of the general conclusion relating to the partition of kinetic energy" (1900a, p. 451). The "Maxwell–Boltzmann doctrine regarding the partition of energy" is the second of two "Nineteenth century clouds over the dynamical theory of heat and light" identified by Lord Kelvin (1901, p. 492). Cloud No. I, "the relative motion of ether and ponderable bodies," only takes up 6 pages in the printed version of Kelvin's famous address, Cloud No. II the remaining 35! We tend to think of equipartition as part of what we now call classical physics but it was certainly not part of what was considered well-established physics at the time. Although we will freely use the term 'classical physics' to contrast it with quantum (and relativity) physics, the controversial status of the equipartition theorem serves as a stark reminder that the old physics of the nineteenth century only came to be called 'classical' with the rise of the new physics of the twentieth (Staley 2005, 2008).

Einstein's light-quantum paper made it clear that blocking the application of equipartition to radiation was not enough to evade the problems that result from the interaction of discrete matter with continuous radiation. To bring out the problem, one only had to apply equipartition to the oscillators interacting with the radiation introduced by Planck. Surely, contemporary theory sanctioned this application! Appealing to equipartition to set the average energy of an oscillator equal to kT and using the relation Planck had derived to find the corresponding expression for the energy density of the radiation at this oscillator's resonance frequency, Einstein arrived at what is now known as the Rayleigh–Jeans law for black-body radiation. Rayleigh (1900b) had found this law in 1900 by applying equipartition directly to the degrees of freedom of the radiation (it is called the Rayleigh–*Jeans* law because James Jeans (1905a, 1905b) corrected a numerical error in Rayleigh's derivation). Rayleigh only expected his law to hold for the "graver modes" (lower frequencies) of the radiation and had, rather arbitrarily, inserted a damping factor to ensure that the energy density decreases exponentially as the frequency increases. Without such a factor, the energy density increases without bound as we move to higher frequencies. This is what Ehrenfest, who had clearly recognized the problem by 1905, later called "the Rayleigh–Jeans catastrophe in the ultraviolet" (Ehrenfest 1911, p. 92, quoted in Klein 1970a, p. 249). The problem, Einstein suggested, could be avoided by adopting the hypothesis that light "consists of a finite number of energy quanta localized at points in space that move without dividing, and can be absorbed or generated only as complete units" (Einstein 1905a, p. 178).

1.2.4   Einstein, fluctuations, and light quanta

The centerpiece of Einstein's 1905 paper was an argument in support of his light-quantum hypothesis based on a consideration of fluctuations in black-body radiation. In the third of three papers on statistical mechanics, in which he independently arrived at many of the results found earlier by Boltzmann and Josiah Willard Gibbs (1902), Einstein (1904) had laid the groundwork for using fluctuations to probe the microstructure of macroscopic phenomena. The most famous application of this technique was his paper on Brownian motion, but it was also central to his work on light quanta, not just in 1905 but in several later papers as well (Einstein 1909a,b, 1914a, 1917a, 1925).

In 1905, Einstein considered black-body radiation in a narrow range in the high-frequency regime where the Planck law reduces to the Wien law. Following Planck, he computed the entropy of this radiation. This allowed him to calculate the change in entropy ΔS for a random fluctuation that would momentarily concentrate the radiation initially occupying a volume V_0 in a subvolume V. Einstein used the relation $\Delta S = k \log W$ to calculate the relative probability W of initial and final state in this fluctuation. Although he called it "Boltzmann's principle," Einstein was the first to use the relation in this way. The principle became central to Einstein's early work on quantum theory, as he realized early on that the canonical ensemble, which he had used in his statistical trilogy to derive the equipartition theorem and a formula for energy fluctuations, does not apply to light quanta. Using Boltzmann's principle, Einstein showed that the probability of the fluctuation he was considering is given by an expression of the form $(V/V_0)^n$, which also gives the probability of a random fluctuation that momentarily concentrates all n molecules of an ideal gas in a container of volume V_0 in the subvolume V. In the case of black-body radiation in the Wien regime, $n = E/h\nu$ (in modern notation), where E is the total energy of the radiation. Einstein concluded that an amount of black-body radiation in a narrow frequency range around ν in the Wien regime of total energy E behaves as if it consists of n quanta of energy $h\nu$.

Einstein's argument supports only part of the light quantum hypothesis as stated above. It establishes that black-body radiation in a narrow frequency range around ν in the Wien regime consists of n *discrete* and *independent* elements of energy $h\nu$. It does *not* establish that these elements are *localized* the way gas molecules are. As we will show in Chapter 3, we find the exact same expression for the probability of the fluctuation in the

Wien regime that Einstein was considering if we take light quanta to be quantized modes of standing waves in the cavity containing the radiation. Rayleigh and Jeans arrived at the law named after them by counting modes of such standing waves. Ehrenfest (1906) indicated and Peter Debye (1910) (see Plate 11) showed in detail how quantizing the energy in these modes leads to the Planck law. Although none of these authors explicitly mentioned this, these standing-wave modes are not localized. Hence, in hindsight, the evidence for the localization of light quanta in Einstein's 1905 paper does not come from the fluctuation argument but from the explanation of the photoelectric effect, the ejection of electrons from a metal plate by radiation of sufficiently high frequency. Somehow the energy of a light quantum gets concentrated at the position of the electron knocked out of the metal plate. It can thus be seen as a form of poetic justice that the light-quantum paper is still widely known as the photoelectric-effect paper and that Einstein won the Nobel prize for his explanation of this effect.

In his first talk at a major physics conference, the 1909 *Naturforscherversammlung*, the annual meeting of the Society of German Physicists and Physicians held that year in Salzburg (cf. Chapter 2, note 20), Einstein (1909b) once again used fluctuations in black-body radiation to argue that light sometimes behaves as a particle and sometimes as a wave. Using the fluctuation formula he had found in 1904 but now deriving it from Boltzmann's principle rather than the canonical ensemble and using the Planck law rather than the Wien law, he showed that the mean square fluctuation of both energy and momentum in a given volume of black-body radiation in narrow frequency range is given by the sum of two terms, one of the form one would expect for particles, the other of the form one would expect for waves (Klein 1964). Neither the 1905 nor the 1909 fluctuation argument convinced the physics community that, under certain circumstances, light behaves as a collection of particles.

In the 1905 light-quantum paper, Einstein, though borrowing elements from Planck's work, had employed the Wien law for black-body radiation rather than the full Planck law. It was only the following year that he addressed the relation between his light-quantum hypothesis and Planck's work. In two papers written that year, Einstein (1906b, 1907) showed that Planck's derivation of his radiation law only works if the energy of his oscillators is actually quantized and not merely discretized for counting purposes. If the energy of an oscillator can take on the full continuum of values, he showed, one arrives at the equipartition value for its average energy, which, as he had shown in 1905, leads directly to the disastrous Rayleigh–Jeans law. If, however, the energy of the oscillator can only take on the values nhv, one arrives at the expression for its average energy that leads to the Planck law. Einstein overstated his case, claiming that Planck's derivation of his black-body radiation law implicitly relied on his own light-quantum hypothesis, but he did show incontrovertibly that it relied on the quantization of the energy of the oscillators.

1.2.5   Lorentz convinces Planck of energy quantization

Planck only accepted that the energy of his oscillators should be quantized in 1908, after a leading physicist of his own generation reached the same conclusion as Einstein (1906b,

1907), albeit via a different route (Kuhn 1987, pp. 196–202). In an influential lecture he gave that year in Rome, Hendrik Antoon Lorentz (1908a, 1909b) (see Plate 5) argued that the electron theory, Lorentz's version of Maxwellian electrodynamics in which electromagnetic fields interact with tiny charge-carrying particles, leads inescapably to the Rayleigh–Jeans law (Kox 2013a). The derivation involved the application of equipartition to the degrees of freedom of the field. Lorentz had come to agree with Jeans that this application was warranted even though it meant that matter and radiation can never be in thermal equilibrium. Following Jeans, he assumed that thermalization was an extremely slow process, which would explain why we never observe matter having lost all its energy to radiation. Lorentz explicitly admitted that, on this view, it was ultimately just a coincidence that all results derived for black-body radiation on the assumption of thermal equilibrium (such as the Stefan–Boltzmann law and the Wien displacement law) happen to agree with all our observations. He also conceded that the available data on black-body radiation clearly supported the Planck law rather than the Rayleigh–Jeans law, but, like Einstein, he recognized that Planck's law could only be had at the expense of profound changes in the theory of electrodynamics. In conclusion, he called for more experiments to decide between the Planck law and the Rayleigh–Jeans law.

Responding to Lorentz's appeal to the experimentalists, two of them, Lummer and Pringsheim (1908), observed that, if the Rayleigh–Jeans law were correct, a metal plate at room temperature would glow in the dark. Lorentz quickly and independently came to that same realization (Lorentz 2008, Doc. 171). And when taken to task by Wien about his Rome lecture, he accepted that the Rayleigh–Jeans law had become untenable. Lorentz was now prepared to start thinking about the consequences of accepting the Planck law. What he found particularly hard to swallow was that a charged oscillator could not emit or absorb an amount of energy less than h times its resonance frequency. In a letter to Lorentz in reply to this objection in the Fall of 1908, Planck for the first time stated unambiguously that the energy of an oscillator of resonance frequency v can only take on the discrete values nhv (Lorentz 2008, Doc. 176). In another letter to Lorentz a little over a year later (Lorentz 2008, Doc. 197), Planck replaced the term 'resonator', which he had been using up to that point but which seemed ill-chosen in view of Lorentz's objection, by the term 'oscillator', which we have been using all along.

By the end of 1908, senior physicists such as Planck, Lorentz, and Wien had thus come to accept two points that Einstein (1906b, 1907) and two other members of a younger generation, Jeans (1905a, 1905b, 1905c) and Ehrenfest (1905, 1906), had been emphasizing for some time. First, the old physics inevitably leads to the Rayleigh–Jeans law. Second, the Planck law, amply confirmed experimentally, calls for radical changes in the old physics.

1.2.6   From Einstein, equipartition, and specific heat to Nernst and the Solvay conference

Einstein (1907) had meanwhile found another area in which quantum ideas could fruitfully be applied, the specific heat of solids at low temperatures. As Maxwell's comments quoted in Section 1.2.3 illustrate, the equipartition theorem was notorious for giving the wrong values for specific heats of both gases and solids. Thus it was only natural for Einstein to use the quantized oscillators of Planck's theory of black-body radiation, with average energies sharply deviating from their equipartition values, to construct a simple model of a solid and check its prediction for the specific heat. The key parameter in Einstein's theory is the ratio $h\nu/kT$. An oscillatory degree of freedom at frequency ν does not contribute appreciably to the specific heat as long as this ratio is much greater than unity. This explained why the specific heat of solids decreases with decreasing temperature, as had been found, for instance, by Heinrich Weber (1872, 1875), one of Einstein's teachers in Zurich. It also provided the solution to Maxwell's puzzle of why the high frequencies of the oscillations presumed to be responsible for the emission of light do not contribute to the specific heat of gases, not even at room temperature.

Einstein's theory drew the attention of low-temperature physicist Nernst, the director of the *Physikalisch-Chemisches Institut* in Berlin (Barkan 1999). Exploiting the vast improvements in cryogenics since the days of Weber, who early on had still used snow as his coolant, Nernst (1911a, 1911b, 1911c) and his collaborators produced striking qualitative confirmation of Einstein's theory (Stone 2013, Chs. 13 and 17). The importance of this work can be gleaned from the following statement by Sommerfeld, made at the *Naturforscherversammlung* in Karlsruhe in September 1911:

> It will always remain a glorious chapter in the history of the first decade of the *Physikalisch-Technische Reichsanstalt* that it has erected one pillar of the quantum theory, the experimental foundations of cavity [black-body] radiation. Perhaps just as much credit deserves to be given to Nernst's institute, which has provided us with the other pillar of the quantum theory, bearing no less weight, with its systematic measurement of specific heats (Sommerfeld 1911a, p. 1060; quoted in Staley 2008, p. 406).

Black-body radiation and the specific heat of solids were the two central themes of the first Solvay conference held in Brussels in late October and early November 1911. This conference brought nascent quantum theory to the attention of a larger subset of the scientific community (Barkan 1993). Nernst had been instrumental in convincing the industrialist Ernest Solvay to finance the conference and to devote it to quantum theory. The conference, which was by invitation only, was attended by most of the major players we encountered above. In the famous group picture taken at the conference (see Plate 9), we see senior figures such as Wien, Planck, Rubens, Lorentz, and Nernst, but also junior ones such as Einstein and Jeans (the most important person missing was Ehrenfest, working in relative isolation in St. Petersburg at the time). In addition, we see physicists who would play an important role in the development of quantum theory in the years to come, such as Ernest Rutherford (Chapter 4) and Sommerfeld (Part Two). Against the wall behind the participants we see a blackboard with the Planck law for black-body radiation and a large sheet with a graph showing the rapid decrease of specific heat as the temperature goes to zero.

1.2.7   Bohr and Rutherford's model of the atom

Two years after the Solvay conference, Bohr's model of the hydrogen atom would shift the focus of quantum theory from black-body radiation and specific heats (both involving the statistics of large collections of quantized oscillators) to spectroscopy and the structure of individual atoms (Darrigol 2002).

In 1911, Bohr (see Plates 15 and 26) had been too junior to be invited to Brussels. He had just earned his doctorate from the University of Copenhagen with a dissertation (Bohr 1911) on the electron theory of metals of Paul Drude (1900) and Lorentz (1905). The dissertation showed that Bohr was thoroughly familiar with the literature on the quantum theories of black-body radiation and specific heats. He clearly recognized that the problems classical theory was facing in these areas would also affect the electron theory of metals. In addition, he showed that classical theory cannot account for either para- or diamagnetism. Eight years later, in a dissertation supervised by Lorentz, Hendrika van Leeuwen (1919) independently arrived at the same result, now known as the Bohr–van Leeuwen theorem. Bohr's prescient diagnosis of what caused these problems for the theory of Lorentz and Drude was that bound as opposed to free electrons do not satisfy the classical laws of mechanics and electrodynamics.

In September 1911, Bohr went to Cambridge for a year of postdoctoral study at the Cavendish Laboratory. Disappointed with J.J. Thomson, the director of the Cavendish, Bohr transferred to Manchester in March 1912 to work with Rutherford (see Plate 12), who had immediately impressed the young Dane when the two men first met in late 1911. On the basis of α-scattering experiments by two members of his group, Hans Geiger (see Plate 12) and Ernest Ernest Marsden (1909), Rutherford (1911) had just published the seminal paper introducing his model of the atom with electrons swirling around a nucleus containing nearly all of the atom's mass. Bohr's project, when he joined Rutherford's group, was to extract more detailed information about atomic structure from α-scattering data. After a few false starts, he was able to account for the energy loss of α-particles traveling through hydrogen gas on the assumption that a hydrogen molecule has two harmonically bound electrons with a resonance frequency calculated from dispersion measurements with the help of the classical dispersion theory of Hermann von Helmholtz, Lorentz, and Drude.[2]

As with many of his later papers, Bohr had great difficulty writing up his results. He was still working on the paper when he left Manchester in the Fall of 1912. He convinced his bride, Margrethe Nørlund, whom he married shortly after his return to Copenhagen, to spend their honeymoon in England so that he could finish his paper and personally hand it to Rutherford, who eventually submitted it for publication (Bohr 1913a). While working on the paper in Manchester in June and July 1912, Bohr had written to his then still fiancée and to his brother Harald that he had "many visions" of unraveling the structure of atoms, yet had also expressed his frustration that these ideas had not yet come to fruition (Aaserud 2013, p. 92).

---

[2] See Duncan and Janssen (2007, sec. 3.1), Jordi Taltavull (2017, Chs. 2–3), and the web resource, *Classical dispersion theory*.

In July 1912, Bohr had given Rutherford a set of notes he had prepared for him on atomic structure. This so-called Rutherford or Manchester memorandum (Bohr 1912) records Bohr's first attempts to construct models for atoms and molecules featuring a quantum condition inspired by Planck. The memorandum contains detailed calculations for simple models of the hydrogen atom, the hydrogen molecule, the helium atom, and a hypothetical helium molecule. All four of these models consist of a single ring of radius a, rotating with frequency v, containing one, two, or four electrons. In the case of the hydrogen and helium atoms, the ring is stabilized by a positively charged nucleus at its center (Bohr was concerned only with mechanical not with radiative stability). In the case of the hydrogen and helium molecules, the ring is stabilized by two positively charged nuclei on either side, placed symmetrically on either side of the plane of the ring, on a line perpendicular to it and going through its center. The conditions for stability gave Bohr a relation between the radius a and the frequency v of his electron rings but he needed an additional condition to determine the values of a and v separately and fix the sizes and frequencies of his electron rings.

To achieve this, Bohr adopted the special hypothesis that the kinetic energy E_{kin} of an electron in one of his rings is proportional to its frequency v. In line with the conclusions he had reached in his dissertation, he emphasized that this hypothesis cannot be derived from classical physics. The only argument he offered in support of it was that Planck and Einstein had successfully used a similar hypothesis to account for various experimental results. He did not elaborate but from his dissertation we know that Bohr had kept abreast of the relevant literature. Contrary to what one might have expected, however, he did not set the proportionality constant K between E_{kin} and v equal to Planck's constant h. He used a value of about $0.6h$ instead.

The Rutherford memorandum does not tell us how Bohr arrived at this value but his reasoning can convincingly be reconstructed (Rosenfeld 1963, p. xxxi). Bohr used the equation $E_{kin} = Kv$ to eliminate the ring radius a from the relation between a and v given by the stability conditions. That gave him an expression for K in terms of v and various constants. He now set the frequency of the electron ring in his model for the hydrogen molecule equal to the resonance frequency he had used in his paper on the energy loss of α-particles traveling through a hydrogen gas. On the Helmholtz–Lorentz–Drude theory of dispersion, this resonance frequency should be equal to the rotation frequency of the two electrons in the model of the hydrogen molecule in the Rutherford memorandum. This line of reasoning led him to set $K \approx 0.6h$.

The Rutherford memorandum contains a list of phenomena Bohr thought he could explain with his models. These are primarily chemical phenomena. Spectral lines are not on the list. The only application for which Bohr provided explicit calculations is the heat of formation of gases. Using his models to calculate the energy of two hydrogen atoms and the energy of one hydrogen molecule, he found a value for the heat of formation of hydrogen gas in the same ballpark as the measured heat of formation of water from hydrogen and oxygen. He also found that the energy of a helium molecule is greater than twice the energy of a helium atom, thus explaining why two helium atoms do not form a helium molecule.

1.2.8   Bohr and Nicholson's theory

In December 1912, a few months after his return to Copenhagen, Bohr came across some papers by John W. Nicholson (see Plate 14). These papers (Nicholson 1911c; 1911d; 1912, discussed in McCormmach 1966) belong to the British tradition of atomic modeling, best known for Thomson's (1904) plum pudding model (Heilbron 1977, pp. 52–63; Kragh 2012, pp. 12–22; Navarro 2012). In this model, electrons were assumed to be moving in rings inside a positively charged sphere. A model along the lines of Bohr's models in the Rutherford memorandum was the 'Saturnian model' proposed by the Japanese physicist Hantaro Nagaoka (1904), inspired by Maxwell's work on the rings of Saturn and published in *Philosophical Magazine* (Kragh 2012, pp. 23–24). As critics were quick to point out, the radiative instability of Nagaoka's model was much worse than that of Thomson's plum-pudding model.

Nicholson's models resembled Nagaoka's—and, unbeknownst to Nicholson, those in Bohr's Rutherford memorandum. Nicholson's models consisted of a ring of electrons rotating around a positively charged nucleus. Bohr, as we saw, adopted this model on the strength of Rutherford's interpretation of Geiger and Marsden's α-scattering data. Nicholson, by contrast, adopted it because he assumed that all mass is electromagnetic.

During the last two decades of the nineteenth century, J.J. Thomson, George F.C. Searle, and Oliver Heaviside had calculated the force a charged sphere experiences from its self-field upon acceleration and thereby shown that the interaction of a charged object with its self-field increases its inertia (Miller 1981, p. 43). This extra inertia was called electromagnetic mass. In the early twentieth century, a number of physicists expected that, ultimately, all mass might be accounted for in terms of electromagnetic mass.[3] A charge distribution's electromagnetic mass is proportional to the energy of its self-field. As the volume occupied by a system of like charges decreases, its potential energy and thereby, via $E = mc^2$, its mass increases.

Since positive charges account for most of the mass in an atom, Nicholson concluded that they had to be concentrated in a small subvolume of the atom. This then is how Nicholson arrived at his nuclear models.

Unlike Bohr, Nicholson was interested in both chemical and spectroscopic applications of his models. He tried to match the frequencies of spectral lines to (overtones of) frequencies of his rotating electron rings. A few years earlier, another British physicist, George A. Schott (see Plate 13), had published detailed calculations in classical electrodynamics showing that models with electron rings, both of the Thomson and of the Nagaoka(–Nicholson–Bohr) variety, can only produce a small number of spectral lines in the optical range of the spectrum and can never produce the converging series of lines seen in many atomic and molecular spectra (Schott 1907). These series satisfied Walter Ritz's (1908b)[4] (see Plate 18 and Appendix B, Section 7) combination principle: the wave number of a spectral line in such a series is the difference between two wave numbers characterized by small integers.[5]

---

[3] In Germany this became the core of the so-called electromagnetic worldview (see Section 1.3.1).

[4] Ritz mostly signed his own name "Walter" but the editor of his collected papers, Pierre Weiss, used "Walther", which seems to have made that the more common spelling (Martinez 2004, p. 4).

[5] See Appendix B for a brief history of spectroscopy.

Taking the lesson of Schott's analysis to heart, Nicholson assumed that the elements of the periodic table, with their exceedingly complicated spectra, are compounds of four simple proto-elements (or "protyles" as he called them), each consisting of a single electron ring, with 2, 3, 4, and 5 electrons, respectively (unlike Bohr, he ruled out rings with just one electron to avoid dipole radiation, the main source of radiative instability). Contrary to Rutherford's model, Nicholson thus assumed that all atoms of the periodic table, including hydrogen, would have several nuclei. Unbound protyles, Nicholson argued, would only be found in highly energetic astrophysical environments. Hence the names he chose for two of them: coronium and nebulium.

Nicholson ran into the same problem as Bohr in the Rutherford memorandum: stability considerations provided a relation between radius and frequency of his electron rings but did not determine the values of these parameters separately. To fix these values, Nicholson matched the frequencies of the rotating electron rings of coronium and nebulium to the frequencies of lines found in the spectra of the solar corona and stellar nebulae (now known to come from ionized oxygen, nitrogen, and iron).

In a paper of June 1912, Nicholson (1912) noted that the value of the angular momentum per electron in his protyles is always close to an integral number times Planck's constant. Nicholson saw this numerical agreement as further confirmation of the existence of his proto elements. His use of Planck's constant was not based on a careful reading of the quantum literature. While Bohr had absorbed developments in Britain as well as on the continent, especially Germany, Nicholson remained firmly ensconced in the British tradition.

Bohr was troubled by the similarity of Nicholson's models to his own and even more so by discrepancies between Nicholson's numerical results and his own calculations in the Rutherford memorandum. Both Bohr and Nicholson had used classical mechanics and electrostatics to find the stability conditions for rings of electrons rotating around positively charged nuclei and then determined the size and rotation frequency of these rings by imposing the condition that the kinetic energy or the angular momentum, two closely related quantities, is proportional to the rotation frequency with a proportionality constant related to Planck's constant (Nicholson on the basis of the frequency of lines in the coronal spectrum, Bohr on the basis of a dispersion frequency in molecular hydrogen). Nicholson, however, found much lower binding energies and much larger radii than Bohr. Bohr eventually pinpointed what lay behind these discrepancies. Nicholson's calculation pertained to highly excited states whereas his own pertained to the ground state. This was the first time Bohr explicitly made this distinction. In the Rutherford memorandum, he had tacitly limited his attention to the ground state.

1.2.9   The Balmer formula and the birth of the Bohr model of the atom

Despite his engagement with Nicholson's theory, Bohr still showed no interest either in excited states or in spectroscopy. As he told Rutherford in a letter of January 31, 1913, he was not calculating "the frequencies corresponding to the lines in the visible spectrum" and continued to focus on "the constitution of the atoms and molecules in their permanent [i.e., ground] state" (Bohr 1972–2008, Vol. 2, p. 580). But Bohr's focus was about to shift. In February 1913, Hans Marius Hansen, a young Danish spectroscopist, alerted him to a formula proposed by a Swiss school teacher, Johann Jakob Balmer (see Plate 17), in a standard text on spectroscopy (Stark 1911). Bohr's personal copy of this book contains a slip of paper with a reference to a paper by H. A. Wilson (1912) with a more general and more suggestive form of the formula,

$$\frac{1}{\lambda} = R\left(\frac{1}{n^2} - \frac{1}{m^2}\right),$$

where we followed modern usage in our choice of letters for the various quantities: $1/\lambda$ (equal to ν/c, the frequency divided by the velocity of light) is the wavenumber of a line in a spectral series satisfying the formula, R is the Rydberg constant, named after the Swedish spectroscopist Johannes Rydberg (see Plate 16), and n and m are integers such that $m > n$ (Hermann 1971, p. 157). For the Balmer series in hydrogen, $n = 2$ and $m = 3, 4, 5, \ldots$.

Bohr later claimed that "as soon as I saw Balmer's formula, the whole thing was immediately clear to me" (Rosenfeld 1963, pp. xxxxix). While this may have been an exaggeration, it is true that, sometime between first laying eyes on the Balmer formula in February and mailing a first version of the first installment of his 1913 trilogy to Rutherford in early March, Bohr recognized that the right-hand side of the Balmer formula could be interpreted as the difference in energy between two quantized orbits in a slightly modified version of the model of the hydrogen atom he had proposed and analyzed in the Rutherford memorandum. First, as he had come to realize while grappling with Nicholson's theory, he had to allow a whole range of quantized orbits rather than just the one ground state, labeled by integers. Second, he had to change the proportionality constant in the quantization condition $E_{\text{kin}} = K\nu$ from $K = 0.6h$ to $K = \frac{1}{2}\tau h$, where τ is an integer labeling the quantized orbits in his model. Identifying the terms on the right-hand side of the formula for the Balmer series as the energies of quantized orbits in his modified model for the hydrogen atom, Bohr could express the Rydberg constant in terms of fundamental constants of nature. Inserting the available values for these constants, he arrived at a value for the Rydberg constant remarkably close to the one found in spectroscopic measurements.

What remained unclear, however, was the mechanism by which the hydrogen atom, as modeled by Bohr, would produce the Balmer lines. The first installment of his trilogy, the paper in which Bohr (1913b) first published his derivation of the Balmer formula, still contains traces of a mechanism Bohr considered before settling on the now familiar mechanism of electrons jumping between orbits. Bohr imagined that radiation was emitted when a hydrogen nucleus captures an electron into one of the quantized orbits labeled by τ. In the Rutherford memorandum, Bohr had ignored the radiative instability of these orbits. In the 1913 trilogy he made it one of the pillars of his theory that electrons do not radiate as long as they stay in the same quantized orbit. He thus guaranteed the radiative stability of these orbits by fiat, i.e., by suspending the laws of classical electrodynamics for them, with the exception of Coulomb's law, which was needed for their mechanical stability. Bohr set the frequency of the radiation in this emission-upon-electron-capture scheme equal to $\nu_\tau/2$, the average of the frequency zero of the electron before capture and the frequency ν_τ of its orbit after capture. Bohr needed this factor of $\frac{1}{2}$ to justify the quantization condition $E_{\text{kin}} = \frac{1}{2}\tau h\nu_\tau$, which he knew led to the correct value of the Rydberg constant. Even with this factor, however, the emission-upon-electron-capture mechanism only allowed Bohr to recover the Balmer formula for the special case that the integer m labeling the initial state of the electron is infinite and the right-hand side of the formula reduces to R/n^2.

To account for the lines in the Balmer series, Bohr had to invent a new mechanism that would produce radiation of the frequency ν given by the Balmer formula for $n = 2$ and $m = 3, 4, 5, \ldots$ The mechanism he came up with is based on two assumptions. First, radiation is emitted only when electrons jump from one orbit to another. Second, the frequency of this radiation is given by the energy difference between these two orbits divided by Planck's constant. This is the famous Bohr frequency condition. The first assumption, as before, suspended most of the laws of electrodynamics for Bohr's electron orbits. The second assumption amounted to another radical departure from these laws. Classically, one would expect a continuum of frequencies if an electron goes from one orbit to another. Bohr, however, assumed that transitions from one orbit to another would result in monochromatic (or, as he called it, homogeneous) radiation. Moreover, the frequency of this monochromatic radiation is different from the orbital frequency of the electron in either the initial or the final orbit, as well as from the average of the two. Not only was this unheard of in classical electrodynamics, it also meant that Bohr had to find a new justification for his quantum condition $E_{\text{kin}} = \frac{1}{2}\tau h\nu_\iota$, where ν_ι, as the subscript indicates, refers to the orbital frequencies ν_n and ν_m, not to the radiation frequency $\nu_{m \to n}$ given by the Balmer formula.

To find this new justification, he started by imposing the more general quantum condition $E_{\text{kin}} = f(\tau)h\nu_\tau$. He then used the form of the Balmer formula to establish that the function f must be linear in τ, i.e., $f(\tau) = c\tau$. To ensure that his quantum theory would merge properly with classical theory in the regime where the latter was expected to remain valid, Bohr now demanded that, for large enough values of N, the difference between the radiation frequency $\nu_{N \to N-1}$ given by the Balmer formula and the orbital frequencies $\nu_N \approx \nu_{N-1}$ be arbitrarily small. He showed that it follows from this requirement that $c = \frac{1}{2}$. This argument can be seen as an application of an embryonic version of what Bohr would later call the correspondence principle (see Section 1.3.4 below).

Severing the relation between orbital frequencies and radiation frequencies (except in the limit of high quantum numbers) was perhaps Bohr's most radical break with classical theory. As Heilbron and Kuhn (1969, p. 263) noted in their classic study of the genesis of the Bohr atom: "That a spectral line of a given frequency must be produced by a charge vibrating at the same frequency was a consequence of electromagnetic theory which even Planck and Einstein had not thought to challenge." As Planck put it in his Nobel lecture:

> That in the atom certain quantized orbits should play a special role could well be granted; somewhat less easy to accept is the further assumption that the electrons moving on these curvilinear orbits . . . radiate no energy. But that the sharply defined frequency

of an emitted light quantum should be different from the frequency of the emitting electron would be regarded by a theoretician who had grown up in the classical school as monstrous and almost inconceivable [*eine ungeheuerliche und für das Vorstellungsvermögen fast unerträgliche Zumutung*] (Planck 1920, p. 19).

Shortly after he published his first papers on wave mechanics, Erwin Schrödinger complained about the same feature in equally strong language. Ever since 1914, he wrote to Lorentz on June 6, 1926, he had considered it to be something "so *monstrous* [*Ungeheuerliches*], that I should like to characterize the excitation of light in this way as really almost *inconceivable* [*undenkbar*]" (Lorentz 2008, Doc. 413, p. 615; English translation in Klein 1967, p. 61).

1.2.10   Einstein and the Bohr model

Despite serious reservations about Bohr's revolutionary model, his contemporaries realized that it could hardly be a coincidence that it produced the Balmer formula with just the right value for the Rydberg constant. As Einstein reflected decades later in his autobiographical notes:

> That this insecure and contradictory foundation was sufficient to enable a man of Bohr's unique instinct . . . to discover the major laws of spectral lines . . . appeared to me like a miracle and appears as a miracle even today. This is the highest form of musicality in the sphere of thought (Einstein 1949, pp. 44–47).

Einstein (1916a, 1917a) incorporated key elements of Bohr's model in the quantum theory of radiation he proposed when he returned to the quantum quandary after completing his general theory of relativity in 1915–6.[6] Today's textbook picture of the Bohr atom is a composite of Bohr's 1913 model and Einstein's 1916–7 theory. Bohr is responsible for the representation of different energy levels by quantized orbits in a miniature solar system and for the radical assumptions that electrons only radiate in transitions from one orbit to another and that the frequency of the emitted radiation is different from the orbital frequencies, yet monochromatic. Einstein took over these radical assumptions but without the picture of a miniature solar system. He remained agnostic as to how the quantized energy states in Bohr's model should be represented. Einstein is responsible for introducing another part of the textbook picture, namely that the transition of an electron from one state to another is accompanied by the emission or absorption of a light quantum in a definite but unpredictable direction. Like most of their contemporaries, Bohr staunchly opposed Einstein's light-quantum hypothesis because it appeared to be irreconcilable with interference phenomena. Bohr took the more conservative point of view that the quantum of energy an electron loses when it jumps from one orbit to another is emitted in the form of a spherical electromagnetic wave.

---

[6] We discuss Einstein's theory, out of chronological order, at the end of Chapter 3.

1.3 The old quantum theory: principles, successes, and failures

Our discussion of the first thirteen years of the quantum theory is centered around the three individuals (Planck, Einstein, and Bohr) responsible for the most important advances during this period. In the twelve years following Bohr's trilogy of 1913, up to the point when a new quantum mechanics emerged, the cast of characters involved in the further development of the theory expands dramatically, and a focus on individual contributors (with two exceptions: Sommerfeld and Ehrenfest) confuses rather than clarifies the conceptual evolution.

We have therefore chosen to organize Part Two of this volume around the structure of the old quantum theory and its confrontation with experimental data. In Chapter 5 we introduce the three central guiding principles of the old quantum theory: the quantization rules of Bohr, Sommerfeld, Schwarzschild, and others, the adiabatic principle championed by Ehrenfest, and Bohr's correspondence principle. In Chapter 6 we examine three prominent examples of the successes of the quantum theory based on these principles: the relativistic fine structure, the explanation of X-ray spectra, and the Stark effect. In Chapter 7, we then illustrate the gradual loss of faith in the theory in the early 1920s by describing three areas in which it failed dramatically to account for the empirical evidence: the complex multiplet structure of spectra, the anomalous Zeeman effect, and the spectrum of helium. We reiterate (see Section 1.1) that not all successes were unqualified successes and that not all failures were complete failures. Though we will remain vigilant in avoiding anachronism, we will draw on hindsight knowledge to answer a question that naturally comes up now that we are in full possession of modern quantum mechanics, viz. why the old quantum theory gave at least partially satisfactory solutions in some areas but failed to do so in others.

It is useful to think of the development of quantum physics in 1914–1925 as institutionally concentrated, with three centers of research playing a key role in bringing together a critical mass of physicists with distinctive as well as overlapping backgrounds. Such a variety of talent turned out to be essential for deciphering the peculiarities of a theory which appeared to defy the precepts of classical physics at almost every turn. These three research centers were, first, the "Sommerfeld (or Munich) School" at the Ludwig Maximilian University of Munich, which fostered the greater part of work on the old quantum theory from 1914 up to the early 1920s (Eckert 1993; Seth 2010), then secondly the Niels Bohr Institute in Copenhagen, founded in 1921 (Robertson 1979), and third, the "Göttingen School", led by Max Born (see Plate 26) (Thorndike Greenspan 2005), which would play a critical role in the transition from the old theory to the matrix and wave mechanics that would emerge in 1925–26.

For the developments covered in the present volume, Sommerfeld's Munich School played by far the largest role. Moreover, Sommerfeld himself proved to be a central figure, providing some of the critical insights in his own work as well as stimulating seminal work in his students, assistants, and associates. The four editions of his *Atombau und Spektrallinien* (Sommerfeld 1919, 1921, 1922a, 1924, cf. Eckert 2013c) constitute invaluable summaries of the evolving state of the field, not only for contemporary workers, but also for historians of physics almost a century later. Although our discussion in Part Two will be organized along conceptual themes, Sommerfeld's role in the development is so significant that we will pause here to give an overview of his career, leading up to the critical year 1915. Toward the end of that year, he introduced the general quantization principles that would guide the old quantum theory for the next decade (Sommerfeld 1915a). Similar biographical information for Planck, Einstein, and Bohr—to the extent that it helps illuminate their early contributions to quantum theory—will be given in due course in Chs. 2–4.

1.3.1   Sommerfeld's path to quantum theory

Arnold Sommerfeld (see Plate 15), born in 1868,[7] belongs to the generation directly between Planck, born in 1858, and Einstein, born in 1879. He was Bohr's senior by almost 17 years. He studied at the Albertus University in his Prussian hometown of Königsberg (now Kaliningrad, Russia) and earned his doctorate in 1891 with a dissertation on the Fourier representation of arbitrary functions. His doctoral adviser, Carl Lindemann, was one of several prominent mathematicians in Königsberg at the time, along with David Hilbert, Adolf Hurwitz, and Hermann Minkowski. This helps explain Sommerfeld's early preference for mathematics. One can nevertheless already detect his partiality for applied mathematics at this early stage: he entered a competition for the best mathematical study of thermal variation in the earth (as a function of depth). The Fourier analysis of the associated data was a direct stimulus for his dissertation topic.

After a year of military service, Sommerfeld moved to Göttingen in 1893, initially as an assistant in crystallography, but gradually falling under the influence of Felix Klein. Their joint four-volume work on the theory of gyroscopes (Klein and Sommerfeld 1897–1910) constitutes a masterpiece of classical mathematical physics, where once again the mathematics is frequently put at the service of engineering problems of contemporary interest (ballistics, torpedoes, ship stabilization, etc.). Sommerfeld's march upward through the German academic establishment took him through appointments at the Clausthal School of Mining (1897–1900), the Technical University of Aachen (1900–1906), and finally, in July 1906, to an appointment (effectively, as Boltzmann's successor, after a gap of several years) to the chair of Theoretical Physics at the Ludwig Maximilian University in Munich.

In Aachen, Sommerfeld's interest had been drawn to the subject of electron theory. As we saw in Section 1.2.8, J.J. Thomson and others had shown that the interaction of a charged sphere with its self-field contributes to its inertia. That this effect accounted for the entire mass of the electron became one of the core tenets of the electromagnetic worldview proposed by Wien (1900b) in a *Festschrift* for Lorentz. Rather than reducing electrodynamics to mechanics, as Maxwell and Heinrich Hertz had attempted, Wien

---

[7] This section owes much to Michael Eckert's (2013b) excellent biography of Sommerfeld.

suggested that mechanics ought to be reduced to electrodynamics (McCormmach 1970, see also Kragh 1999, Ch. 8).

Max Abraham, a student of Planck, became the leading proponent of this electro-magnetic program. Abraham (1902, 1903) modeled the electron as a rigid spherical surface charge distribution without any Newtonian mass. Lorentz (1904) adapted this model for his own purposes, subjecting Abraham's electron to a microscopic version of the Lorentz–FitzGerald contraction (Janssen and Mecklenburg 2007). The two models give slightly different formulas for the velocity dependence of the electron's mass. In both cases, however, mass increases sharply with velocity. Sommerfeld's (1904a, 1904b, 1905) contributions to this burgeoning field were important enough to be discussed in a seminar on electron theory organized by Hilbert, Minkowski, and others in Göttingen in 1905 (Eckert 2013b, pp. 137–141, Pyenson 1985, Ch. 5).

Measurements of the deflection of beta rays, already known to consist of high-speed electrons, by external electromagnetic fields were expected to be decisive between the various electron models and the theories they were part of (Miller 1981). Experiments of this kind by Walter Kaufmann (1906) appeared to support Abraham's theory. Discussing Kaufmann's results at the *Naturforscherversammlung* in Stuttgart in 1906, Planck (1906b, p. 761) argued that what was at stake in these experiments was nothing less than a fundamental choice between the basic postulate of Abraham's theory (i.e., all physics can be reduced to electrodynamics) and the basic postulate of what Planck called the "Lorentz–Einstein theory" (i.e., no absolute velocities can ever be detected). Planck made it clear that he preferred the latter. In the discussion following this talk, Sommerfeld, two years shy of his 40th birthday, brought the house down when he quipped: "I suspect that the gentlemen under forty will prefer the electrodynamical postulate, while those over forty will prefer the mechanical-relativistic postulate" (ibid.).

It was not until a few months after this conference that Sommerfeld first read the paper in which the 26-year old Einstein (1905d) derived Lorentz's formula for the velocity dependence of mass from the relativity principle without any assumptions about the shape or nature of the electron. Sommerfeld was no instant convert to relativity (Sommerfeld 2000, Doc. 103). It was Minkowki's space-time reformulation of the theory that won him over. When Minkowski died in January 1909, Sommerfeld saw to it that his lecture "Space and time" at the *Naturforscherversammlung* in Cologne in 1908 was published posthumously (Minkowski 1909). Two papers by Sommerfeld (1910a, 1910b) on Minkowski's four-dimensional formalism the following year played an important role in getting special relativity accepted by the physics community. Max Laue (see Plate 20), who in 1909 had transfered from Berlin and Planck to Munich and Sommerfeld, enshrined this four-dimensional approach in the first textbook on relativity (Laue 1911).

At the same conference in Cologne where Minkowski gave his space-time lecture, Alfred Bucherer (1908) presented the results of beta-ray deflection experiments *à la* Kaufmann that seemed to support the Lorentz–Einstein formula. Although these and other experiments would convince the physics community over the next few years of the correctness of this formula, later reanalysis showed that none of them were accurate enough to distinguish between the different contenders (Zahn and Spees 1938). As we will see in Section 1.3.5, Sommerfeld's work on the old quantum theory would result in a much more accurate, and convincing, demonstration of the correctness of the relativistic formula for the velocity dependence of mass.

Sommerfeld made his first foray into quantum theory with his so-called "*h*-theory", in which Planck's constant was presumed to play a role in the energetics of beta and gamma radiation, X-ray production from cathode rays, and eventually even in an explanation of the photoelectric effect that eschewed Einstein's light quanta. In the first two cases, the critical phenomenon was the emission of electromagnetic radiation when a charged particle (in both cases, an electron) undergoes either acceleration or deceleration, as required by Maxwellian electrodynamics. In the context of X-ray production, cathode ray electrons impinging on an anti-cathode (positively charged electrode) would undergo *Bremsstrahlung* or 'braking radiation'—a term coined by Sommerfeld (1909)—resulting in the emission of polarized and anisotropic electromagnetic radiation, which was termed the "primary" radiation, to be distinguished from a separate secondary X-ray component which was not polarized, isotropic, and containing discrete frequency components characteristic of the composition of the anti-cathode. In the case of beta and gamma radiation, it had been observed that the emission of beta particles from radioactive materials was almost always accompanied by gamma radiation, which was assumed to be due to the acceleration of electrons leaving the atom (the mechanism of acceleration being unknown).

In 1911, Sommerfeld invoked Planck's constant in an attempt to explain the energetics of gamma-ray emissions accompanying beta radiation (Sommerfeld 1911b) and the anisotropic (noncharacteristic) X-rays emitted when cathode rays strike a metal target (Sommerfeld 1911a). The subject of X-rays was of particular relevance in Munich, where Sommerfeld's experimental counterpart, Wilhelm Röntgen, the discoverer of said rays, continued to work intensively on their properties. It was primarily in the hope of theoretical assistance in deciphering the properties of X-rays, that Röntgen had secured the chair in Munich for Sommerfeld, having to overcome significant opposition from none other than Lindemann, Sommerfeld's doctoral supervisor in Königsberg (Eckert 2013b, p. 144).

The role played in Sommerfeld's "*h*-theory" by Planck's constant (as a unit of action = energy times time) was to constrain the time duration of acceleration (in the beta-gamma case) or deceleration (in the X-ray case) by equating it to Planck's constant divided by the corresponding energy gain (or loss). None of this made any sense, as subsequent developments were to show,[8] and the X-ray experiments planned by Sommerfeld in 1912 to quantitatively test his *h*-hypothesis were in any event derailed by Laue's use of the apparatus for the far more consequential discovery of crystal X-ray diffraction.[9] As it

---

[8] For example, the emission of beta and gamma radiation from radioactive nuclei are independent events, with the gamma radiation the result of a previous beta transition leaving the final nucleus in an excited quantum state, which later drops to the ground state with the emission of a gamma photon. The gamma radiation is *not* associated with the acceleration of the beta electron. The temporal relation of beta and gamma radiation in radioactive decay remained controversial for a considerable time, as witnessed by the title of a paper by Lise Meitner (see Plate 24) from 1925: "The gamma radiation of the actinium sequence and the proof that the gamma rays are only emitted after a completed radioactive decay" (Meitner 1925, cf. Sime 1996, p. 94).

[9] See Eckert (2015a, pp. 11-12; 2015b) for a detailed discussion of these developments.

turned out, the animated discussions over Sommerfeld's quantum of action approach to X-ray phenomena at the first Solvay conference (Sommerfeld 1912) did not presage any further fruitful elaboration of these ideas. By the time of the second Solvay conference in October 1913, devoted to "The Structure of Matter," the focus was almost exclusively on the remarkable results obtained since 1911 with X-ray crystal interference.

Sommerfeld's theoretical work was always strongly linked to the press of new experimental discoveries, and given that his call to Munich was primarily due to Röntgen's desire for theoretical support of his X-ray investigations, the motivation for his work on the *h*-hypothesis in relation to X-ray phenomena is clear. Another experimentalist who would come to be closely connected with Sommerfeld (and the work of the "Sommerfeld school") was Friedrich Paschen (see Plate 2), working in Tübingen. Sommerfeld had already been in communication with Paschen in 1904, at a time when the still mysterious gamma rays were interpreted by some (including Paschen) as highly energetic electrons (in other words, as super-energetic beta rays). By 1912, Paschen's experimental focus had moved to spectroscopy, specifically the baffling variety of anomalous Zeeman splitting patterns displayed by the spectra of various atomic systems under the influence of a strong magnetic field. The specific splitting pattern initially found by Pieter Zeeman (1896) (see Plate 8), and immediately explained by Lorentz (1897) on the basis of his classical electron theory (Kox 1997), had proven to be the exception rather than the rule for the vast majority of magnetically influenced spectra studied in the two decades after Lorentz's work. In 1912, however, Paschen and his assistant Ernst Back (see Plate 24) discovered that when the applied magnetic field becomes very strong, the complicated Zeeman types tend to merge together and the original simple Lorentz triplet structure reappears. This startling "Paschen–Back effect" became the stimulus for Sommerfeld to begin an intensive study of the Zeeman effect in atomic spectroscopy, culminating in a paper on the Zeeman effect for anisotropically bound electrons (Sommerfeld 1913).

This, of course, is just around the time that Bohr in Copenhagen was beginning to appreciate the value of spectra for studying the internal dynamics of atoms. Well after the Bohr theory had seen the light of day, however, Sommerfeld's work on spectral lines, which was further stimulated by the demonstration of line splitting in the presence of an electric field by Johannes Stark (see Plate 10),[10] continued to be based entirely on classical theory. This was presumably because there already existed, at least for the Zeeman effect, the sophisticated and phenomenologically successful coupled-oscillator theory of the Göttingen theoretical physicist Woldemar Voigt (see Plate 19). Voigt's theory could not only reproduce the bewildering variety of anomalous Zeeman splittings, but even incorporate the transition to normal triplet structure observed by Paschen and Back (cf. note 29 below). Sommerfeld devoted considerable effort to understanding, and streamlining, this Voigt theory, resulting in another paper on the Zeeman effect (Sommerfeld 1914).

---

[10] Stark informed Sommerfeld of his discovery of this effect in a letter of November 21, 1913 (Sommerfeld 2000, Doc. 204). For historical discussion of the discovery of the effect by Stark (1913a) and, independently, Antonino Lo Surdo (1913), see Hermann (1965a), Leone, Paoletti and Robotti (2004), Kox (2013b), and Duncan and Janssen (2015).

1.3.2   Quantum conditions: Planck, Sommerfeld, Ishiwara, Wilson, Schwarzschild, and Epstein

Bohr's groundbreaking trilogy appeared over the course of the Summer and Fall of 1913. By early September Sommerfeld was already communicating with Bohr, complimenting him on the success of the theory in accounting for the Rydberg constant in terms of Planck's constant, and enquiring whether Bohr planned to extend the theory to the Zeeman effect, a topic with which Sommerfeld, as we saw above, was seriously engaged at the time. By the Spring of 1914, we can see a shift in Sommerfeld's interests towards the problem of atomic structure. He realized that previous purely classical attempts to treat this problem were almost surely inadequate and that Bohr's ideas would have to be taken into account. By May 1914, Sommerfeld and his assistant Wilhelm Lenz (who had arrived in February) were lecturing on the treatment of both Zeeman and Stark effects both from a classical (Voigt) and from a quantum (Bohr) standpoint. Bohr himself presented his latest results (mostly concerning the spectra of hydrogen and helium) at Munich's famous Wednesday colloquium on July 15, 1914 (Eckert 2013b, p. 199).

A fortnight after Bohr's colloquium, Austria–Hungary declared war on Serbia, and by the following week Europe was at war. The forty-five year old Sommerfeld was deemed unnecessary for war service and was able to continue his academic activity. In the Winter semester (October 1914–February 1915), he offered a special seminar in Munich on the "Zeeman effect and spectral lines" (Eckert 2013b, p. 200). At this point, Sommerfeld was convinced that Bohr's treatment of the Balmer series held the key to any dynamical understanding of spectral lines. As in the case of Einstein's relativity theory, it thus took Sommerfeld a while to recognize the promise of Bohr's new theory. Once he did, however, his contributions greatly advanced Bohr's theory, just as his contributions to relativity had greatly advanced Einstein's.

On February 22, 1915, he wrote to Wien that he had "beautiful new results" concerning the Bohr elucidation of the Balmer series (Sommerfeld 2000, Doc. 214). He continued to give tantalizing hints of progress along these lines to various colleagues throughout the year, but the public unveiling of his extension of the Bohr theory would not come until a lecture "On the Theory of the Balmer Series" on December 6 to the Bavarian Academy of Sciences (Sommerfeld 1915a).[11] The delay seems to have been due to Sommerfeld's unsuccessful attempts in the interim to arrive at a quantitatively successful treatment of the Stark effect on the basis of his new understanding of the substructure of the Balmer lines (Eckert 2013b, p. 206; see note 12 below).

The possibility of elliptical electron orbits had been discussed by Bohr (1914b) in a paper on the effect of electric and magnetic fields on spectral lines, but only as a consequence of the distorting effect of an applied electric field; the undisturbed electron orbits were still assumed circular and the selection of stationary states therefore involved only a single quantum number n (which could, for circular orbits, be associated with the

---

[11] Cf. the introductions by Eckert (2013a, 2014), in German and English respectively, to reprints and translations of this paper and its sequel (Sommerfeld 2013, 2014a, 2014b).

angular momentum as in the treatment of the Balmer series in hydrogen). The (small) elliptical distortion of each of the original circular orbits by a weak electric field of (small) size $\mathcal{E}$ would shift the energy difference between two stationary states by $\mathcal{E}(f(n_1) - f(n_2))$.

For Sommerfeld, the existence of mechanically stable elliptical orbits of continuously variable eccentricity made it very difficult to understand the sharp spectral lines observed both in the absence and presence of an external electric field. Unless one artificially restricted the allowed stationary states to exactly circular orbits, the energy in a Keplerian system would necessarily depend on both the (independently variable) angular momentum and eccentricity of the orbiting particle. Imposing a quantization condition on the angular momentum only would still leave a continuously variable degree of freedom in the eccentricity, making the appearance of states with sharply defined energy (and sharp spectral lines from the transition between such states) incomprehensible.

In his presentation to the Bavarian Academy in December 1915, Sommerfeld addressed this problem head-on, providing a new supplementary quantization condition which would select a discrete subset of (nonzero) eccentricity values for the quantized Kepler system, while maintaining the angular momentum quantization introduced by Bohr. Quite naturally, for a system of two degrees of freedom (an electron following a Keplerian orbit in a plane), the complete specification of the motion required two independent quantization conditions. Sommerfeld's great contribution was to rephrase the quantization of angular momentum in terms of a phase integral of the momentum p_ϑ canonically conjugate to the orbit angle ϑ over a complete cycle of ϑ, setting this integral equal to an integer multiple n of Planck's constant. A similar condition for the phase integral for the other degree of freedom (in polar coordinates) in a Kepler orbit—the radial distance r—then led to the introduction of a second quantum number n', and a quantization condition for the eccentricity ϵ of the orbit in terms of n and n'. Sommerfeld's new version of quantization was directly inspired by ideas of Planck, going back to 1906, in which the quantization of action was given primacy over that of energy. The selected motions of a one-dimensional oscillator were associated with the slicing of its mechanical phase space, spanned by the position q and the momentum q, into separate domains with areas (in units of action) exactly equal to Planck's constant h. These ideas had received considerable attention at the first Solvay conference in 1911, which Sommerfeld had attended.

The original set of circular Bohr orbits (characterized by $n' = 0$, $\epsilon = 0$) were now supplemented by a new set of allowed noncircular orbits with $\epsilon \neq 0$ and $n' \neq 0$. The additional richness of allowed quantum states in the Balmer series remained invisible because the energy of these doubly quantized orbits still only depended on the single integer $n + n'$. It was perfectly clear to Sommerfeld (and had been for the better part of 1915[12]) that the application of an electric field would lead to different energies for orbits of varying eccentricity but identical values for $n + n'$, thereby resulting in the splitting

---

[12] In the published version of his December talk, Sommerfeld (1915a, p. 426) noted that "I have already presented these matters a year ago in a lecture, but delayed their publication with the intention of applying to the elucidation of the Stark effect, among other matters. This intention was frustrated in the meantime . . . by the difficulty of applying the quantum conditions to nonperiodic orbits."

of spectral lines observed in the Stark effect. Sommerfeld, to put it in modern terms, had thus hit upon the idea of degeneracy and of lifting such degeneracy through the application of external fields (Duncan and Janssen 2015, p. 227). At this point, however, he lacked the technical apparatus necessary to apply his phase integral approach to a nonperiodic system, which the Kepler motion of an electron would become once an external electric field was applied.

Sommerfeld was not alone in 1915 in ferreting out the generalization of the quantization procedures of Planck and Bohr that would prove to be so fruitful in the coming years. Around the same time, and independently, similar ideas were put forward by Jun Ishiwara (1915) in Japan and by William Wilson (1915), working at King's College in London. Moreover, in late 1915, Planck was intensely occupied with extending the phase space slicing he had introduced in 1906 for a one-dimensional periodic system to (periodic) systems with several degrees of freedom. He summarized his results in a paper published early the following year (Planck 1916). As we shall see in Section 5.1.1, Planck's new procedure is equivalent (though not obviously so) to the mathematically precise, and maximally general, form the quantization conditions would soon assume, via the formalism of action and angle variables, for general conditionally periodic systems, at the hands of Karl Schwarzschild (1916) (see Plate 21), best known today for having found the first exact solution of the field equations of general relativity.

Schwarzschild's contributions to quantum theory, shortly before his untimely death on May 11, 1916, bring up an interesting issue with regards to the formal preparation in classical mechanics that theorists concerned with the development of quantum physics at this juncture could rely on.[13] Today it is commonly supposed that the generation of Planck, Sommerfeld, and Einstein was completely at home in the sophisticated reformulation and extension of Lagrangian mechanics we now refer to as Hamilton–Jacobi theory. In fact, understanding of and proficiency in these techniques was almost entirely restricted to those physicists primarily concerned with celestial mechanics.[14] Schwarzschild, who had studied astronomy in the 1890s in Strasbourg and Munich, was an expert in this area, where the use of Hamilton–Jacobi methods and the action-angle variables exploited by Charles-Eugène Delaunay (1860–1867), Henri Poincaré (1892–1899), Carl Vilhelm Ludwig Charlier (1902–1907) and others were indispensable (Nakane 2015).

Schwarzschild made the connection between these techniques from celestial mechanics and Sommerfeld's phase integrals in a letter to his Munich colleague of

---

[13] In Appendix A we provide an extensive review of classical mechanics, including those aspects of the canonical formalism that are critical for understanding the progression of ideas in the old quantum theory.

[14] Boltzmann is a striking exception to this rule. His student in Vienna, Paul Ehrenfest, was not, as is illustrated by the following anecdote (see also Section 1.3.3). In his acceptance speech upon receiving the Lorentz medal, John H. Van Vleck (1974, p. 9) recalled that when Ehrenfest visited the University of Minnesota in 1924 and attended a talk by Van Vleck on the "Correspondence Principle for Absorption," in which Hamilton–Jacobi techniques were prominently employed, Ehrenfest was "surprised at my being so young a man—the lengthy formulas for perturbed orbits in my publication on the three-body problem of the helium atom had given him the image of a venerable astronomer making calculations in celestial mechanics" (Duncan and Janssen 2007, p. 627).

March 1, 1916 (Sommerfeld 2000, Doc. 240). In his response eight days later, Sommerfeld, for all his mathematical sophistication, had to admit that he was not familiar with these techniques (Sommerfeld 2000, Doc. 243; cf. Duncan and Janssen 2015, p. 232). This exchange between Sommerfeld and Schwarzschild, still on war duty and suffering the ravages of a fatal illness, provided a critical impetus for Paul Epstein's (1916b) (see Plate 20) derivation of the Stark effect splittings.[15]

Like Sommerfeld, Planck was unfamiliar with these techniques from celestial mechanics. He did not recognize, for instance, that his phase space slicing approach, which seemed rather ad hoc and was certainly awkward to implement in a general way, was exactly equivalent to a reformulation of the problem in terms of action-angle variables (see Section 5.1.1).

The explanation of the Stark effect was one of the major successes of the old quantum theory (Section 6.3). Yet, as Epstein (1916b, p. 507) and Sommerfeld (1919, pp. 502–503) both recognized, it also highlighted a serious problem for the theory: which orbits are allowed depends on the coordinates in which the quantization conditions are imposed. The choice of coordinates does not affect the energy levels found but it does affect the shape of the orbits (Duncan and Janssen 2014, sec. 3).

1.3.3 Ehrenfest and the adiabatic principle

The second guiding principle that played a fundamental role in the conceptual development of the old quantum theory is the adiabatic principle, associated primarily with the name of Paul Ehrenfest (see Plate 8).[16]

Ehrenfest was born in Vienna in 1880, where he spent his childhood and secondary school years, before enrolling at the University of Vienna in 1899. His studies in Vienna began with a concentration in chemistry but soon drifted towards physics, under the influence of Ludwig Boltzmann, and were interspersed with visits to Göttingen to attend lectures of Voigt and Stark in physics, and Klein and Hilbert in mathematics, among others. In 1903, Ehrenfest also visited Leyden where he attended Lorentz's lectures on radiation theory. His doctoral degree from Vienna was granted in 1904 (with a dissertation on classical hydrodynamics). That same year, he married Tatiana Alexeyevna Afanassjewa, a physics and mathematics student from Russia he had met in Göttingen. In 1907, the couple moved to St. Petersburg, where Ehrenfest secured a teaching position at the Polytechnic Institute. In 1912, he landed a permanent position as Lorentz's successor in Leyden.

Ehrenfest had absorbed thoroughly the statistico-mechanical approach (Darrigol 2018) to thermal phenomena that Boltzmann had championed in the teeth of vigorous

---

[15] Essentially the same results for the Stark splittings were obtained independently by Schwarzschild (1916) himself, though with a minor error.

[16] As in the case of Michael Eckert's (2013b) superb biography of Sommerfeld, we have a beautiful account of Paul Ehrenfest's life and contributions to physics (up to the year 1920) in Martin Klein's (1970a) *Paul Ehrenfest: The Making of a Theoretical Physicist*. For further discussion of the adiabatic principle, see Navarro and Pérez (2004, 2006), Pérez (2009), and Duncan and Pérez (2016).

opposition from both anti-atomists such as Wilhelm Ostwald or even committed atomists such as his Vienna colleague Josef Loschmidt. In October 1906, a month after Boltzmann's suicide, Ehrenfest was invited by Felix Klein to give a lecture in Göttingen. In this lecture, Ehrenfest addressed various objections to Boltzmann's statistical interpretation of the second law of thermodynamics. Klein was so impressed with the lecture that he asked Ehrenfest to write the article on the conceptual foundations of statistical mechanics for the *Encyclopedia of the Mathematical Sciences* (of which Klein was an editor) that Boltzmann had finally agreed to write after Klein had threatened to ask his critic Zermelo instead (Klein 1970a, pp. 81–83). The preparation of this article, which Ehrenfest wrote jointly with his wife, took a full five years. It was published in 1911 (Ehrenfest and Ehrenfest-Afanassjewa 1911, cf. note 24 in Chapter 2).

Ehrenfest (1911) then returned to Planck's theory of black-body radiation on which he had published some insightful papers before (Ehrenfest 1905; 1906; see Sections 1.2.3, 2.3, and 3.4.2). In this 1911 paper, he tried to identify the irreducible conceptual ingredients necessary for arriving at the Planck law for the spectral distribution of black-body radiation (Norton 1993). Just as Boltzmann had assumed that equal infinitesimal volumes d^3p of the momentum phase space of a massive gas particle should be assigned equal a priori weight in the statistical analysis of a collection of such particles, Ehrenfest posited the existence of a function $\gamma(\nu, E)$ such that $\gamma(\nu, E)dE$ would give the a priori weight for electromagnetic modes of frequency ν in the energy interval $(E, E + dE)$. It could easily be shown (basically by a Doppler-effect argument) that the ratio E/ν was invariant under a reversible adiabatic compression of the black-body radiation in an enclosure. Ehrenfest concluded that, for such a compression to leave the entropy constant, the weight function determining the probability of a configuration would have to take the form $G(E/\nu)$, i.e., it would have to be a function solely of the "adiabatic invariant" E/ν. He was further able to show that the requirement that the Planck law hold for all frequencies required this function to have only discrete support, being nonzero only when E/ν were integer multiples of an elementary value (which, of course, was just Planck's constant h).

Ehrenfest arrived in Leyden in the Fall of 1912 and although occupied with numerous bureaucratic and academic tasks (including the preparation of his inaugural lecture on special relativity; Ehrenfest 1913a), he continued to ponder the significance of the adiabatic invariance of the mysterious ratio of energy to frequency, which had played such a critical role in understanding the immunity of the classical second law of thermodynamics to the intrusion of the quantum in black-body radiation (Klein 1970a, pp. 260–263). He had already seen the connection of the adiabatic invariance to quantization in his black-body work. It was clear to him that any integrally quantized quantity (and therefore not subject to continuous variation) would have to be essentially "stuck" at its initial quantized value during slow (infinitesimal) variations of the parameters of the system (e.g., the volume of the black-body enclosure). If the regular laws of mechanics were to continue to hold during such variation, this suggested immediately that the appropriate quantities for quantization would have to be just those that were mechanically guaranteed to be invariant under such variations.

Ehrenfest found what he needed in a theorem of Boltzmann (discovered independently by Rudolf Clausius and Kalman Szily) that asserted that, under slow changes[17] of the parameters of a mechanical system undergoing periodic motion, the integral of the kinetic energy of the system over a single period is time invariant. This means that the average kinetic energy $\overline{K}$ times the period T (or, equivalently, $\overline{K}$ divided by the frequency ν) is time invariant. The energy E in the E/ν ratio appearing in the black-body case should thus really be restricted to the average kinetic energy $\overline{K}$ only. This had the added virtue of resolving the puzzling issue of ambiguity in the zero of the potential energy part in any mechanical system, which would seem to be incompatible with a fixed quantization of the total energy.[18] In many cases, such as the harmonic oscillator, a classical virial theorem ensures the equality (more generally, proportionality) of average kinetic and potential energy, so the quantization of the total energy follows automatically from the quantization of the kinetic energy part. In particular, the quantization of energy $E = nh\nu$ for Planckian oscillators implies that the adiabatic invariant $\overline{K}/\nu$ should assume the quantized values $\frac{1}{2}nh$.

In his characteristically lucid fashion, Ehrenfest laid all of this out in a letter to his Russian friend Abram Joffe in February 1913 (Klein 1970a, p. 261). He also set about applying his new quantization condition to the rotational energy of diatomic molecules. This allowed him to calculate the rotational contribution to the specific heat of diatomic gases. A diatomic molecule with moment of inertia I and angular velocity $\omega = 2\pi\nu$ has rotational kinetic energy

$$\frac{1}{2}I\omega^2 = \frac{1}{2}nh\nu = \frac{1}{2}n\hbar\omega,$$

where $\hbar$ is our modern notation for $h/2\pi$. It follows that the angular momentum of a diatomic molecule is quantized: $L = I\omega = n\hbar$. Ehrenfest's (1913b) paper was submitted in May 1913. It is clearly independent of Bohr's (1913b) similar quantization of angular momentum in the first paper of the trilogy and is based on completely different arguments. The underlying motivation for Ehrenfest's quantization condition, based on the Boltzmann–Clausius–Szily adiabatic theorem, was laid out in another paper submitted to the Amsterdam Academy in November that same year (Ehrenfest 1913c).

The appearance of Sommerfeld's papers in early 1916 generalizing Bohr's treatment of hydrogenic (one-electron) atoms with circular electron orbits to the general Keplerian case (including relativistic effects) provoked a complicated response from Ehrenfest. On the one hand, he had clearly received the Bohr theory with a great deal of antipathy, calling it "completely monstrous" (*ganz kanibalisch*) in a letter to Sommerfeld in May 1916 (Sommerfeld 2000, Doc. 254; quoted in Klein 1970a, p. 286, and in Kragh 2012, p. 91). On the other hand, he realized that the successes of Bohr, and their extension to

---

[17] Here "slow" means that the fractional change in the parameters over one period of the motion is much less than one.

[18] In his letter to Joffe cited below, Ehrenfest credited his wife Tatiana with insisting on the importance of eliminating this ambiguity.

the relativistic fine structure by Sommerfeld, could hardly be pure accident. Moreover, he had realized immediately (and pointed out in his letter to Sommerfeld) the connection between Sommerfeld's phase integrals and his adiabatic invariants. In fact, for a one-dimensional periodic system, Sommerfeld's phase integral

$$\oint p\,dq = \oint p\dot{q}\,dt = 2 \oint \frac{p^2}{2m}\,dt$$

(where $\oint$ gives the integral over a single period and where we used that $p = m\dot{q}$) is equal to (twice) Boltzmann's time integral of the kinetic energy and should therefore, according to Ehrenfest, be quantized with the value nh, just as Sommerfeld had assumed.

By June 1916, Ehrenfest (1916a) had prepared a paper for the Amsterdam Academy, in which he explained in detail the connection between the adiabatic principle and Sommerfeld's phase integral approach to quantization. An extended version was published in *Annalen der Physik* (Ehrenfest 1916b). In this paper, Ehrenfest formulated the "adiabatic hypothesis" in a particularly concise way:

> Under reversible adiabatic transformation of a system, (quantum-theoretically) "allowed" motions are always changed into "allowed" motions (Ehrenfest 1916b, p. 328).

The utility of this (deceptively simple) principle for quantum theory was immediately clear. If the appropriate quantization rules had been established (and, presumably, empirically confirmed) for one type of system, then the appropriate quantization rules could be inferred for any new system related to the first by an adiabatic transformation (if periodicity of motion could be maintained throughout the transformation). At a deeper level, the association of adiabatic invariants with quantized quantities showed the way to a more general approach to quantization, which did not rely on the particularities of a given mechanical system with a specified energy function.

In Ehrenfest's 1916 papers, the complete connection with the Sommerfeld approach, generalized by Schwarzschild to any conditionally periodic system describable by action-angle variables, was not yet at hand. Ehrenfest could only assert the adiabatic invariance of the total kinetic energy of a system, which implied the quantization of a sum of phase integrals for the general case of a system of more than one degree of freedom (although in certain cases, the presence of additional conservation laws, such as angular momentum, allowed one to "detach" and quantize individual phase integrals, as Sommerfeld had done in his treatment of central motion). The final step would follow promptly, with the proof of Jan Burgers (1917a, 1917b, 1917c), one of Ehrenfest's students in Leyden, that the *individual* action variables for conditionally periodic systems were indeed adiabatic invariants. Ehrenfest later confessed to Burgers that he was "horribly ignorant and inept" in the "theory of integration for canonical systems" (Klein 1970a, p. 291). We see once again that even physicists supremely well prepared in classical physical theory, as Ehrenfest undoubtedly was, were typically unversed in the intricacies of Hamilton–Jacobi theory and its elaborations, at least in 1916 (cf. note 14 above).

Although the pragmatic and empirically oriented Sommerfeld was not at first particularly receptive to the implications or use of the adiabatic principle (it only received adequate attention in the third edition of *Atombau und Spektrallinien*), both Einstein and Bohr realized the profound character of the result. Unlike the Bohr–Sommerfeld quantization conditions, the adiabatic principle would survive the transition from the old to the new quantum theory basically intact. It played an important role in the old quantum theory. Redubbed the "principle of mechanical transformability", it became, with the correspondence principle, a foundational pillar on which Bohr based much of his reasoning in his massive monograph "On the Quantum Theory of Line Spectra" (Bohr 1918).

1.3.4 The correspondence principle from Bohr to Kramers, Born, and Van Vleck

The correspondence principle, the third of our "guiding principles" underlying the development of the old quantum theory in the decade 1915–1925, only received its present name in 1920 (in a lecture of Bohr's in Berlin[19]). Previously, Bohr had not used a special name for his attempts to "to trace the analogy between the quantum theory and the ordinary theory of radiation as closely as possible" (Bohr 1918, p. 4). Nowadays, the correspondence principle is sometimes identified with the obvious requirement that the predictions of quantum theory match those of classical physics for those phenomena where classical methods give an empirically verified description. Bohr employed the term in a far more specific way, to indicate that the fusion of the physics of quantized *mechanical* systems (e.g., electrons orbiting nuclei) with the completely classical Maxwellian electromagnetic theory (which Bohr adhered to even after the Compton effect had convinced most physicists of the need for a quantized electromagnetic field) placed serious constraints on the description of processes in which classical electromagnetic radiation was emitted or absorbed by quantized material systems.

Of course, the existence of a "classical limit" in which quantum and classical behaviors merge had already been apparent purely within the realm of electromagnetic phenomena, in the transformation of the Planck formula for black-body radiation into the classical Rayleigh–Jeans form for the "graver modes" (low frequencies) of the radiation. The first application of Bohr's formal-analogy/correspondence arguments specifically at the intersection of quantized matter and continuous radiation appears in the first paper of the 1913 trilogy, in the final argument for the quantization condition selecting the stationary states of given principal quantum number N (see Section 1.2.9). To bridge the gap between discrete and continuous behavior, Bohr used that, for very high values of the quantum number N, the electron energy E_N and orbit frequency ν_N vary almost continuously from one value of N to the next, so one almost has a classical description for matter as well as radiation. It made sense in this domain that the frequency $\nu_{N \to N-1}$ of the

---

[19] The most explicit definition given by Bohr can be found in the Appendix to Part III of Bohr (1918), which appeared in 1922 (Bohr 1922a). For insightful discussion of the correspondence principle, see Darrigol (1992, Part B). The more recent literature on the correspondence principle includes Fedak and Prentis (2002, and, building on their paper, Bokulich 2008), Rynasiewicz (2015), and Jähnert (2016).

emitted radiation in a transition from the Nth to the $(N-1)$th orbit should agree (at least asymptotically) with that expected on the basis of classical Maxwellian theory, namely the orbital frequency ν_N itself. Given the Bohr frequency condition, $E_N - E_{N-1} = h\nu_{N\to N-1}$, and $\nu_{N\to N-1} \approx \nu_N \approx \nu_{N-1}$, the desired asymptotic agreement is found with precisely the quantization condition introduced earlier on the basis of an analogy with Planck's second theory (in which resonators absorb energy continuously but emit it only in units of $h\nu$) and with the reinterpretation that immediately follows in terms of a quantized angular momentum (as suggested by earlier work by Nicholson).

Once the full canonical apparatus for conditionally periodic systems had been introduced into quantum theory in 1916 by Schwarzschild, and enthusiastically adopted by Sommerfeld and his student Epstein (see Section 1.3.2), Bohr realized that the simple arguments of his 1913 trilogy could be substantially extended and applied in a wider context. In particular, the reappearance of classical radiative behavior in the limit of large quantum numbers had implications not only for the frequency of the emitted radiation, but also for its intensity. A bound charged particle executing periodic motion around a stationary positive nucleus, as in hydrogen atoms, would classically emit dipole radiation at all the frequencies corresponding to the Fourier analysis of its coordinate $\vec{r}(t)$. The intensity of the associated radiation at any one of these frequencies would, moreover, be proportional to the square of the coefficient of the corresponding sinusoidal term in the Fourier expansion. The calculation of these quantities would be greatly facilitated by the application of the action-angle techniques that physicists working on the quantum theory in the mid-1910s were rapidly absorbing.

By 1918, in his monograph on the quantum theory of line-spectra, Bohr was fully converted to this approach and correspondence-principle arguments (still phrased in terms of a "close relation between the spectrum calculated on the quantum theory and that to be expected on ordinary electrodynamics," Bohr 1918, p. 27) were reexpressed in terms of Fourier expansions in which the desired intensity amplitudes were (in principle) calculable functions of the quantized action variables of the system. The Dutch physicist Hendrik "Hans" Kramers (see Plate 25), in his doctoral thesis written under Bohr's supervision, was even able to compute analytically, in a technical tour-de-force, the Fourier amplitudes for the electron motion both in the relativistic case and for the Stark effect (Kramers 1919).[20] An application that proved especially significant for deciphering line spectra in the late 1910s and early 1920s was the use of correspondence-principle arguments to "deduce" selection rules. Transitions between states corresponding to Fourier components (of the motion of the electron in the initial state) that vanished were presumed to have zero intensity, in other words, to be "forbidden". This principle was applied even in the regime of low quantum numbers—clearly taking us away from the original motivation of correspondence-principle arguments.

Another, considerably less cogent, application of the correspondence principle came in Bohr's "second atomic theory" of 1920–21, in which Bohr attempted to build up (hence the term "Aufbau theory") multi-electron atoms by adding electrons one by

---

[20] For discussion of Kramers' life and work, see Dresden (1987).

one to an initially isolated positively charged nucleus, and applying correspondence-principle reasoning to identify the possible final orbits that each electron could occupy after dropping to the lowest possible energy. In hindsight, it is easy to see that Bohr's correspondence-principle "derivation" (never supported by actual quantitative calculations) of the electronic shell structure of the elements of the periodic table (and thereby their chemical properties) was invalid. A quantum theory of atoms built with electrons of the kind Bohr was envisioning at the time—distinguishable, spinless, not subject to the Pauli exclusion principle—does not give the distinctive periodic shell arrangements that Bohr claimed to have detected by waving the "magic wand" (as Sommerfeld (1919, p. 403) famously called it) of the correspondence principle.

A further deepening—or better, wholesale transformation—of the correspondence principle was carried out by Born, Kramers, and Van Vleck in 1924 (Duncan and Janssen 2007). In their hands, the "formal analogy" between classical theory and quantum mechanics became a precise mathematical prescription, in which the classical-mechanical equations of motion, in the action-angle framework, were converted to quantum equations by the simple expedient of replacing differential equations containing derivatives with respect to a (continuous) action variable by discrete difference equations, reflecting the quantized character of the action variables in the quantum theory. This procedure, used in tandem with Einstein's stochastic radiation theory (see Section 1.2.10), would lead to the quantum dispersion theory of Kramers, a direct precursor of matrix mechanics. These remarkable developments will be discussed in detail in Volume Two of our book.

1.3.5   The old quantum theory's winning streak: fine structure, Stark effect, X-ray spectra

The initial success of the Bohr theory of 1913 in accounting for the Balmer lines of hydrogen and ionized helium was not immediately followed by dramatic leaps in the understanding of atomic structure. There was, as we have seen, a hiatus of two years in which little was accomplished. Only with the deeper understanding of quantization that developed over the course of 1915 and was codified in its most general form in the Spring of 1916 by Schwarzschild, was further progress possible. But when it came, progress was dramatic and impressive. The basic tenets of the Bohr theory were confirmed in quantitatively successful explanations of the fine structure of the hydrogen and helium spectra, in an understanding of line splittings in X-ray spectra over an enormous range of atomic numbers, and in the treatment of the effect of strong electric fields on optical spectra (the Stark effect). All this was accomplished during the first half of 1916.

It can be argued that, for the remaining nine years of its persistence, the old quantum theory would never again experience such spectacular success. That the advances in these three areas were all associated with Arnold Sommerfeld established the dominance of the Munich school in teasing out the relation between spectroscopy and atomic structure. This prominence would be maintained later by the appearance of successive editions of *Atombau und Spektrallinien* (Atomic Structure and Spectral Lines) (Sommerfeld 1919, 1921, 1922a, 1924). Sommerfeld's treatise provided comprehensive and up-to-date reviews of the theoretical situation and became known as the bible of atomic physics (Eckert 2013b, pp. 255–256).

With hindsight, we can see that these first great successes of the old quantum theory (post-Bohr trilogy) were partly the result of a fortunate choice of problems and of some remarkable numerical "accidents" resulting in the cancellation of physical effects that were either unknown or improperly treated in the theory. In the first place, the quantization rules were being applied to one of the few dynamical situations in which the results would turn out to agree quantitatively with the new quantum theory that emerged in 1925. These were situations in which the charged particle interacting with the electromagnetic field (and hence giving rise to the observed spectra) was moving in a force field that to a high level of approximation was a central Coulomb field.[21] This was guaranteed in "hydrogenic" systems in which a single electron orbited a single positively charged nucleus, i.e., in hydrogen or singly ionized helium. But it was also approximately true in X-ray spectra, where the binding of an inner electron to a positive nucleus (especially of high charge, i.e., large Z) is dominated by the Coulomb attraction of the nucleus and where the electrostatic repulsion of the other electrons plays only a minor role.

Another way in which the problems chosen by Sommerfeld *et al.* at this time (i.e., 1915–6) were particularly apposite, given the state of the theory, was that the essential features of the problem did not involve magnetic effects. The reason this was critical is that the existence of electron spin, with the concomitant appearance of an intrinsic magnetic moment for the electron, was completely unknown at this point. There was simply no way, in the absence of this information, to arrive at an adequate understanding of such spectral features as the complex multiplet structure and the anomalous Zeeman effect that depend critically on the magnetic properties of the electron. In the Stark effect, by contrast, the applied electric field couples directly to the position of the electron and ignores the magnetic moment that comes with the spin of the electron.

As Dirac's (1928a, 1928b) two-part paper on the relativistic theory of the electron would make clear, the relativistic fine structure of the spectrum also critically depends on the spin of the electron. It thus seems quite baffling that Sommerfeld was able to derive essentially the correct result in a spin-less theory in 1916! Here we come to the second critical component of the early successes of the old quantum theory—the presence of remarkable, and often totally fortuitous, numerical accidents resulting in the cancelation of compensating errors in calculations based on quantized orbits of spin-less electrons. The premier example of this type of serendipity can be found in Sommerfeld's calculation of the fine structure of "hydrogen and hydrogen-like lines" in Part I of his article "On the quantum theory of spectral lines," submitted to *Annalen der Physik* in July 1916 (Sommerfeld 1916a). Sommerfeld had already realized that the peculiar degeneracy (equality of energy) of circular and elliptical orbits emerging from his quantization of both radial and angular degrees of freedom in the Kepler problem

---

[21] The other dynamical system in which the quantization procedures of the old quantum theory would give results essentially consonant with modern quantum mechanics is, of course, the harmonic oscillator.

depended on (*i*) the exact Coulomb character of the potential energy field (which would be altered in the presence of an external electric field, as in the Stark effect), and (*ii*) the use of nonrelativistic mechanics in describing the kinetic energy of the electron. The effect of an applied electric field had already been treated successfully by Schwarzschild and Epstein in the Spring of 1916 with the help of the new techniques of quantized action variables for conditionally periodic systems (see Section 1.3.2). Sommerfeld concentrated instead on working out the consequences for spectroscopy of the inclusion of relativistic effects in the description of the Kepler motion of an electron moving in a purely (as in the case of hydrogen or ionized helium) or approximately (as in the binding of inner electrons responsible for X-ray lines) Coulomb field. The inclusion of relativistic effects could be accomplished by a fairly simple generalization of his earlier nonrelativistic treatment: even though the resulting motion was now only conditionally periodic (involving two distinct periods, resulting in a precession of the perihelion, the point of the electron's orbit closest to the nucleus), the radial and angular quantization conditions could be imposed directly, without the necessity of a complicated detour through the action-angle formalism, as in the case of the Stark effect.

The formula obtained by Sommerfeld (1916a)[22] for the total relativistic energy of a bound electron in a stationary state turns into the formula obtained by Dirac twelve years later if we replace Sommerfeld's angular quantum number n by $j + \frac{1}{2}$, where j is the total angular momentum of the electron (orbital plus spin). In the new quantum mechanics, the quantum number $j + \frac{1}{2}$ assumes exactly the range of integer values $(1, 2, 3, \dots)$ that the purely orbital one n did in the old quantum theory. That Sommerfeld's formula gave perfectly correct results was thus a consequence of a cancellation of three effects: (*i*) the incorrect association of $n = 1, 2, 3, \dots$ states in the old theory with electrons which in fact had orbital angular momentum $0, 1, 2, \dots$; (*ii*) the ignorance of a factor of one-half for the spin contribution to the total angular momentum; (*iii*) a peculiar factor of $\frac{1}{2}$ (now called a Maslov index; Gutzwiller 1990, p. 211) which is missing in the old quantum theory in the quantization condition for certain systems (a well-known example being the zero-point energy in the harmonic oscillator).

Sommerfeld's formula gave corrections to the original Bohr energy levels which separated the energy of circular and elliptical orbits by amounts proportional to $(Z\alpha)^2$, where Z is the nuclear charge and $\alpha \equiv e^2/\hbar c$. As Sommerfeld was to point out in the first edition of *Atombau und Spektrallinien*, this "fine-structure constant" represented a "confluence of the three primary strands of modern research in theoretical physics: electron theory [through the electron charge e], quantum theory [through Planck's constant h], and relativity theory [through the speed of light c]" (Sommerfeld 1919, p. 370).

The presence of a squared factor of Z, the nuclear charge, meant that the line splittings induced by relativistic effects would be larger in helium than in hydrogen. Accordingly,

---

[22] The final formula was actually obtained by Lenz (Eckert 2013a, pp. 48–49). Sommerfeld obtained a series expansion of the energy in powers of the fine-structure constant. In a letter of March 7, 1916 (Sommerfeld 2000, Doc. 242), Lenz derived and communicated to Sommerfeld a simple analytic expression, which can conveniently be compared with Dirac's result.

the first quantitatively impressive confirmation of the correctness of Sommerfeld's formula did come from the measurements of the splitting of lines in ionized helium by Paschen (1916), Sommerfeld's experimental colleague in Tübingen. The agreement between Sommerfeld's predictions and Paschen's measurements (at roughly the 1% level) was certainly encouraging, but what convinced Sommerfeld that he was on the right track were measurements of the L-doublet splittings in X-ray spectra (corresponding to the relativistic splitting in energy of inner electrons of principal quantum number 2) over an enormous range of atomic numbers (from mid-sized elements with $Z = 47$ up to uranium with $Z = 92$; see Sommerfeld 1916b and Section 6.2). These measurement were made by the British physicist Henry G. J. (Harry) Moseley (1913, 1914) (see Plate 22), who was lost to science when he was killed in action during the Battle of Gallipoli in Turkey on August 10, 1915.[23]

In the first edition of *Atombau und Spektrallinien*, Sommerfeld (1919, Ch. 5, sec. 6) emphasized the significance of the fine-structure results in providing direct empirical confirmation of the velocity dependence of mass required by special relativity as opposed to that predicted by Abraham's purely electromagnetic theory (see Section 1.3.1). Experiments with cathode rays by Kaufmann and others aimed at settling the issue had proved extremely difficult: such methods simply could not compete with the precision of spectroscopic measurements. After showing that the kinematics implied by the Abraham theory gives a fine-structure splitting 20% lower than that measured by Paschen (and inferred from the X-ray results), Sommerfeld comments

> *In summary we must conclude that the absolute* [Abraham] *theory suffers shipwreck on the shores of the spectroscopic facts, and must definitively cede its previously asserted prominence to relativity theory* (Sommerfeld 1919, p. 376, emphasis in the original).

An important aspect of these early spectroscopic successes of the old quantum theory was the issue of selection rules. The quantization conditions allowed one to select a discrete set of allowed electron orbits and to calculate the corresponding energies. The observed phenomena, however, were *spectral lines*, each of which was associated with a pair of stationary states, corresponding to an initial and a final orbit, with the frequency of the line determined by the energy difference between the two via Bohr's frequency condition. A priori, one would therefore expect a multiplicity of NM lines in a transition from an orbit of principal quantum number N to an orbit of principal quantum number M, once the degeneracy of both sets of orbits has been lifted, either by relativistic effects or, additionally, in the case of the Stark effect, by the presence of an external electric field. In the case of the relativistic fine structure, however, fewer lines were seen, suggesting the existence of "selection rules" forbidding transitions between certain pairs of lines. In rather ad hoc fashion, Sommerfeld had originally (in his 1916 papers) argued for certain restrictions on changes in quantum number in radiative transitions. By the time of the first edition of *Atombau und Spektrallinien*, he had come to accept arguments of

---

[23] For more on Moseley's life and work, see Heilbron (1974, 1966).

the Polish physicist Wojciech (Adalbert in German) Rubinowicz (1918a, 1918b) (see Plate 23), based on conservation of energy and angular momentum, that suggested that the appropriate rule was to restrict transitions to those in which the angular momentum quantum number changed by at most unity, i.e., $\Delta n = 0, \pm 1$. This reduced the number of lines observed in a transition from $N = 4$ to $M = 3$ from twelve to nine. Before long, Sommerfeld had become aware of arguments (presented in Bohr's (1918) monograph on spectral lines) based on the correspondence principle that indicated a more stringent constraint, viz. $\Delta n = \pm 1$, reducing the number of allowed lines from nine to five, a multiplicity confirmed by more precise measurements. In the Stark effect, selection rules based on correspondence-principle arguments also played an important role (see Section 6.3).

1.3.6   The old quantum theory's luck runs out: multiplets, Zeeman effect, helium

Our experience with the evolution of physical theories since the scientific revolution has led us to expect that the most successful theories at any given time share two important features: (*i*) the ability to provide a quantitatively accurate description of a reasonably wide range of phenomena; (*ii*) a conceptual cohesion that makes it possible to express its fundamental tenets in a reasonably simple mathematical form. We do not expect any such theory to remain permanently immune to the accumulation of more detailed and accurate information about the phenomena it covers or to remain pertinent in the event that an entirely new class of phenomena becomes relevant. A highly successful theory, such as the Newtonian theory of gravity, can continue to provide a robust and accurate description of the phenomena in its domain (terrestrial and celestial gravitational phenomena) over several centuries, yet is still likely to be superseded eventually. Even though in many situations the new theory might be practically equivalent to the old, there could be a conceptual gulf between them, as illustrated by the case of the Newtonian theory of gravity and its successor, Einstein's general theory of relativity.

The old quantum theory lacked the conceptual cohesion of these gravitational theories. It remained an awkward amalgam of Newtonian mechanics, Maxwellian electrodynamics, and quantum rules. Moreover, the range of phenomena for which it gave a quantitatively successful description was extremely limited. In at least one sense, however, it dealt with a broader range of systems than Newtonian gravity. The latter was tightly focused on just one system—our own solar system. The old quantum theory borrowed the techniques developed in celestial mechanics to deal with the perturbing effects of one celestial body on another to deal with the behavior of a variety of atoms exposed to a variety of external fields. The relevant data were not positions of celestial bodies but frequencies, intensities, and polarizations of radiation emitted by these atoms under various circumstances. This made the theory extremely vulnerable to precisely the empirical data that were critical for its initial successes, namely, the extraordinary wealth of information, provided with increasingly high precision by spectroscopists during the first quarter of the twentieth century. All in all, it should therefore not be too surprising that the lifetime of the old quantum theory was little more than a decade.

The failures of the old quantum theory actually already began to appear shortly after its guiding principles had been formulated. They can be traced to two basic problems. First, there was a critical piece of information missing, which, had it been available to atomic theorists in the decade before 1925, would certainly have improved the understanding of atomic structure, even within the limited framework of the old quantum theory. This missing piece was the existence of intrinsic electron spin and its concomitant magnetic moment. In the absence of this concept, it was simply impossible to arrive at a proper understanding of either the complex multiplet structure of line spectra or the closely related problem of the splitting behavior of these multiplets under the influence of an external magnetic field (i.e., the anomalous Zeeman effect).[24] Second, the whole space-time picture of dynamics at the atomic level, in terms of electrons following well-defined orbits, was fundamentally in error, as would become clear once the new mechanics of Heisenberg and Schrödinger emerged.[25]

The quantitative successes of the old theory, despite these basic flaws, are due to a series of mathematical accidents that can be traced back to the extremely simple character of the dynamics of a single point particle subject to a harmonic or Coulomb potential. These accidents evaporate the moment one has a more complicated system, one, for example, with more than a single electron orbiting a nucleus, with the noncentral inter-electron interactions properly taken into account. The old quantum theory was therefore doomed to "suffer shipwreck" (to use Sommerfeld's colorful phrase) once it attempted to confront the problem of the binding energy of the neutral helium atom with just two electrons. This failure, unlike those encountered in multiplet theory or the Zeeman effect, did not depend on electron spin. Using Hartree–Fock methods in modern quantum mechanics, we find that the ground-state binding energy of two distinguishable electrons without spin to a nucleus of charge +2 is the same, in first approximation, as that of two indistinguishable electrons with spin.

The intricate story of the complex multiplet structure of optical spectra is related in Section 7.1.[26] We have already seen that Sommerfeld realized that the degeneracy of stationary states of an electron in the hydrogen atom (or singly-ionized helium) would be lifted by relativistic effects. In late 1916, he turned his attention to multi-electron atoms most similar in behavior to hydrogen, the alkali metals of the first group of the periodic table, where the single outer "valence" electron could be regarded as orbiting a core (i.e., nucleus plus inner electrons) of unit positive charge. Sommerfeld (1916c) pointed out that the quantized energy levels of the valence electron in such atoms could be calculated by assuming that the total electrostatic potential experienced by this electron from the core could be approximately regarded as a Coulomb potential e^2/r modified by faster falling powers of r ($1/r^2$, $1/r^3$, etc.), as a consequence of the core charge (nuclear

---

[24] Ignorance, until 1924, of Pauli's exclusion principle, had similarly perfidious consequences for any attempt to explain the periodic optical and chemical properties of the elements.

[25] As illustrated by a comment by Kramers (1923, p. 339) quoted toward the end of Section 7.4, there already was a strong suspicion by early 1923 that the motion of the electrons in these orbits might not satisfy the laws of classical mechanics.

[26] The history of quantitative spectroscopy, starting with Fraunhofer, is reviewed in Appendix B.

plus inner electrons) being spread out and not point-like. In other words, Sommerfeld explicitly took into account the "screening" effect of the inner electrons on the nuclear electrostatic force exerted by the nucleus on the outer electron.

The net effect of these modifications was similar to that of the relativistic treatment of the kinetic energy but quantitatively much larger. It produced a splitting in energy of the N circular and various elliptical orbits possible for principal quantum number N, leading to N distinct terms (or energy levels). In this way, Sommerfeld arrived at a qualitative understanding of the existence of the various optical series spectra (sharp, principal, diffuse, etc.[27]) and, in particular, of the mathematical form of the Balmer-formula-like expressions introduced by Rydberg to describe these series in the final decade of the nineteenth century.

It had already been established in the nineteenth century that the spectral series lines of the alkali metals in the first group of the periodic table were narrow doublets, and those of the alkaline earths in the second group narrow triplets. With the advent of the Rowland grating in the 1880s, wavelengths could be measured to a thousandth of an Ångstrom (i.e., to seven significant digits), so there could be no doubt about the existence of this additional structure. The multiplicity of lines here, however, went beyond that expected by Sommerfeld's "dissection" of the degenerate stationary states. For example, for alkali metals, while the s terms (orbits of angular momentum $n - 1$) were "simple" (there was just a single stationary state), those for any orbit with $n > 1$ (i.e., the $p, d, f, \ldots$ terms, etc.) were all narrow doublets. Adding relativistic effects to those induced by screening would shift the term energies slightly, but could not suddenly result in a doubling of levels for the orbits of angular momentum $n > 1$. For alkaline earths, terms other than s appeared as either singlets or narrow triplets. This overabundance of possible states clearly posed a serious problem to the whole Bohr planetary picture of the atom, even when supplemented by the sophisticated machinery of action-angle variables and the inclusion of relativistic effects.

Sommerfeld's first serious response to the crisis posed by the complex multiplets of series spectra was characteristically pragmatic and empirically based. It was clear that some additional degree of freedom was needed to account for the additional term structure exposed by the multiplets as well as for new selection rules that were emerging. These new selection rules appeared to regulate transitions between the subcomponents of terms which, according to the tenets of the original Bohr theory, should have corresponded to single, unique quantized orbits. Sommerfeld (1920) added an "inner quantum number" to the original pair of radial and angular quantum numbers. This inner quantum number was not associated with any specific physical mechanism, although Sommerfeld speculated that it perhaps corresponded to some form of "hidden rotation". The only type of rotational dynamics comprehensible to atomic theorists at this time was orbital angular momentum—the concept of spin was still five years in the future.

[27] See Appendix B.

In late 1921, a specific mechanical model was concocted by Sommerfeld's new student, Werner Heisenberg (1922). In this "core model" the (orbital) angular momentum of the valence electron (in alkali metal atoms) and that of the rest ("core") of the atom— i.e., the nucleus and inner electrons—were quantized separately and at *half-integral* values. The angular momentum of the core was set equal to $\hbar/2$, irrespective of the (also half-integral) angular momentum of the valence electron. The core angular momentum was forced to be either completely aligned or anti-aligned with the valence angular momentum. This gave two distinct states of different energy, due to the difference in interaction of the magnetic moment of the core with the magnetic field generated by the current loop arising from the orbit of the outer valence electron.

By yet another one of those fortuitous accidents from which the old quantum theory so frequently profited,[28] the resulting energy splitting worked out to be almost identical to the modern spin-orbit splitting, with the core angular momentum of magnitude $\hbar/2$ playing the role that would eventually be taken over by electron spin. Heisenberg's model was therefore, at least for the alkali metal group, quantitatively successful. Over the next two years, however, it became increasingly clear (as was emphasized, in particular, by Alfred Landé (see Plate 24), a student of Sommerfeld working in Frankfurt at the time) that the ad hoc dynamical assumptions of the model were incompatible with elements of classical mechanics (e.g., the Larmor theorem), which still formed an integral part of the old quantum theory.

The second failure of the old quantum theory that we have selected for special attention arises in the treatment of the anomalous Zeeman effect (Sections 7.2–7.3). We already touched on the prehistory of this topic in Section 1.3.1. In 1896, Zeeman observed first the broadening, then (with more careful measurements) the splitting of the sodium D lines under the influence of an imposed magnetic field. A month later, Lorentz had devised an explanation for magnetic spectral splitting based on the effect of a magnetic field on harmonically bound negatively charged "mobile ions" (the term "electron" had not yet been introduced). The monochromatic radiation (at the frequency of the harmonic motion) emitted by the ions absent a magnetic field would, according to Lorentz, be split into three components, associated with the three motions into which the three-dimensional motion of an ion could be separated: motion along the direction of the magnetic field (with no change in frequency, as the magnetic force on a charged particle is zero for motion along the direction of the field), and radiation displaced equally above and below the original frequency, corresponding to clockwise or anti-clockwise motion around the direction of the field (Kox 1997).

Unfortunately, this "normal Lorentz triplet" turned out to be the exception rather than the rule when, within a few years, spectroscopists began to study the effect of a magnetic field on a wide variety of spectral lines. For example, the two sodium D lines were found to split into four and six components respectively under the influence of a

---

[28] In this case, physicists were ignorant of two factors that happened to cancel: the gyromagnetic ratio 2 of the electron and the Thomas precession correction factor of 1/2.

magnetic field. Even more complicated "anomalous Zeeman splittings" were found in spectral lines associated with elements towards the center of the periodic table.

For two decades following Zeeman's discovery, a considerable amount of work was done to extract the phenomenological regularities exhibited in an increasingly vast empirical data base of Zeeman splittings. For example, the splittings for elements in the same column of the periodic table exhibited similar patterns (Preston's law). As we already noted in Section 1.3.1, the Göttingen physicist Woldemar Voigt succeeded in developing a complex and elegant generalization of Lorentz's analysis of harmonically bound electrons coupled to a magnetic field, from which much of the phenomenology could be reproduced by a suitable choice of coupling parameters. This theory, however, was completely classical. It was basically an ingenious magnetic extension of the Helmholtz–Lorentz–Drude dispersion theory.[29] After the Bohr theory appeared, it was clear that the whole problem had to be rethought in the context of Coulomb-bound electrons orbiting a central positive nucleus. The first convincing treatment of the Zeeman effect within the framework of the old quantum theory was given by Sommerfeld (1916d) and, independently, the Dutch physicist Peter Debye, who had studied with Sommerfeld in Aachen (see Eckert 2013a, p. 52). Unfortunately, their result was simply a confirmation of Lorentz's original result: individual spectral lines should undergo a transition to a normal Lorentz triplet once a magnetic field was applied, a conclusion which remained valid even when relativistic effects were included (as the normal Zeeman effect simply appeared superimposed on the relativistically split lines).

The history of the anomalous Zeeman effect in the old quantum theory between 1916 and 1926, when the matter was finally resolved with the introduction of electron spin and the use of the new methods of quantum mechanics, is an extraordinarily complex and convoluted one. Our account of the period 1916–1923 in Section 7.3 is undoubtedly the most intricate and demanding part of this volume. It is impossible to present here a readable abbreviated account of all the twists and turns of the story without doing violence to the details, which, in this case, are critically important. To give the reader some guidance, however, it will be helpful to distinguish two important phases of the theoretical odyssey followed by the chief practitioners of quantum theory in this period.

Between the first efforts of Debye and Sommerfeld in 1916 and early 1922, the main focus was on establishing a comprehensive description of the numerous empirical regularities exhibited by the anomalous splittings. A typical example was the further elaboration of Runge's hypothesis (dating from 1907) that, for weak magnetic fields, the observed anomalous Zeeman splittings were rational multiples of the normal Lorentz splitting. A number of rules were then evolved for associating the denominators of these rational fractions with the type of complex spectral multiplet undergoing the Zeeman splitting. All of this led, by the beginning of the 1920s, to a fairly complete phenomenological understanding of a wide range of spectral "Zeeman types".

---

[29] See Section 7.2.2, and the web resource *Voigt Theory*, for an account of the Voigt theory of the Zeeman effect.

From 1922 on, the emphasis shifted to attempts to build explicit models of atomic structure in which the electron orbits would respond to applied magnetic fields in the ways indicated by this vast phenomenology. The core model of Heisenberg (1922), discussed above in connection with the problem of complex multiplets, was the premier effort along these lines. Not only did it provide (for alkali metals) a quantitatively successful account of the complex multiplet structure (*sans* magnetic field), it also correctly reproduced the transition from the weak-field anomalous splitting to the strong-field limit, in which, as Paschen and Back (1912) had discovered a decade earlier, the normal Lorentz triplet structure reemerged. However, Heisenberg's model came under almost immediate attack (initially from Landé and Bohr, later from Pauli and others) for internal dynamical inconsistencies. Over the next three years, no dynamically consistent atomic model was found that could completely avoid the conceptual difficulties of the core-model approach. However, the sustained involvement of leading physicists with this problem was to lead directly to Pauli's discovery (via an ingenious application of the adiabatic principle) of the exclusion principle, as well as to the introduction of electron spin, first by Ralph Kronig, who decided not to publish the idea after failing to convince Pauli and others of its merits (Jammer 1966, pp. 146–148), and then, independently, by George Uhlenbeck and Samuel Goudsmit (1925).[30] Even though ultimately a failure for the old quantum theory, the anomalous Zeeman effect thus turned out to be an extremely fruitful failure.

The final topic we have selected as an example of failure in the old quantum theory is the problem of helium (Section 7.4). Specifically, we examine the inability of the old quantum theory to come to terms with the spectroscopy of a neutral helium atom or, even more basically, the binding energy of its ground state. In the Rutherford memorandum and in the second part of the 1913 trilogy, Bohr had advanced a simple picture of a helium atom in its ground state. The two electrons would rotate on the same circular orbit, in the same sense, at opposite locations on the circle, with each electron carrying one unit $\hbar$ of angular momentum. The single ionization energy (i.e., the energy needed to remove one electron, leaving the other in the lowest quantum level) for this configuration was found to be 28.9 electron volts, clearly incompatible with early measurements of James Franck (see Plate 26) and Gustav Hertz (1913), which gave about 20 eV (Gearhart 2017). The experimental measurements would be improved and eventually settle, in 1919, at about 25 eV. By that time, the Bohr picture for the ground state of helium had already been criticized and discarded for several other reasons, such as the disagreement of the model with dispersion measurements and the absence of paramagnetic properties for helium (which the magnetic moment of two similarly rotating electrons would imply).

The other feature of helium that had bothered spectroscopists for some time was the existence of two disconnected sets of terms (quantum levels), between which radiative transitions were not observed. Initially this had even led to the assumption that there were two distinct elements involved: helium and parhelium, later renamed orthohelium and parahelium. This idea was abandoned after all efforts to separate the elements chemically

---

[30] These developments will be covered in Volume Two of our book.

had failed. What lies behind this peculiar feature of the spectrum of helium according to the new quantum theory is that the dominant radiative transitions in atoms are electric dipole transitions, which do not affect the electron spin, so that states in which the spins are aligned (orthohelium) only connect radiatively with each other, as do states in which the spins are anti-aligned (parahelium).

Ignorant as they were of spin, the theorists of the old quantum theory devised an explanation for helium's peculiar spectrum in terms of different spatial orientations of the two electron orbits. Screening calculations by Sommerfeld and Landé aimed at estimating the quantized energy levels for the measured para- and orthohelium terms suggested that the two electron orbits should be tilted with respect to each other for parahelium states but coplanar for the orthohelium ones. By the early 1920s, these arguments led to the Bohr–Kemble[31] model for the ground state of helium (by this time known to be a parahelium state) in which the two electrons occupied similar almost circular orbits with one tilted with respect to the other at an angle of approximately 60 degrees.

Heroic calculations in 1921–22 by Kemble's student John H. Van Vleck (1922a, 1922b) at Harvard and Bohr's assistant Kramers (1923) in Copenhagen both led to the result that the single ionization energy was only around 20 eV, which is just about as far off from the correct value 24.6 eV as Bohr's 1913 result, albeit from the other side.[32] A detailed examination by Born and Heisenberg (1923) of the term energies for excited states of helium also ended in failure.

1.3.7 Born taking stock

Max Born's *Vorlesungen über Atommechanik* (Lectures on Atomic Mechanics), written in 1924 and published in early 1925, can be regarded as a final summing up of the methods and achievements of the old quantum theory. On the very last page, Born (1925, p. 341) concedes that the principles of the quantum theory—namely, calculation of the motion according to the rules of classical mechanics, followed by selection of stationary states by quantization of action variables—were only in agreement with experiment for systems involving a single electron. Already for the next simplest case, that of two electrons bound to a helium nucleus, the theory failed. The failure in this case went to the core of the conceptual inadequacy of the model, its reliance on an entirely incorrect spatio-temporal dynamics based on particles having well-defined positions at any given time and following well-defined orbits.

---

[31] Edwin Kemble had been the first American physicist to write a predominantly theoretical dissertation on the old quantum theory (Duncan and Janssen 2007, p. 562).

[32] Adiabatic considerations were critical in these calculations. One could imagine the final orbits to arise from initially independent motions of the electrons with the electrostatic repulsion of the electrons turned off, and then examine the change in energy as the repulsion is adiabatically reintroduced.

Such difficulties ensured that Born was keenly aware of the provisional character of the old quantum theory. As he wrote in the preface of his book:

> [T]he work is deliberately conceived as an attempt . . . to ascertain the limits within which the present principles of atomic and quantum theory are valid and . . . to explore the ways by which we may hope to proceed . . . [T]o make this program clear in the title, I have called the present book "Vol. I"; the second volume is to contain a closer approximation to the "final" atomic mechanics . . . The second volume may, in consequence, remain for many years unwritten. In the meantime let its virtual existence serve to make clear the aim and spirit of this book (Born 1925, p. v).

As we shall see in Volume Two of our book, a correct understanding of helium, both qualitatively and quantitatively, followed within two years with the development of Heisenberg's matrix mechanics and Schrödinger's wave mechanics, in combination with the new ideas of electron spin and the Pauli exclusion principle. The same aggregation of new concepts would likewise provide the resolution of the other two failures of the old quantum theory highlighted in Chapter 7, the complex multiplet structure (due to the spin-orbit interaction) and the anomalous Zeeman effect (due to the combined effect of spin-orbit interactions and the direct interaction of the double magnetic moment of the electron with the applied magnetic field). The sequel to Born's 1925 book could thus be written much sooner than its author anticipated. It was published only five years later (Born and Jordan 1930).

Part I

Early Developments

2

Planck, the Second Law of Thermodynamics, and Black-body Radiation

2.1 The birthdate of quantum theory?

It is often said that quantum theory was born in Berlin on December 14, 1900. In the meeting that day of the German Physical Society (*Deutsche Physikalische Gesellschaft*), Max Planck (1900d) presented a derivation of a new law for the spectral distribution of black-body radiation, i.e., for the way its energy is distributed over frequency. Black-body radiation is the heat radiation given off by perfect absorbers—bodies that absorb all radiation that falls on them and do not reflect any. All bodies at finite temperature give off heat radiation. This is electromagnetic radiation emitted at all frequencies, strongest in the infrared when the temperature is below or at room temperature, strongest in the visible and the ultraviolet at higher temperatures. As was first shown by Gustav Kirchhoff (1860), the ratio of a body's power of emission to its power of absorption is the same for all bodies and depends only on the body's temperature and the frequency of the radiation emitted or absorbed. It follows that the power of emission is the same for all perfect absorbers, i.e., for all black bodies. The energy per unit volume in black-body radiation is thus a universal function of the frequency ν and the temperature T. In the late nineteenth century, several physicists tried to find this universal formula. Planck eventually found it by cleverly combining two older formulas that worked well in different frequency regimes. Figure 2.1 gives a graphical representation of Planck's new formula. Planck (1900c) unveiled it in a meeting of the German Physical Society on October 19, 1900. This new formula was in excellent agreement with the experimental data and Planck immediately set to work to derive it from first principles. In the meeting of December 14, 1900, he presented for the first time the derivation he had found.

This derivation crucially involves the relation $\varepsilon = h\nu$, where ε is an energy, ν is a frequency, and h is what is now known as *Planck's constant*. Although Planck (1900d, p. 40) referred to this relation as "the most essential point of the whole calculation," he had little to say about its interpretation. We are now accustomed to reading it as

Constructing Quantum Mechanics. Anthony Duncan and Michel Janssen, Oxford University Press (2019).
© Anthony Duncan and Michel Janssen. DOI: 10.1093/oso/9780198845478.001.0001

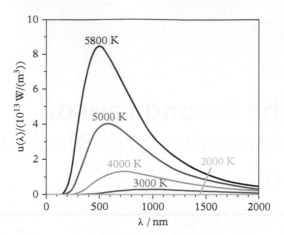

Figure 2.1 *The distribution of energy over wavelength (visible light is roughly in the 390–700 nm range) for black-body radiation at various temperatures (well above room temperature). The equivalent distribution of energy over frequency for different temperatures T is given by Planck's expression (2.24) below for ρ_v, the energy of the radiation per unit volume in the frequency interval $(v, v + dv)$, as a function of T.*

expressing energy quantization. For a modern reader it conjures up the idea that light of frequency v comes in discrete packages of energy hv. Although this is how one of the early readers of Planck's papers—Paul Ehrenfest (1905) (see Section 2.6)—interpreted them, this is definitely not what their author had in mind. The idea of light quanta was not proposed and championed until five years later by Albert Einstein (1905a).[1] It met with fierce resistance for almost twenty years. Planck applied the equation $\varepsilon = hv$ to what he called resonators, which we can think of as charged harmonic oscillators (charges on a spring) with resonance frequency v, interacting with black-body radiation.[2] Even when applied to these resonators, however, it is not at all clear whether Planck thought of the equation as expressing a restriction of the possible values of their energies to integral multiples of hv. Planck needed the equation $\varepsilon = hv$ to make the derivation of his new black-body radiation law work. Whatever he thought about this equation and the new constant h, there is no indication whatsoever that he realized in 1900 that they would require a complete overhaul of the physics of his day.

By the turn of the century, Planck had been working on black-body radiation for several years in connection with attempts to provide an electromagnetic foundation for the second law of thermodynamics. In Section 2.3, we will examine his efforts in this area more closely. Here we just give a brief overview. Between 1897 and 1899,

---

[1] Ehrenfest (1906) only argued that Planck had implicitly assumed that radiation comes in discrete lumps in the derivation of his black-body radiation law—he did not endorse the idea.

[2] In 1910, Planck would drop the term 'resonator' and start using the term 'oscillator' instead (see Section 3.4.2).

Planck presented a lengthy paper in five installments, entitled "On irreversible radiation phenomena," to the Prussian Academy of Sciences in Berlin (Planck 1897–1899). Following Massimiliano Badino (2015, sec. 3.2), we refer to this paper as Planck's pentalogy.

Planck approached the problem of finding a formula for the spectral distribution of the energy in black-body radiation in a rather roundabout way. He considered a resonator at frequency v interacting with the radiation in a narrow frequency interval around v. He was able to derive a simple relation between the average energy U_v of the resonator and the energy density ρ_v of the radiation. This meant that he had an expression for ρ_v once he had one for U_v. To find an expression for U_v, in turn, all Planck had to do was to specify the entropy S_v of a resonator as a function of U_v. He could then use the standard relation $dS/dU = 1/T$ from thermodynamics and solve for U_v as a function of T. The task of finding the black-body radiation law thus boiled down to choosing an expression for the entropy of a resonator as a function of its average energy and justifying that choice.[3]

In the final installment of his pentalogy, presented to the Berlin Academy in May 1899, Planck reported that he had only been able to find one function S_v of U_v compatible with the second law of thermodynamics. This function led directly to the law for black-body radiation found experimentally by Friedrich Paschen (1896–1897) in Hannover[4] and theoretically by Wilhelm (Willy) Wien (1896) in Berlin. Now remembered as the Wien law, it was often called the Paschen–Wien law in the closing years of the nineteenth century. Although Planck could not derive the Paschen–Wien law from the second law of thermodynamics—which, after all, would have required him to prove and not merely conjecture the uniqueness of the entropy function leading to it—he certainly put it on a more secure theoretical footing than Wien had done.

The Paschen–Wien law agreed well with the experimental data available at the time. Many of these data were collected at the Imperial Physical and Technical Institution (*Physikalisch-Technische Reichsanstalt* or PTR for short) in Berlin-Charlottenburg. Founded in 1887, the PTR can be seen as the German forerunner of the American National Bureau of Standards founded in 1901. Right around the time of the publication of the final installment of the pentalogy, however, experimentalists at the PTR began to cast doubt on the Paschen–Wien law, though Paschen (1899a, 1899b) continued to stand

---

[3] Reflecting on this research decades later, Planck wrote: "Back then a considerable number of physicists had turned their attention to the problem of the energy distribution in the normal [read: black-body] spectrum, both from the theoretical and the experimental side. But they all only looked in the direction of trying to express the intensity of the radiation [ρ_v] as a function of the temperature T, whereas I suspected that a deeper connection could be found in the dependence of the entropy S on the energy U. Since the meaning of the entropy concept was not yet given the recognition it deserves, nobody cared about the method I used and I could afford to do my calculations at leisure and thoroughly, without having to worry about being interrupted or overtaken by anybody" (Planck 1943, p. 261; repeated verbatim in Planck 1949, p. 392). As we will see, Planck exaggerated the lack of interest of his contemporaries in his approach but it is true that his approach stood apart from theirs.

[4] In 1901, Paschen (see Plate 2) became Full Professor in Tübingen, where in the 1910s and early 1920s he did important measurements of the fine structure and the Zeeman effect (see Chapters 6 and 7). Paschen's colleagues in Hannover in the late 1880s and early 1890s were Heinrich Kayser and Carl Runge (Mehra and Rechenberg 1982a, p. 164, Kangro 1976, p. xiv). Kayser, Paschen, and Runge did important work on line spectra (see Appendix B).

by it. The first to report deviations were Otto Lummer and Ernst Pringsheim (1899a, 1899b, 1900).[5] In a paper published in March 1900, Planck (1900b) followed Paschen in attributing these deviations to experimental errors. In this same paper, he withdrew his earlier conjecture that the second law of thermodynamics uniquely picks out the entropy function that leads to the Paschen–Wien law—or, as it was now starting to get called, the Planck–Wien law. However, he immediately supplied some new considerations, which in conjunction with the second law once again led to the entropy function for the Paschen–Wien–Planck law. Given his new derivation of this entropy function, Planck (1900b, p. 720) wrote, his confidence in the Wien law had only been strengthened.

Then, in October 1900, Heinrich Rubens (see Plate 3) and Ferdinand Kurlbaum (1900) informed him that they had found unmistakable deviations from the Wien law in the infrared. Planck (1900c) now used the wiggle room the second law of thermodynamics left him to cook up a new expression for the entropy of his resonators as a function of their energy. This new function leads to a law for the spectral distribution of the energy in black-body radiation that interpolates between the Wien law at high frequencies and what today is known as the Rayleigh–Jeans law at low frequencies (cf. note 37 below). In Section 2.4, we show in detail how Planck arrived at this new law for black-body radiation.

Planck's new law agreed remarkably well with the experimental data. At this point, however, it was little more than an inspired guess. He still had to justify his new choice for the function $S_\nu(U_\nu)$. In papers presented in December 1900 and January 1901, Planck (1900d, 1901a) provided such a justification. It is based on a definition of entropy introduced by Ludwig Boltzmann in a paper on the kinetic theory of gases. This is where Planck introduced the crucial relation $\varepsilon = h\nu$. It will be useful to preview this argument, which we will examine in detail in Section 2.5, and indicate what role the relation $\varepsilon = h\nu$ plays in it.

Boltzmann (1877) suggested that the entropy of a system in a certain macrostate is proportional to the logarithm of the number of different microstates that realize that macrostate.[6] The equilibrium state of a closed system, the state with maximum entropy, is then identified as the macrostate with the largest number of possible microstates. The number of microstates corresponding to the equilibrium state will be overwhelmingly larger than the number of microstates for any other macrostate. Boltzmann could thus rephrase the second law of thermodynamics, which says that a closed system will evolve toward its state of maximum entropy, as the statement that a system will tend to evolve toward the macrostate with the largest number of microstates and, once there, will tend not to wander too far from it.

---

[5] The experimentalists who provided the data on black-body radiation are not nearly as well known as the theorists who drew far-reaching conclusions from them. Channeling Shakespeare and Stoppard, Allan Franklin (1986, p. 1) drew attention to the neglect of experiment in this case by noting that the announcement "Lummer and Pringsheim are dead" would be about as shocking to physicists or historians and philosophers of physics as the announcement "Rosencrantz and Guildenstern are dead" in Hamlet.

[6] Boltzmann himself used the terms "state distribution" (*Zustandsverteilung*) and "complexion" to refer to macrostates and microstates, respectively.

Planck was actually the first to write down the now familiar formula, $S = k \log W$, carved into Boltzmann's tombstone (see Plate 6), which expresses this definition of the entropy S in terms of the number W of microstates. In particular, Planck was the first to identify the proportionality constant k as a constant of nature, which is the same regardless of whether we are dealing with a container filled with gas or a container filled with black-body radiation. Given its origins in Boltzmann's work, it is only natural that this constant came to be known as *Boltzmann's constant*, but Boltzmann only introduced it once (Boltzmann 1883, p. 90; see Darrigol 2018, p. 264) and never used it again (cf. Planck 1958, Vol. 3, p. 126, p. 265).

The now famous 1877 paper in which Boltzmann introduced the definition of entropy that Planck turned to in 1900 was not well-known at the time. What makes it all the more remarkable that Planck used it to derive an expression for the entropy of his resonators is that he strongly opposed the underlying view of the second law of thermodynamics as a statistical law (see Section 2.3). On this view, promoted by Boltzmann and James Clerk Maxwell, nature does not strictly preclude the decrease in entropy in isolated systems, it just makes it exceedingly unlikely. Planck's view was that the second law is an exact law, which does not allow exceptions, no matter how improbable. In fact, as the title of his pentalogy, "On irreversible radiation phenomena," indicates, what had drawn Planck to the subject of black-body radiation in the first place was that he had hoped to find an electrodynamic foundation for this strict version of the second law. Boltzmann (1897a, 1897b, 1898) immediately recognized that this attempt would fare no better than attempts to derive it from mechanics. Planck conceded that his original approach failed but adjusted it to get around Boltzmann's objections. Only in 1914, eight years after Boltzmann's suicide, did Planck come to accept that the second law is a statistical law (Needell 1988, p. xxxviii). In 1900, however, this did not stop him from borrowing some of Boltzmann's techniques for dealing with molecules in a gas and applying them to his resonators.

Planck found the average entropy S of a single resonator at frequency ν by computing the entropy S_N of a set of N such resonators and dividing the result by N. To find S_N, he determined in how many ways the total amount of energy E of these N resonators can be distributed over its members. Proceeding along the same lines as Boltzmann in a similar calculation for gas molecules, Planck divided the total energy into energy elements of size ε to get a finite count. Boltzmann assumed that ε was small yet large enough for there to be many molecules with energies in the interval $(E, E + \varepsilon)$. Assuming that the number of molecules with energies in such intervals varies continuously with E, he replaced sums over ε by integrals over E. In this way, ε dropped out of his final result (Kuhn 1987, p. 50). Planck, however, only arrived at the expression for S that he knew ahead of time would lead to his new black-body radiation law if he set $\varepsilon = h\nu$. In that case, the energy element ε does not drop out of the final result.

In Section 2.6, which concludes this chapter, we address the question of how Planck interpreted the relation $\varepsilon = h\nu$. Basically, it can be viewed in two different ways. The first option is to assume that the energy of a resonator at frequency ν can only take on the values $nh\nu$, where n is an integer. In that case the energy of a resonator is clearly quantized. The second option is to allow the full continuum of values for the energy of a resonator but to assume that all values *between* $nh\nu$ and $(n + 1)h\nu$ belong to the same microstate when counting the number of ways the energy can be distributed over these resonators.

Einstein (1906b, p. 203) eventually showed that the second option is not viable (see Section 3.4.1). If the resonator energy can take on a continuum of values, he elaborated another three years later, it would mean that, for large ν, microstates with appreciably different energies are lumped together in the count (Einstein 1909b, pp. 494–495). Incredible as it may seem, Einstein appears to have been the first to put his finger on this problem (see Section 3.4.3).

One can find snippets in Planck's writings suggesting that he was thinking in terms of this second option, interpretating the relation $\varepsilon = h\nu$ in terms of energy bins rather than energy parcels.[7] In the course of the derivation of December 1900, for instance, he wrote that "[i]f the ratio thus calculated [E/P in our notation] is not an integer, we take for P an integer in the neighborhood [of E/P]" (Planck 1900d, p. 40). This caveat, however, does not return in the more widely read version of his derivation published a month later in *Annalen der Physik* (Planck 1901a). It was not until 1908 that Planck explicitly endorsed the view that the equation expresses energy quantization, only to retreat again a few years later, when he presented what came to be known as "Planck's second theory", in which resonators absorb energy continuously but emit energy only in units of $h\nu$ (Planck 1913, pp. 190–191; cf. Chapter 3, note 78, and Section 5.1.1).

Over the past few decades, historians of science have carefully reexamined Planck's work on black-body radiation. This reexamination was triggered by *Black-Body Theory and the Quantum Discontinuity, 1894–1912*, an influential book by Thomas S. Kuhn (1978), famous for *The Structure of Scientific Revolutions* (Kuhn 1962). Before Kuhn, the received view was that Planck, the prototype of a reluctant revolutionary, had grudgingly, and perhaps not entirely consistently, accepted in December 1900 that his new black-body radiation law required the quantization of the energy of his resonators. This is the picture, for instance, emerging from the account of Planck's work given by the eminent historian of modern physics, Martin J. Klein (1962, 1963a, 1977). In his 1978 book, building on this work by Klein and a book on the early history of black-body theory and experiment by Hans Kangro (1970), Kuhn presented the diametrically opposed view that Planck did not quantize the energy of his resonators at all. Although historians of physics still disagree about the interpretation of various aspects of Planck's work (for instance, about how exactly his 1900–1 derivation of the entropy of his resonators is supposed to work) and even though the textual evidence is more ambiguous than Kuhn made it sound, there is broad consensus that Kuhn's revisionist interpretation is much closer to the truth than the standard interpretation, articulated most forcefully by Klein.[8]

---

[7] This terminology was suggested to us by Ted Jacobson (private communication).

[8] The most important contributions to the reevaluation of Planck's work on black-body radiation triggered by Kuhn's book are (in roughly chronological order): Klein's review of Kuhn's book (Klein, Shimony, and Pinch 1979, pp. 430–434); the (unpublished) dissertation on Planck by Allan A. Needell (1980, supervised by Klein) as well as his introduction to a reprint of Planck's *The Theory of Heat Radiation* (Needell 1988); an article reflecting on the controversy by Peter Galison (1981); Kuhn's (1984) response to criticisms of

2.2 Early work on black-body radiation (1860–1896)

A decade after the formulation of the first two laws of thermodynamics by Rudolf Clausius (1850) and William Thomson (1851), who became Lord Kelvin in 1892, Gustav Kirchhoff (1860) applied thermodynamics to heat radiation, which he conceived of as vibrations of a mechanical ether, the medium through which light waves and related waves in the infrared and the ultraviolet were thought to propagate. Kirchhoff showed that the ratio of a body's power of emission to its power of absorption must be a universal function of the temperature T of the body and the frequency ν of the radiation emitted and absorbed (or, equivalently, its wavelength λ). It does not depend on the body's shape, chemical composition, or other properties. It follows that all perfectly absorbing or black bodies at the same temperature T emit radiation with the same energy-over-frequency distribution. This result posed both an experimental and a theoretical challenge.[9]

The experimental challenge was to measure this universal spectral distribution of the energy in black-body radiation at different temperatures. A small hole in a box with walls kept at temperature T could serve as the source of black-body radiation. Such a box, after all, is an example of a perfect absorber. Black-body radiation was therefore also called, especially in Germany where most of this research was done,[10] cavity radiation (*Hohlraum Strahlung*). The light was separated into different frequencies with a prism and then the energy at different frequencies was measured. Starting in the 1880s, a new instrument, the bolometer, invented by the American astronomer and physicist Samuel Pierpont Langley (1881), greatly improved the accuracy of these measurements (Loettgers 2003; Lombardi 2003).

The theoretical challenge was to derive Kirchhoff's universal function—i.e., the formula for the intensity ρ of black-body radiation as a function of frequency ν and temperature T—from first principles. An important first step was the Stefan–Boltzmann law, which says that the energy per unit volume emitted per unit time by a black body at temperature T at all frequencies combined is equal to σT^4, where the proportionality constant σ is now known as the Stefan–Boltzmann constant. This law put an important constraint on candidates for ρ as a function of ν and T. Integrated over all frequencies, any such candidate had better give a result proportional to the fourth power of the absolute temperature. The law was first found experimentally by Joseph Stefan (1879) in Vienna. Five years later, Boltzmann (1884), who had been Stefan's student and was appointed professor at the University of Vienna himself in 1869, derived it from

---

his book, reprinted as an afterword in its second edition (Kuhn 1987); and subsequent analyses of Planck's work by Olivier Darrigol (1992, 2000b, 2001), Jochen Büttner, Jürgen Renn, and Matthias Schemmel (2001), Clayton A. Gearhart (2002), and Massimiliano Badino (2009, 2015). Of these more recent commentators, Gearhart is closest to Klein and Darrigol (1992, pp. 76–77) is closest to Kuhn.

[9] See Kangro (1976). For the experimental work, see Ch. 4 on 1890–1895 and Ch. 7 on 1896–1899. For the theoretical work, see Ch. 5 on Wien and Ch. 6 on Planck. For work specifically at the PTR in Berlin, see Hoffmann (2001).

[10] For discussion of contributions from British physicists, such as Balfour Stewart and John Tyndall, see Kangro (1976, Ch. 1; see also pp. 113–116 for discussion of Planck's reading of Tyndall).

a combination of thermodynamics and electrodynamics (Segrè 1980, Appendix 1, p. 301). When Boltzmann published this derivation, it was still not generally accepted, especially not on the European continent, that light is an electromagnetic wave, even though Maxwell (1861–1862) had already come to this conclusion more than twenty years earlier. It was only when Heinrich Hertz (1888a, 1888b) succeeded in generating electromagnetic waves of much longer wavelength than light in the laboratory and demonstrated that they behave in exactly the same way as visible light in reflection, refraction, diffraction, and interference experiments that Maxwell's field theory of electromagnetism, including his electromagnetic theory of light, displaced older action-at-a-distance theories on the continent.

Since Hertz, like Maxwell, died young,[11] it fell to the Dutch physicist Hendrik Antoon Lorentz to further develop Maxwell's theory. His counterpart in Britain was Joseph Larmor (Darrigol 1991, 2000a). Lorentz (1892, 1895) deviated from Maxwell in two important ways. First, he made a sharp distinction between matter and ether, the medium carrying electric and magnetic fields including those constituting light waves. In Lorentz's view, ether and matter were completely different substances, which accordingly were subject to entirely different laws. Newtonian mechanics had jurisdiction over matter, Maxwell's equations over the ether. Maxwell had derived his equations for electric and magnetic fields using various mechanical analogies (Lazaroff-Puck 2015) and never questioned that these laws were compatible with Newtonian mechanics. Lorentz, however, conceived of the ether as completely stationary, which means that Newton's third law cannot apply to it. Electric and magnetic fields can set matter in motion but matter cannot set the ether in motion. Lorentz recognized this but did not see it as a problem. "It is true," he conceded,

> that this conception would violate the principle of the equality of action and reaction...but nothing, as far as I can see, forces us to elevate that principle to the rank of a fundamental law of unlimited validity (Lorentz 1895, p. 28; discussed in Janssen 2003, pp. 34–36).

Lorentz saw electric and magnetic fields as disturbances in a new non-mechanical substance governed by new nonmechanical laws. This made it all the more interesting to study these new entities in the 1890s. It also renewed and reinvigorated interest in the problem of black-body radiation in this period.

Lorentz's second innovation, in line with his commitment to an atomistic view of matter, was to introduce the concept of small positively and negatively charged particles present in all matter and responsible for the interaction between ether and matter.[12] Given the sharp distinction Lorentz drew between the two, matter could not drag along

[11] Maxwell died in 1879 at age 48, Hertz in early 1894 just shy of his 37th birthday. Later that same year, Hertz's teacher, Hermann von Helmholtz, who had tried to connect Maxwell's theory with continental theories of electromagnetism since the 1870s, also died.

[12] For detailed analysis of this move into microphysics, see Buchwald (1985).

any ether as it moved through this medium. The only way in which the two could interact was for charged particles to generate fields in the ether and for fields, in turn, to exert forces on charged particles. Lorentz introduced these particles—first called "ions", later "electrons"—well before 1897, around which time J. J. Thomson and others succeeded in identifying so-called cathode rays (rays going from cathode to anode in vacuum tubes) as streams of particles soon to be called electrons.[13]

To return now to black-body radiation, Wilhelm Wien (1893) (see Plate 1),[14] following up on Boltzmann's prescient combination of thermodynamics and electrodynamics, found another important constraint on Kirchhoff's universal function for the spectral distribution of its energy. He showed that the intensity ρ depends on the temperature T only via the product λT of wavelength and temperature (or, equivalently, the ratio ν/T of frequency and temperature). For any T, the intensity ρ will have a maximum at some wavelength λ_{max}. Let λ_{max_1} be the wavelength where ρ peaks at T_1 and λ_{max_2} be the wavelength of the peak at T_2. As long as the dependence of ρ on λ is some power of λ multiplied by some function of λT, Wien's result entails that $\lambda_{max_1} T_1 = \lambda_{max_2} T_2$.[15] This means that the peak of the spectral distribution of black-body radiation shifts to lower wavelengths (higher frequencies) as the temperature increases. This fits with our experience. When an iron bar is heated it goes from glowing red to glowing blue as the temperature rises. This shift in the position of the peak is what Lummer and Pringsheim (1899b, p. 219) first called the "displacement law". Wien's more general result—that ρ only depends on T via λT or ν/T—also came to be called the (Wien) displacement law and played an important role in Planck's work on black-body radiation.

Illustrating how controversial the application of thermodynamics to electromagnetic radiation still was at the time, Wien (1930, pp. 16–17) recalled in his autobiography that Hermann von Helmholtz, his former teacher and his boss at the PTR in 1893, initially thought that radiation could not be treated thermodynamically. He also recalled that Helmholtz had told him that Lord Kelvin, a good friend of Helmholtz's, had referred to the application of thermodynamics to radiation as "[t]hermodynamics . . . going mad" (Wien 1930, p. 17).[16] Wien (1894) defended his approach in another paper on the topic the following year.

---

[13] Buchwald and Warwick (2001) edited a collection of essays on the discovery of the electron. Lorentz contributed to this discovery by showing that his "electron theory"—as it came to be called—could account for the splitting of spectral lines in a magnetic field discovered in 1897 by his countryman Pieter Zeeman after whom the effect is named (see Section 7.2.1). In 1902, Zeeman and Lorentz shared the Nobel prize for discovering and explaining this effect, respectively (Kox 1997). The Zeeman effect, especially the more complicated patterns of splittings in the so-called anomalous Zeeman effect subsequently found, went on to play an important role in the development of quantum physics (see Section 7.3).

[14] For discussion of Wien's contributions to black-body theory, see Kangro (1976, Ch. 5)

[15] Let $\rho(\lambda, T) = \lambda^k f(\lambda T)$. If this function has a maximum for $\lambda = \lambda_{max}$, then $\partial \rho / \partial \lambda = k\lambda^{k-1} f(\lambda T) + \lambda^k f'(\lambda T)T = 0$ or $kf(x) + xf'(x) = 0$ (with $x \equiv \lambda T$) for $\lambda = \lambda_{max}$. It follows that the product $\lambda_{max} T$ is the same for all temperatures T.

[16] Quoted by Mehra and Rechenberg (1982a, p. 30, note 31).

Two years later, Wien (1896) proposed a concrete candidate for Kirchhoff's universal function satisfying the constraints coming from the Stefan–Boltzmann law and the Wien displacement law.[17] Following a suggestion by the Russian physicist Vladimir Alexandrovich Michelson (1887),[18] Wien found this formula for the distribution of the energy of black-body radiation over different wavelengths by connecting it to the well-known Maxwell–Boltzmann distribution of the energy of a gas over different velocities of its molecules (Maxwell 1860; Boltzmann 1872). In an ideal gas in thermal equilibrium at temperature T the fraction $f(v)\,dv$ of molecules (mass m) per unit volume with speeds between v and $v + dv$ is given by (in modern notation)

$$f(v) = C(T)v^2 e^{-\frac{1}{2}mv^2/kT}, \qquad (2.1)$$

with $C(T) \equiv 4\pi\,(m/2\pi kT)^{3/2}$. Wien imagined black-body radiation emitted by such a gas. He assumed that the amount of radiation energy at a certain wavelength λ is proportional to the number of gas molecules emitting radiation of that wavelength. He added the rather dubious assumption that the wavelength λ at which a gas molecule emits radiation is a function only of its velocity v. Given those assumptions, he argued, the spectral distribution of the energy density in black-body radiation takes the form

$$\rho(\lambda, T) = F(\lambda)e^{-f(\lambda)/T}, \qquad (2.2)$$

where $F(\lambda)$ and $f(\lambda)$ are two yet to be determined functions. Appealing to his displacement law, he set $f(\lambda)/T = C_2/\lambda T$. Appealing to the Stefan–Boltzmann law, he set $F(\lambda) = C_1/\lambda^5$. The constants C_1 and C_2 are parameters to be fitted to the black-body radiation curves obtained experimentally. Inserting these choices for $F(\lambda)$ and $f(\lambda)$ into Eq. (2.2), we arrive at

$$\rho(\lambda, T) = (C_1/\lambda^5)\,e^{-C_2/\lambda T}. \qquad (2.3)$$

A law of the same form but with one more free parameter had already been proposed on experimental grounds by Friedrich Paschen (1896–1897) (see Plate 2). Instead of C_1/λ^5, Paschen had C_1/λ^α. His experiments suggested a value for α of about 5.5 (Kangro 1976, p. 85). This value was close enough to 5 for Eq. (2.3) to become known as the Paschen–Wien law.

The energy $\rho(\lambda, T)d\lambda$ per unit volume in a narrow wavelength interval $(\lambda, \lambda + d\lambda)$ can be rewritten as the energy $\rho(v, T)dv$ per unit volume in the corresponding frequency interval $(v, v - dv)$. Using that $\lambda = c/v$ (with c the velocity of light) and $d\lambda/dv = -c/v^2$ and

---

[17] Inspection of Fig. 2.3 suffices to see at least qualitatively that these constraints are satisfied. With increasing temperature, the area under the curve sharply increases and the peak moves to lower wavelengths.

[18] See Kangro (1976, pp. 30–36) for discussion of the contributions to black-body theory by Michelson (1887, 1888), not to be confused with the Polish-born but American physicist Albert A. Michelson.

absorbing c into new parameters $c_1 \equiv C_1/c^4$ and $c_2 \equiv C_2/c$, we can rewrite the Paschen–Wien law (2.3) in terms of frequency:

$$\rho(v, T) = c_1 v^3 e^{-c_2 v / T}. \tag{2.4}$$

For the next three years, the Paschen–Wien law held up well as black-body radiation measurements were done with ever increasing accuracy over ever widening ranges of frequencies and temperatures.

2.3   Planck, the second law of thermodynamics, and black-body radiation (1895–1899)

In June 1896, Wien left Berlin to take up a professorship in Aachen and Planck replaced him as "resident theoretician" (*Haustheoretiker*) at the PTR (Jungnickel and McCormmach 1986, Vol. 2, p. 254). Planck had succeeded Kirchhoff in Berlin in 1889 (after Boltzmann had turned down the position) and become Director of a new Institute for Theoretical Physics, created especially for him. Though he would spend the rest of his career in Berlin, he had earned his doctorate in Munich—the year that Maxwell died and Einstein was born—with a dissertation on the second law of thermodynamics (Planck 1879). In later reminiscenses, he complained that none of the leading German physicists at the time, notably Helmholtz, Kirchhoff, and Clausius, had shown any interest in his dissertation (Planck 1943, p. 257).

As we mentioned in Section 2.1, Planck believed in the strict validity of the second law. Initially, he even believed that the second law was incompatible with the existence of atoms. By 1896, Planck had joined Boltzmann in his attack on Mach's anti-atomism (Heilbron 1986, p. 45) but in the conclusion of one of his early papers, he wrote:

> The second law of the mechanical theory of heat, consistently developed, is incompatible with the assumption of finite atoms. It can therefore be foreseen that the further development of the theory will lead to a battle between these two hypotheses [the second law as Planck understood it and the atomic hypothesis] in which one of them will perish (Planck 1882, p. 475; discussed by Kuhn 1987, pp. 23–24, and Darrigol 1992, pp. 22–23).

Planck went on to say that, although it would be premature to predict the outcome of this battle, he saw signs "that atomic theory despite its great success, will ultimately have to be abandoned in favor of the assumption of continuous matter" (ibid.).

In a footnote to the first sentence of the passage quoted above,[19] Planck referred to a famous thought experiment in Maxwell's book *Theory of Heat* (Maxwell 1871, p. 328–329). Maxwell had first described this thought experiment in a letter to Peter

---

[19] This footnote is missing from the reprint of Planck (1882) in Planck (1958, pp. 134–163).

Guthrie Tait of December 11, 1867 (Klein 1970b). Consider a container filled with an ideal gas at a certain uniform temperature. The velocity distribution of the molecules in this gas will be given by the Maxwell–Boltzmann distribution for that temperature. Now imagine a partition dividing the container into two halves and a small hole in that partition fitted with a door that can be opened and shut rapidly and with negligible effort to allow molecules to pass from one side to the other, one at the time. Maxwell now introduced a doorkeeper, who was to open and close the door so as to trap relatively fast molecules on one side of the partition and relatively slow molecules on the other. In this way, the elusive doorkeeper, immediately called a "demon" by William Thomson and known ever since as "Maxwell's demon" (Klein 1970b, p. 86), could raise the temperature of one half of the gas and lower the temperature of the other half without doing any appreciable work. By continuing to do so, the demon could move heat from a cold to a hot reservoir without doing work, in violation of the second law of thermodynamics in the formulation of Clausius (1850).

The demon was only window dressing for Maxwell. The point of his thought experiment was that there is an admittedly tiny probability that for a short period of time the redistribution of slow and fast molecules over the two halves of the container that would result from the actions of the demon would happen spontaneously. As Maxwell summarized the upshot of his thought experiment in a letter to John William Strutt (Lord Rayleigh) of December 6, 1870: "The second law has the same degree of truth as the statement that, if you throw a tumblerful of water into the sea, you cannot get the same tumblerful of water out again" (Klein 1970b, p. 86).

Planck not only disagreed with this interpretation of the second law, he took a dim view of the underlying kinetic theory of gases. In 1891, for instance, in a lecture at the annual meeting of the Society of German Natural Scientists and Physicians (*Gesellschaft Deutscher Naturforscher und Ärzte*), held that year in Halle,[20] he remarked: "Anyone who has studied the works of Maxwell and Boltzmann ... will scarcely be able to escape the impression that the remarkable physical insight and mathematical skill exhibited in conquering these problems is inadequately rewarded by the fruitfulness of the results gained" (Planck 1891, p. 373; quoted and discussed by Kuhn 1987, p. 22).

Given this attitude toward atomism and the kinetic theory of gases, it is hardly surprising that Planck turned to electromagnetism and the ether to find a foundation for the second law of thermodynamics. In this connection, he studied idealized miniature versions of the electric circuits that Hertz had used to generate and detect electromagnetic waves. He published a pair of papers analyzing the *radiation damping* of such Hertzian resonators, the damping of the oscillation of the resonator as a result of its interaction with the radiation (Planck 1895, 1896). He quickly discovered that the results do not depend on the detailed structure of the resonators. For our purposes, it is convenient to think of them as charged harmonic oscillators. For one thing, this connects Planck's resonators to the charged particles responsible for the emission and absorption of light in

---

[20] Starting in 1822, such a meeting, known as the *Naturforscherversammlung*, was held in a different city in the German-speaking world every year.

Lorentz's theory, even though Planck, given his reservations about atomism at the time, did not make that connection (Darrigol 1992, pp. 30–31, pp. 37–38).

The key point for Planck was that radiation damping of a resonator is a *conservative* process, unlike damping due to friction or to the electrical resistance of a current-carrying wire, both resulting in a loss of heat to the environment. In radiation damping, the sum of the energy in the resonator and the energy in the radiation always remains constant. In Planck's view, this made radiation damping a promising candidate for the derivation of the second law of thermodynamics from fundamental processes. As he put it in the introduction of his second paper on the topic:

> The study of conservative damping appears to me to be of fundamental importance because it shows the possibility of a general explanation of irreversible processes with the help of conservative forces—a problem which confronts the theoretical research in physics more urgently every day (Planck 1896, p. 470, quoted and discussed by Mehra and Rechenberg 1982a, p. 34).

He immediately added, however, that the elaboration of this idea would have to wait for another occasion. In the conclusion of his first paper on radiation damping, Planck mentioned in passing that his analysis also throws light on Kirchhoff's law about the universality of the ratio of power of emission and power of absorption (Planck 1895, p. 458). This is the only reference to black-body radiation in these two papers.

Planck carried out this program of finding an electromagnetic foundation for the second law of thermodynamics in what would become a five-part paper entitled "On irreversible radiation phenomena" (Planck 1897–1899, Pts. I–V).[21] As part of the program of this pentalogy, Planck hoped to derive the law for the spectral distribution of black-body radiation. He already made this clear in the introduction of the first installment:

> The resonator, however, also influences the "coloring" of the wave exciting it, if one takes this to mean the distribution of the total intensity over the various simply periodic oscillations contained in it, and it is to be expected that it will somehow produce here too a certain evening out among the intensities of different colors, from which important conclusions could then be drawn about the energy distribution in the stationary [i.e., equilibrium] state of the radiation (Pt. I, p. 496).

Planck's subsequent work would prove this expectation wrong but eventually he would nonetheless find a way to derive the law for black-body radiation within the general framework he developed.

Planck's original idea for deriving irreversibility from the interaction between resonators and radiation also turned out to be wrong. On the face of it, the idea seems plausible enough. When an electromagnetic plane wave of frequency v hits a resonator

---

[21] Our discussion of this paper follows Darrigol (1992, Ch. 3, pp. 39–56). See also Kuhn (1987, Ch. 4) and Badino (2015, sec. 3.2, pp. 45–75, "The Pentalogy").

with a resonance frequency close to ν, the resonator will absorb energy from this wave and reemit part of it in the form of an outgoing concentric spherical wave. We never witness the inverse of this process, i.e., a concentric wave converging on the resonator at its center, which is then reemitted as a plane wave. Planck thus seemed to have found a conservative yet irreversible process.

Boltzmann (1897b) quickly put his finger on the flaw in Planck's reasoning.[22] Maxwell's equations, like the equations of Newtonian mechanics, are time symmetric: if they allow the scenario of an incoming plane wave reemitted as an concentric wave, they also allow the time-reversed scenario of an incoming concentric wave reemitted as a plane wave. The difference is that the latter calls for delicate initial conditions not normally realized.

Planck came to accept this criticism and looked for ways to rule out those special initial conditions (Planck 1897–1899, Pt. II; Pt. III, p. 511, p. 531). Inspired by Boltzmann's hypothesis of "molecular chaos", he adopted the hypothesis of "natural radiation" (Planck 1897–1899, Pt. IV, sec. 25, pp. 556–558; Pt. V, sec. 9, pp. 571–573; cf. Darrigol 1992, pp. 40–43). The radiation in which a resonator is immersed will be a superposition of waves of different frequencies moving in different directions. The hypothesis of "natural radiation" says that the phases of all these different waves are uncorrelated.[23] Reflecting their different views on irreversibility, Planck and Boltzmann looked upon their respective hypotheses in fundamentally different ways. For Boltzmann, "molecular chaos" implied that it is *highly unlikely* that the state of all the molecules in a gas is such that the entropy of the gas decreases. For Planck, "natural radiation" implied that such processes as converging concentric waves are *strictly ruled out*. A reinterpretation of Boltzmann's hypothesis of "molecular chaos" along these lines provided Planck with a way to reconcile the existence of atoms with the second law of thermodynamics: Nature somehow prohibits scenarios such as those encountered in Maxwell's demon argument (Planck 1900a, pp. 619–620; cf. Darrigol 1992, pp. 53–54).

Planck was now finally ready to give his electromagnetic derivation of the second law of thermodynamics. In the last and longest installment of his pentalogy, submitted in May 1899—and reprised, for the most part verbatim, in a paper submitted to *Annalen der Physik* in November 1899 (Planck 1900a)—he showed that the entropy, suitably defined, of a system consisting of resonators and electromagnetic radiation can never decrease. The reactions of Helmholtz and Kelvin to Wien's work mentioned in Section 2.2 may serve as a reminder that the application of thermodynamics to electromagnetic radiation was still controversial in the 1890s. Planck thus explicitly defended the position that one should ascribe both entropy and temperature to radiation (Planck 1897–1899, Pt. V, Secs. 20–21, pp. 592–594; 1900a, Secs. 20–21, pp. 112–115; 1900b, pp. 721–722).

[22] See Darrigol (1992, pp. 39–41) and Badino (2015, pp. 51–60) for discussion of the ensuing controversy between Boltzmann (1897b, 1897c, 1898) and Planck (1897–1899, Pt. II, Pt. IV, sec. 26, pp. 558–559).
[23] For a more careful definition of "natural radiation", see the web resource *Classical dispersion theory*.

Planck's proof that the combined entropy of resonators and radiation never decreases is closely analogous to Boltzmann's (1872) proof of his famous *H*-theorem. Oddly, Planck did not mention the *H*-theorem or refer to Boltzmann in this context. Yet, it will be instructive to explore this analogy and to bring out an important disanalogy between the two cases.[24]

Boltzmann (1872) showed that, barring what he took to be highly improbable initial states, any distribution of the velocities of the molecules in a gas will over time evolve toward the Maxwell–Boltzmann distribution (see Eq. 2.1). He proceeded in two steps. First, he derived an equation, now known as the *Boltzmann equation*, for the time evolution of the velocity distribution $f(\vec{v}, t)$ as a result of elastic collisions of pairs of molecules. This equation told him that the time derivative of $f(\vec{v}, t)$ vanishes if the velocity distribution is the Maxwell–Boltzmann distribution. Hence, the Maxwell–Boltzmann distribution is *an* equilibrium distribution.

The second step was to prove that the Maxwell–Boltzmann distribution is the *only* equilibrium distribution. To this end, Boltzmann introduced the quantity *H* for an arbitrary velocity distribution $f(\vec{v}, t)$. He defined it as

$$H \equiv \int d^3 v f(\vec{v}, t) \ln f(\vec{v}, t).$$ (2.5)

Using the Boltzmann equation for $\partial f(\vec{v}, t)/\partial t$, Boltzmann showed that the time derivative of *H* is always less than or equal to zero: $dH/dt \leq 0$. This result is known as the *H*-theorem. The only distribution for which $dH/dt = 0$ is the Maxwell–Boltzmann distribution. In other words, the *H*-function monotonically decreases and reaches its only minimum for the Maxwell–Boltzmann distribution. Boltzmann had thus demonstrated that this is indeed the only equilibrium distribution. He had actually accomplished a good deal more. The function *H*, which never increases, is directly related to the entropy, which never decreases. In fact, Boltzmann showed that $- H$ has all the properties of entropy. His derivation of the *H*-theorem thus amounted to a derivation of the second law of thermodynamics from a conservative and time-reversible process (the elastic collision of molecules) *and some statistical assumptions*. The paper five years later (Boltzmann 1877) that Planck would draw on for the derivation of his new black-body radiation law in December 1900, was written to clarify these statistical assumptions in response to incisive criticism of the *H*-theorem by Boltzmann's Vienna colleague Joseph Loschmidt (1876).

---

[24] The first to explore this analogy and to put his finger on the disanalogy was Boltzmann's student Paul Ehrenfest (1905, 1906). For discussion, see Klein (1970a, pp. 232–233, pp. 238–239), Kuhn (1987, Chs. 3 and 6), Darrigol (1992, p. 44, p. 51, p. 55), and Badino (2015, p. 60, p. 74). Kuhn and Darrigol refer to Planck's result as the "electromagnetic *H*-theorem". A classic discussion of Boltzmann's *H*-theorem can be found in the famous article in the *Encyclopedia of the Mathematical Sciences* on the foundations of statistical mechanics by Ehrenfest and his wife Tatiana Ehrenfest-Afanassjewa (1911). For more recent discussion, see Klein (1970a, Secs. 6.3–6.9, pp. 100–128), Brown, Myrvold, and Uffink (2009) and, presenting a view opposing that of Klein and Brown *et al.*, Badino (2011).

Twenty years later, Boltzmann (1896, 1897a) had to defend his H-theorem again, this time against the criticism of one of Planck's former students, Ernst Zermelo (1896a, 1896b). Zermelo tried to refute the H-theorem by appealing to a theorem by the French mathematician Henri Poincaré (1890, pp. 67–72). According to this theorem, any mechanical system will eventually (even though this may take longer than the age of the universe) return to a state arbitrarily close to the state it is currently in. This is now called *Poincaré recurrence*. In the opening paragraph of his reply, Boltzmann (1896, p. 773), sounding somewhat exasperated, pointed out that he and others had already explained many times before that the second law of thermodynamics is a statistical law and that the H-theorem should therefore not be treated as a theorem in mechanics (Badino 2015, pp. 50–51).

Planck (1897–1899, Pt. I, p. 494) referred to the exchange between Boltzmann and Zermelo in the first part of his pentalogy. He explicitly put forward his electromagnetic approach to deriving the second law as an alternative to Boltzmann's. He had to answer Boltzmann's (1897b, 1897c, 1898) criticisms of this approach[25] before he could finally take a stab at such a derivation. When he finally did, he modeled his derivation on Boltzmann's derivation of the H-theorem, but then, oddly, failed to acknowledge its originator.[26]

Intuitively, the mechanism behind Planck's electromagnetic version of the H-theorem is the following. Consider a plane wave of frequency v traveling through a cavity containing many resonators with resonance frequency v, sufficiently far away from each other to make any direct influence they might have on one another completely negligible. After a period of time, the resonators will have scattered the plane wave in such a way that throughout the cavity the radiation will no longer be traveling in one direction but in all directions. In other words, the scattering will make the radiation homogeneous and isotropic. This corresponds to an increase in entropy of the radiation. Planck introduced expressions for the entropy of radiation and resonators modeled on Boltzmann's H-function and derived an equation for the time evolution of the total entropy that has a structure similar to the equation Boltzmann derived for the time evolution of H. Just as Boltzmann could show that $dH/dt \leq 0$, Planck could show that $dS_{tot}/dt \geq 0$, where S_{tot} is the sum of the entropy of the resonators and the radiation (Planck 1897–1899, Pt. V, secs. 17–18, p. 584–589; 1900a, secs. 17–18, pp. 104–109).

This finally brings us to the disanalogy between the two cases alluded to above (see note 24). The collision processes studied by Boltzmann drive the velocity distribution of the gas molecules to the equilibrium Maxwell–Boltzmann distribution. By contrast, the scattering processes studied by Planck do not change the frequency distribution of the

[25] Among other things, Boltzmann (1897c) pointed out that the derivation of the second law that Planck proposed was just as vulnerable to Zermelo's recurrence objection as the H-theorem (Badino 2015, p. 53).

[26] This is not meant to denigrate Planck's result, which according to one of the commentators whose analysis of the electromagnetic H-theorem we are following "might be regarded as Planck's greatest theorem" (Darrigol 1992, p. 48).

radiation at all (Ehrenfest 1905, p. 1303; 1906, pp. 528–529)![27] Contrary to what Planck expected when he embarked on his program of giving an electromagnetic derivation of the second law, a resonator only interacts with radiation in a narrow range around its resonance frequency. The interaction will smoothen out the radiation at that frequency (i.e., make it homogeneous and isotropic), but there will be no transfer of energy from one frequency band to another.[28] In the preceding paragraph, we considered radiation and resonators at one particular frequency. Now imagine radiation and resonators at many different frequencies. The interaction of resonators and radiation will smoothen out the radiation at all frequencies but it will preserve whatever distribution of energy over the different frequencies we started out with.[29] Another mechanism is needed to change some arbitrary initial distribution to the universal equilibrium distribution of black-body radiation. Fortunately, there is no need to analyze that additional mechanism in detail. We can just imagine the whole system of radiation and resonators to be put in contact with a heat bath, in this case the walls of the cavity kept at some fixed temperature T (see Figure 2.2).

As long as the system of resonators and radiation at all different frequencies is enclosed in a container with perfectly reflecting walls, its overall equilibrium state could be one in which the equilibrium states of subsystems consisting of resonators and radiation at the same frequency correspond to different temperatures (Ehrenfest 1905, pp. 1309–1310). Once the walls, kept at temperature T, can absorb radiation at one frequency and emit it at another, all these subsystems will rapidly evolve to the universal black-body equilibrium distribution at one and the same temperature T. To ensure that the system of resonators and radiation reaches thermal equilibrium, Planck thus implicitly assumed the second law of thermodynamics. He could not derive the approach to equilibrium from his electromagnetic H-theorem (cf. note 24). What he could derive was that, for all frequencies considered separately, the total entropy of resonators and radiation at that frequency can never decrease.

Imagining the resonators and radiation at all frequencies to be in contact with a heat bath at temperature T, however, Planck did produce a derivation of a formula for black-body radiation that he had hoped to find as a by-product of his electromagnetic derivation

---

[27] Modern quantum electrodynamics removes the disanalogy: Photon–photon scattering will in the long run drive the spectral distribution of radiation to the universal black-body equilibrium distribution just as collisions between molecules will drive the velocity distribution of the molecules in a gas to the Maxwell–Boltzmann distribution.

[28] Planck acknowledged this in the conclusion of his book on heat radiation as well as in his Nobel lecture (Planck 1906a, p. 220; Planck 1920, p. 123). In the conclusion of the book he suggested that collisions between resonators could change the distribution of energy over frequency. Earlier in the book, Planck (1906a, p. 66) had introduced a "speck of soot" [*Kohlenstäubchen*] to accomplish this same goal. Peter Debye (1910, p. 1428 and p. 1432) also introduced a speck of soot as the conduit through which the energy of the radiation reaches its equilibrium spectral distribution.

[29] Ehrenfest (1906, p. 529, note) quotes the same passage from the first installment of the pentalogy that we quoted above in which Planck suggested that the interaction of the radiation with the resonators would also change the frequency distribution of its energy. Ehrenfest rejects this suggestion. He argues that the energy in the radiation will not be redistributed over different frequencies, even if there is overlap between the narrow frequency ranges in which resonators at neighboring frequencies interact with the radiation.

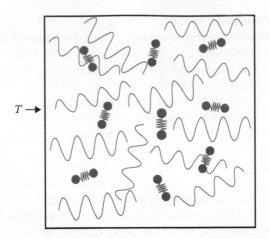

Figure 2.2 *Resonators at frequencies ν_1 and ν_2 (with $\nu_1 < \nu_2$) interacting with radiation at those same frequencies in a cavity with walls kept at temperature T absorbing and reemitting radiation.*

of the second law. For this derivation, Planck had to combine a standard relation from thermodynamics with two results of his pentalogy. The standard relation is

$$\frac{dS}{dU} = \frac{1}{T}, \tag{2.6}$$

where S is the entropy and U is the energy.[30] In the case of radiation, Planck *defined* the temperature through this relation (Planck 1897–1899, Pt. V, p. 594; 1900a, p. 115).

In addition, Planck put to good use a simple relation he had found between the energy density ρ_ν of natural radiation at ν and the average energy U_ν of a resonator at that same frequency immersed in this radiation

$$\rho_\nu = \frac{8\pi \nu^2}{c^3} U_\nu. \tag{2.7}$$

It had taken Planck considerable effort to derive this result (Planck 1897–1899, Pt. V, p. 575; 1900a, p. 93).[31]

---

[30] When an amount of heat ΔQ is added to a system, it is converted into internal energy ΔU and/or work $\Delta W = p\Delta V$ (where p is the pressure and ΔV is the change in volume). If there is no work, the relation $dQ = dW + dU = pdV + dU$ reduces to $dQ = dU$. Using the thermodynamical definition of entropy $dS = dQ/T$ (i.e., turning the inexact differential dQ into an exact one), one finds $dS/dU = 1/T$.

[31] For a modern derivation of this result, see the web resource *Classical dispersion theory*. A condensed version of such a derivation can be found in Darrigol (2009, Appendix, pp. 164–165). For a reconstruction of Planck's own derivation, see Darrigol (1992, pp. 43–44, "The fundamental equation," and p. 49).

The final piece of the puzzle was Planck's choice of an expression for the entropy S_v of a resonator at frequency v as a function of U_v (Planck 1897–1899, Pt. V, p. 585; 1900a, p. 104):

$$S_v = -\frac{U_v}{av} \ln \frac{U_v}{ebv},$$ (2.8)

where a and b are constants and e is the base of the natural logarithm. This choice of S_v leads directly to the Wien law for ρ_v. From Eqs. (2.6) and (2.8), it follows that[32]

$$\frac{1}{T} = \frac{dS_v}{dU_v} = -\frac{1}{av} \ln \frac{U_v}{bv}.$$ (2.9)

Solving for U_v, we find

$$U_v = bve^{-av/T}.$$ (2.10)

Using Eq. (2.7), we arrive at

$$\rho_v = \frac{8\pi bv^3}{c^3} e^{-av/T},$$ (2.11)

which is the Wien law (2.4) with the constants $(8\pi b/c^3)$ and a replacing c_1 and c_2, respectively.

How did Planck find the critical expression (2.8) for S_v as a function of U_v? First, it follows from the Wien displacement law that S_v must be a function of U_v/v. In a footnote to the paper in which he first introduced the function $S_v(U_v)$ leading to his new law for black-body radiation, Planck (1900c, p. 36) stated explicitly that the displacement law is equivalent to assuming that $S_v = f(U_v/v)$ (with f some arbitrary function). In the paper on his new law in *Annalen der Physik* early the following year, Planck (1901a, pp. 310–311) gave a proof of this statement. Second, comparison of Eqs. (2.5) and (2.8) suggests that Planck modeled S_v on Boltzmann's H function. Other than that, Planck justified Eq. (2.8) (and a similar expression for the entropy of the radiation) by noting that no other choice he tried worked. "I have made repeated efforts," he confessed, "to modify or generalize [Eq. (2.8) for S_v] in such a way that all well-founded electromagnetic and

---

[32] Note that

$$\frac{dS_v}{dU_v} = \frac{d}{dU_v}\left(-\frac{U_v}{av}\ln\frac{U_v}{bv} + \frac{U_v}{av}\right)$$

$$= -\frac{1}{av}\ln\frac{U_v}{bv} - \frac{U_v}{av}\left(\frac{bv}{U_v}\frac{1}{bv}\right) + \frac{1}{av}.$$

The last two terms cancel each other.

thermodynamical laws are still satisfied but I have not succeeded" (Planck 1897–1899, Pt. V, p. 596; 1900a, p. 118; cf. Darrigol 1992, p. 49). He briefly discussed the generalization

$$S_v = -\frac{U_v}{f(v)} \ln \frac{U_v}{\varphi(v)} \tag{2.12}$$

of Eq. (2.8), where $f(v)$ and $\varphi(v)$ are arbitrary functions. Proceeding the way we got from Eq. (2.8) for S_v to Eq. (2.11) for ρ_v, we find that the generalization Eq. (2.12) of the former leads to the generalization

$$\rho_v = \frac{8\pi v^2 \varphi(v)}{e\, c^3} e^{f(v)/T} \tag{2.13}$$

of the latter. Planck rejected the more general form (2.12) for S_v on the grounds that the Wien displacement law and the Wien radiation law (2.4), respectively, require that $f(v)$ and $\varphi(v)$ in Eq. (2.13) are linear in v as in Eq. (2.11). Finally, he explained:

> If one attempts to start from a energy distribution law of a form different from [Eq. (2.13)] and work backwards to an expression for the entropy, one always finds contradictions to the theorems proved [in an earlier section of Planck's paper] about entropy increase (Pt. V, p. 596; 1900a, p. 118).

These considerations suggest that Planck found Eq. (2.8) for the entropy of a resonator, at least in part, by working backwards from the Wien law (Klein 1970a, p. 223).

Planck was confident that what he, despite considerable effort, had been unable to find did not exist: an alternative to expression (2.8) for the entropy of a resonator compatible with the second law of thermodynamics. As far as Planck was concerned, the Wien law, which follows directly from this expression for the entropy, and the second law of thermodynamics stand or fall together:

> I believe that it must therefore be concluded that the definition for the entropy of radiation, and also the Wien distribution law for the energy that goes with it, is a necessary consequence of applying the principle of entropy increase to the electromagnetic theory of radiation, and that the limits of validity of this law, should there be any, therefore coincide with those of the second law of thermodynamics. Further experimental test of this law naturally acquires all the more fundamental interest for this reason (Planck 1897– 1899, Pt. V, p. 597; 1900a, p. 118; cf. Klein 1977, p. 5).

Further evidence of Planck's strong confidence in his derivation of the Wien law is provided by the concluding section of both the fifth part of his pentalogy and its reprise in the *Annalen*. In this section, he proposed a system of natural units based on the values of the velocity of light, Newton's gravitational constant, and the two constants a and b in the Wien law (Planck 1897–1899, Pt. V, sec. 26, pp. 599–600; 1900a, sec. 26, pp. 120–122).

2.4 From the Wien law to the Planck law: changing the expression for the entropy of a resonator

Right after deriving the Wien law in his paper of May 1899, Planck (1897–1899, Pt. V, p. 595) reported that the law's "validity to at least a very good approximation" had recently been established in on-going experiments on black-body radiation.[33]

Figure 2.3 is an example of the kind of graph in which the results of these experiments were presented. It shows experimental and theoretical plots of the intensity of the radiation against its wavelength for various temperatures. The theoretical plots are based on the Wien law with a choice of values for its two free parameters (C_1 and C_2 in Eq. (2.3)) that give the best fit for the range of wavelengths and temperatures probed in the experiment. The most recent results cited by Planck in May 1899 had been published in February by Lummer and Pringsheim (1899a) and in April by Paschen (1899a). While Lummer and Pringsheim reported small deviations from the Wien law, especially for long wavelengths, Paschen continued to find agreement with the law he himself had first proposed (see Section 2.2).

By the time Planck submitted the minimally reworked version of his May 1899 paper to *Annalen der Physik* in November 1899, Lummer and Pringsheim (1899b) had just published another paper reporting further deviations from the Wien law, as had their PTR colleague Rubens (1899). Planck (1900a, p. 117) added a footnote to the page proofs of his *Annalen* paper acknowledging these findings but did not change the sentence in the main text proclaiming the validity of the Wien law "to at least a very good approximation" nor the paragraph quoted above stating that the Wien law and the second law of thermodynamics stand or fall together.

In March 1900, Planck (1900b) published another paper in *Annalen*, in which he reexamined his derivation of the Wien law given that its unrestricted validity had been called into question, even though Paschen's (1899b) latest measurements, published in December 1899, continued to support it. One outcome of this reexamination was that Planck had to reverse himself on an important point. Contrary to what he had confidently asserted the year before, the second law of thermodynamics in and of itself does *not* uniquely determine the expression for the entropy of a resonator as a function of its energy.[34] In fact, the only constraint the second law puts on this function is that its second-order derivative cannot be positive:

$$\frac{d^2 S_\nu}{dU_\nu^2} = -f(U_\nu),\tag{2.14}$$

where f can be any non-negative function (Planck 1900b, pp. 730–731). On the basis of more specific (though ultimately unwarranted) assumptions about his resonators, Planck

---

[33] See Kangro (1976, pp. 167–180) for detailed analysis of the experimental situation in 1899.

[34] Both Lummer and Pringsheim (1899b, cf. Kangro 1976, p. 180) and Wien (1900a, p. 38) complained that Planck had not proven this uniqueness. For Planck's reaction, see Planck (1900d, p. 43).

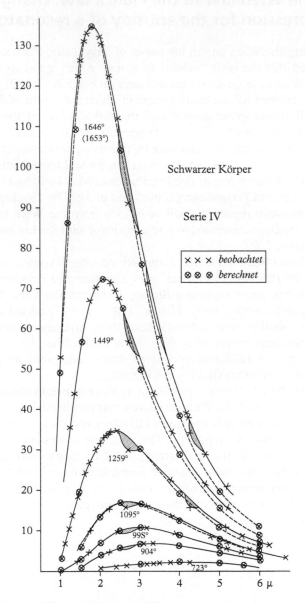

Figure 2.3 *Plots of the intensity of black-body radiation against wavelength at different temperatures from a paper of November 1899 by Otto Lummer and Ernst Pringsheim (1899b, p. 217; also reproduced in Kangro 1976, p. 176). Observed (beobachtet) values are indicated by the symbol ×; calculated (berechnet) values by ⊗. The calculated values are based on the Wien law (2.3) with a choice of C_1 and C_2 giving the best overall fit with the data.*

(1900b, p. 732) was able to determine the form of $f(U_\nu)$ and could replace Eq. (2.14) by the more specific

$$\frac{d^2 S_\nu}{dU_\nu^2} = -\frac{\alpha}{U_\nu}. \tag{2.15}$$

Using the Wien displacement law, he could set $\alpha = 1/a\nu$ (1900b, p. 733). Planck was thus led straight back to the Wien law: Differentiating Eq. (2.9) for the first derivative of the entropy function (2.8) leading to the Wien law, we find

$$\frac{d^2 S_\nu^{\text{Wien}}}{dU_\nu^2} = \frac{d}{dU_\nu} \left(-\frac{1}{a\nu} \ln \frac{U_\nu}{b\nu} \right) = -\frac{1}{a\nu U_\nu}. \tag{2.16}$$

Understandably, given this new derivation, Planck (1900b, p. 720) announced in the introduction of his paper that his confidence in the Wien law had only been strengthened, "even though some of the grounds on which it is based have shifted somewhat."

In the following months, however, experimental support for the Wien law continued to erode. In a paper based on their presentation at the *Naturforscherversammlung* (cf. note 20) in Aachen in September 1900, Lummer and Pringsheim (1900, p. 171) concluded that "the Wien–Planck spectral equation does not represent the black radiation measured by us in the range from 12μ to 18μ [i.e., in the mid-infrared]" (Kangro 1976, pp. 195–196). Deducing values of the parameters C_1 and C_2 from data at different frequencies and temperatures, they found a "wandering" (Kangro 1976, p. xv) of these constants and concluded, contrary to what Planck (1897–1899, Pt. V, pp. 599–600; 1900a, pp. 120–122) had suggested, that especially the constant C_2 "in the Wien–Planck equation should not be treated as a 'natural constant'" (Lummer and Pringsheim 1900, p. 172).

Meanwhile, in June 1900, Lord Rayleigh (1900b) had published a short article in *Philosophical Magazine* in which he proposed an alternative to the Wien law (Klein 1970a, pp. 235–236; Kangro 1976, pp. 190–192).[35] Unlike Planck, Rayleigh did not take the detour via the entropy of resonators to derive his black-body radiation law. He used a powerful but highly controversial result from statistical mechanics, the *equipartition theorem*,[36] and applied it directly to the oscillations in the ether making up electromagnetic radiation according to the physics of the day. He found that the energy density ρ_ν of black-body radiation is proportional to $\nu^2 T$.[37] Rayleigh realized that

---

[35] Various other alternatives to the Wien law were proposed, notably by Max Thiesen (1900) and by Lummer and Eugen Jahnke (1900). For discussion, see Kangro (1976, Ch. 8, pp. 184–186 (Thiesen), pp. 192–194 (Lummer and Jahnke)).

[36] See Eqs. (3.3)–(3.8) in Section 3.2 for a derivation and Section 1.2.3 for discussion of the controversy surrounding the theorem.

[37] As we will see in Section 3.3.1, Einstein (1905a) pointed out that this is a direct consequence of Planck's Eq. (2.7) as long as the equipartition theorem can be used to set the average energy U_ν of Planck's resonators (which have two degrees of freedom) equal to kT. The resulting black-body radiation law, $\rho_\nu = (8\pi/c^3) \nu^2 kT$, is now known as the Rayleigh–Jeans law. This is because James Jeans (1905a, 1905b) corrected a mistake of a factor 8 in Rayleigh's (1905a) derivation of the proportionality constant (see Section 3.2, Eqs. (3.9)–(3.15)). For further discussion of the origin of the Rayleigh–Jeans law, see Kuhn (1987, pp. 144–152) and Pais (1982, pp. 373–374).

this law could not possibly hold for all values of ν: The energy would increase without limit as the frequency ν would go from the infrared into the visible and the ultraviolet. This is what Ehrenfest (1911, p. 92) would later call the *ultraviolet catastrophe* (Klein 1970a, p. 249).[38] Rayleigh (1900b) only wanted to apply the equipartition theorem to the "graver modes" of the radiation, i.e., to the modes with wavelengths/frequencies in the infrared. To arrive at a viable law for the entire spectrum, Rayleigh, rather arbitrarily from a theoretical standpoint, multiplied the expression $c_1 \nu^2 T$ by an exponential function $e^{-c_2\nu/T}$ (where c_1 and c_2 are constants to be fitted to the data). Lummer and Pringsheim (1900, p. 179) found that Rayleigh's law did much better than the Wien law for longer wavelengths but much worse for shorter ones.

It was Rubens (see Plate 3) who finally convinced Planck that the Wien law had become untenable (Planck 1949, pp. 393–394; 1943, pp. 262–263). On October 7, 1900, he and his wife visited Planck at his house and Rubens told his colleague that his latest measurements with Kurlbaum agreed with Rayleigh's law at long wavelengths (Hettner 1922, p. 1036; cf. Kangro 1972, pp. 46–47).[39] That same night apparently, Planck used the wiggle room the second law of thermodynamics still left him to accommodate Rubens and Kurlbaum's findings by cooking up a suitable function $f(U)$ for the right-hand side of Eq. (2.14) for d^2S/dU^2. Rubens compared the resulting new black-body radiation law to his and Kurlbaum's data and found that they agreed to within a few percent.[40] The following year, new data published by Rubens and Kurlbaum (1901) (see Fig. 2.4) and Paschen (1901) confirmed Planck's law. It has enjoyed excellent agreement with the data ever since.

Rubens and Kurlbaum (1900) submitted the paper with their latest results to the Berlin Academy on October 25, 1900. Six days earlier, Kurlbaum had given a talk about this work at a meeting of the German Physical Society. That same meeting saw the public unveiling of Planck's new law. After reviewing his derivation of the Wien law of March 1900 (see Eqs. (2.14)–(2.16) above), Planck (1900c, p. 36) changed Eq. (2.14) to

$$\frac{d^2 S_\nu}{dU_\nu^2} = -\frac{\alpha}{U_\nu(\beta + U_\nu)}. \tag{2.17}$$

[38] The phrase Ehrenfest actually used was "the Rayleigh–Jeans catastrophe in the ultraviolet." Although he only introduced this colorful phrase in 1911, he had already identified the problem in an earlier paper: "For higher frequencies, [the Rayleigh–Jeans law] must be rejected because it leads to infinitely large intensities for $\nu = \infty$" (Ehrenfest 1906, p. 531).

[39] Kangro (1976, pp. 198–200) mentions a letter he received from Hettner in which Hettner told him that he got this information directly from Planck when he went to visit Planck while preparing an article on Rubens for *Die Naturwissenschaften* (Hettner 1922).

[40] Planck's recollections as to when exactly Rubens informed him of these findings vary. In one version, Rubens told him one or two days after his visit of October 7 (Kangro 1972, p. 47, Hettner 1922, p. 1036). In two very similar passages written in the 1940s, however, Planck (1943, p. 263; 1949, p. 394) wrote that Rubens visited him the morning after he first presented his new law in the meeting of the German Physical Society of October 19 and told him at that point that "during the night he had carefully compared my new formula with his data and had found satisfactory agreement everywhere."

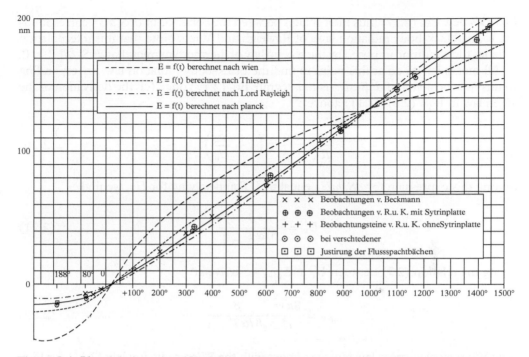

Figure 2.4 *Plot of the intensity of black-body radiation at a single frequency deep in the infrared against temperature from a paper submitted to* Annalen der Physik *in February 1901 by Heinrich Rubens and Ferdinand Kurlbaum (1901). The figure clearly shows that for long wavelengths the intensity varies linearly with the temperature, in agreement with both the Planck law and the Rayleigh–Jeans law (which fails for shorter wavelengths). The data, however, are in sharp disagreement with the Wien law (Gearhart 2002, p. 180; Stone 2013, p. 9).*

If the Wien law were correct, $d^2 S/dU^2$ is proportional to $-1/U_\nu$ (see Eq. (2.16)). However, the proportionality of U_ν to T for low frequencies, which was proposed by Rayleigh and confirmed by Rubens and Kurlbaum, suggests that $d^2 S/dU^2$ is proportional to $-1/U_\nu^2$ instead. After all, if $U_\nu \propto T$, the standard relation (2.6) gives

$$\frac{dS_\nu}{dU_\nu} = \frac{1}{T} \propto \frac{1}{U_\nu},$$
(2.18)

from which it follows that

$$\frac{d^2 S_\nu}{dU_\nu^2} \propto -\frac{1}{U_\nu^2}.$$
(2.19)

Planck's new Eq. (2.17) thus interpolates between the behavior of d^2S/dU^2 correspond-ing to the black-body radiation laws of Rayleigh and Wien. Note that this interpolation satisfies the condition imposed by the second law of thermodynamics that $d^2S_\nu/dU_\nu^2 \leq 0$.

All that is left to do at this point to find the black-body radiation law corresponding to this new behavior of d^2S/dU^2 is to integrate Eq. (2.17) once (cf. the derivation of the Wien law in Eqs. (2.6)–(2.11) above):[41]

$$\frac{dS_\nu}{dU_\nu} = \frac{\alpha}{\beta} \ln\left(1 + \frac{\beta}{U_\nu}\right). \tag{2.20}$$

Setting this expression for dS_ν/dU_ν equal to $1/T$ and solving for U_ν, we find

$$U_\nu = \frac{\beta}{e^{\beta/\alpha T} - 1}. \tag{2.21}$$

Using the relation (2.7) between U_ν and ρ_ν, we arrive at[42]

$$\rho_\nu = \frac{8\pi\nu^2}{c^3} \frac{\beta}{e^{\beta/\alpha T} - 1}. \tag{2.22}$$

The Wien displacement law and agreement with the Wien radiation law (2.11) for high frequencies demand that the ratio β/α and β itself are proportional to ν. With malice aforethought we set

$$\alpha = k, \quad \beta = h\nu, \tag{2.23}$$

as k and h will turn out to be Boltzmann's constant and Planck's constant, respectively. When we substitute these definitions into Eq. (2.22), we arrive at the Planck law for black-body radiation in its now familiar form:

$$\rho_\nu = \frac{8\pi}{c^3} \frac{h\nu^3}{e^{h\nu/kT} - 1}. \tag{2.24}$$

---

[41] One readily verifies that differentiating Eq. (2.20) reproduces Eq. (2.17):

$$\frac{d}{dU_\nu}\left[\frac{\alpha}{\beta} \ln\left(1 + \frac{\beta}{U_\nu}\right)\right] = \frac{\alpha}{\beta} \cdot \frac{1}{1 + \frac{\beta}{U_\nu}} \cdot -\frac{\beta}{U_\nu^2} = -\frac{\alpha}{U_\nu^2 + \beta U_\nu}.$$

[42] Planck (1900c, p. 37) actually stated his new law in terms of wavelength rather than frequency (cf. Eq. (2.3) for the Wien law):

$$\rho_\lambda = \frac{C\lambda^{-5}}{e^{c/\lambda T} - 1},$$

where, as usual, C and c are constants to be fitted to the data.

Unsurprisingly, given that Planck started from the interpolated Eq. (2.17), the Planck law reduces to the Wien law (cf. Eq. (2.11)) for high frequencies (and/or low temperatures), when $e^{h\nu/kT} \gg 1$,

$$h\nu \gg kT: \qquad \rho_\nu \approx \frac{8\pi}{c^3} h\nu^3 e^{-h\nu/kT}, \tag{2.25}$$

and to the Rayleigh–Jeans law (cf. note 37) for low frequencies (and/or high temperatures), when $e^{h\nu/kT} \approx 1 + h\nu/kT$,

$$h\nu \ll kT: \qquad \rho_\nu \approx \frac{8\pi h\nu^3}{c^3} \frac{kT}{h\nu} = \frac{8\pi}{c^3} \nu^2 kT. \tag{2.26}$$

Note that Planck's constant drops out in Eq. (2.26).

2.5 Justifying the new expression for the entropy of a resonator

To justify his new black-body radiation law, Planck had to justify the choice of the entropy S_ν of a single resonator as a function of its average energy U_ν that lead to the new law. To find $S_\nu(U_\nu)$ for this new law, we need to integrate Eq. (2.20) for dS_ν/dU_ν. Inserting the expressions (2.23) for α and β into Eq. (2.20), we can rewrite this equation as

$$\frac{dS_\nu}{dU_\nu} = \frac{k}{h\nu} \ln\left(1 + \frac{h\nu}{U_\nu}\right). \tag{2.27}$$

Integration of this equation gives:

$$S_\nu = k\left\{\left(\frac{U_\nu}{h\nu} + 1\right) \ln\left(\frac{U_\nu}{h\nu} + 1\right) - \frac{U_\nu}{h\nu} \ln\frac{U_\nu}{h\nu}\right\}, \tag{2.28}$$

as one readily verifies.[43] For high frequencies, the average energy U_ν of a resonator will be much smaller than $h\nu$. In that case the first term becomes negligible relative to the second, and Eq. (2.28) reduces to Eq. (2.8) for $S_\nu(U_\nu)$, which leads to the Wien law.

---

[43] Differentiating Eq. (2.28) reproduces Eq. (2.27)

$$\frac{dS_\nu}{dU_\nu} = \frac{k}{h\nu} \ln\left(\frac{U_\nu}{h\nu} + 1\right) + \frac{k}{h\nu} - \frac{k}{h\nu} \ln\frac{U_\nu}{h\nu} - \frac{k}{h\nu}$$

$$= \frac{k}{h\nu} \ln\left(\frac{\frac{U_\nu}{h\nu} + 1}{\frac{U_\nu}{h\nu}}\right) = \frac{k}{h\nu} \ln\left(1 + \frac{h\nu}{U_\nu}\right).$$

As Planck recalled in his Nobel lecture:

> Even if the radiation formula should prove to be absolutely correct, it would as just a happily guessed interpolation formula only be of limited value. On the very day that I formulated this law, I therefore began to devote myself to the task of investing it with true meaning, and that issue led me automatically to a consideration of the connection between entropy and probability and thus to Boltzmann's trend of ideas; until after some weeks of the most strenuous work of my life, light came into the darkness (Planck 1920, p. 125; repeated, almost verbatim, in Planck 1949, pp. 394–395; see also Planck 1943, p. 264).[44]

Although it took some time to work out the details, Planck quickly hit upon the basic idea behind the derivation of Eq. (2.28) for the resonator entropy that he would present two months later. Only one week after presenting his new black-body radiation law, he wrote to his Berlin colleague Otto Lummer:

> If the prospect of a theoretical derivation of the radiation law should exist at all, which I naturally assume, then, in my opinion, this can only be the case if it is possible to derive the expression for the probability of a radiation state, and this, you see, is given by the entropy ... According to my formula, the entropy of the resonator would come to

$$S = \alpha \log \frac{(\beta + U)^{\beta+U}}{U^U}$$

> and this form very much recalls expressions occurring in the probability calculus.[45] After all, in the thermodynamics of gases, the entropy S is the log of a probability magnitude, and Boltzmann has already stressed the close relationship between the function X^X, which enters the theory of combinatorials, and the thermodynamic entropy (Planck to Lummer, October 26, 1900).

This letter already contains several elements of Planck's (1900d) presentation at the German Physical Society on December 14, 1900. He announced this presentation in a letter a full month in advance: "I now also have a theory [to back up the new

---

[44] This passage is quoted and discussed, for instance, by Klein (1970a, p. 225) and Kuhn (1987, pp. 97–98).

[45] Differentiating this formula for S twice, we recover an expression for d^2S/dU^2 equivalent to Eq. (2.17). The first derivative is:

$$\frac{dS}{dU} = \alpha \frac{d}{dU}\left((\beta + U)\log(\beta + U) - U\log U\right) = \alpha\left(\log(\beta + U) + 1 - \log U - 1\right).$$

The second derivative is:

$$\frac{d^2S}{dU^2} = \alpha \frac{d}{dU}\left(\log(\beta + U) - \log U\right) = \alpha\left(\frac{1}{\beta + U} - \frac{1}{U}\right) = -\frac{\alpha\beta}{U(U+\beta)}.$$

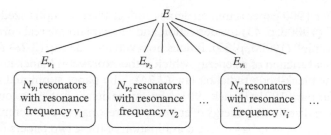

Figure 2.5 *Distributing the total energy E over groups of resonators at different frequencies: an amount of energy E_{ν_i} goes to the N_{ν_i} resonators at frequency ν_i (i = 1, 2, ...). The distribution $\{E_{\nu_1}, E_{\nu_2}, ...\}$ is subject to the constraint $E_{\nu_1} + E_{\nu_2} + ... = E$.*

black-body radiation formula] which I will lecture on here at the Physical Society in four weeks" (Planck to Wien, November 13, 1900).[46]

As Planck (1900d, p. 43) noted in his December paper, "the whole deduction is based on the single theorem that the entropy [S] of a system of resonators with given energy is proportional to the logarithm of the total number [W] of possible complexions for the given energy." Planck (1900d, p. 39) cited Boltzmann (1877) for this idea.[47] Figure 2.5 illustrates how Planck intended to proceed.

Start with a group of a large number N_{ν_i} resonators all at the same resonance frequency ν_i. Planck (1900d, pp. 39–40) considered a large number of such groups, one for every frequency interval, and asked what the equilibrium distribution of the total energy E over the different frequencies would be. In general, the group of N_{ν_1} resonators at ν_1 will get an amount E_{ν_1}, the group of N_{ν_2} resonators at ν_2 will get an amount E_{ν_2}, etc. Any such distribution $\{E_{\nu_1}, E_{\nu_2}, ...\}$ will be subject to the constraint

$$\sum_{i=1}^{\infty} E_{\nu_i} = E. \tag{2.29}$$

To find the equilibrium distribution $\{E_{\nu_1}, E_{\nu_2}, ...\}_{\text{equilibrium}}$, Planck would have had to find the distribution $\{E_{\nu_1}, E_{\nu_2}, ...\}$ that maximizes W and thereby the entropy $S = k \ln W$, where W is the number of ways in which the distribution $\{E_{\nu_1}, E_{\nu_2}, ...\}$ can be realized. As it turns out, Planck only had to determine W for an *arbitrary* distribution $\{E_{\nu_1}, E_{\nu_2}, ...\}$. The Wien displacement law provided him with a convenient shortcut for the cumbersome process of finding the one maximizing W (Kuhn 1987, p. 241).

---

[46] These letters from Planck to Lummer and Wien are quoted and discussed by Jungnickel and McCormmach (1986, Vol. 2, pp. 261–262; see also Darrigol 1992, pp. 67–68).

[47] Planck may have been alerted to this paper by Boltzmann's reference to it in his *Lectures on Gas Theory* (Boltzmann 1896–1898, Vol. 1, sec. 6, p. 58). In that same section, Boltzmann is criticizing some of Planck's work and there is a reference to this section in the introduction of Planck (1900a, p. 75; cf. Gearhart 2002, p. 184, note 42).

The December 1900 paper is thin on details and, as Planck emphasized, was intended only to "sketch" (1900d, p. 43) the derivation and "to explain the real core of the theory as clearly as possible" (1900d, p. 39). It does not even include Eq. (2.28) for the entropy of a resonator as a function of its energy, which is the expression Planck needed to derive to justify his new black-body radiation law (2.24). The paper submitted to *Annalen der Physik* in early January 1901 (Planck 1901a) has more details but leaves out others (such as the basic idea of distributing energy over groups of resonators at different frequencies). The reconstruction below is based on a combination of these two papers (Planck 1900d, 1901a).[48] We have enhanced Planck's own notation to bring out more clearly how he managed to derive Eq. (2.28).

Let W_{v_i} be the number of ways in which the energy E_{v_i} can be distributed over N_{v_i} resonators. W, the number of ways in which the distribution $\{E_{v_1}, E_{v_2}, \ldots\}$ can be realized, is the product of these numbers

$$W = \prod_{i=1}^{\infty} W_{v_i}. \tag{2.30}$$

The entropy S of all resonators will accordingly be the sum of the logarithms of these numbers:

$$S = k \ln W = k \ln \left(\prod_{i=1}^{\infty} W_{v_i} \right) = \sum_{i=1}^{\infty} k \ln W_{v_i} \tag{2.31}$$

In other words, S is the sum of the entropies

$$S_{N_{v_i}} \equiv k \ln W_{v_i} \tag{2.32}$$

of the various groups of N_{v_i} resonators at v_i.

To find the entropy of a single resonator at v_i, the quantity from which the black-body radiation law is derived, Planck (1901a, p. 308) set (we drop the subscript i at this point)

$$S_v = \frac{S_{N_v}}{N_v}, \tag{2.33}$$

on the argument that the time average of one resonator can be replaced by an average over a group of many such resonators (Planck 1900d, p. 38, and—more clearly—Planck 1901a, p. 307).[49]

---

[48] See Nauenberg (2016, Appendix A, pp. 717–718) for an alternative reconstruction of Planck's adaptation of Boltzmann's methods to derive his black-body radiation law. Nauenberg ignores the "sketch" of Planck's argument in the December 1900 paper.

[49] See Section 3.2 for discussion of the conditions under which these two kinds of averages can be set equal to one another.

To find S_{N_ν}, Planck had to count the number of ways W_ν in which the energy E_ν can be distributed over N_ν resonators. To get a finite count he borrowed a trick from Boltzmann.[50] He imagined that the energy E_ν comes in a large number P_ν of small parcels of energy ε_ν (1900d, p. 40; 1901a, p. 308):

$$E_\nu = P_\nu \, \varepsilon_\nu. \qquad (2.34)$$

Given Planck's discretization (not necessarily quantization!) of the energy E_ν, the number W_ν is now the large but finite number of ways in which P_ν parcels of energy ε_ν can be distributed over N_ν resonators. Planck (1900d, p. 41; 1901a, p. 309) appealed to standard combinatorics to set W_ν equal to:[51]

$$W_\nu = \frac{(P_\nu + N_\nu - 1)!}{P_\nu!(N_\nu - 1)!}. \qquad (2.35)$$

Since both P_ν and N_ν are large numbers, we can neglect the term -1 in both the numerator and the denominator of Eq. (2.35) and use Stirling's formula, $N! \approx N^N$, for the various factorials:

$$\frac{(P_\nu + N_\nu - 1)!}{P_\nu!(N_\nu - 1)!} \approx \frac{(N_\nu + P_\nu)^{N_\nu + P_\nu}}{N_\nu^{N_\nu} P_\nu^{P_\nu}}. \qquad (2.36)$$

Planck did not give a source for either Eq. (2.35) or Eq. (2.36) but the letter to Lummer quoted above suggests that he found both in Boltzmann (1877, pp. 181–182).[52]

More than a decade later, in a paper co-authored with Leyden's most famous low-temperature physicist, Ehrenfest gave an exceedingly simple argument establishing Eq. (2.35) (Ehrenfest and Kamerlingh Onnes 1914; 1915 cf. Klein 1970a, pp. 255–256). Let P_ν dots (•) and $N_\nu - 1$ bars (|) represent the energy elements and the resonators, respectively. The bars actually represent dividers between these resonators, which is why there is one less bar than there are resonators. Any sequence of P_ν dots and $N_\nu - 1$

---

[50] See Klein (1977, pp. 16–18) and Gearhart (2002, pp. 181–185) for clear expositions of the relevant passage in Boltzmann (1877, sec. 1, pp. 167–186). See Kuhn (1987, pp. 102–110) for a clear explanation of how the problem Boltzmann was addressing differs from the problem Planck was addressing. Boltzmann was interested in finding the most probable way of dividing a given amount of energy over a large number of individual molecules. Planck was interested in finding the most probable way of dividing a given amount of energy over a large number of groups of resonators, each containing a large number of individual resonators. He was not interested in finding the most probable way of subdividing the fraction of the energy in one particular group of resonators over the individual resonators in that group.

[51] Following Rosenfeld (1936), Klein (1962, p. 474) and others have suggested that Planck originally hit upon the derivation of Eq. (2.28) reconstructed here by working backwards from this combinatorial expression. The letter to Lummer of October 26, 1900, quoted above, can be seen as supporting this suggestion.

[52] Boltzmann (1877, p. 181) considered the "binomial coefficient ... $\mathcal{J} = \binom{\lambda + n - 1}{\lambda}$." The binomial coefficient is defined as $\binom{n}{k} \equiv \frac{n!}{k!(n-k)!}$. Eq. (2.35) can thus also be written as $W_\nu = \binom{P_\nu + N_\nu - 1}{P_\nu}$. Boltzmann (1877, p. 182) proceeded to use the Stirling approximation for $\mathcal{J}$.

bars (e.g., the one starting with • • •|• •||•| ...) represents a particular distribution of P_ν energy elements over N_ν resonators. The number of dots before the first bar gives the energy (in multiples of ε_ν) of the first resonator in this particular distribution, the number of dots between the first and the second bar the energy of the next resonator, etc. The number of dots after the last bar gives the energy of the N_ν^{th} and final resonator. If the bars and dots were all *distinguishable* from one another, there would be $(P_\nu + N_\nu - 1)!$ different sequences of this kind. Since, however, the bars are *indistinguishable* from one another, as are the dots, this number still needs to be divided by $(N_\nu - 1)!$ and by $P_\nu!$, which results in Eq. (2.35).

Using Eq. (2.36) for W_ν in Eq. (2.32) (with ν instead of ν_i), we arrive at (Planck 1901a, p. 309):

$$S_{N_\nu} \approx k \{ (N_\nu + P_\nu) \ln (N_\nu + P_\nu) - N_\nu \ln N_\nu - P_\nu \ln P_\nu \}. \qquad (2.37)$$

If we add and subtract $k(N_\nu + P_\nu) \ln N_\nu$ and regroup terms, we can rewrite this as:

$$S_{N_\nu} \approx k \left\{ (N_\nu + P_\nu) \ln \left(1 + \frac{P_\nu}{N_\nu} \right) - P_\nu \ln \frac{P_\nu}{N_\nu} \right\}. \qquad (2.38)$$

Dividing by N_ν, we obtain an expression for the average entropy S_ν of a single resonator with resonance frequency ν:

$$S_\nu = \frac{S_{N_\nu}}{N_\nu} \approx k \left\{ \left(1 + \frac{P_\nu}{N_\nu} \right) \ln \left(1 + \frac{P_\nu}{N_\nu} \right) - \frac{P_\nu}{N_\nu} \ln \frac{P_\nu}{N_\nu} \right\}. \qquad (2.39)$$

The average energy U_ν of such a resonator is equal to E_ν/N_ν. Combining this with Eq. (2.34) for E_ν, we have $U_\nu = P_\nu \varepsilon_\nu / N_\nu$. Substituting

$$\frac{U_\nu}{\varepsilon_\nu} = \frac{P_\nu}{N_\nu} \qquad (2.40)$$

in Eq. (2.39), we arrive at (Planck 1901a, p. 309):

$$S_\nu \approx k \left\{ \left(1 + \frac{U_\nu}{\varepsilon_\nu} \right) \ln \left(1 + \frac{U_\nu}{\varepsilon_\nu} \right) - \frac{U_\nu}{\varepsilon_\nu} \ln \frac{U_\nu}{\varepsilon_\nu} \right\}. \qquad (2.41)$$

This result holds for any distribution $\{E_{\nu_1}, E_{\nu_2}, \ldots\}$. What is special about the equilibrium distribution $\{E_{\nu_1}, E_{\nu_2}, \ldots\}_{\text{equilibrium}}$ is that the Wien displacement law holds. This is why Planck did not have to do any maximization. As Planck (1901a, pp. 310–311) proved explicitly in his January *Annalen* paper, the Wien displacement law entails that S_ν must be a function of U_ν/ν. In the course of this proof, he wrote the displacement law in a simple form—the formula for the spectral distribution of black-body radiation must be of the form:

$$\rho_v = \frac{v^3}{c^3} f(T/v). \tag{2.42}$$

If S_v is a function of U_v/v, it follows that, for the equilibrium distribution, ε_v in Eq. (2.41) has to be proportional to v. Calling the proportionality constant h, Planck thus introduced the relation

$$\varepsilon_v = hv. \tag{2.43}$$

With the help of this relation, Eq. (2.41), for the special case of the equilibrium distribution, turns into Eq. (2.28) for the entropy of a resonator as a function of its energy that leads directly to Planck's new black-body radiation law (2.24) (Planck 1901a, p. 312).

2.6   Energy parcels or energy bins?

In his December 1900 paper, Planck (1900d, p. 40) stated that "assuming [the energy] E to be composed of a well-defined number of equal parts and use thereto the constant of nature h" was "the most essential point of the whole calculation." Planck is right to emphasize the central role of the relation $\varepsilon = hv$. If we let ε go to zero, Planck's new black-body radiation law (2.24) reduces to the highly problematic Rayleigh–Jeans law (2.26). It is clear that Planck realized that the constant h introduced a new element into physics. What is not clear is how he interpreted that new element at the time. Even his later reminiscences, written long after it had become evident that his black-body radiation law calls for energy quantization, remain ambiguous on this point. In an oft-quoted letter of October 7, 1931, for instance, Planck wrote to the American physicist Robert W. Wood:

> You recently expressed the wish ... that I should describe from a psychological point of view the considerations which led me to propose the hypothesis of energy quanta ... Briefly summarized, what I did can be described as simply an act of desperation. By nature I am peacefully inclined and reject all doubtful adventures. But by then I had been wrestling unsuccessfully for six years (since 1894) with the problem of equilibrium between radiation and matter and I knew that this problem was of fundamental importance to physics; I also knew the formula that expresses the energy distribution in normal spectra. A theoretical derivation therefore had to be found at any cost, no matter how high ... The two laws [of thermodynamics], it seems to me, must be upheld under all circumstances. For the rest, I was ready to sacrifice every one of my previous convictions about physical laws. Boltzmann had explained how thermodynamic equilibrium is established by means of a statistical equilibrium, and if such an approach is applied to the equilibrium between matter and radiation, one finds that the continuous loss of energy into radiation can be prevented by assuming that energy is forced, at the outset, to remain together in certain quanta. This was a purely formal assumption and I really did not give it much thought except that, no matter what the cost, I must bring about a positive result (Planck to Robert W. Wood, October 7, 1931).[53]

---

[53] The translation of the entire letter is given in Hermann (1971, pp. 23–24).

This letter has to be treated with extreme caution. Several clauses in it ("an act of desperation," "no matter what the cost") and the statement that he had been willing to give up all accepted physical laws except those of thermodynamics suggest that Planck was well aware at the time that his energy elements would wreak havoc on physics as he knew it. This is not supported by his writings in 1900–1. It is also hard to reconcile with his stubborn refusal a few years later to accept the implications of his new radiation law spelled out by Einstein and others (see Section 3.4). Moreover, the argument that without energy quanta there would be "a continuous loss of energy into radiation" was not part of Planck's reasoning in 1900 at all. Einstein (1905a) and Ehrenfest (1906) were the first to recognize that the introduction of energy quanta could be used to escape this dire consequence of the equipartition theorem. Yet, despite taking such liberties with the historical record, Planck still qualified the statement of his hypothesis of energy quanta by adding that it was "a purely formal assumption," which he "did not give ... much thought."

As we indicated in Section 2.1, the relation $\varepsilon = h\nu$ can be interpreted in (at least) two ways, in terms of *energy parcels* or in terms of *energy bins*. We now take a closer look at these two interpretations. On the first interpretation, a resonator with characteristic frequency ν can only absorb energy from the ambient radiation in *parcels* of $h\nu$. For that to be true, it seems, at least one of the following two conditions must hold. The resonator's energy is quantized, i.e., it can only take on the values $nh\nu$ ($n = 0, 1, 2, \ldots$), or the radiation's energy is quantized, which amounts to Einstein's light-quantum hypothesis. The first alternative corresponds to Lorentz's reading of Planck. "[A]ccording to Planck," Lorentz (1903, p. 668) wrote, "the energy that is stored up in a resonator cannot increase or diminish by gradual changes, but only by whole 'units of energy' [$h\nu$]." The second alternative corresponds to Ehrenfest's reading of Planck. In 1903, while still a student in Vienna, Ehrenfest visited Leyden in 1903 and attended Lorentz's lectures on the theory of radiation (Klein 1970a, p. 232). In 1905, the year after he obtained his degree, he published the incisive study of Planck's research program that we already referred to in Section 2.3. One of the hypotheses Ehrenfest (1905, p. 1313) identified as underlying Planck's new theory of late 1900 was "that the radiant energy of the various colors consists of minuscule energy particles of magnitude [$h\nu$]."[54] We know that Planck resisted Lorentz's reading until 1908 (see Section 3.4.2). We know that he resisted Ehrenfest's reading, which amounts to Einstein's light-quantum hypothesis, until the early 1920s. This makes it doubtful he would have subscribed to either one in 1900–1.

The interpretation of $\varepsilon = h\nu$ in terms of bins rather than parcels offers a way to make sense of Planck's reasoning that requires neither the quantization of the energy of the resonators nor the quantization of the energy of the radiation. Once again consider the idea that a resonator with characteristic frequency ν can only absorb energy from the ambient radiation in parcels of $h\nu$. The total energy E_ν of a group of N_ν of such

---

[54] These (mis)readings of Planck by Lorentz and Ehrenfest are quoted and discussed by both Klein (1970a, pp. 231–234) and Kuhn (1987, pp. 138–139). Ehrenfest had not read Einstein's (1905a) paper proposing the light-quantum hypothesis at this point.

resonators can then only be distributed over the individual resonators in P_ν such parcels (where $E_\nu = P_\nu h\nu$). Ehrenfest's clever combinatorial argument shows that there are $(N_\nu + P_\nu - 1)!/P_\nu!(N_\nu - 1)!$ ways of doing this. In any such distribution, some number w_0 of resonators will end up with energy 0, w_1 with $h\nu$, w_2 with $2h\nu$, etc. Any set $\{w_n\}$ of occupation numbers of these energy states will have to satisfy the constraints

$$\sum_{n=0}^{P_\nu} w_n = N_\nu, \qquad \sum_{n=1}^{P_\nu} w_n(nh\nu) = E_\nu, \qquad (2.44)$$

which simply express that the total number of resonators is N_ν and that their total energy is E_ν. Any set $\{w_n\}$ satisfying these constraints specifies one way in which the resonators can be divided over the allowed energy states. We can use the same combinatorial argument as before to convince ourselves that there are $(N_\nu + P_\nu - 1)!/P_\nu!(N_\nu - 1)!$ ways of distributing N_ν resonators over P_ν energy states under the constraints (2.44).

Switching from thinking in terms of *energy parcels distributed over resonators* to thinking in terms of *resonators distributed over energy states* prepares us for the alternative interpretation of the relation $\varepsilon = h\nu$ in terms of energy bins. We can change the interpretation of w_n from "the number of resonators with *exactly* the energy $nh\nu$" to "the number of resonators with energies *between* $nh\nu$ to $(n + 1)h\nu$" without changing the count of the total number of ways in which E_ν can be distributed over N_ν resonators. Since h is small, we will only make a small error—or so it would seem[55]—if we assume that all values between $nh\nu$ to $(n + 1)h\nu$ can be set equal to $nh\nu$. On this interpretation, energy is *not* quantized (neither the energy of the resonators nor the energy of the radiation) and what the relation $\varepsilon = h\nu$ expresses is that, for whatever reason, all that matters for Planck's counting purposes is how many resonators there are in each energy bin of size $h\nu$.

When Planck wrote his papers of December 1900 and January 1901, was he thinking in terms of parcels or bins? One can point to passages in these papers in support of either answer.

First, as we saw in Section 2.6, Planck did his derivation in terms of distributing energy parcels over resonators and not in terms of distributing resonators over energy states or energy bins. That suggests parcels rather than bins. So does the clause "assuming [the energy] E to be composed of a well-defined number of equal parts" in the December paper, which is echoed in the clause "assuming that energy is forced, at the outset, to remain together in certain quanta" in the letter to Wood. Yet, in the same paragraph of the December paper in which Planck stated this assumption, he also wrote that "dividing E_ν by ε_ν we get the number P_ν of energy elements which must be divided over the N_ν resonators. If the ratio thus calculated is not an integer, we take for P_ν an integer in the neighborhood" (Planck 1900d, p. 40; subscripts ν added). This qualification of the strict validity of $\varepsilon_\nu = h\nu$, however, does not return in the January paper.

---

[55] Einstein (1906b, p. 203; 1909b, pp. 494–495) showed that this is not true (see Sections 3.4.1 and 3.4.3 for discussion).

There is clear evidence that Planck was fully aware of the equivalence of the two ways distinguished above of doing his counting—distributing parcels over resonators or distributing resonators over bins (Kuhn 1987, pp. 128–130). Even though the occupation numbers w_n introduced above do not appear anywhere in either his December or his January paper, they are featured prominently in the section of Boltzmann's paper on which Planck modeled his derivation (Boltzmann 1877, cf. note 50, see pp. 175–176 for the constraints given in Eq. (2.44) above). Moreover, in another paper on his new black-body radiation law, written later in 1901, Planck (1901d, pp. 633–634) presented Boltzmann's original problem—true to Boltzmann's own intentions (Kuhn 1987, p. 50)—in terms of molecules distributed over energy bins, wrote that a "very similar procedure" can be used to find the entropy of his resonators, and then proceeded, without pausing to explain the difference, to present his own problem in terms of energy parcels distributed over resonators (Planck 1901d, p. 637).

There is one passage in Planck's January 1901 paper in particular that suggests that he was thinking in terms of bins rather than parcels. This is also as close as Planck came in these papers to offering an interpretation of the new element his new theory brought into physics. In a short subsection of his January paper, Planck explained that the hypothesis on which his whole derivation is based is that the number W_ν in the formula $S_{N_\nu} = k \log W_\nu$ is given by the number

> of all possible complexions [Boltzmann's term for a particular realization of a certain distribution] formed by distribution of the energy U_ν among the N_ν resonators; or, in other words, [the hypothesis that] any given complexion is just as probable as any other. Whether this is actually the case in nature can, in the final analysis, only be shown by experience. But should experience decide in its favor it will be possible to make inferences in the other direction, from the validity of the hypothesis to the particular nature of the vibrations of the resonators, i.e., to the nature of the "equivalent and as far as their sizes are concerned comparable original regions [*Spielräume*]," to use the expression of J. v. Kries.[56] In the current state of affairs it would seem to be premature to pursue this line of thought any further (Planck 1901a, p. 309; subscripts ν added).

What Planck appears to be saying here is that the Planck law, if shown to be empirically valid, tells us something about which ranges of energy should be considered equiprobable.

In 1901, as the last sentence of the quotation above shows, Planck thought that it was premature to pursue this idea any further. However, in his 1906 *Lectures on the Theory of Heat Radiation*, in which he systematically presented the results of his extensive research on the subject, he would develop a way to give meaning to his constant h along these lines. Instead of considering energy bins of size $h\nu$ to be equiprobable, he now considered areas of the phase space spanned by the position q and the momentum p of his resonators to be equiprobable (Planck 1906a, sec. 150, pp. 154–156; see Section 5.1.1). As Planck points out, the motion of a resonator, conceived of as a mass m on a spring with

---

[56] Planck cites a page in a book by Johannes von Kries (1886, p. 36) on probability theory.

spring constant k (and resonance frequency $\nu = \sqrt{k/m}/2\pi$), will be represented by an ellipse in the pq-plane, the size of which is determined by its energy $U = (p^2/2m) + kq^2/2$. The area between successive ellipses for the values $U = nh\nu$ ($n = 1, 2, \ldots$) is equal to h.

The original choice between (a) parcels, in which case energy (of resonators, radiation, or both) would have to be quantized, and (b) bins, in which case energy need not be quantized, now turns into a choice between (a) restricting the possible orbits in phase space of the resonators to these ellipses and (b) allowing phase space orbits in between these ellipses as well. Einstein (1906b, 1907, see Section 3.4.1) showed that only the former quantized option is viable. In his 1906 book, Planck, however, clearly chose the latter continuous option (Kuhn 1987, pp. 128–129). To the phrase "the number of resonators with energy of a given magnitude," he now added parenthetically: "better: which lie within a given region" (Planck 1906a, p. 151). A few pages later, he talked about "the probability that the energy of a resonator lies between U and $U + \Delta U$" (Planck 1906a, p. 155).

In 1908, at least in part as a result of an influential lecture by Lorentz that year at a conference in Rome (see Section 3.4.2), Planck came to accept that the energy of his resonators really is quantized, i.e., that the only possible states are the ones corresponding to the set of ellipses $U = nh\nu$ in phase space. The first unequivocal statement to this effect by Planck can be found in a letter to Lorentz of October 7, 1908: "The energy of the resonator at a given instant is $gh\nu$ (g a whole number or zero)" (Lorentz 2008, Doc. 176).[57]

A few years later, Planck (1911a, 1911b) once again changed his mind, publishing what quickly became known as Planck's "second theory", in which a resonator can absorb energy continuously but emit energy only in whole quanta of $h\nu$ once it has accumulated an amount of energy that is itself an integral multiple of $h\nu$ (Kuhn 1987, pp. 235–254; Needell 1988, pp. xxx–xxxi; cf. Section 5.1.1). This is the theory Planck defended in the second edition of *The Theory of Heat Radiation* (Planck 1913).[58]

Planck's views on the meaning of h and the meaning of the relation $\varepsilon = h\nu$ thus underwent considerable change in the first decade after their introduction. It would therefore be unwise to attribute a definite view to Planck in late 1900 and early 1901. However, the passage about which complexions are equiprobable quoted above makes it tempting to speculate that Planck thought that h and $\varepsilon = h\nu$ had something to do with how to carve up the energy of resonators with various resonance frequencies into equiprobable *energy bins*. This can be seen, as we indicated above, as the germ of later work on phase space quantization by Planck himself as well as by Arnold Sommerfeld

[57] This letter is quoted and discussed by Kuhn (1987, p. 198) and Gearhart (2002, p. 192). We will return to it in Section 3.4.2.

[58] A new feature of Planck's second theory that was to become a permanent fixture of quantum physics is the zero-point energy of the harmonic oscillator. Planck added a term $\frac{1}{2}h\nu$ to Eq. (2.21) for the average energy U_ν of his resonators (Giulini and Straumann 2006, p. 125).

and others (see Section 5.1).[59] Similarly, we can look upon Planck's analysis of how to distribute *energy parcels* over resonators as the germ of later work on the statistics of indistinguishable particles by Ehrenfest (Klein 1959), Władysław Natanson (1911), and others, including, once again, Planck himself (Darrigol 1988, 1991).

There is an oft-repeated tale that, sometime in late 1900, Planck took his seven-year-old son Erwin, who would eventually be executed for his role in the assassination attempt on Adolf Hitler in 1944, for a walk and told the youngster that he had just made the greatest discovery in physics since Newton (Kuhn 1987, p. 113). After expressing skepticism about this story, the German historian of physics Armin Hermann received a letter from the Göttingen physicist Robert W. Pohl[60] corroborating it. The younger Planck and Pohl had been friends. In his letter to Hermann of June 1972, Pohl also identified the momentous discovery Planck allegedly told his son about. It was not energy quantization. Pohl told Hermann:

> On one of our boat rides, Erwin spontaneously said to me, "Father knew, in his own words, that his discovery of the new natural constant had the same significance as that of Copernicus." Therefore, after [Max] Planck's death [in 1947], I took the trouble to see that the constant with its numerical value was placed on his gravestone [in Göttingen] (Kuhn 1987, p. 285, note †44).

As a matter of fact, Planck had found *two* new constants at the end of 1900, h and k, which appropriately adorn the tombstones of Planck and Boltzmann, respectively (see Plates 6 and 7). As Planck (1920, p. 126; 1943, p. 265) emphasized on several occasions, Boltzmann had used the ratio R/N of the ideal gas constant and Avogadro's number instead of the constant k. Planck was the first to recognize that the constant k applies to black-body radiation as well. As it turned out, a cavity filled with black-body radiation allowed for a much more accurate determination of k (and hence, Avogadro's number) than a vessel filled with the molecules of a gas.

Unsurprisingly, given his proposal for a set of universal constants the year before (Planck 1897–1899, Pt. V, pp. 599–600; 1900a, pp. 120–122; cf. Section 2.3), Planck put great emphasis on these constants, which could be determined with considerable accuracy from the wealth of data on black-body radiation gathered during the preceding years. The numerical evaluation of these constants takes up the concluding sections of both the December 1900 and the January 1901 paper (Planck 1900d, pp. 44–45; 1901a, pp. 312–313). The values Planck arrived at are:

---

[59] The prominence of this approach in the late 1910s and early 1920s (see Part Two) may have colored the account Planck gave in his Nobel lecture in 1920 of what he had done twenty years earlier: "While this constant [h] was absolutely indispensable to the attainment of a correct expression for entropy—*for only with its aid could be determined the magnitude of the 'elementary region[s]'* [Elementargebiete] *or 'range[s]'* [Spielräume] *of probability, necessary for the statistical treatment of the problem*—it obstinately withstood all attempts at fitting it, in any suitable form, into the frame of the classical theory" (Planck 1920, p. 12, our emphasis). The italicized clause clearly suggests bins rather than parcels.

[60] See Mehra and Rechenberg (1982a, p. 293) for a concise summary of Pohl's career in physics.

$$h = 6.55 \cdot 10^{-27} \text{ erg} \cdot \text{sec}, \qquad k = 1.346 \cdot 10^{-16} \text{ erg/deg}.$$

These values are remarkably close to the modern values (in the units used by Planck):

$$h = 6.626070 \cdot 10^{-27} \text{ erg} \cdot \text{sec}, \qquad k = 1.380649 \cdot 10^{-16} \text{ erg/deg}.$$

One can readily understand why Planck took special pride in these results.[61]

In May 1901, Planck (1901c) presented an addendum to his pentalogy on irreversible radiation phenomena (Planck 1897–1899), in which he made the changes to his electromagnetic foundation of the second law of thermodynamics necessitated by the replacement of the entropy function (2.8), which leads to the Wien law, by the entropy function (2.28), which leads to the Planck law. It would be another thirteen years before Planck announced in public for the first time that he had come to accept Boltzmann's statistical interpretation of the second law. Planck (1914), by now Rector of the University of Berlin, made this announcement in an address commemorating the founding of the university. He delivered this address on August 3, 1914, the day Germany declared war on France (Needell 1988, p. xxxviii).

---

[61] Around the same time, Planck (1901b) wrote a separate paper focusing on Boltzmann's constant k and using its value, determined on the basis of black-body radiation measurements, to find new values for related constants such as Avogadro's number and the electron charge.

3

Einstein, Equipartition, Fluctuations, and Quanta

3.1 Einstein's annus mirabilis

In May 1905, Einstein wrote a letter to his friend Conrad Habicht. Together with Maurice Solovine, Einstein and Habicht formed a discussion group in Bern that its three members mockingly referred to as the Olympia Academy. "I promise you four papers," Einstein told Habicht, the first of which "deals with radiation and the energy characteristics of light and is very revolutionary" (Einstein 1987–2018, Vol. 5, Doc. 27). This is the paper introducing light quanta. It was submitted to *Annalen der Physik* in March and published in June (Einstein 1905a). The second is the dissertation that earned Einstein his doctorate in physics from the University of Zurich (Einstein 1905b).[1] It was published in slightly revised form the following year (Einstein 1906a). The third is the paper on Brownian motion, submitted to the *Annalen* in May 1905 and published in July (Einstein 1905c). The fourth, he told Habicht, "is only a rough draft at this point and is an electrodynamics of moving bodies by the use of a modification of the theory of space and time" (Einstein 1987–2018, Vol. 5, Doc. 27). This is the paper introducing what came to be known as the special theory of relativity. Submitted in June 1905 and published in September, it is the third paper by Einstein in Vol. 17 of the *Annalen* (Einstein 1905d). The dissertation, these three papers, plus a short addendum to the third, introducing his signature equation $E = mc^2$ (Einstein 1905e), made 1905 Einstein's *annus mirabilis*, his year of miracles (Stachel 2005).

---

[1] This was Einstein's third attempt to obtain his doctorate. The first attempt had been with Heinrich Weber, his most important physics professor at the Swiss Federal Polytechnic in Zurich (renamed the *Eidgenössische Technische Hochschule* or ETH in 1911), where Einstein had been a student from 1896 to 1900, in a program training high school teachers in physics and mathematics. This attempt was aborted before it had begun in earnest in the Summer of 1901. In November of that same year, he submitted a dissertation to Alfred Kleiner at the University of Zurich. This dissertation, which does not survive, was rejected. In 1905, Einstein submitted a new one to Kleiner and this time it was accepted. For further discussion, see the editorial note, "Einstein's dissertation on the determination of molecular dimensions" (Einstein 1987–2018, Vol. 2, pp. 170–182), Renn (2000, pp. 116–117, pp. 120–122), and Uffink (2006, pp. 65–67).

Constructing Quantum Mechanics. Anthony Duncan and Michel Janssen, Oxford University Press (2019).
© Anthony Duncan and Michel Janssen. DOI: 10.1093/oso/9780198845478.001.0001

Einstein did all of this while working at the Swiss Patent Office in Bern, where he had been appointed patent examiner third class in June 1902. This job required his attendance, if perhaps not his full attention, eight hours a day, six days a week. Despite his stunning research productivity in 1905, Einstein must have continued to perform his duties at the patent office to the satisfaction of his employer. In 1906, he earned himself a promotion. He became patent examiner second class.

Einstein was right to reserve the label "very revolutionary" for the light-quantum paper. Whereas the relativity paper can be seen as the culmination of developments in nineteenth-century electrodynamics, the light-quantum paper constitutes a sharp break with that same tradition. Special relativity, accordingly, was accepted by the physics community within a few years, while the light-quantum hypothesis faced staunch opposition for almost two decades. In the letter dated June 12, 1913, in support of their (successful) proposal to offer Einstein a prestigious position in the Prussian Academy, the Berlin physicists Planck, Nernst, Rubens, and Emil Warburg wrote: "That he may sometimes have missed the target in his speculations, as for example, in his hypothesis of light quanta, cannot really be held against him" (Einstein 1987–2018, Vol. 5, Doc. 445; quoted, e.g., in Pais 1982, p. p. 382).[2]

Einstein's light-quantum hypothesis went against one of the main tenets of the physics of his day. Early in the nineteenth century, on the basis of interference phenomena, Thomas Young and Augustin Jean Fresnel had revived Christiaan Huygens' idea that light consists of waves rather than particles, as Isaac Newton and his eighteenth-century followers had maintained. By the middle of the nineteenth century the triumph of the new wave theory over the old particle theory had been complete. During the second half of the century, light waves had been identified as electromagnetic waves (Maxwell 1861–1862; Hertz 1888a, 1888b).

Special relativity left no room for the ether, the medium in which these electromagnetic waves were supposed to propagate. As we shall see in later chapters, however, it remained common throughout the 1910s and early 1920s to talk about light as waves in the ether, even for physicists championing special relativity, such as Arnold Sommerfeld. More generally, special relativity, far from undercutting Maxwellian electrodynamics, including its electromagnetic wave theory of light, showed that Newtonian mechanics had to be modified to resolve the conflict between the theories of Newton and Maxwell that had opened up in the 1890s in the optics and electrodynamics in moving systems (Janssen and Stachel 2004).

By contrast, Einstein's suggestion that light sometimes behaves as a collection of particles called into question the validity of Maxwell's equations. Moreover, it was completely unclear how the light-quantum hypothesis could be reconciled with interference phenomena. In a paper in which the experimental confirmation of a formula derived from the light-quantum hypothesis was announced, the hypothesis itself was thus called "a bold, not to say reckless, hypothesis" that "flies in the face of the thoroughly established facts of interference" (Millikan 1916b, p. 355; see Section 3.3.3). It took considerable

---

[2] See Kirsten and Körber (1975, pp. 201–208) for transcription and facsimile of this letter.

intellectual courage to put these ideas in print and one can understand why Einstein gave his paper a very cautious title: "On a heuristic point of view concerning the production and transformation of light."

One wonders whether the paper would even have been accepted for publication in the venerable *Annalen der Physik* had it been Einstein's first submission to the journal. The editor of the *Annalen* during this period was the Leipzig physicist Paul Drude, whose work on the electron theory of metals and on optical dispersion theory later became important for the development of quantum theory. Planck had been Associate Editor since 1895 and took over as Editor after Drude's suicide in 1906. There was no formal refereeing process in those days. Once authors had established their credentials with a few publications, they did not have to worry too much about getting subsequent papers published.[3] By 1905, Einstein had been publishing in the *Annalen* for several years, even though he had not produced anything as groundbreaking as the papers of his miracle year. In addition to a number of book reviews and two papers he later dismissed as "my two worthless beginner's works,"[4] he had published three papers on the foundations of statistical mechanics. Given this track record, it is understandable that Planck, despite his own strong reservations, allowed Einstein's revolutionary light-quantum paper to be published in the *Annalen*.

3.2   The statistical trilogy (1902–1904)

Before we turn to the light-quantum paper, we need to take a look at Einstein's early papers on the foundations of statistical mechanics, now often referred to as the statistical trilogy (Einstein 1902, 1903, 1904).[5]

The trilogy built on Boltzmann's (1896–1898) *Lectures on Gas Theory*, which Einstein had already studied carefully a few years earlier. It is unclear how much of Boltzmann's other work he was familiar with at this point. He certainly did not know any of the work by the American theoretical physicist Josiah Willard Gibbs. The German translation of Gibbs's (1902) book, *Elementary Principles in Statistical Mechanics*, only appeared in 1905, two years after its author's death. As Max Born (1949, p. 164) noted decades

---

[3] Drude, however, insisted on some changes when in August 1905 Einstein submitted his dissertation for publication in the *Annalen*, which consequently was not published until the following February (Einstein 1906a).

[4] Einstein to Johannes Stark, December 7, 1907 (Einstein 1987–2018, Vol. 5, Doc. 66). For discussion of these papers, see the editorial note, "Einstein on the nature of molecular forces" (Einstein 1987–2018, Vol. 2, pp. 3–8), Norton (2006, pp. 84–85) and Kox (2014, pp. 105–107).

[5] For analysis of these three papers, see the editorial note, "Einstein on the foundations of statistical physics" (Einstein 1987–2018, Vol. 2, pp. 41–55), Pais (1982, Ch. 4), Kuhn (1987, pp. 171–179), Gearhart (1990), Renn (2000), Uffink (2006), Norton (2006, 2008), Rynasiewicz and Renn (2006), and Renn and Rynasiewicz (2014). Kuhn (1987) describes the trilogy as "a series of three brilliant papers" (p. 171). Uffink (2006), while recognizing the trilogy's remarkable achievements, puts his finger on various questionable assumptions and dubious steps in its arguments. We cover Einstein's trilogy only to the extent that it is relevant for understanding Einstein's subsequent work on the light-quantum hypothesis.

later: "The similarity is quite amazing." Einstein (1911a, p. 176) himself later wrote that he would not have published these papers had he known of Gibbs's work.

Although in many places the trilogy duplicates earlier work by Boltzmann and Gibbs, it set the stage for much of Einstein's work in 1905 and beyond. It put in place some of the foundations for this work and, perhaps even more importantly, it prepared him for seeing the limitations of these foundations. It is also where Einstein developed the idea of using fluctuations to probe the microstructure of macroscopic systems.

As we saw in Chapter 2, Planck believed that the second law of thermodynamics holds strictly, which implies that the laws of nature do not allow fluctuations. Maxwell and Boltzmann disagreed: fluctuations do occur and can sometimes even be observed macroscopically. In one of his responses to Zermelo (see Section 2.3), Boltzmann noted in passing that

> for small particles in a gas, motions have been observed which can result from the fact that in such cases the pressure on a part of their surface no longer negligible compared to their total surface is sometimes a little larger, sometimes a little smaller (Boltzmann 1896–1898, p. 572).

More than two decades earlier, in a lecture for the British Association for the Advancement of Science, Maxwell had already traced this idea back to the ancient atomists:

> When Lucretius wishes us to form a mental representation of the motion of atoms, he tells us to look at a sunbeam shining through a darkened room ... and to observe the motes which chase each other in all directions through it. This motion of the visible motes, he tells us, is but a result of the far more complicated motion of the invisible atoms which knock the motes about (Maxwell 1873, Vol. 2, p. 373).

Einstein realized that, under special circumstances, fluctuations not only manifest themselves macroscopically but provide detailed information about the microscopic constituents of the macroscopic systems we can directly observe.

The most famous example of such use of fluctuations is Einstein's 1905 paper on Brownian motion.[6] Like Lucretius's motes, particles suspended in a liquid, small but large enough to be visible under a microscope, are hit by the molecules of the liquid from all sides producing net unbalanced forces making these particles move irregularly. Einstein could predict the average distance traveled by a particle as a result of these collisions. His prediction was confirmed within a few years by the French physicist Jean Perrin (1913, cf. Nye 1972). This provided direct evidence for the reality of atoms.

---

[6] For discussion of this and other papers on Brownian motion, see the editorial note, "Einstein on Brownian motion" (Einstein 1987–2018, Vol. 2, pp. 206–222), Pais (1982, Ch. 5), Stachel (2005, 73–84), Renn (2005), and Kox (2014). For expositions of this paper aimed at physics undergraduate students, see Cheng (2013, Ch. 2, pp. 20–28) and, in more detail, Kennedy (2012, Ch. 3, pp. 56–104).

Fluctuations also take center stage in Einstein's work on quantum theory, both in the 1905 light-quantum paper and in several other papers that we will examine in detail in this chapter (Einstein 1909a, 1909b in Section 3.4.3; Einstein 1916b in Section 3.6.2).[7] In these papers, Einstein considered energy fluctuations and (in 1909 and 1916) momentum fluctuations in black-body radiation. Boltzmann and Gibbs only applied statistical mechanics to gases. In the statistical trilogy, Einstein tried to make sure that the formalism he was developing would be flexible enough to deal with other systems, such as, in particular, a cavity filled with black-body radiation.

In the first installment of the trilogy, Einstein (1902) considered an arbitrary mechanical system with n degrees of freedom with a Hamiltonian $H(q_i, p_i)$ defined on its phase space spanned by the generalized coordinates $q_i = (q_1, \ldots, q_n)$ and their conjugate momenta $p_i = (p_1, \ldots, p_n)$.[8] Following Boltzmann, he assumed that $H(q_i, p_i) = E$ is the only conserved quantity of the system. This is equivalent to another assumption introduced by Boltzmann, which was named the *ergodic hypothesis* by Paul Ehrenfest and Tatiana Ehrenfest-Afanassjewa (1911) in their review article on statistical mechanics.[9] A system is ergodic if it eventually passes through every point of its energy shell in phase space, i.e., if it eventually goes through all values of (q_i, p_i) for which the Hamiltonian has the value E.[10]

The following simple argument shows that there is a close connection between ergodicity and energy being the only conserved quantity (but see note 9). If there were other conserved quantities besides the energy, the energy shell could be divided up into parts corresponding to different values for those other conserved quantities and the system would always be found in one of those parts. Conversely, if the system were always found in a particular part of the energy shell, we could introduce a conserved quantity that has one value in that part and different values elsewhere.

Einstein considered an ensemble of infinitely many copies of a system with energies in the narrow range between E and $E + \Delta E$ (with $\Delta E \ll E$). He recovered the result, already found by Boltzmann and Maxwell, that the only distribution of copies of the system over its energy shell that stays the same over time is one that is uniform in phase space. In that case the number of copies dN with coordinates and momenta between (q_i, p_i) and $(q_i + dq_i, p_i + dp_i)$ is proportional to the volume element $dq_1 \ldots dp_n$ in phase space:

---

[7] In his report for the first Solvay congress in 1911 on his 1907 quantum theory for the specific heats of solids (see Section 3.5), Einstein likewise considered energy fluctuations in his simple model of a solid (Einstein 1914a, pp. 342–343). In 1925, he considered the energy fluctuations predicted by his quantum theory of the ideal gas (Einstein 1925, sec. 8, pp. 8–9).

[8] See the Appendix, Section A.1.3, for an introduction to Hamiltonian mechanics.

[9] Strictly speaking, it follows from ergodicity that energy is the only conserved quantity but the converse is not necessarily true. For discussion, see Darrigol (2018, p. 126–128).

[10] Einstein's reliance on the ergodic hypothesis was criticized a few years later by Leonard Ornstein (1910), a student of Lorentz (Uffink 2006, p. 50). Shortly after the publication of the review article by the Ehrenfests (1911), it was shown mathematically by Michel Plancherel (1913) and Arthur Rosenthal (1913) that there can be no strictly ergodic systems. It is widely assumed, however, that any system of physical interest is at least *quasi-ergodic*, which means that it eventually gets *arbitrarily close* to every point of its energy shell in phase space. For a brief discussion of the early history of the ergodic hypothesis, see Brush (1967, pp. 168–177, 181–183).

$$dN = A\,dq_1 \ldots dq_n dp_1 \ldots dp_n, \quad E \leq H(q_i, p_i) \leq E + \Delta E \tag{3.1}$$

where the proportionality constant A serves to normalize the number dN. This is known as the *micro-canonical ensemble*.

The systems of interest to Einstein, however, exchange energy with their surroundings: think of a gas or black-body radiation in contact with a heat bath. On the assumption that it is small compared to its environment, such a system can be described by what is known as the *canonical ensemble*, in which the energy is no longer restricted to a narrow range around a fixed value E but is exhibiting a Maxwell–Boltzmann distribution. The relative number dN of copies of the system with coordinates and momenta between (q_i, p_i) and $(q_i + dq_i, p_i + dp_i)$ is now given by (in modern notation)

$$dN = A' e^{-\beta H(q_i, p_i)} dq_1 \ldots dq_n dp_1 \ldots dp_n, \tag{3.2}$$

where $\beta \equiv 1/kT$. Einstein equated the relative number dN of systems in a certain state in such a canonical ensemble with the relative probability dW of finding a single system in that state at any given time (Uffink 2006, p. 40). This allowed him to replace the average energy of a single system over time by the average energy of the infinitely many systems in the canonical ensemble at an arbitrary fixed time. Using this assumption, he proceeded to prove the equipartition theorem for the time average of the energy stored in the various degrees of freedom of the system (Einstein 1902, pp. 421–428; see also Einstein 1904, pp. 357–358).[11]

We present a modern version of this proof. Consider a term of the form aq_k^2 in the Hamiltonian $H(q_i, p_i)$, i.e., a term quadratic in one of the generalized coordinates $(q_1, \ldots, q_n)$, and representing the entire dependence of the Hamiltonian on the variable q_k. The same argument works for terms quadratic in one of the conjugate momenta $(p_1, \ldots, p_n)$. The Hamiltonian $H = (p_x^2 + p_y^2 + p_z^2)/2m$ of a free particle of mass m has three such terms; the Hamiltonian $H = (p^2/2m) + \frac{1}{2}m\omega^2 q^2$ of a one-dimensional harmonic oscillator with characteristic angular frequency ω has two. The canonical ensemble average (indicated by an overbar) of a term of this form is:

$$\overline{aq_k^2} = \frac{\displaystyle\int_{-\infty}^{\infty} \ldots \int_{-\infty}^{\infty} dq_1 \ldots dp_n \, aq_k^2 A e^{-\beta H(q_i, p_i)}}{\displaystyle\int_{-\infty}^{\infty} \ldots \int_{-\infty}^{\infty} dq_1 \ldots dp_n \, A e^{-\beta H(q_i, p_i)}} \tag{3.3}$$

(here we have dropped the prime on A' in Eq. (3.2)). The integrals in the numerator and the denominator, except for the ones over q_k, cancel each other and we are left with

---

[11] Einstein initially left open the form of the parameter β in Eq. (3.2), which in his notation is $2h$, and showed in the course of proving the equipartition theorem that it must be equal to $1/kT$, or in his notation $2h = 1/2\kappa T$ with $2\kappa = k$ (Einstein 1902, pp. 427–428; Uffink 2006, pp. 39–41).

$$\overline{aq_k^2} = \frac{\int_{-\infty}^{\infty} dq_k \, aq_k^2 e^{-\beta aq_k^2}}{\int_{-\infty}^{\infty} dq_k e^{-\beta aq_k^2}}. \tag{3.4}$$

Defining the denominator as Z_k, now called a partition function or partition sum, and using the standard integral $\int_{-\infty}^{\infty} dx e^{-ax^2} = \sqrt{\pi/a}$, we obtain

$$Z_k \equiv \int_{-\infty}^{\infty} dq_k e^{-\beta aq_k^2} = \sqrt{\frac{\pi}{\beta a}}. \tag{3.5}$$

Since

$$-\frac{dZ_k}{d\beta} = \int_{-\infty}^{\infty} dq_k \, aq_k^2 e^{-\beta aq_k^2}, \tag{3.6}$$

we have

$$-\frac{d\ln Z_k}{d\beta} = -\frac{1}{Z_k}\frac{dZ_k}{d\beta} = \overline{aq_k^2}. \tag{3.7}$$

From Eqs. (3.5) and (3.7), it follows that

$$\overline{aq_k^2} = -\frac{d}{d\beta}\ln\sqrt{\frac{\pi}{\beta a}} = -\frac{1}{2}\frac{d}{d\beta}\left(\ln(\pi/a) - \ln\beta\right) = \frac{1}{2\beta} = \tfrac{1}{2}kT. \tag{3.8}$$

This is the equipartition theorem. Every degree of motion—more precisely: every term quadratic in q_k or p_k in the Hamiltonian $H(q_i, p_i)$—has the same average amount of energy of $\tfrac{1}{2}kT$.

When the equipartition theorem is applied to the degrees of freedom in electromagnetic radiation, we arrive at the highly problematic Rayleigh–Jeans law for black-body radiation (see Section 2.4). As we will see in Section 3.3.1, Einstein derived this law from what we now think of as classical physics[12] by applying the equipartition theorem not to the radiation itself but to the resonators introduced by Planck interacting with the radiation. However, Rayleigh (1900b), Jeans (1905a, 1905b) and Lorentz (1909b, see Section 3.4.2) derived this law by applying equipartition directly to the radiation. These derivations played an important role in the reception of Planck's theory of black-body radiation and Einstein's light-quantum hypothesis. We therefore insert a modern version of this application of the equipartition theorem at this point.

---

[12] For careful discussion of the origin of this use of the term "classical", see Staley (2005, 2008) and Gooday and Mitchel (2013).

Consider a cubical box of dimensions $L \times L \times L$, with conducting walls, so that the electric field vanishes whenever x, y, or z are equal to 0 or L. The electric field in the possible standing wave modes in the box will be of the form

$$E \propto \sin\left(\frac{n_x \pi x}{L}\right) \sin\left(\frac{n_y \pi y}{L}\right) \sin\left(\frac{n_z \pi z}{L}\right) \sin(2\pi \nu t), \qquad (3.9)$$

where (n_x, n_y, n_z) are whole positive integers 1, 2, 3,... This field satisfies the wave equation

$$\frac{\partial^2 E}{\partial x^2} + \frac{\partial^2 E}{\partial y^2} + \frac{\partial^2 E}{\partial z^2} = \frac{1}{c^2}\frac{\partial^2 E}{\partial t^2}, \qquad (3.10)$$

where c is the speed of light. Inserting Eq. (3.9) into Eq. (3.10), we find

$$\left(n_x^2 + n_y^2 + n_z^2\right)\frac{\pi^2}{L^2} = \frac{4\pi^2 \nu^2}{c^2}. \qquad (3.11)$$

Introducing the vector $\vec{n} \equiv (n_x, n_y, n_z)$ (with positive integer components), we have:

$$|\vec{n}| = \frac{2L\nu}{c}. \qquad (3.12)$$

The vector $\vec{n}$ occupies an octant in three-dimensional space, so the number of standing wave modes in a shell $(|\vec{n}|, |\vec{n}| + d|\vec{n}|)$ is

$$\frac{1}{8} \cdot 4\pi |\vec{n}|^2 d|\vec{n}| \cdot 2 = \frac{1}{8} \cdot 4\pi \left(\frac{4L^2\nu^2}{c^2}\right)\left(\frac{2Ld\nu}{c}\right) \cdot 2 = \frac{8\pi L^3 \nu^2 d\nu}{c^3} \qquad (3.13)$$

(the factor of 2 is there to take into account that each mode has two independent polarizations). Rayleigh (1900b) initially only showed that the number of modes is proportional to ν^2 and did not bother to calculate the proportionality constant. When five years later he did (Rayleigh 1905a), he omitted a factor of $\frac{1}{8}$, effectively allowing positive and negative values for the mode numbers n_x, n_y, n_z, even though the corresponding fields are not independent modes (they differ at most by a sign from the modes in the positive octant). Rayleigh's error was corrected by Jeans (1905a; 1905b; cf. Chapter 2, note 37).

Equipartition says that each mode will have an average energy of kT, with the electric and the magnetic field each contributing $\frac{1}{2}kT$ (the two contributions turn out to be equal for electromagnetic radiation *in vacuo*). Hence the total energy per unit volume in a frequency interval $(\nu, \nu + d\nu)$ should be

$$\rho_{\text{class}} d\nu = \frac{1}{L^3} \cdot \frac{8\pi L^3 \nu^2 d\nu}{c^3} \cdot kT \qquad (3.14)$$

and the spectral distribution of black-body radiation should be given by

$$\rho_{\text{class}}(\nu, T) = \frac{8\pi}{c^3} kT\nu^2. \tag{3.15}$$

This is the law we found in Section 2.5 as the low-frequency limit of the Planck law (see Eq. (2.24) and Eq. (2.26)). Recall that Rayleigh (1900b) only proposed this law for low frequencies (what he called the "graver modes") and rather arbitrarily multiplied the expression $c_1 T\nu^2$ he found by $e^{-c_2\nu/T}$ (with c_1 and c_2 constants to be fitted to the data). Also recall that Planck proposed his new law because of deviations from the Wien law found by his experimental colleagues, *not* because he realized that classical theory led to what Ehrenfest (1911, p. 92) would later call the ultraviolet catastrophe. In fact, Einstein (1905a) and Ehrenfest (1906) were the first to point out that classical theory leads to this disastrous law for the spectral distribution of black-body radiation.

After this intermezzo on equipartition and the Rayleigh–Jeans law, we return to Einstein's statistical trilogy. In the third installment, Einstein (1904, pp. 359–360) derived another now standard result for the canonical ensemble, a formula for the mean square fluctuation of the energy

$$\overline{\Delta E^2} \equiv \overline{(E - \overline{E})^2} = \overline{E^2} - \overline{E}^2 = kT^2\frac{d\overline{E}}{dT}. \tag{3.16}$$

Einstein did not know at this point that this formula can already be found in Gibbs. As we did with the equipartition theorem, we present a concise modern derivation of this fluctuation formula. Defining the denominator of Eq. (3.3) as the partition sum Z and setting $H(q_i, p_i) = E$, we obtain

$$Z \equiv \int_{-\infty}^{\infty} \cdots \int_{-\infty}^{\infty} dq_1 \ldots dp_n A e^{-\beta E}. \tag{3.17}$$

We find Eq. (3.16) for $\overline{\Delta E^2}$ by setting two expressions for $d^2\ln Z/d\beta^2$ equal to one another. On the one hand,

$$\frac{d^2 \ln Z}{d\beta^2} = \frac{d}{d\beta}\left(\frac{1}{Z}\frac{dZ}{d\beta}\right) = \frac{1}{Z}\frac{d^2 Z}{d\beta^2} - \frac{1}{Z^2}\left(\frac{dZ}{d\beta}\right)^2 = \overline{E^2} - \overline{E}^2, \tag{3.18}$$

where, in the last step, we used that

$$-\frac{1}{Z}\frac{dZ}{d\beta} = \overline{E}, \qquad \frac{1}{Z}\frac{d^2 Z}{d\beta^2} = \overline{E^2}. \tag{3.19}$$

On the other hand,

$$\frac{d^2 \ln Z}{d\beta^2} = \frac{d}{d\beta} \left(\frac{1}{Z} \frac{dZ}{d\beta} \right) = -\frac{d\overline{E}}{d\beta} = kT^2 \frac{d\overline{E}}{dT}, \tag{3.20}$$

where, in the last step we used that, since $\beta = 1/kT$, $d\beta = -dT/kT^2$.

Combining Eqs. (3.18) and (3.20), we arrive at Eq. (3.16) for the mean square fluctuation of the energy in the canonical ensemble. Note that this fluctuation formula is much more general than the equipartition theorem (3.8), which only holds for systems with Hamiltonians consisting of a sum of terms quadratic in q's or p's.

Much of Einstein's effort in the second and third installments of his statistical trilogy went into attempts, not all of them successful, to make his statistical formalism for dealing with macroscopic systems independent of the detailed dynamics of their microscopic constituents. The distinction between coordinate and momentum variables turned out to be immaterial and with it the distinction between kinetic and potential energy. The Hamiltonian could be replaced by some other conserved quantity not necessarily representing the system's energy. The equations of motion could accordingly be more general than Hamilton's equations. The statistical trilogy thus shows a gradual relaxation of the mechanical underpinnings of the formalism Einstein sought to develop.

In the concluding section of the third installment, Einstein (1904, pp. 360–362) used the fluctuation formula (3.16) to find a rough estimate of the fluctuations in the energy of black-body radiation in a cavity of dimensions λ_{max}, the wavelength of the peak in the spectral distribution of the radiation's energy (Rynasiewicz and Renn 2006, pp. 13–14).

Using the plausible estimate that in a volume of those dimensions the energy would fluctuate between zero and its mean value, he set the mean square fluctuations equal to the square of the mean energy $\overline{E}$:

$$\overline{\Delta E^2} \approx \overline{E}^2. \tag{3.21}$$

Using the Stefan-Boltzmann law to set

$$\overline{E} = (\sigma T^4/c)\, V = (\sigma/c)\, T^4 \lambda_{\mathrm{max}}^3 \tag{3.22}$$

and inserting this into Eq. (3.16), he arrived at

$$\overline{\Delta E^2} = kT^2 \frac{d}{dT}\left((\sigma/c)\, T^4 \lambda_{\mathrm{max}}^3 \right) = (k\sigma/c)\, T^5 \lambda_{\mathrm{max}}^3. \tag{3.23}$$

where we used that $\lambda_{\mathrm{max}} T$ is a constant. Einstein, it seems, failed to take into account the temperature dependence of λ_{max}, which would explain why he found an extra factor of 4 on the right-hand side of Eq. (3.25) below. According to Eqs. (3.21) and (3.22), $\overline{\Delta E^2}$ should be equal to

$$\overline{E}^2 = (\sigma/c)^2\, T^8 \lambda_{\mathrm{max}}^6. \tag{3.24}$$

It follows that

$$\lambda_{\max} T = \left(\frac{kc}{\sigma}\right)^{1/3}. \tag{3.25}$$

Inserting values for k, σ, and c, we find that $(kc/\sigma)^{1/3} = .26\,\mathrm{K}\cdot m$. Einstein arrived at $(4kc/\sigma)^{1/3} = .42\,\mathrm{K}\cdot m$ instead. This value is still of the same order of magnitude as the value $\lambda_{\max} T = .293\,\mathrm{K}\cdot m$ found from measurements of black-body radiation, which inspired confidence that the formalism Einstein had developed in the statistical trilogy was applicable to black-body radiation. This confidence, however, would not last.

The central argument of the light-quantum paper, completed one year after the final installment of the statistical trilogy, also involves fluctuations in black-body radiation. The light-quantum paper, however, makes little use of the formalism developed in these earlier papers. In 1902–1904, Einstein used the canonical ensemble to give him the probabilities he needed to derive the equipartition theorem (3.8) and the fluctuation formula (3.16). In the light-quantum paper, he started calculating relative probabilities using the equation $W = e^{\Delta S/k}$ instead.

He obtained this equation by inverting the relation $S = k\ln W$ (cf. Eq. (2.31)). In fact, it was Einstein (1905a, p. 187) who coined the term "Boltzmann's principle" for this relation, first written down in this form by Planck (see Section 2.1). He alluded to this expression for the entropy in the introduction of his 1904 paper, mentioning both Boltzmann and Planck's theory of black-body radiation, in which the expression was put to good use (see Section 2.5). Yet, he did not explicitly write down the equation $S = k\ln W$, nor gave references to the relevant papers of Boltzmann and Planck. And although he claimed to have found an analogous expression for the entropy in his own formalism, where W is replaced by some volume in phase space, the connection between the two expressions remained unclear (Uffink 2006, p. 48). In the light-quantum paper, Einstein invoked the relation $S = k\ln W$ without any attempt to connect it to the phase space needed to define the canonical ensemble. As we will see, he had good reason to sever this connection (Norton 2006, 2008).

3.3 The light-quantum paper (1905)

3.3.1 Classical theory leads to the Rayleigh–Jeans law

Einstein opened his 1905 light-quantum paper[13] by pointing to a tension in contemporary theory: It treats matter as discrete, with a large but finite number of degrees of freedom, but electric and magnetic fields as continuous, with an infinite number of

---

[13] This paper is discussed in many places. See, e.g., the editorial note, "Einstein's early work on the quantum hypothesis" (Einstein 1987–2018, Vol. 2, pp. 134–148), Stachel (2005, pp. 167–176), Klein (1963b), Dorling (1971), Pais (1982, Ch. 19), Kennedy (2012, pp. 39–55), Norton (2006, 2008), and Darrigol (2014).

degrees of freedom. He conceded that the explanations based on continuous fields of phenomena that boil down to the observation of time averages of these fields—such as diffraction, reflection, refraction, and dispersion phenomena—were so successful that they would probably never be replaced by anything else. He went on to suggest, however, that phenomena involving the emission and absorption of light—such as black-body radiation and the "production of cathode rays by ultraviolet light" (better known today as the photoelectric effect)—might be better explained by a theory in which radiation, like matter, is discrete. This led him to introduce the following hypothesis, to be examined and argued for in the rest of the paper:

> [I]n the propagation of a light ray emitted from a point source, the energy is not distributed continuously over ever-increasing volumes of space, but consists of a finite number of energy quanta localized at points in space that move without dividing, and can be absorbed or generated only as complete units (Einstein 1905a, p. 178).

Note that Einstein's hypothesis is not just that radiation is discrete but also that its discrete components are independent of one another and localized in space.

One way to bring out the tension between discrete matter and continuous radiation is to apply the equipartition theorem to radiation. If this application were allowed, a system consisting of both matter and radiation could never reach equilibrium. Since continuous radiation would have infinitely many more degrees of freedom than discrete matter, the former would drain all the energy from the latter. The easy way out of this conundrum is to deny that equipartition applies to the radiation's degrees of freedom. To show that the problem can not so easily be circumvented, Einstein turned to the relation that Planck[14] had derived between the energy density ρ_ν of black-body radiation at frequency ν and the average energy U_ν of an resonator with that same frequency ν as its resonance frequency:

$$\rho_\nu = \frac{8\pi\nu^2}{c^3} U_\nu. \tag{3.26}$$

(see Eq. (2.7)). Applying the equipartition theorem to the two degrees of freedom of the resonator, Einstein inserted

$$U_\nu = kT \tag{3.27}$$

(writing Boltzmann's constant k as the ideal gas constant R divided by Avogadro's number N) into this equation and arrived at what we now recognize as the Rayleigh–Jeans law (2.26):

---

[14] Einstein (1905a, p. 180) cites the reprise in *Annalen der Physik* (Planck 1900a, p. 93) of the last installment of Planck's pentalogy on irreversible radiation phenomena in the Proceedings of the Berlin Academy (Planck 1897–1899, Pt. V, p. 575). Two letters of April 1901 to his fiancé, Mileva Marić, show that Einstein was reading Planck's papers in the *Annalen* early on (Einstein 1987–2018, Vol. 1, Docs. 96 and 97).

$$\rho_v = \frac{8\pi}{c^3} v^2 (R/N) T. \tag{3.28}$$

Einstein did not cite Rayleigh (1900b) at this point and Jeans (1905a, 1905b) only corrected the numerical mistake in Rayleigh's derivation of the proportionality constant between ρ_v and $v^2 T$ a couple of months after Einstein submitted his light-quantum paper (see Eq. (3.13) above and note 37 in Chapter 2). In his biography of Einstein, Abraham Pais (1982, p. 372) thus calls Eq. (3.28) the Rayleigh–Einstein–Jeans law.

If Eq. (3.28) were correct, ρ_v does not reach a maximum for a certain frequency v_{max} but continues to get bigger as the frequency goes up. This is why Rayleigh (1900b) originally added, in a purely ad hoc fashion, a factor of $e^{-cv/T}$. Without such a factor, the law entails what Ehrenfest (1911) later called the ultraviolet catastrophe (see Sections 2.4 and 3.2). As Einstein (1905a, p. 182) noted, in the continuous theory of radiation the distribution of energy over matter and radiation can never reach equilibrium. All energy will end up in the radiation. Though he only coined the term later, Ehrenfest (1905) drew attention to this dire consequence of the prevailing theory around the same time as (and independently of) Einstein.

3.3.2   Einstein's argument for light quanta: fluctuations in black-body radiation at high frequencies

After devoting sec. 1 of his paper to this argument against the continuous conception of radiation, Einstein (1905a, secs. 2–6) spent the next five sections developing an ingenious argument in support of his discrete alternative conception. He derived a formula for the small probability of a large fluctuation in high-frequency black-body radiation that has the exact same form as the formula for the corresponding fluctuation in an ideal gas.[15]

In sec. 2, Einstein (1905a, p. 182) introduced the Planck law, citing Planck's (1901a) paper of January 1901 in the *Annalen* (see Section 2.5). He wrote it as:

$$\rho_v = \frac{\alpha v^3}{e^{\beta v/T} - 1}. \tag{3.29}$$

So instead of the constants h and k introduced by Planck (see Eq. (2.24)), Einstein used α and β, related to h and k via

$$\alpha = \frac{8\pi h}{c^3}, \qquad \beta = \frac{h}{k}. \tag{3.30}$$

---

[15] For earlier analyses of this argument, see, for instance, Dorling (1971) and Norton (2006, 2008), who calls it "Einstein's miraculous argument."

For low frequencies, $e^{\beta v/T} \approx 1 + \beta v/T$, and Eq. (3.29) reduces to

$$\rho_v = \frac{\alpha v^3}{\beta v/T} = \frac{\alpha}{\beta} v^2 T, \tag{3.31}$$

which is the Rayleigh–Jeans law (3.28) as long as

$$\frac{\alpha}{\beta} = \frac{R}{N} \frac{8\pi}{c^3}. \tag{3.32}$$

Inserting numerical values for α, β, R, and c (for which Einstein used the letter L at this point), he confirmed that this relation reproduces the accepted value for Avogadro's number N. For high frequencies, as Einstein (1905a, p. 186) reminded his readers at the beginning of sec. 4, $e^{\beta v/T} \gg 1$, and the Planck law (3.29) reduces to the Wien law,

$$\rho_v = \alpha v^3 e^{-\beta v/T}, \tag{3.33}$$

which, with the identification of α and β in Eq. (3.30), is just Eq. (2.25).

In sec. 3, Einstein introduced the entropy density $\varphi(\rho_v, v)$ of radiation with energy density ρ_v at frequency v. The entropy S of radiation at all frequencies enclosed in a volume V is given by[16]

$$S = \int_0^\infty dv \varphi(\rho_v, v) V. \tag{3.34}$$

Following Wien (1894), explicitly cited at this point, and Planck (1900a, p. 115), cited in sec. 1 of his paper, Einstein assigned both entropy and temperature to radiation (cf. our discussion in Sections 2.2–2.3) and posited the basic relation $dS/dU = 1/T$ from standard thermodynamics between the two, except that he replaced entropy and energy by entropy and energy *densities*:

$$\frac{\partial \varphi(\rho_v, v)}{\partial \rho_v} = \frac{1}{T}. \tag{3.35}$$

Einstein noted that one could use this relation either to determine ρ_v starting from an expression for φ or to determine φ starting from an expression for ρ_v. In Chapter 2, we saw Planck follow the former route to find both the Wien law (2.11) and the Planck law (2.24).[17] Einstein (1905a, sec. 4) followed the latter route. The steps in Einstein's

---

[16] Einstein used lower case v for volume. We use upper case V to make it easier to distinguish between the volume V and the frequency v.

[17] In the case of the Planck law, Planck started from an expression for the *second* derivative of the entropy with respect to the energy (see Eq. (2.17)). Planck did so not in terms of the energy and entropy density of the radiation itself but in terms of the energy U_v and entropy S_v of resonators interacting with the radiation.

derivation of an expression for φ correspond to the steps in Planck's derivations of expressions for U_ν (compare Eqs. (3.33)–(3.38) to Eqs. (2.8)–(2.10)). Einstein just reversed the order of these steps.

Einstein focused on black-body radiation in a narrow frequency interval in the Wien regime of high frequencies. Solving for T in Eq. (3.33) for ρ_ν, we obtain:

$$\frac{1}{T} = -\frac{1}{\beta\nu} \ln \frac{\rho_\nu}{\alpha\nu^3}. \tag{3.36}$$

Eq. (3.35) then tells us that

$$\frac{\partial\varphi}{\partial\rho_\nu} = -\frac{1}{\beta\nu} \ln \frac{\rho_\nu}{\alpha\nu^3}. \tag{3.37}$$

Integrating this expression, we find (cf. Chapter 2, note 32):

$$\varphi(\rho_\nu, \nu) = -\frac{\rho_\nu}{\beta\nu} \ln \frac{\rho_\nu}{e\alpha\nu^3}. \tag{3.38}$$

Einstein now introduced the energy E_ν and entropy S_ν of the radiation in the frequency interval $(\nu, \nu + d\nu)$ in some volume V:

$$E_\nu \equiv \rho_\nu V d\nu, \qquad S_\nu \equiv \varphi(\rho_\nu, \nu) V d\nu. \tag{3.39}$$

From these definitions and Eq. (3.38), it follows that:

$$S_\nu = -\frac{E_\nu}{\beta\nu} \ln \frac{E_\nu}{e\alpha\nu^3 V d\nu}. \tag{3.40}$$

The change in entropy if the volume taken up by the radiation goes from V_0 to V while the energy E_ν remains constant is thus given by

$$\Delta S_\nu = S_\nu - S_\nu^0 = \frac{E_\nu}{\beta\nu} \ln \frac{V}{V_0}. \tag{3.41}$$

It is at this point that Einstein (1905a, sec. 5) introduced Boltzmann's principle, $S = (R/N) \ln W$. His promise to give a general justification of the principle in a separate paper was not fulfilled until several years later.[18] In the 1905 paper, he basically only offered a plausibility argument for the logarithmic dependence of entropy on probability: to find the entropy of a system consisting of several parts we need to *add* the entropies of the

---

[18] For discussion of the relevant passages in Einstein (1909a, 1910), see Klein (1974) and the editorial note, "Einstein on critical opalescence" (Einstein 1987–2018, Vol. 3, pp. 283–285).

subsystems; to find the probability of the state of the system we need to *multiply* the probabilities of the states of the subsystems.

Einstein used Boltzmann's principle to calculate the (tiny) probability that, due to a random fluctuation, an amount of black-body radiation at frequency v in the Wien regime initially taking up a volume V_0 momentarily gets concentrated in a subvolume $V < V_0$. According to Boltzmann's principle, the entropy decrease in this process is proportional to the logarithm of the relative probability W of the final state with respect to the initial state:

$$\Delta S_v = S_v - S_v^0 = (R/N)\ln W. \tag{3.42}$$

Rewriting Eq. (3.41) as

$$\Delta S_v = (R/N)\ln\left(\frac{V}{V_0}\right)^{\frac{E_v}{R\beta v/N}}, \tag{3.43}$$

and combining it with Eq. (3.42), Einstein inferred that

$$W = \left(\frac{V}{V_0}\right)^{\frac{E_v}{R\beta v/N}} = \left(\frac{V}{V_0}\right)^{\frac{E_v}{hv}}, \tag{3.44}$$

where in the last step we used Eq. (3.30) to replace Einstein's combination of constants $R\beta/N$ by the familiar Planck constant h (Einstein would only start doing so in 1909; see Section 3.4.3).

On the face of it, Einstein had no right to combine Eqs. (3.42) and (3.43) (Norton 2006, p. 87). Eq. (3.43) gives the difference in entropy between two equilibrium states, whereas Eq. (3.42) gives the difference in entropy between an initial equilibrium state and a final state that is the result of a random fluctuation. This final state thus need not be an equilibrium state. It definitely cannot be an equilibrium state *at the same temperature* as the initial state. After all, the Stefan–Boltzmann law, $u = (\sigma T^4)/c$, tells us that the energy *density* of black-body radiation is fixed by the temperature. Hence, if the energy density increases, as it must when the volume decreases while the energy remains constant, the temperature must also increase.

In combining Eqs. (3.42) and (3.43), Einstein implicitly made the assumption that the final state can be treated as an equilibrium state *at some higher temperature T'*. He did not have to worry about the value of T'. As Eq. (3.40) shows, he wrote the entropy of black-body radiation in the Wien regime in terms of V and E and not in terms of, say, V and T. Treating the state resulting from the fluctuation Einstein considered as an equilibrium state can be justified by the observation that the vast majority of all possible microstates for a macrostate with volume $V < V_0$ and energy E belong to the equilibrium (macro)state with that smaller volume, that same energy and a higher temperature.[19]

The final step in Einstein's argument was to compare the probability of this particular fluctuation in black-body radiation in the Wien regime, given by Eq. (3.44), to the probability that, due to a random fluctuation, all n molecules of an ideal gas initially occupying a volume V_0 would momentarily be found in a subvolume $V < V_0$. Since the molecules all move independently of one another, this probability is given by

$$W = \left(\frac{V}{V_0} \right)^n. \tag{3.45}$$

Comparing the exponents in Eqs. (3.44) and (3.45), Einstein concluded:

> [M]onochromatic radiation of low density (within the range of validity of Wien's radiation formula) behaves thermodynamically as if it consisted of [n] mutually independent energy quanta of magnitude [$h\nu$] (Einstein 1905a, p. 191).

Note that only the discreteness and the independence of the energy quanta are mentioned in this passage. The light-quantum hypothesis, as stated in the introduction of his paper, also says that light quanta are point-like and localized (Einstein 1905a, p. 178). The analogous behavior between black-body radiation in the Wien regime and an ideal gas when it comes to fluctuations of the kind considered by Einstein certainly suggests that light quanta possess all three properties: discreteness, independence, and localization. A closer look at Einstein's argument, however, reveals that while it supports discreteness and independence, it does not support localization.

It turns out that one also arrives at Eq. (3.44) if we treat the n quanta of energy $h\nu$ into which the total energy E in the container with black-body radiation can be divided *not* as n point particles but as the excitation energy of n modes of standing waves in the narrow frequency interval $(\nu, \nu + d\nu)$. These are the modes counted by Rayleigh (1900b) and Jeans (1905a, 1905b) in their derivations of the radiation law named after them (see Section 3.2) and quantized by Ehrenfest (1906) and Peter Debye (1910) (see Section 3.4.2). In the Wien regime, the total number of modes in this frequency interval will be much larger than n. This means that there will be (except very rarely) at most one quantum of energy $h\nu$ allocated to any one of them. To determine the entropy $S = k \ln W$ of this system, we need to evaluate W, the number of ways in which the n quanta of energy $h\nu$ can be distributed over the available modes in the frequency interval $(\nu, \nu + d\nu)$. The number of modes is proportional to the volume of the container (see Eq. (3.13) above). As a result of the fluctuation considered by Einstein, this volume is effectively reduced from V_0 to V. There is still at most one quantum of energy per mode, so we can set the ratio of the numbers W before and after the fluctuation equal to $(V/V_0)^n$. This is exactly the same ratio as in Eq. (3.45), yet instead of a system of localized particles, we

---

[19] For more detailed discussion of this issue, see the web resource *The (first) Einstein fluctuation theorem.*

are now considering a system of *nonlocalized* standing-wave modes. In other words, the fluctuation that Einstein thought clearly demonstrated the particle-like behavior of light is also compatible with wave-like behavior, provided the waves correspond to discretely occupied standing wave modes of radiation in a cavity.[20] Einstein's fluctuation argument thus does not support the clause in the statement of the light-quantum hypothesis in the introduction of Einstein's paper that these energy elements are "localized at points in space ... move without dividing, and can be absorbed or generated only as complete units" (Einstein 1905a, p. 178). Support for that part of the hypothesis only comes from its application to the three phenomena discussed in secs. 7–9 of the paper, most importantly the photoelectric effect.

Before turning to these applications, Einstein calculated the average energy of his light quanta, $\overline{E}_{\text{light quanta}}$, for the hypothetical case that the Wien law (3.33) holds for all frequencies. To this end, he divided the total energy of the radiation in some volume V,

$$\int_0^\infty \rho_\nu^{\text{Wien}} d\nu\, V = \alpha V \int_0^\infty d\nu \nu^3 e^{-\beta\nu/T}, \tag{3.46}$$

by the total number of quanta in that volume,

$$\int_0^\infty \frac{\rho_\nu^{\text{Wien}} d\nu}{h\nu} V = \frac{\alpha V}{h} \int_0^\infty d\nu \nu^2 e^{-\beta\nu/T}. \tag{3.47}$$

Using the standard integral $\int_0^\infty dx\, x^n e^{-ax} = n!/a^{n+1}$ with $a \equiv \beta/T = h/kT$, he found that

$$\overline{E}_{\text{light quanta}} = h \frac{\int_0^\infty d\nu \nu^3 e^{-a\nu}}{\int_0^\infty d\nu \nu^2 e^{-a\nu}} = h\frac{3!/a^4}{2!/a^3} = \frac{3h}{a} = 3kT. \tag{3.48}$$

As Einstein noted without further comment, this is twice the equipartition value for the mean energy of a molecule in an ideal gas. So, even if the Wien law were accurate over the entire spectrum of black-body radiation, light quanta do not behave in quite the same way as molecules in an ideal gas.[21]

In his statistical trilogy (see Section 3.2), Einstein had derived the equipartition theorem from the canonical ensemble (see Eqs. (3.3)–(3.8)) and he had applied the canonical ensemble to black-body radiation (see Eqs. (3.21)–(3.25)). By the time he wrote the light-quantum paper, Einstein may already have realized that this application is unwarranted. This would explain why it apparently did not bother him that, despite the analogy he sought to establish between black-body radiation in the Wien regime and

---

[20] See the web resource *The (first) Einstein fluctuation theorem* for a more detailed exposition of the argument in this paragraph.

[21] If we use the Planck law (3.29) rather than the Wien law (3.33) to calculate $\overline{E}_{\text{light quanta}}$, the result, it turns out, changes only slightly, from $3kT$ to about $2.7kT$.

an ideal gas, the mean energy of his light quanta is twice the equipartition value for the molecules in an ideal gas. More generally, it would explain why Einstein, in his light-quantum paper, hardly made use of the results he had derived for the canonical ensemble in the statistical trilogy and relied almost exclusively on Boltzmann's principle instead. Four years later, Einstein had *definitely* realized that the canonical ensemble cannot be used to describe black-body radiation. Einstein (1909a, pp. 188–189) now calculated the mean square energy fluctuation in black-body radiation using the same formula he had derived in 1904 on the basis of the canonical ensemble (see Eqs. (3.16)–(3.20) above). Before applying this formula to black-body radiation, however, he rederived it avoiding the canonical ensemble altogether and using Boltzmann's principle instead (see Section 3.4.3, Eqs. (3.61)–(3.71) below).

We suggested above that it must have taken considerable intellectual courage on Einstein's part to publish the light-quantum hypothesis. His ingenious fluctuation argument, which takes up five of the nine sections of the light-quantum paper, was probably what, more than anything else, gave him this courage. The argument shows that the discreteness of radiation is forced upon us by the validity of the Wien law at high frequencies, a well-confirmed empirical result, and Boltzmann's principle, which Einstein took to be a fundamental principle of statistical physics that applies to both matter and radiation. Though in hindsight we can see that the fluctuation argument only establishes the discreteness and the independence of light quanta in the Wien regime, not their localization, it was only natural for Einstein to assume the latter as well. Light thus seemed to exhibit particle-like behavior in these fluctuations. A satisfactory theory of light would have to do justice to such behavior as well as to the more familiar wave-like behavior. In his papers on light quanta, Einstein mostly refrained from speculation about what such a theory might look like and limited himself to identifying constraints it would have to satisfy. Using well-established theoretical and experimental results, he tried to find "inductively secure fixed point[s] from which to carry on" (Rynasiewicz and Renn 2006, p. 6). What accordingly set Einstein apart from his contemporaries in physics was not some iconoclastic vision for revolutionary new theories but his confidence in this way of taking advantage of lessons learned from Nature, no matter how contradictory they appeared to be, coupled to his extraordinary skill and perseverance in following this strategy.

The fluctuation argument, convincing as it may have been to Einstein, did not sway his colleagues. Maybe this was simply because no argument for particle-like behavior of light would have at the time. It may also have been, at least in part, because the analogy between an ideal gas and black-body radiation at the heart of the argument was neither perfect nor new.

We have already seen that the analogy is not perfect. Its most serious shortcoming, undoubtedly, is that the similarity of the expressions for W in Eqs. (3.44) and (3.45) does nothing to establish the localization of light quanta. Nobody, however, seems to have spotted that problem at the time, even though first Ehrenfest (1906) and then Debye (1910) showed that quantizing (nonlocalized) modes of the radiation leads to the Planck law. Two other shortcomings, however, would have been readily apparent. First, the analogy only works for the high frequencies in the Wien regime. Second, even for those frequencies, the average energy of a light quantum is twice that of a molecule in an ideal gas.

Einstein's analogy was not new either. In 1896, Wien found his law for the distribution of the energy of black-body radiation over different frequencies explicitly drawing on a shaky analogy to the Maxwell–Boltzmann law for the distribution of molecules over different velocities in an ideal gas (see Section 2.2). Readers of Einstein's paper in 1905 may well have remembered this dubious genetic link between the Wien distribution and the Maxwell–Boltzmann distribution. It may therefore have looked to them as if Einstein was simply bringing out in a slightly different way a superficial analogy that Wien had relied on in 1896 and that Planck had already done away with when he rederived the Wien law in 1899 (see Section 2.3).

For Einstein, the similarity between the Wien distribution and the Maxwell–Boltzmann distribution was anything but superficial. A little over a decade later, he opened another paper on light quanta (which will be covered in Section 3.6) with the observation that "[t]he formal similarity between the chromatic distribution curve for thermal radiation and the Maxwell velocity-distribution law is too striking to have remained hidden for long" (Einstein 1917a, p. 63).

Hence, even though his readers may not have taken the analogy between black-body radiation and an ideal gas all that seriously, Einstein himself most certainly did. "[I]t seems reasonable," he wrote at the end of sec. 6 of the 1905 light-quantum paper, "to investigate whether the laws of generation and conversion of light are also so constituted as if light consisted of such energy quanta" (Einstein 1905a, p. 191).

3.3.3 Evidence for light quanta: the photoelectric effect

In the last three sections of the light-quantum paper (secs. 7–9), Einstein showed how the light-quantum hypothesis can be used to explain three phenomena that are extremely puzzling from the point of view of the wave theory: Stokes's rule about fluorescence, the photoelectric effect, and the ionization of gases by ultraviolet light. It is in the explanation of these phenomena that Einstein used the part of the light-quantum hypothesis that says that the energy quanta $h\nu$ are "localized at points in space ... move without dividing, and can be absorbed or generated only as complete units" (Einstein 1905a, p. 178).

Stokes's rule says that the frequency of light given off by a fluorescent substance is never greater than the frequency of the light shone on that substance.[22] On the light-quantum hypothesis, the energy of light is proportional to its *frequency*, which makes Stokes's rule a direct consequence of energy conservation. The energy coming out can never be greater than the energy going in: $E_{out} \leq E_{in}$. Since according to the light-quantum hypothesis $E_{out} = h\nu_{out}$ and $E_{in} = h\nu_{in}$, it follows that $\nu_{out} \leq \nu_{in}$. In the wave theory of light, energy is proportional not to frequency but to the square of the *amplitude* of the associated wave, so it is possible to have $E_{out} \leq E_{in}$ while $\nu_{out} > \nu_{in}$. That the light-quantum hypothesis makes the energy of light proportional to its *frequency* also

---

[22] Stokes discovered but never published this rule. William Thomson first announced it in 1883, giving Stokes full credit (Stone 2013, pp. 76–77).

plays a central role in the other two applications of the light-quantum hypothesis. In the case of the photoelectric effect, by far the most famous of the three, it leads to a simple quantitative relation that can be checked empirically.[23]

When light of sufficiently high frequency falls on a metal surface, electrons are released from the metal, even when the intensity of the light is very low. Up to a certain threshold frequency, however, light, even at high intensity, will not release any electrons from the metal. This phenomenon is known as the photoelectric effect. In the late 1880s, Hertz, in the course of the experiments that would finally convince the scientific community that light consists of electromagnetic waves (see Section 2.2), had noticed that light can get metals to emit cathode rays. These rays were identified as streams of electrons in the late 1890s (see Chapter 2, note 13). So what Hertz had observed ten years earlier was light knocking electrons out of metals. Hertz thus not only demonstrated that light behaves as an electromagnetic wave, he also discovered the effect that would later be seen as strong evidence of its particle behavior.[24]

Einstein had already recognized the importance of the photoelectric effect four years before he published his light-quantum paper. A letter to Mileva Marić, his fiancé, of April 30, 1901, contains a remark that can be seen as a first step toward the light-quantum hypothesis: "Recently the idea came to me that in the production of light perhaps a direct conversion of the energy of motion into light occurs" (Einstein 1987–2018, Vol. 1, Doc. 102). Presumably, Einstein used the term "direct conversion" to contrast this picture of the production of light with the picture used by Planck in which light waves of frequency ν are generated by resonators oscillating at that frequency. A month later, on May 28, 1901, Einstein sent Marić another letter in which he rhapsodized about "a marvelous article" he had just read by Philipp Lenard (1900) "on the production of cathode rays by ultraviolet light" (Einstein 1987–2018, Vol. 1, Doc. 111). Einstein's enthusiasm is readily understandable given that the process Lenard was investigating is just the inverse of the process of "direct conversion" that Einstein was beginning to contemplate (Rynasiewicz and Renn 2006, pp. 31–32). More generally, these snippets from Einstein's correspondence with Marić suggest that the light-quantum paper had a long gestation period.

According to the light-quantum hypothesis, shining light on a metal surface only results in the release of electrons if the frequency ν of the light is such that its energy $h\nu$ is greater than the minimum energy that an electron needs to escape the metal. Let $E_{\text{esc}}^{\text{min}}$ denote the *minimum escape energy*. Any surplus energy will go into the kinetic energy of the escaped electron. Let $E_{\text{kin}}^{\text{max}}$ denote the *maximum kinetic energy* an escaped electron can acquire. The light-quantum hypothesis predicts a simple relation between $E_{\text{kin}}^{\text{max}}$ and $E_{\text{esc}}^{\text{min}}$:

$$E_{\text{kin}}^{\text{max}} = h\nu - E_{\text{esc}}^{\text{min}}. \tag{3.49}$$

---

[23] This sets the photoelectric effect apart from the other two phenomena and helps explain why it played a more prominent role in the development of quantum physics (Kao 2015).

[24] See Pais (1982, pp. 379–382) for a concise history of the photoelectric effect from 1887 to 1915.

To test this relationship, an experimental setup is used in which the metal surface is the positive plate of a capacitor producing an electric field of adjustable strength that opposes the motion of the escaped electron. The electric potential is cranked up to the value V_{stop}, the so-called stopping potential, the minimum value needed to ensure that no electrons reach the negative plate of the capacitor. The potential energy of an electron in this field, $-eV_{\text{stop}}$ (where e is the absolute value of the electron charge), equals minus its maximum kinetic energy. Substituting eV_{stop} for $E_{\text{kin}}^{\max}$ in Eq. (3.49), we find that the stopping potential satisfies the simple relation

$$V_{\text{stop}} = (h/e)\nu - (E_{\text{esc}}^{\min}/e). \tag{3.50}$$

A plot of V_{stop} against ν should thus be straight line with a slope h/e. The photoelectric effect thus allows us to determine the ratio of these two constants of nature.

In the light-quantum paper, Einstein cited two papers by Lenard (1902, 1903) on the photoelectric effect but it was not clear whether the plot of V_{stop} against ν for his results gave a straight line. Decisive confirmation of the linear relation in Eq. (3.50) allowing for an accurate measurement of Planck's constant (given the value for the electron charge) was only found in late 1915 and published in early 1916, by Robert A. Millikan (1916a, 1916b), then at the University of Chicago. Figure 3.1 shows a plot from one of Millikan's papers. The result he reported for Planck's constant, with an accuracy of about half a percent, was $h = 6.57 \cdot 10^{-27}\,\text{erg}\cdot\text{sec}$ (Millikan 1916b, p. 388). This agrees within the margin of error with the value $6.55(03) \cdot 10^{-27}$ that Planck extracted from measurements

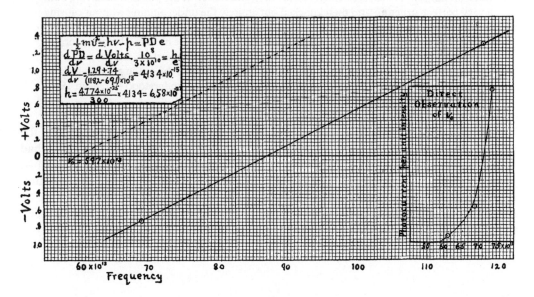

Figure 3.1 *The experimental verification of Einstein's formula for the photoelectric effect: Plot of stopping potential against frequency for sodium (Millikan 1916a, p. 24).*

of black-body radiation and gets us a little closer to the modern value of $6.6261 \cdot 10^{-27}$ (cf. Section 2.7).

In his autobiography, Millikan recalled his "complete verification of the validity of Einstein's equation" [Eq. (3.50)] and added that this had

> proved simply and irrefutably I thought, *that the emitted electron that escapes with the energy gets that energy by the direct transfer of units of energy from the light to the electron* and hence scarcely permits of any other interpretation than that which Einstein had originally suggested, namely that of the semi-corpuscular or photon theory of light itself (Millikan 1950, pp. 101–102; italics in the original).[25]

Millikan was 82 when he published his autobiography and his recollections of how he viewed his confirmation of Eq. (3.50) for the photoelectric effect in 1916 differ sharply from what he wrote more than three decades later.[26]

We already quoted his earlier characterization of the light-quantum hypothesis as "reckless" (Millikan 1916b, p. 355). Toward the end of the same paper Millikan opined that "the semi-corpuscular theory by which Einstein arrived at his equation seems at present to be wholly untenable" (ibid., p. 383, quoted, e.g., in Stuewer 1975, pp. 73–74). In his book *The Electron* published the following year, he again strictly separated the equation from the theory behind it:

> Despite then the apparently complete success of the Einstein equation, the physical theory of which it was designed to be the symbolic expression is found so untenable that Einstein himself, I believe, no longer holds to it, and we are in the position of having built a very perfect structure and then knocked out entirely the underpinning without causing the building to fall. It stands complete and apparently well tested, but without any visible means of support. These supports must obviously exist, and the most fascinating problem of modern physics is to find them. Experiment has outrun theory, or, better, guided by erroneous theory, it has discovered relationships which seem to be of the greatest interest and importance, but the reasons for them are as yet not at all understood (Millikan 1917, p. 230; quoted, e.g, in Stuewer 2014, p. 156).

Rather than dismantling the scaffold on which he had built his equation for the photoelectric effect, as Millikan suggests in this passage, Einstein (1916a, 1916b) had actually used it to build a new quantum theory of radiation (see Section 3.6). This had given him renewed confidence that he was on the right track with his light-quantum hypothesis. Millikan's dim view of Einstein's light quanta, however, continued to be shared by many physicists in the late 1910s and early 1920s. When Einstein was awarded the 1921 Nobel Prize (which he officially received in 1922), it was "for his discovery of

---

[25] Quoted, e.g., by Franklin (2013, p. 592) and Stuewer (2014, p. 143).

[26] Stuewer (2014, p. 143) presents this as an example of what he fittingly calls "Millikan's philosophy of history: if the facts don't fit your theory, change the facts."

the law of the photoelectric effect," not for the theory of light quanta from which he had derived this law (Franklin 2013, p. 588).[27]

3.4   Black-body radiation and the necessity of quantization

3.4.1   The quantization of Planck's resonators

Given that Einstein (1905a, p. 182) mentioned Planck's law for black-body radiation in his light-quantum paper and cited Planck's (1901a) paper in the *Annalen* in which it was derived, why did he not make more extensive use of it? Our discussion in Section 3.3.2 already provided part of the answer: The fluctuation argument that forms the centerpiece of the light-quantum paper only works with the Wien law. Another part of the answer is that, in 1905, Einstein still did not know what to make of Planck's work (Klein 1963b, sec. 6, pp. 81–85).

If the later recollections of Michele Besso, his most important sounding board at the time, are to be trusted, Einstein may have suppressed some criticism of Planck in his light-quantum paper. Einstein and Besso met as students in Zurich and remained lifelong friends. From 1902 to 1908, they were colleagues at the Swiss Patent Office in Bern. Their discussions earned Besso the lone acknowledgment in Einstein's (1905d) relativity paper.[28] More than two decades later, in a letter of January 17, 1928 to his by then famous friend, Besso made the following intriguing remark:

> As far as I'm concerned, I was your audience in the years 1904 and 1905; if, in the drafting of your publications on the quantum problem, I deprived you of some of your glory, in exchange I provided you with a friend in Planck (Speziali 1979, p. 141).[29]

The final clause of this statement in all likelihood refers to the relativity paper. Planck had been one of relativity's earliest and strongest advocates (Goldberg 1976). The glory that would have been Einstein's had it not been for Besso's intervention is presumably Einstein's share in the credit for introducing energy quantization (Darrigol 2014, p. 124; Stone 2013, p. 73). What was the criticism of Planck that Besso convinced Einstein to suppress?

Einstein probably considered using the argument against the continuous conception of radiation in sec. 1 of the light-quantum paper (see Section 3.3.1) as an argument against Planck's theory of black-body radiation as well (Darrigol 1988, p. 57; Rynasiewicz and Renn 2006, pp. 30–31). Recall that the argument was based on the relation Planck had derived between the energy density ρ_ν of black-body radiation at frequency ν and the average energy U_ν of a resonator with resonance frequency ν (see Eq. (2.7) and

---

[27] See Friedman (2001, pp. 126–140) and Elzinga (2006) for accounts of how Einstein won the Nobel Prize.
[28] The Einstein–Besso correspondence was published in a volume edited by Speziali (1979). See Bernstein (1991) for a review of this volume.
[29] This letter may actually not have been sent (Bernstein 1991, p. 155).

Eq. (3.26)). Planck had derived this relation using standard electrodynamics. Appealing to the equipartition theorem to set

$$U_\nu = kT,$$ (3.51)

Einstein used Planck's relation between ρ_ν and U_ν to show that (in modern terms) classical theory leads to the Rayleigh–Jeans law and the ultraviolet catastrophe. This also shows that Planck's theory of black-body radiation contradicts the theory it was supposedly based on. Planck's theory, after all, gives an expression for U_ν,

$$U_\nu = \frac{h\nu}{e^{h\nu/kT} - 1}$$ (3.52)

(see Eqs. (2.21) and (2.23)), that only reduces to Eq. (3.51) in the limit of low frequencies. That this amounts to a serious objection to Planck's theory must have been perfectly clear to Einstein in 1905, but he did not drive the point home. The omission may have been on Besso's advice. It may also have been because Einstein could not put his finger on where exactly the difference between the two conflicting expressions (3.51) and (3.52) for U_ν was coming from. In the introduction of his next paper on the light-quantum hypothesis, dated March 1906, Einstein did pronounce on the relation between his hypothesis and Planck's theory of black-body radiation. After a brief summary of the 1905 light-quantum paper, Einstein began the next paragraph confessing that, when he wrote that paper,

> it seemed to me as though Planck's theory of radiation in some sense formed a contrast to my work. New considerations . . . have made it clear to me that the theoretical foundation on which Planck's radiation theory rests differs from the foundation that would result from Maxwell's theory and the electron theory, and indeed differs exactly in that Planck's theory implicitly makes use of the light-quantum hypothesis (Einstein 1906b, p. 199).

Einstein overstated his case here. What he went on to show in this paper is that Planck's law for black-body radiation follows if the energy of the resonators interacting with the radiation is assumed to be quantized.[30] This is a far cry from showing that Planck implicitly relied on the quantization of the energy of the radiation itself.

---

[30] Einstein only showed that quantization of the energy of the resonators is *sufficient* to derive Planck's law. Jeans (1910) appears to have been the first to try and prove the converse: that quantization is *necessary* to derive Planck's law. A rigorous proof of this necessity was subsequently provided by Ehrenfest (1911) and Poincaré (1911, 1912). For careful analysis of these papers by Jeans, Ehrenfest, and Poincaré, see Norton (1993). On Poincaré's excursion into quantum theory, see also McCormmach (1967).

This was clearly recognized, for instance, by Max Laue (see Plate 20), Planck's assistant from 1905 to 1909.[31] In a letter to Einstein of June 2, 1906, Laue wrote that he agreed that "radiant energy can only be emitted and absorbed in certain finite quanta" as long as it is understood that the quantization is "not a characteristic of electromagnetic processes in a vacuum, but of the absorbing or emitting material." This, Laue argued, was in the spirit of the heuristic viewpoint of the 1905 light-quantum paper that "radiation does not consist of light quanta ... but only behaves during energy exchange with matter as though it did." "By the way," he added, "I have never discussed your heuristic viewpoint with my boss" (Einstein 1987–2018, Vol. 5, Doc. 37; cf. Kuhn 1987, p. 189). As we will see, the boss resisted even Laue's cautious reformulation of Einstein's conclusion.

Einstein (1906b) reached his conclusion by considering the entropy of a Planck resonator. Using a general formula from the second installment of his statistical trilogy (Einstein 1903), he derived two expressions for this entropy, one on the assumption that the energy of a resonator is continuous and one on the assumption that it is quantized (Einstein 1906b, pp. 200–202). Suppressing the details of the derivations, he concluded that with continuous energy one finds an expression for the entropy of a resonator that leads to the Rayleigh–Jeans law for black-body radiation, whereas with quantized energy one finds an expression that leads to the Planck law.[32] In a subsequent paper, dated November 1906, Einstein (1907, pp. 180–183) similarly derived two expressions for the average energy of a Planck resonator, one if its energy is continuous and one if it is quantized. This time he presented the derivations in detail, referring to the second part of the statistical trilogy only for the general formalism in which these calculations are done.

If the resonator's energy E is continuous, its average energy U_ν is given by the canonical ensemble average

$$\overline{E} = \frac{\displaystyle\int_0^\infty dE\, E\, e^{-E/kT}}{\displaystyle\int_0^\infty dE\, e^{-E/kT}}. \tag{3.53}$$

We already calculated this ensemble average when we proved the equipartition theorem in Section 3.2 (see Eqs. (3.3)–(3.8)). The energy of a resonator, $E = (p^2/2m) + \frac{1}{2}m\omega^2 q^2$, has a term quadratic in the momentum p and a term quadratic in the coordinate q. Eq. (3.53) thus gives $\overline{E} = kT$.

---

[31] Laue had been so impressed by Einstein's (1905d) relativity paper that he resolved to pay him a visit in Bern. It is unclear whether this first visit took place in August 1906 or August 1907 (Einstein 1987–2018, Vol. 5, Doc. 57, note 11). Apparently expecting Einstein to be on the faculty of the University of Bern, Laue was surprised to find him at the Patent Office instead and initially walked right past the unassuming patent clerk (Hoffmann 1972, p. 84). On September 2, 1907, Laue wrote to Jakob Laub, who was working with Einstein on a paper on the electrodynamics of moving media (Einstein 1987–2018, Vol. 2, pp. 503–507, Vol. 5, p. 640): "This fellow is a revolutionary. In the first two hours of conversation he overturned all of mechanics and electrodynamics, and this on the basis of statistics" (Einstein 1987–2018, Vol. 3, p. xxviii).

[32] Kuhn (1987, pp. 182–184) supplies some of the details of the derivation of these two expressions for the resonator entropy.

If the resonator's energy E is quantized, the probability distribution of a canonical ensemble of such resonators will be sharply peaked at the values $E = nh\nu$ where n is a nonnegative integer. The relative number of copies of the system with energies between E and $E + dE$ is then given by (cf. Eq. (3.2)):

$$dN = C\omega(E)e^{E/kT} dE, \tag{3.54}$$

where $\omega(E)$ is given by[33]

$$\omega(E) = \sum_{n=0}^{\infty} \delta(E - nh\nu). \tag{3.55}$$

As in the statistical trilogy, Einstein equated the relative number dN of resonators with an energy between E and $E + dE$ with the relative probability of finding a single resonator with an energy in that interval at any given time (cf. the comments following Eq. (3.2)). The average energy U_ν of the resonator is then given by the canonical ensemble average

$$\overline{E} = \frac{\int_0^\infty dE\, \omega(E)E e^{-E/kT}}{\int_0^\infty dE\, \omega(E)e^{-E/kT}} = \frac{\displaystyle\sum_{n=0}^{\infty} nh\nu e^{-nh\nu/kT}}{\displaystyle\sum_{n=0}^{\infty} e^{-nh\nu/kT}}. \tag{3.56}$$

This average energy can be written as a derivative of the logarithm of the partition sum $Z \equiv \sum e^{-nh\nu\beta}$ (with $\beta \equiv 1/kT$) in the denominator of Eq. (3.56) (cf. Eqs. (3.5)–(3.6)):

$$\overline{E} = -\frac{1}{Z}\frac{dZ}{d\beta} = -\frac{d\ln Z}{d\beta}. \tag{3.57}$$

The partition sum has the form of the geometric series $1 + x + x^2 + \ldots = 1/(1 - x)$ with $x \equiv e^{-h\nu\beta}$. Hence

$$Z = \frac{1}{1 - e^{-h\nu\beta}}. \tag{3.58}$$

Inserting this expression for Z into Eq. (3.57), we find that

$$\overline{E} = \frac{d}{d\beta}\ln\left(1 - e^{-h\nu\beta}\right) = \frac{h\nu e^{-h\nu\beta}}{1 - e^{-h\nu\beta}}. \tag{3.59}$$

---

[33] Here we help ourselves to the Dirac delta function, which Einstein, of course, did not have at his disposal in 1906. It was not introduced until two decades later (Dirac 1927).

Setting $U_\nu = \overline{E}$ and $\beta = 1/kT$, we recover Eq. (3.52),

$$U_\nu = \frac{h\nu}{e^{h\nu/kT} - 1},\tag{3.60}$$

which, in conjunction with Planck's relation (3.26) between ρ_ν and U_ν, gives the Planck law (3.29) for ρ_ν (cf. Eqs. (2.21)–(2.24)).

From these calculations of the average energy of a resonator in Einstein's paper of November 1906 one can draw the same conclusions that Einstein drew from his calculations of the entropy of a resonator in his paper of March 1906. As he explained in the introduction of this March 1906 paper, he had come to the conclusion that Planck's theory complements rather than contradicts his own theory. Planck, he argued, had tacitly made two assumptions. First,

> [t]he energy of an elementary resonator can only assume values that are integral multiples of $(R/N)\beta\nu$ [Einstein still wrote $h\nu$ in this cumbersome way]; by emission and absorption, the energy of a resonator changes by jumps of integral multiples of $(R/N)\beta\nu$ (Einstein 1906b, p. 202).

Since this assumption, Einstein continued, is at odds with Maxwell's theory of electrodynamics, it undermines the derivation of relation (3.26) between ρ_ν and U_ν needed to get from Eq. (3.60) for U_ν to the Planck law for ρ_ν. This is why a second assumption is needed, namely that this relation continues to hold even though Planck's derivation of it is no longer valid. "In my opinion," Einstein concluded,

> the above considerations do not at all disprove Planck's theory of radiation; rather, they seem to me to show that with his theory of radiation Mr. Planck introduced into physics a new hypothetical element: the hypothesis of light quanta (Einstein 1906b, p. 203).

As Rynasiewicz and Renn (2006, pp. 30–31) point out, Planck could hardly be expected to fall for "the ruse [of trying to get him] to accept the light-quantum [hypothesis] by offering him credit for its invention."

In a letter to Einstein of July 6, 1907, largely written in response to Einstein's (1906c) review of the first edition of Planck's (1906a) *Lectures on the Theory of Heat Radiation*, Planck did not give an inch on the quantization of radiation and still stopped short of conceding that at least the energy of his resonators is quantized. "I am not looking for the meaning of the elementary quantum of action (light quantum) in the vacuum," he wrote, "but at the places of absorption and emission, and assume that vacuum processes are *exactly* described by Maxwell's equations" (Einstein 1987–2018, Vol. 5, Doc. 47). Planck, it seems, was not ready to accept that Einstein had pulled the rug out from under his derivation of relation (3.26) between ρ_ν and U_ν.

In the closing paragraph of his letter, Planck tried to make a conciliatory gesture of his own. Surely, he suggested, defending the relativity principle was more urgent than debating the "rather old question" of the meaning of the quantum of action. "As long as

not have been truly black bodies, at least not for short wavelengths (Lorentz 1909b, pp. 339–340). As far as theory was concerned, his assessment was that

> [i]f one compares the theories of Planck and Jeans, one finds that both have their merits and their drawbacks. Planck's theory is the only one that gives a formula in agreement with the results of experiments, but we can adopt it only by altering profoundly our fundamental conceptions of electromagnetic phenomena. One sees this immediately as soon as one takes into consideration that a single electron moving in an arbitrary way emits radiation of all possible wavelengths. Clearly, it is impossible to apply to such an electron the hypothesis of energy elements the size of which depends on their frequency. Jeans's theory, by contrast, obliges us to attribute to chance the presently inexplicable agreement between observation and the laws of Boltzmann and Wien (Lorentz 1909b, p. 341).

Lorentz then cheerfully concluded that "fortunately, one may hope that new experimental determinations of the radiation law will permit us to decide between the two theories" (Lorentz 1909b, p. 341). Lorentz was severely criticized, if not downright ridiculed, for these pronouncements. In a letter of May 18, 1908, Wien told Sommerfeld he was deeply disappointed that Lorentz had seen fit to defend the utterly untenable Rayleigh–Jeans law in his Rome lecture. "This time," Wien summed up the situation, "Lorentz has not shown himself to be a leader in physics" (Sommerfeld 2000, Doc. 132; cf. Kuhn 1987, p. 192). The day before, Wien had written directly to Lorentz. Although he had great respect for his Dutch colleague, he did not pull any punches in that letter either. "I don't believe anybody who has ever done experimental work on radiation," he wrote, "would grant you that there is even the slightest possibility that Jeans's theory could be in agreement with experience" (Lorentz 2008, Doc. 170). As if the point needed further amplification, he added:

> That a cavity, even when it is left standing at some fixed temperature for an arbitrarily long time, will not eventually contain ultraviolet radiation of great intensity, let alone X-rays [as it should if the Rayleigh–Jeans law were correct], is something that I am more firmly convinced of than I am of the principle of energy conservation or any law of nature, including [sic] the Pythagorean theorem (Lorentz 2008, Doc. 170).

Lorentz's Rome lecture came in for sharp criticism from experimentalists as well. In a short paper, entitled "On the Jeans–Lorentz radiation formula," submitted June 1, 1908 to *Physikalische Zeitschrift* and published in the issue of July 15, Lummer and Pringsheim countered Lorentz's suggestion that their black body may not have absorbed enough radiation to qualify as such with measurements showing that, at least in the visible range, it absorbed more than 99.999% of the radiation that fell on it (Lummer and Pringsheim 1908, p. 450). They conceded that it would be much more difficult to measure the degree of absorption for longer wavelengths. They thus found it ironic that Lorentz did *not* question the blackness of their black body for those wavelengths. Their most devastating objection, however, was that, if the (Rayleigh–Einstein–)Jeans–Lorentz formula were correct, "steel at room temperature would have to emit a bright light" (Lummer and Pringsheim 1908, p. 449).

To his credit, Lorentz quickly accepted these criticisms.[40] In a lengthy reply to Wien dated June 6, 1908, he granted Wien that the Rayleigh–Jeans law is untenable, even though he was still reluctant to accept the notion of energy elements underlying Planck's law. What he found especially puzzling was that, according to Planck's law, resonators at high frequencies do not take up *any* energy from the radiation they are exposed to (cf. Section 2.6). In this letter, Lorentz also worked out an example equivalent to the one in the paper Lummer and Pringsheim (1908) had just submitted. Lorentz's letter mentions Lummer and Pringsheim but not their response to his Rome lecture. "I have gained clarity through the following simple consideration," Lorentz told Wien:

> According to Jeans's theory, black-body radiation of a particular wavelength is simply proportional to the temperature; and would thus have to be five times less at $15°$ C[elsius] [≈ 288K] than at $1200°$ [≈ 1473K, i.e., a temperature roughly 5 times higher]. A polished silver plate, which, let's say, reflects 93% of the yellow light that falls on it and absorbs 7%, should thus have an emission power for yellow light equal to $1.4[= 7 \div 5]$% of the emission power of a white-hot [black] body.[41] At $15°$ C it should thus be visible in the dark (Lorentz 2008, Doc. 171).

As he promised Wien in this letter, Lorentz added a note to the published version of his Rome lecture, in which he included the observation that, according to the Rayleigh–Jeans law, a silver plate at room temperature should glow in the dark (Lorentz 1909b, p. 342). Since he cited Lummer and Pringsheim (1908) for the confirmation of the blackness of their black body, it may not have been clear to his readers that he had thought of this example independently of theirs.[42]

Lorentz's Rome lecture and the reactions to it finally convinced those in the physics community interested in black-body radiation (still a rather small minority) of two basic points that Einstein, Ehrenfest, and Jeans had been trying to get across for some time. First, classical theory inevitably leads to the empirically disastrous Rayleigh–Jeans law. Secondly, the empirically successful Planck law can only be had at the cost of significant modifications of classical theory.

Even Planck himself finally came around, at least to some extent (Kuhn 1987, pp. 196–202). In a letter to Lorentz of October 7, 1908,[43] he unequivocally stated for the first time that the only values the energy of his resonators can take on are integer multiples of $h\nu$. Planck was responding to a letter from Lorentz of August 2, 1908, which is no longer extant. Planck's reply, however, makes it clear that Lorentz raised the same question he

---

[40] Kox (2013a, p. 158) cites a letter from Lorentz to Zeeman of August 20, 1908 that clearly shows Lorentz's embarrassment about the position he had taken in Rome (Lorentz 2018, Doc. 87).

[41] Kirchhoff had demonstrated that the ratio of emission and absorption power of any physical object is a universal function of frequency and temperature (cf. Section 2.2).

[42] Two decades later, in a lecture in Berlin commemorating Lorentz shortly after his death, Planck (1928, p. 347) correctly attributed the example of a silver plate glowing in the dark to the deceased. However, Planck changed his mind when the experimentalist Rudolf Ladenburg told him afterwards that the example was due to Lummer and Pringsheim and added a note to this effect in the printed version of his lecture (Planck 1928, p. 349).

[43] We already quoted from this letter in Section 2.6 (see note 57).

had raised in his letter to Wien of June 6: How can it be that resonators at high frequency do not absorb any energy from the ambient radiation? "A resonator," Planck admitted,

> will be set into vibration by the field. But—and now comes the essential point—this excitation does not correspond to the simple known law of the pendulum; rather there exists a certain threshold: the resonator does not respond at all to very small excitations; if it responds to larger ones, it does so only in such a way that its energy is a multiple of the energy element $h\nu$, so that the instantaneous value of the energy is always represented by such an integral multiple.
> In sum, I might therefore say, I make two assumptions:
>
> 1) the energy of a resonator at a given instant is $gh\nu$ (g a whole number or 0);
>
> 2) the energy emitted and absorbed by a resonator during an interval containing many billion oscillations (and thus also the average energy of a resonator) is the same as it would be if the usual pendulum equation applied.

These two assumptions do not seem to me to be incompatible (Lorentz 2008, Doc. 176; translation from Kuhn 1987, pp. 197–198).

Well over a year later, on January 7, 1910, Planck wrote in another letter to Lorentz:

> Of course, you are entirely right to say that such a resonator no longer deserves its name,[44] and that has moved me to strip it of its honorific title and call it by the more general name "oscillator" (as I also did in the enclosed draft) (Lorentz 2008, Doc. 197; quoted by Kuhn 1987, pp. 200–201).

The "enclosed draft" is for an article submitted to the *Annalen* about a week later (Planck 1910). What Planck used to call resonators are consistently called oscillators in this paper.

Lorentz's Rome lecture had one unfortunate consequence, which Lorentz, had he known about it, would have deeply regretted. Lorentz's public criticism of Planck's derivation of his radiation law—recall the conclusion that "we can adopt [Planck's theory] only by altering profoundly our fundamental conceptions of electromagnetic phenomena" (Lorentz 1909b, p. 341)—provided welcome ammunition to those out to derail the effort to award Planck the 1908 Nobel Prize for his work on black-body radiation (Nagel 1982, see also Heilbron 1986, pp. 23–26, and Stone 2013, Ch. 14, "Planck's Nobel Nightmare"). Another decade would pass before Planck finally won the Nobel Prize "in recognition of the services he rendered to the advancement of Physics by his discovery of energy quanta." In 1919, he was awarded the prize for 1918. Wien

---

[44] Planck is responding to a letter from Lorentz of July 30, 1909. Although this letter does not survive, two partial drafts for it do. In these drafts, Lorentz raises some objections to Planck's use of his resonators but he does not object to the *term* "resonator" (Lorentz 2008, Docs. 194 and 195).

had already received this honor "for his discoveries regarding the laws governing the radiation of heat" in 1911.[45]

3.4.3   Einstein's 1909 Salzburg lecture: fluctuations and wave–particle duality

After the paper of January 1907, part of which we discussed in Section 3.4.1 and to which we will return in Section 3.5.1, Einstein did not publish on the problem of radiation for two years. In January 1909, he submitted a paper to *Physikalische Zeitschrift* with the title "On the present status of the radiation problem" (Einstein 1909a). In September, he presented some of the results of this paper in a lecture he gave at the *Naturforscherversammlung*, held that year in Salzburg (Einstein 1909b, cf. note 20 in Chapter 2).[46]

In between these two events, Einstein was appointed Associate Professor (*Außerordentlicher Professor* or *Extraordinarius*) of physics at the University of Zurich. As Robert Schulmann (1993, 2014, pp. 428–429) has argued, Einstein had already set his sights on a return to Zurich when he left for Bern and the Patent Office in 1902. Einstein had graduated from the Zurich Polytechnic in 1900 with a teacher's certificate but had decided shortly thereafter that, instead of becoming a high-school teacher, he wanted to pursue an academic career. This explains the attempts he made in 1901 to obtain his doctorate, first with Weber at the Polytechnic, then with Kleiner at the University (see note 1). In late 1901, a Dean at the University of Zurich first floated the idea of establishing an *Extraordinarius* position at the university to relieve Kleiner, its only *Ordinarius* (Full Professor) in physics, of some of his teaching duties (Schulmann 1993, pp. 18–19). No position was created at that point but it was agreed that promising graduates should be encouraged to take the necessary steps to qualify for such a position if and when it materialized. This is how Kleiner ended up grooming Einstein for his return to Zurich (Schulmann 1993, p. 17).[47]

In 1905, Einstein cleared the first hurdle, finally obtaining his doctorate with Kleiner. The next step in the German academic system was (and is) *Habilitation*. This requires a *Habilitationsschrift*, which can be thought of as a second dissertation. In 1907, Einstein submitted a collection of his papers as such to the University of Bern but this was deemed unacceptable. In 1908, he therefore submitted a new document especially written for the occasion: "Consequences for the constitution of radiation of the energy distribution law of black-body radiation" (Einstein 1987–2018, Vol. 5, p. 621). This *Habilitationsschrift*, which unfortunately does not survive, was accepted. Einstein now advanced to the position of Instructor (*Privatdozent*), the lowest rung on the academic ladder. For two semesters Einstein taught in this capacity at the University of Bern. He had to do so in his spare time as the position carried no regular salary. A *Privatdozent* could only

[45] See www.nobelprize.org/nobel_prizes/physics/.
[46] For discussion of Einstein's Salzburg lecture, see Klein (1964, pp. 5–15) and Stone (2013, pp. 136–140).
[47] Stone (2013, p. 24) suggests that Kleiner hindered rather than helped Einstein's cause but he only looked at part of the documentary evidence on which Schulmann (1993) based his account.

collect student fees for courses taught outside the regular curriculum. That same year, 1908, Kleiner was appointed to a two-year term as rector of the University of Zurich and the plan to create an *Extraordinarius* position finally came to fruition (Schulmann 1993, p. 23). Despite Kleiner's influence, however, the university's first choice was Friedrich Adler, the son of the prominent social-democrat Victor Adler. Einstein moved into first place only after Adler, recognizing that he could not hold a candle to Einstein, withdrew his candidacy.[48] It was not until May 1909 that the appointment, effective that Fall, was finalized. October 1909 thus marks the end of Einstein's days as a patent clerk and the proper beginning of his academic career.

So, technically, Einstein was still a patent clerk when he took the podium in Salzburg in September 1909 for his very first conference talk. However, he was about to join the ranks of the physicists in his audience, most of whom laid eyes on the man who was changing the face of their discipline for the first time. Planck chaired the session and was probably responsible for getting Einstein this prestigious speaking engagement. Senior figures such as Wien, Rubens, and Sommerfeld were in attendance. So were Laue and several other physicists of Einstein's own generation whom we will meet later: James Franck, Paul Epstein, Rudolf Ladenburg, Fritz Reiche, and Max Born (Fölsing 1993, p. 771, note 81).

Einstein's talk was titled "On the development of our views concerning the nature and constitution of radiation" (Einstein 1909b). Planck must have been pleased that the speaker spent the first half of his lecture reviewing the changes in our conception of radiation brought about by special relativity, such as the abandonment of the ether. Right at the beginning, however, Einstein already made it clear that more drastic changes were on the horizon. As he wrote in the introduction of the published version of the Salzburg lecture:

> [T]here is an extensive group of facts concerning radiation that show that light has certain fundamental properties that can be understood much more readily from the standpoint of Newton's emission theory than from the standpoint of the wave theory. It is therefore my opinion that the next phase of the development of theoretical physics will bring us a theory of light that can be interpreted as *a kind of fusion of the wave and emission theories*. To give reasons for this opinion and to show that a profound change in our views on the nature and constitution of light is imperative is the purpose of the following remarks (Einstein 1909b, pp. 482–483; emphasis added).

After covering the changes wrought by special relativity, Einstein reminded his audience of some of the facts alluded to in the passage quoted above. These were phenomena such as the photoelectric effect, which are baffling from the point of view of the wave theory but just what one would expect on the basis of the light-quantum hypothesis

---

[48] Einstein returned the favor during the First World War, when Adler was awaiting trial for his assassination in November 1916 of the Austrian prime minister Count Stürgkh. On behalf of a number of Zurich physicists, Einstein drafted a petition to the Austrian authorities to spare Adler's life. Adler was handed a death sentence but on appeal it was commuted to eighteen years in prison. He was pardoned right after the war (Galison 2008, Janssen and Mecklenburg 2007, pp. 106–107, 121–122, note 59, Sigmund 2017, pp. 91–99).

(pp. 490–491). Einstein then turned to black-body radiation, amplifying a point he had already made in 1906, when he first discussed the relation between his own light-quantum hypothesis and Planck's theory of black-body radiation in print.

In the Wien regime, where $h\nu \gg kT$, the mean energy $\overline{E}$ of a resonator (the term still used in 1909) with a characteristic frequency ν is *much smaller* than $h\nu$ (Einstein 1906b, p. 203; in 1909 Einstein finally replaced his own clumsy $(R/N)\beta$, by Planck's h). In his Salzburg lecture, Einstein elaborated on this observation in two ways (Kuhn 1987, p. 186). First, he gave a numerical example. For $T = 1700K$ and $\lambda = .5\mu$, the ratio $h\nu/\overline{E}$ is equal to $6.5 \cdot 10^7$ (1909b, p. 494). Second, he spelled out why this amounts to a devastating objection against the idea that the energy of these resonators can take on a continuum of values. In Section 2.6, we argued that Planck initially vacillated between two interpretations of the relation $\varepsilon = h\nu$. The relation could be interpreted, it seemed, either in terms of *energy parcels* or in terms of *energy bins*. Einstein conclusively demonstrated that the latter is not a viable option. To derive the formula for the entropy of a resonator corresponding to his new law for black-body radiation, Planck (1900d, 1901a) had to count the number of microstates W (complexions in Boltzmann's terminology) to be inserted into the general expression $k \ln W$ for the entropy (see Section 2.5). If we think of $\varepsilon = h\nu$ as giving the size of an energy bin and focus on resonators with characteristic wavelength $\lambda = .5\mu$ at temperature $T = 1700K$, Einstein observed, we are counting complexions

> as if the energy of the resonators can only assume the value zero, the $6.5 \cdot 10^7$-fold value of its mean energy, or a multiple of this. It is clear that if we proceed in this manner, we use for the calculation of the entropy only a vanishingly small part of those energy distributions that we must consider as possible according to the foundations of the [classical] theory. Thus, according to the foundations of the theory, the number of these complexions is not an expression for the probability of the state in Boltzmann's sense. In my opinion, to accept Planck's theory means plainly to reject the foundations of [classical] radiation theory (Einstein 1909b, pp. 494–495).

As in 1906, Einstein drew the conclusion that Planck's theory implicitly relied on the light-quantum hypothesis, once again not clearly distinguishing between the quantization of the resonators and the quantization of light itself (see Section 3.4.1). This time, however, Einstein did explicitly recognize that while the hypothesis is sufficient to derive the Planck law, it might not be necessary (cf. note 30). "Would it not be possible," he asked,

> to replace the hypothesis of light quanta by another assumption that would also fit the known phenomena? . . . [W]ould it not be possible to retain at least the equations for the propagation of radiation and only conceive of the elementary processes of emission and absorption in a new way? (Einstein 1909b, p. 495).

To answer these questions, Einstein turned to the same "fixed point" strategy he had used in the 1905 light-quantum paper (Rynasiewicz and Renn 2006, p. 6; cf. Section 3.3.2). In 1905, he showed that it follows straight from the validity of the Wien law for black-body radiation at high frequencies and Boltzmann's principle that there must be fluctuations in which light behaves as a collection of discrete independent quanta (even if he did not establish, as we saw in Section 3.3.2, that these quanta behave as point-like particles). In 1909, both in the Salzburg lecture and in the paper published earlier that year, he could do even better. He now showed that it follows directly from the validity of the Planck law for black-body radiation at all frequencies and Boltzmann's principle that there must be fluctuations in which light exhibits *both* wave-like and particle-like behavior at the same time.

Einstein was quite explicit in his Salzburg lecture about the strategy he was adopting:

> To clarify these matters, we will try to proceed in the opposite direction than that taken by Mr. Planck in his radiation theory. We consider Planck's radiation formula as correct[49] and ask ourselves whether some conclusion about the constitution of radiation can be inferred from it (Einstein 1909b, p. 495–496).[50]

In the paper published earlier that year, Einstein (1909a, secs. 6–8, pp. 188–191) had used Planck's law and Boltzmann's principle to derive formulas for both energy and momentum fluctuations in black-body radiation that were the sum of a wave term and a particle term. In the Salzburg lecture he only covered momentum fluctuations—without derivations, which he would present in full the following year (Einstein and Hopf 1910). He imagined a system reminiscent of the one he had considered in his 1905 paper on Brownian motion (and, in analyzing the system, put the methods developed in that paper to good use). This system consisted of a plate or a mirror, which can move only in the direction perpendicular to its surface, immersed in black-body radiation of a given temperature, which is exerting radiation pressure on it.

In their paper the following year, Einstein and Ludwig Hopf, a student of Sommerfeld (Eckert 2013b, p. 174), considered this same system albeit for a different purpose: they used it for yet another proof that classical theory inescapably leads to the Rayleigh–Jeans law. Six years later, Einstein (1916b) considered a very similar system in the context of the new quantum theory of radiation he proposed that year (Einstein 1916a). He now replaced the mirror immersed in black-body radiation by molecules immersed in black-body radiation emitting and absorbing light quanta while transitioning between quantized energy states. We will cover the argument of this 1916 paper in Section 3.6.2. For our purposes here, it suffices to derive Einstein's 1909 formula for the mean square fluctuation of the *energy* in black-body radiation satisfying the Planck law.

---

[49] As further evidence for the empirical adequacy of Planck's formula, Einstein (1909b, p. 495) noted that it gives a value for the electron charge that agrees well with values found more recently on the basis of scintillation counting of α-particles (Rutherford and Geiger 1908; Regener 1909).

[50] As Klein (1964, p. 9) put it: "The key to his reasoning was his reversal of Planck's procedure. Instead of trying to derive the distribution law from some more fundamental starting point, he turned the argument around. Planck's law had the solid backing of experiment; why not assume its correctness and see what conclusions it implied as to the structure of radiation? Einstein had already done just this sort of thing in 1905 when he based his argument for the granular structure of radiation on the consequences of Wien's radiation law... This time he applied the method to Planck's law itself with equally impressive results." The title of his lost 1908 *Habilitationsschrift*, "Consequences for the constitution of radiation of the energy distribution law of black-body radiation," suggests that he adopted the same strategy in that document.

In the final installment of his statistical trilogy, Einstein had already derived a formula for the mean square energy fluctuation and applied it to black-body radiation (see Section 3.2, Eqs. (3.16)–(3.25)). However, the canonical ensemble for which he had derived this formula cannot be used to describe black-body radiation (see the discussion following Eq. (3.48) and Section 3.4.1, especially Eqs. (3.53)–(3.60)). This would explain why, in 1909, Einstein rederived the formula, avoiding the canonical ensemble and using Boltzmann's principle instead (Einstein 1987–2018, Vol. 2, p. 146).[51]

Consider the instantaneous fluctuation

$$\varepsilon \equiv \Delta E = E - \overline{E} \tag{3.61}$$

of the energy E from its equilibrium value $\overline{E}$ in a small subvolume V of a box with perfectly reflecting walls filled with black-body radiation. Let the volume of the rest of the box be V_0 (see Fig. 3.2).[52] The corresponding instantaneous fluctuation of the energy in V_0 is

$$-\varepsilon = \Delta E_0 = E_0 - \overline{E_0}. \tag{3.62}$$

The mean square fluctuation of the energy in V is given by

$$\overline{\Delta E^2} = \frac{\int_{-\infty}^{\infty} \varepsilon^2 W(\varepsilon)\,d\varepsilon}{\int_{-\infty}^{\infty} W(\varepsilon)\,d\varepsilon}, \tag{3.63}$$

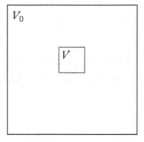

Figure 3.2 *A box with perfectly reflecting walls that is filled with black-body radiation and divided into two subvolumes, a small volume V and a large volume V_0 (the volume of the rest of the box).*

---

[51] As Pais (1982, p. 73) notes, the formula for the mean square energy fluctuation "had been derived independently by Gibbs and Einstein, using in essence the same method. In 1909, Einstein gave a new derivation, this one all his own." Our presentation of Einstein's (1909a, pp. 188–189) new derivation follows Klein (1964, pp. 10–11).

[52] Einstein (1909a, p. 188) did not specify how the total volume of the system is to be divided into a small and a large volume, which he designated by v and V, respectively. Our specific choice was made with malice aforethought (see note 53).

where the probability $W(\varepsilon)d\varepsilon$ of any instantaneous fluctuation ε is given by Boltzmann's principle:

$$W(\varepsilon)d\varepsilon = e^{\Delta S(\varepsilon)/k}d\varepsilon. \tag{3.64}$$

We thus need to calculate the change in entropy

$$\Delta S(\varepsilon) = (S + S_0) - \overline{(S + S_0)} \tag{3.65}$$

of the whole system that accompanies the energy fluctuation $\varepsilon = \Delta E = -\Delta E_0$. A Taylor expansion of Eq. (3.65) around the equilibrium value $\overline{S} + \overline{S_0}$ gives:

$$\Delta S(\varepsilon) = \left(\frac{\overline{\partial S}}{\partial E} - \frac{\overline{\partial S_0}}{\partial E_0} \right) \varepsilon + \frac{1}{2} \left(\frac{\overline{\partial^2 S}}{\partial E^2} + \frac{\overline{\partial^2 S_0}}{\partial E_0^2} \right) \varepsilon^2 + O(\varepsilon^3). \tag{3.66}$$

Given that $\overline{\partial S/\partial E} = 1/T$, $\overline{\partial S_0/\partial E_0} = 1/T_0$ and $T = T_0$ (the two systems are in thermal equilibrium), the term of order ε vanishes. For the term of order ε^2, we note that

$$\overline{\left(\frac{\partial^2 S}{\partial E^2} \right)} = \overline{\frac{\partial}{\partial E} \left(\frac{1}{T} \right)} = \overline{\frac{\partial}{\partial T} \left(\frac{1}{T} \right) \overline{\left(\frac{\partial T}{\partial E} \right)}} = -\frac{1}{T^2 C_V}, \tag{3.67}$$

where we used that the heat capacity at constant volume is defined as

$$C_V \equiv \frac{d\overline{E}}{dT}. \tag{3.68}$$

We similarly have $\overline{\partial^2 S_0/\partial E_0^2} = -1/(T_0^2 C_{V_0})$. Since $C_{V_0} \gg C_V$, the second term in the factor multiplying ε^2 in Eq. (3.66) is negligible compared to the first. To order $O(\varepsilon^2)$, Eq. (3.66) thus reduces to

$$\Delta S(\varepsilon) = -\frac{\varepsilon^2}{2T^2 C_V}. \tag{3.69}$$

Inserting this result into Eqs. (3.63)–(3.64) and introducing the abbreviation $a \equiv 1/(2kT^2 C_V)$, we find:

$$\overline{\Delta E^2} = \frac{\displaystyle\int_{-\infty}^{\infty} \varepsilon^2 e^{-a\varepsilon^2}\, d\varepsilon}{\displaystyle\int_{-\infty}^{\infty} e^{-a\varepsilon^2}\, d\varepsilon} = \frac{\frac{1}{2a}\sqrt{\pi/a}}{\sqrt{\pi/a}} = \frac{1}{2a}. \tag{3.70}$$

We thus arrive at:

$$\overline{\Delta E^2} = kT^2 C_V = kT^2 \frac{d\overline{E}}{dT}, \tag{3.71}$$

where in the last step we used Eq. (3.68). This formula for the mean square energy fluctuation is exactly the same as the one Einstein had derived in 1904 using the canonical ensemble (cf. Eq. (3.16)).

We now apply Eq. (3.71) to black-body radiation. The mean energy $\overline{E_{(\nu,V)}}$ of black-body radiation of energy density $\rho(\nu, T)$ at temperature T in a narrow frequency range $(\nu, \nu + \Delta\nu)$ in a small subvolume V of the box containing the radiation is given by

$$\overline{E_{(\nu,V)}} = \rho(\nu, T) V \Delta\nu. \tag{3.72}$$

To find the mean square fluctuation $\overline{\Delta E^2_{(\nu,V)}}$ of the energy in the interval $(\nu, \nu + \Delta\nu)$ in the subvolume V, we insert Eq. (3.72) into the general formula (3.71):

$$\overline{\Delta E^2_{(\nu,V)}} = kT^2 \frac{d\overline{E_{(\nu,V)}}}{dT} = kT^2 \frac{\partial\rho(\nu, T)}{\partial T} V \Delta\nu. \tag{3.73}$$

Choosing a specific law for the spectral distribution $\rho(\nu, T)$ of black-body radiation, we find the corresponding formula for $\overline{\Delta E^2_{(\nu,V)}}$. Using the Rayleigh–Jeans law (2.26), we find:

$$\overline{\Delta E^2_{(\nu,V)}}_{\text{Rayleigh-Jeans}} = \frac{c^3}{8\pi\nu^2} \frac{\overline{E_{(\nu,V)}}^2_{\text{Rayleigh-Jeans}}}{V\Delta\nu}. \tag{3.74}$$

Hence, for the Rayleigh–Jeans law, $\overline{\Delta E^2_{(\nu,V)}}$ is proportional to the square of the mean energy. This is the *signature of waves* (Norton 2006, p. 71).[53]

Using the Wien law (2.25), we find:

$$\overline{\Delta E^2_{(\nu,V)}}_{\text{Wien}} = h\nu \overline{E_{(\nu,V)}}_{\text{Wien}}. \tag{3.75}$$

Hence, for the Wien law, $\overline{\Delta E^2_{(\nu,V)}}$ is proportional to the mean energy itself. This is the *signature of particles* (ibid.). For independent light quanta of frequency ν randomly

---

[53] Lorentz (1916, p. 59) later derived Eq. (3.74) from classical electromagnetic theory. The equivalent of Eq. (3.74) for a simple model, proposed by Ehrenfest (1925), was derived by Max Born, Werner Heisenberg and Pascual Jordan (1926, Ch. 4, sec. 3) as a prelude to their attempt to derive Eq. (3.77) below from matrix mechanics. In this simple model, waves on a string play the role of electromagnetic waves in some container and the mean square fluctuation of the energy in a small segment of this string replaces the mean square fluctuation of the energy in a small subvolume of this container. For a reconstruction of the calculations for this simple model, see Duncan and Janssen (2008, sec. 4.1, pp. 651–656).

entering and leaving the subvolume V, with an average number of quanta $\bar{n}$ and average energy $\overline{E_{(v,V)}} = \bar{n}h\nu$, the average fluctuation in the energy, $\overline{\Delta E_{(v,V)}}$, will be $\sqrt{\bar{n}}h\nu$. The root mean square fluctuation will thus be

$$\overline{\Delta E^2_{(v,V)}} = \bar{n}(h\nu)^2 = h\nu\overline{E_{(v,V)}}, \tag{3.76}$$

which is indeed of the form of Eq. (3.75).

Finally, using the Planck law (2.24), we find (Einstein 1909a, p. 189):

$$\overline{\Delta E^2_{(v,V)\,\text{Planck}}} = \frac{c^3}{8\pi v^2}\frac{\overline{E_{(v,V)}}^2_{\text{Planck}}}{V\Delta v} + h\nu\overline{E_{(v,V)}}_{\text{Planck}}. \tag{3.77}$$

Hence, for the Planck law, $\overline{\Delta E^2_{(v,V)}}$ has both a wave and a particle term. Given that the Planck law was originally obtained as an interpolation between the Wien and the Rayleigh–Jeans law (see Section 2.4), it is perhaps not so surprising that Eq. (3.77) combines Eqs. (3.74) and (3.75). It remains striking, however, that Einstein, using essentially only Boltzmann's principle, was able to show that Planck's law for black-body radiation implies what we would now call the wave–particle duality of light.[54]

For the mean square fluctuations of *momentum* in black-body radiation satisfying the Planck law, Einstein (1909a, p. 190) found an expression very similar to Eq. (3.77) for the mean square fluctuations of its energy. Once again, the formula consists of two terms, one pointing to particles, the other pointing to waves. These fluctuation considerations formed the basis for Einstein's opinion, stated at the beginning of the lecture, that "the next phase of the development of theoretical physics will bring us a theory of light that can be interpreted as a kind of fusion of the wave and emission theories" (Einstein 1909b, pp. 482–483).[55]

As he told his audience in Salzburg, it was not yet clear what such a theory would look like. The remarkable formulas he found for the mean square fluctuations of energy and momentum in black-body radiation "offer few formal clues on which to build a theory" (Einstein 1909b, p. 499). "Imagine that the diffraction and interference phenomena are not yet known," Einstein continued, but that we would have a fluctuation formula of the

---

[54] Einstein's wave–particle duality of 1909, however, is different from the wave–particle duality associated with Bohr's complementarity principle. According to the latter, depending on the experimental context, light should show either wave or particle behavior. Measuring the mean square fluctuation of the energy in a narrow frequency interval in a small subvolume of a cavity with black-body radiation, however, one would expect to confirm Eq. (3.77) and thus observe wave and particle behavior in one and the same experiment (Duncan and Janssen 2008, p. 635).

[55] Interviewed by Kuhn and George Uhlenbeck for the *Archive for the History of Quantum Physics* (AHQP) in 1962 (Kuhn 1967, pp. 100–102), Reiche still remembered that he "was very much impressed by the appearance of the second term in the fluctuation formula" (Hermann 1971, p. 68). "Though it is of course," Reiche added, "a rather indistinct proof of 'photons'. I remember of course that people were opposed and tried to find another reason or tried to give the formula another form" (ibid.).

form of Eq. (3.74) for the Rayleigh–Jeans law. "Who would have sufficient imagination," he asked rhetorically, "to construct the wave theory on such a basis?" (Einstein 1909b, p. 499). Yet, Einstein could not resist speculating on the nature of a theory fusing the wave and particle theories of light. He suggested that light quanta might be singularities in the electromagnetic field, like point charges, then immediately downplayed the importance of this suggestion:

> [N]o importance should be attached to such a picture as long as it has not led to an exact theory. All I wanted is briefly to indicate with its help that the two structural properties (the undulatory structure and the quantum structure) simultaneously displayed by radiation according to the Planck formula should not be considered as mutually incompatible (Einstein 1909b, p. 500).

This cautionary remark concludes (the published version of) the talk.

Einstein had ended the paper of January 1909, in which he first presented these fluctuation arguments, on a more confident note. In the final section of the paper, the speculations at the end of the Salzburg lecture were presented in more detail. Einstein argued that we should be looking for some non-linear generalization of Maxwell's equations (or Maxwell–Lorentz equations as they were often called back then) and listed some of the constraints such a generalization would have to satisfy (such as Lorentz invariance and compatibility with the Rayleigh–Jeans law for low frequencies). In the concluding paragraph of this paper, Einstein sounds optimistic about the prospect of finding this new theory:

> I have not yet succeeded in finding a system of equations fulfilling these conditions … The variety of possibilities does not seem so great, however, for one to have to shrink from this task (Einstein 1909a, p. 193).

One factor that may help explain the change in tone between January and September 1909 is an exchange of letters Einstein had in between with the most important physicist *not* present in Salzburg: Lorentz. Einstein did not meet the great Dutch physicist in person until February 1911, when he gave a talk in Leyden (Kox 1993, p. 46). But on March 30, 1909, he sent Lorentz a copy of his January paper with a cover letter highlighting the momentum fluctuations implied by the Planck law and the proposed modification of the Maxwell–Lorentz equations in the final section of the paper (Einstein 1987–2018, Vol. 5, Doc. 146). Two weeks later, on April 13, Einstein followed up with a postcard expressing his admiration for Lorentz's proof in his Rome lecture, a version of which he had just read, that classical theory inevitably leads to the Rayleigh–Jeans law (Einstein 1987–2018, Vol. 5, Doc. 149). Another three weeks later, on May 6, Lorentz sent Einstein a long letter in which he expressed strong agreement on several points— and, more generally, made it clear that he had great respect for Einstein and had been following his work with great interest—but also raised some serious objections to the notion of light quanta (Einstein 1987–2018, Vol. 5, Doc. 153). One of his objections was that light quanta could not be point-like, as that would be incompatible with interference occurring over appreciable distances. Lorentz also told Einstein that he had tried to modify Maxwell's equations himself but had found that "as soon as one makes even the slightest change ... the greatest difficulties arise." Einstein was delighted to receive this letter from Lorentz. "I admire this man like no other," he wrote to his collaborator Laub (cf. note 31), "I might say I love him."[56] On May 23, 1909, he sent Lorentz a detailed reply (Einstein 1987–2018, Vol. 5, Doc. 163). At this point, however, the correspondence breaks off.[57] Einstein clearly failed to convince Lorentz. Lorentz's reservations about the way in which Einstein tried to combine wave and particle behavior of light in a nonlinear generalization of the Maxwell–Lorentz theory may help explain the more cautious tone he adopted in Salzburg.

Another factor was undoubtedly that Planck was chairing the session. Using (if not abusing) his privilege as chair, Planck was the first to respond to Einstein's lecture. He remained unconvinced. "According to the latest considerations of Mr. Einstein," he remarked,

it would be necessary to conceive the free radiation in vacuum, and thus the light waves themselves, as atomistically constituted, and hence to give up Maxwell's equations. This seems to me to be a step which, in my opinion is not yet necessary (Einstein et al. 1909, p. 825).

In the discussion that followed, Johannes Stark came to Einstein's defense. Stark was the only physics professor at the time who was sold on the idea of light quanta (Stuewer 2014, p. 147). How else could it be explained, Stark asked, that X rays could travel "up to 10 m [and] still achieve concentrated action on a single electron?" (Einstein et al. 1909, p. 826).[58] Even though nobody had an answer to Stark's question, the audience appears to have been on Planck's side. Asked decades later whether Einstein's Salzburg lecture "had a great effect," one of the physicists who had attended the talk as a young man, recalled: "No. You see the chairman of the meeting was Planck, and he immediately said that it was very interesting but that he did not quite agree with it. And the only man who seconded at that meeting was Johannes Stark. You see, it was too far advanced" (Hermann 1971, p. 68).[59]

---

[56] Einstein to Jakob Laub, May 19, 1909 (Einstein 1987–2018, Vol. 5, Doc. 161). For discussion of the relationship between Einstein and Lorentz, see Kox (1993).

[57] For more detailed discussion of this exchange between Einstein and Lorentz in 1909, which can also be found in Lorentz (2008, Docs. 183, 185, 189 and 190), see Stone (2013, pp. 131–135).

[58] Stark claimed that experiments on the production of X rays by cathode rays and vice versa supported the light quantum hypothesis, a claim challenged by Sommerfeld (Stuewer 1975, pp. 31–37; see also Wheaton 1983, pp. 120–132, and Eckert 2013b, p. 173). A Japanese physicist visiting colleagues in Britain and Germany reported that "Stark . . . was propounding his 'Lichtquantentheorie' . . . The Germans say he is full of phantasies, which may be partly true" (Hantaro Nagaoka to Ernest Rutherford, February 22, 1911; quoted in Stuewer 1975, p. 37).

[59] Interview with Paul Epstein by John Heilbron in 1962 for the AHQP. See note 55 for Reiche's reaction to Einstein's lecture.

3.5 The breakdown of equipartition and the specific heat of solids at low temperatures (1907–1911)

Missing from Einstein's 1909 lecture in Salzburg was a topic that would take center stage in the development of quantum physics for the next couple of years: the specific heat of solids at low temperatures. And missing from the audience in Salzburg was the man responsible for putting the spotlight on this topic, the physical chemist Walther Nernst.[60]

Einstein proposed his quantum theory of specific heats at the end of 1906. It takes up the second part (pp. 184–190) of his paper, "Planck's theory of radiation and the theory of specific heat" (Einstein 1907). In the first part (pp. 180–184), Einstein showed that the equipartition theorem must break down for the resonators that Planck had used in the derivation of his black-body radiation law (see Section 3.4.1). To recover Planck's law, the average energy of these resonators had to deviate sharply from its equipartition value kT. Einstein showed that one arrives at the desired non-equipartition values by quantizing the energy of these resonators (see Eqs. (3.54)–(3.60)).

The equipartition theorem was notorious for giving the wrong values for the specific heats of both solids and gases. In the introduction to this volume, we already quoted some celebrated passages from the work of Maxwell, Kelvin, and Rayleigh drawing attention to these and related problems (see Section 1.2.3). It was thus only natural for Einstein to use the quantized resonators from black-body theory to construct a simple model of a solid and check its prediction for the specific heat.[61] Contrary to what the equipartition theorem predicted, but in agreement with the experimental data, Einstein found that the specific heat goes to zero as the temperature goes to zero (see Fig. 3.3).

In Einstein's model, a solid is taken to be a lattice with N harmonic oscillators (where N is chosen to be Avogadro's number), oscillating *independently* with the same resonance frequency ν. If the energy of these oscillators could take on the full continuum of possible values, their average energy, as Einstein had shown in the first part of his paper (see Eq. (3.53)), would be $\overline{E} = 3kT$ and every oscillator would contribute

$$\frac{d\overline{E}}{dT} = 3k \tag{3.78}$$

---

[60] For the history of the problem of the specific heat of solids and the work of Einstein and Nernst in this area, see, e.g., Klein (1965), Pais (1982, Ch. 20, pp. 389–401), Kuhn (1987, pp. 210–220), Stone (2013, Ch. 13, pp. 103–110 (Einstein); Ch. 17, pp. 141–148 (Nernst)), and editorial material in Einstein (1987–2018, Vol. 2, pp. 142–143, Vol. 3, pp. xxi–xxviii).

[61] In his lecture at the *Naturforscherversammlung* in Karlsruhe from which we have already quoted in the introduction (see Section 1.2.6), Sommerfeld observed that Kelvin's (1901) "Cloud No. II," related to equipartition, was responsible for the problems classical theory was facing in two areas, black-body radiation and specific heats. "Summarizing what we have learned from radiation and specific heats about the general question of energy distribution," he quipped, "we can say with a variation on Schiller: 'One should weigh degrees of freedom, not count them.'" (Sommerfeld 1911a, p. 1061; the famous German poet and playwright had said this about votes).

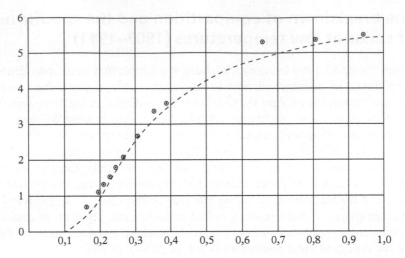

Figure 3.3 *Plot of specific heat C_V against $x \equiv kT/h\nu$ (Einstein 1907, p. 186).*

to the specific heat. The specific heat of the N oscillators making up Einstein's solid would then be $C_V = 3Nk = 3R$ (where R is the ideal gas constant).

This is the value given by a law for specific heats named after the two French physicists who first formulated it, Pierre Dulong and Alexis Petit (1819). The two of them measured the specific heat for a dozen metals and sulfur at room temperature and found that the product of their specific heat per gram and their atomic mass (a quantity first introduced by John Dalton in 1808) was a constant. The atoms of all elements, they concluded, have the same heat capacity. In modern terms, the Dulong–Petit law says that the specific heat of a mole of any solid is equal to $3R$. Boltzmann (1871) was the first to show that this result follows directly from the equipartition theorem. A mole of a solid consists of a lattice with N atoms or molecules. As long as each atom or molecule contributes six terms to the Hamiltonian that are quadratic in either a component of its momentum or a component of its position, the average energy of N such atoms or molecules is given by $\overline{E} = 3NkT$ (cf. Eqs. (3.3)–(3.8)). The specific heat, in that case, is $3R$, as it should be according to the Dulong–Petit law.

As early as the 1840s, however, deviations from the Dulong–Petit law had been observed. One of the pioneers of experiments on the specific heat of solids at low temperatures was Einstein's teacher at the Zurich Polytechnic, Heinrich Weber. Using a calorimeter filled with snow, he found that the specific heat of diamond dropped by a factor of 3 if he lowered the temperature from $200°$ to $0°C$ (Weber 1872). Three years later, using dry ice (solid carbon dioxide) and other coolants, Weber got the temperature down to $-100°C$ and measured the change in the specific heat of diamond, graphite, boron, and silicon in going from $-100°$ to $1000°C$, finding a variation by a factor 15 in the case of diamond (Weber 1875, cf. Pais 1982, p. 391). These experiments landed Weber the appointment in Zurich. Some of Weber's data are included in his student's paper on specific heat (Einstein 1907, p. 190).

Einstein studied with Weber in the late 1890s.[62] It is unclear whether, as a student, Einstein learned about the experiments with which Weber had made a name for himself. However, two letters to Mileva Marić of March 1901 strongly suggest that by that time he was aware of Weber's results. In these letters Einstein already speculated about the connection between "electric resonators" (of the kind considered by Planck in his work on black-body radiation) and deviations from the Dulong–Petit law for specific heats (Einstein 1987–2018, Vol. 1, Docs. 93 and 94).

To explain the deviation of the specific heat of solids at low temperature from its equipartition Dulong–Petit value, Einstein assumed that the energy of the oscillators in his simple model of a solid is quantized and can only take on values that are integral multiples of $h\nu$. In that case, the average energy of an oscillator is no longer the equipartition value $\overline{E} = 3kT$ but is given by (cf. Eq. (3.59)):

$$\overline{E} = \frac{3h\nu}{e^{h\nu/kT} - 1} = 3kT \left(\frac{h\nu/kT}{e^{h\nu/kT} - 1} \right). \tag{3.79}$$

The specific heat of a solid consisting of N such quantized oscillators is given by:

$$C_V = N\frac{d\overline{E}}{dT} = 3R\frac{(h\nu/kT)^2\, e^{h\nu/kT}}{\left(e^{h\nu/kT} - 1\right)^2}. \tag{3.80}$$

This can be written as $3R$ times the function

$$f(x) \equiv \frac{e^{1/x}}{x^2 \left(e^{1/x} - 1\right)^2}, \tag{3.81}$$

where $x \equiv kT/h\nu$. Figure 3.3 shows the plot of C_V against x given in Einstein's paper. For large x, we have:

$$f(x) \approx \frac{1 + \dfrac{1}{x}}{x^2 \left(1 + \dfrac{1}{x} + \dfrac{1}{2x^2} - 1\right)^2} = 1 + O\left(\frac{1}{x^2}\right). \tag{3.82}$$

Hence, if x goes to ∞, $f(x)$ goes to 1 and the expression for C_V in Eq. (3.80) reduces to $3R$. This means that, as long as the resonance frequency ν is somewhere in the infrared, the specific heat will only start to deviate from its equipartition value at very low temperatures.

---

[62] In the early 1880s, Nernst, who would produce experimental confirmation of Einstein's theory of specific heat, had also taken classes with Weber (Barkan 1999, pp. 30–31, p. 115).

For small x, we have

$$f(x) \approx \frac{e^{1/x}}{x^2 \left(e^{1/x}\right)^2} = \frac{1}{x^2} e^{-1/x}. \tag{3.83}$$

Hence, if $x = kT/h\nu$ goes to 0, $f(x)$ also goes to 0. Contrary to the constant value predicted by the equipartition theorem, the specific heat of a solid consisting of quantized oscillators goes to zero as the temperature goes to zero. Moreover, if the resonance frequency of an oscillator is relatively high (say in the ultraviolet), that oscillator will only start to contribute to the specific heat at high temperatures. So Einstein's simple model could explain both why the specific heats of solids like diamond drop well below the equipartition value at low temperature and why the degrees of freedom of the electrons assumed to be present in atoms to account for the emission of light do not contribute to the specific heat at all.

Einstein's theory of specific heats did not attract any attention at first, which helps explain why he did not mention it in his Salzburg lecture in 1909. The following year, however, Nernst and his collaborators confirmed Einstein's prediction that the specific heat goes to zero as the temperature goes to zero (Nernst 1910, 1911a, 1911b, 1911c, Nernst, Koref and Lindemann 1910; Nernst and Lindemann 1911).[63] This result fit with the "heat theorem" that Nernst (1906) had first proposed a few years earlier (Klein 1965, p. 176). Now known as the third law of thermodynamics, this "theorem" says that the entropy at absolute zero is a constant, which can be set equal to zero. At the time, however, Nernst's "theorem" was, in fact, just a conjecture, which had already proved to be very useful—Nernst's Berlin colleague Fritz Haber relied on it in his work on extracting nitrogen from the air—but still lacked a proper foundation (Stone 2013, pp. 144–145). One consequence of the "theorem", as Nernst pointed out in lectures at Yale in 1906 published several years later (Klein 1965, p. 176), was "that the atomic [i.e., specific] heats in the solid state decrease greatly at low temperatures" (Nernst 1913, p. 63). It was no surprise therefore that Nernst was drawn to Einstein's theory, which likewise predicted such a decrease and might thus provide the means to shore up his general heat theorem. The agreement between theory and experiment was not perfect. The experiments showed that, when the temperature goes to zero, the specific heat does not drop off as fast as it should according to Eq. (3.80) (Pais 1982, p. 397). Nernst thought that this discrepancy was due to the unrealistic assumption that the quantized oscillators in Einstein's model (the individual atoms in the solid) all have the same oscillation frequency. Einstein (1911b) tried to develop a more realistic model but it was Peter Debye (1912) (see Plate 11) who developed the model that fit the experimental data much better (Pais 1982, p. 397).[64]

---

[63] Frederick Lindemann later became scientific advisor to Winston Churchill.

[64] In Debye's model, the thermal energy is calculated as a sum over contributions from lattice sound waves of all possible frequencies and directions, up to the high-frequency (short-wavelength) cutoff imposed by the minimum separation of the atoms (lattice spacing). The result, as in the calculation of the total energy density in black-body radiation, is that the low-temperature behavior of the energy is softened from exponential to power vanishing.

Despite verifying (at least qualitatively) Einstein's formula for specific heats, Nernst had strong reservations about the quantum theory from which the formula was derived. His attitude toward Einstein's theory, however, was not nearly as negative as Millikan's when he verified Einstein's formula for the photoelectric effect in 1916 (see Section 3.3.3). In a lecture delivered at the Prussian Academy on January 26, 1911, to honor the birthday of Emperor Wilhelm II the following day, Nernst said:

> At this time, the quantum theory is essentially a computational rule, one may well say a rule with most curious, indeed grotesque, properties. However, in the hands of Planck, when it comes to radiation, and Einstein, when it comes to molecular mechanics, it has borne such rich fruits and admits, as I have indicated above, so many other applications, that there is now a research duty to take a position on it from as many angles as possible and subject it to experimental tests (Nernst 1911a, p. 86; partly quoted in Pais 1982, p. 399).

The year before, Nernst had made a stop in Zurich on a trip to Lausanne to pay Einstein, whom he had not met before, a visit and tell him about the specific heat experiments (Einstein 1987–2018, Vol. 3, p. xxii). On March 16, 1910, Einstein mentioned this visit in a letter to his collaborator Laub and wrote enthusiastically: "The quantum theory is established, as far as I am concerned. My predictions concerning specific heats seem to be brilliantly confirmed" (Einstein 1987–2018, Vol. 5, Doc. 199). Nernst had been impressed with Einstein. "[W]e can be very happy to have found such an original young thinker," he wrote to Arthur Schuster on March 17, 1910, calling the young Associate Professor at the University of Zurich a reincarnation of Boltzmann, a "Boltzmann redivivus" (Barkan 1999, p. 183). Nernst's visit did not go unnoticed in Zurich. In an interview decades later, the Hungarian chemist George de Hevesy, an assistant at what in 1910 was still called the Polytechnic (it was renamed the ETH the following year), recalled that Nernst's visit "made Einstein famous. Einstein came as an unknown man to Zurich. Then Nernst came, and people in Zurich said, 'This Einstein must be a clever fellow, if the great Nernst comes so far from Berlin to Zurich to talk to him' " (Kuhn 1987, p. 215).[65]

This may have been a coincidence, but a mere five weeks after Nernst's visit, Einstein was offered a full professorship at the German university in Prague (Stone 2013, p. 150). Einstein readily accepted this promotion from *Extraordinarius* to *Ordinarius* (cf. Section 3.4.3). The following year, in April 1911, Einstein, his wife, and their two young sons moved to Prague. There his star continued to rise and in July 1912, he returned to Zurich, now as a full professor at the ETH. A couple of months before Einstein's return to his alma mater, Weber died. Illustrating the strained relationship with his former teacher, Einstein, in a letter to a close friend, called this "a good thing for the Polytechnic."[66] Einstein did not stay long at the ETH either. Even before his move back to Zurich, in April 1912, Einstein had visited Berlin to talk to Nernst, Haber, and others about moving

---

[65] Interview with Hevesy by Kuhn and Emilio Segrè in 1962 for the AHQP.
[66] Einstein to Heinrich Zangger, after June 5, 1912 (Einstein 1987–2018, Vol. 5, Doc. 406, quoted in Fölsing 1993, p. 79, and Stone 2013, p. 152).

to Berlin. Nernst had already started to explore the possibility of bringing Einstein to Berlin in July 1910. In 1913, these efforts finally bore fruit. In July of that year, Planck and Nernst went to Zurich to offer Einstein a salaried position at the Prussian Academy with no teaching obligations. Einstein accepted and, in March 1914, arrived in Berlin to take up his new position. This is where he would stay until the Nazi takeover of 1933 forced him to leave Germany.

The attention his 1907 quantum theory of specific heat was belatedly receiving in the early 1910s was an important factor in Einstein's meteoric rise from Associate Professor at the University of Zurich in 1909 to salaried member of the Prussian Academy in Berlin in 1914. Nernst was largely responsible for this attention. Not only did his work lead to an exponential growth of papers on the theory of specific heats during the years 1911–1913 (Kuhn 1987, p. 217, Figs. 2a and 2b), Nernst also convinced the Belgian industrialist Ernest Solvay to make available generous funds to support a conference in Brussels devoted to the fledgling quantum theory. The five-day conference, which was held October 30–November 3, 1911, brought together the cream of the crop in physics from Europe (see Plate 9).[67] Invitations were sent out in June 1911. When Einstein received his invitation, his interest was already shifting from quantum theory to a relativistic theory of gravity. On October 21, 1911, shortly before the conference, he wrote to Besso: "When the witches' Sabbath in Brussels is over ... I will again be my own man" (Einstein 1987–2018, Vol. 5, Doc. 296).

While for many attendees, the Solvay conference was their first encounter with the conundrums of quantum theory, Einstein learned nothing new. He gave a masterful presentation focusing on specific heats (Einstein 1912, 1914a) but was dismissive about the conference in his comments afterwards. "The whole business would have been a delight for the diabolical Jesuit fathers," he wrote to Zangger on November 15, 1911 (Einstein 1987–2018, Vol. 5, Doc. 305). In a letter to Besso of December 26, 1911, he compared the whole affair to the "lamentation over the ruins of Jerusalem" (Einstein 1987–2018, Vol. 5, Doc. 331). These statements also reflect that Einstein was starting to lose confidence that progress could be made on the quantum problem. "The h disease looks ever more hopeless," he wrote to Lorentz on November 23, 1911 (Einstein 1987–2018, Vol. 5, Doc. 313).

Einstein's loss of confidence is nicely illustrated by an anecdote about his last days in Prague. In his biography of Einstein, Philipp Frank, Einstein's successor in Prague, relates the following probably apocryphal story (Gordin 2020):

> Einstein's office ... overlooked a park with beautiful gardens and shady trees ... [H]e learned that it was a park belonging to the insane asylum of the province of Bohemia. The people walking in the garden were inmates of this institution ... When I first went to Prague, Einstein showed me this view, explained it to me, and said playfully: "Those are the madmen who do not occupy themselves with the quantum theory" (Frank 1947, p. 98).

---

[67] For discussion, see, e.g., Barkan (1993, 1999, Ch. 11, pp. 181–207), Staley (2008, Ch. 10, pp. 397–422), and Stone (2013, Ch. 18, pp. 149–159).

This ushered in "three years of skepticism" about quantum matters, as Olivier Darrigol (2014, Sec. 8, pp. 131–133) has called it. In a paper co-authored with Otto Stern, his colleague first in Prague and then in Zurich, he even gave a derivation (though, by his own admission, not a strict one) of Planck's law for black-body radiation without any quantization, using the zero point energy of the harmonic oscillator introduced in Planck's second theory instead (Einstein and Stern 1913, pp. 556–559).[68] During this period, Einstein focused more and more on his new theory of gravity (Janssen and Renn 2015). It was only in 1916, after he completed his general theory of relativity, that he made his next great contribution to quantum theory.

3.6 Einstein's quantum theory of radiation (1916)

On August 11, 1916, Einstein sent an exuberant postcard to his friend Besso:

> A beautiful light has dawned on me about radiation absorption and emission... An astonishingly simple derivation, I dare say, *the* derivation of the Planck formula. A thoroughly quantized affair (Einstein 1987–2018, Vol. 8, Doc. 250).

Einstein had published this derivation in July 1916 (Einstein 1916a). As can be inferred from a postcard to Besso of August 24, he sent Besso an offprint of this paper (Einstein 1987–2018, Vol. 8, Doc. 251). This second postcard also mentions a follow-up paper (Einstein 1916b), which was published in a special issue of the Communications (*Mitteilungen*) of the Zurich Physical Society in honor of Kleiner, Einstein's dissertation adviser and mentor (see Section 3.4.3). This paper was reprinted in *Physikalische Zeitschrift* in March 1917 (Einstein 1917a). This is the version typically cited today and we will also refer to it.

This 1916–17 paper consists of two parts. In the first part, Einstein (1917a, secs. 1–3) reprised the "astonishingly simple derivation" of Planck's black-body radiation law of his July 1916 paper. The crucial premises of this derivation are fundamental assumptions about the exchange of energy between molecules and radiation. In his August 1916 paper, Einstein added assumptions about the exchange of *momentum* between molecules and radiation. As Einstein (1917a, p. 77) noted in the conclusion of the paper, it was important to consider momentum as well as energy given the intimate connection between the two that had been revealed by special relativity.

Einstein (1917a, secs. 4–7) used the analysis of the momentum exchange between molecules and radiation in the second part of his paper to give yet another derivation of the Planck law, this one based on the analysis of the momentum exchange between molecules and radiation. This derivation provides a stringent consistency check on Einstein's assumptions about the momentum exchange between molecules and radiation.

---

[68] In this same paper Einstein and Stern also considered the temperature-dependence of the specific heat of molecular hydrogen. See Gearhart (2010) for a detailed history of this problem, which would not be solved until 1926, after the arrival of the new quantum mechanics.

If the assumptions are changed ever so slightly, the Planck law no longer obtains. As Einstein announced in the introduction of his paper: "It will become apparent that we shall only arrive at a theory free from contradictions, if we consider [the elementary processes of absorption and emission] to be perfectly directional; this embodies the main result of the subsequent discussion" (Einstein 1917a, p. 65).[69]

3.6.1   New derivation of the Planck law

In secs. 1–2 of the paper, Einstein (1917a) laid out the basic tenets of his new quantum theory of radiation and its interaction with matter. As far as radiation is concerned, he adopted a strict particle view, realizing full well that the theory he was proposing was only preliminary and did not do justice to the wave aspects of light. His model of matter can be seen as a more abstract version of Bohr's (1913b) model for the hydrogen atom (see Chapter 4). A molecule can only exist in a number of discrete states Z_n (where 'Z' stands for *Zustand*, i.e., state) with internal energy ε_n. Einstein was careful not to associate such states with electron orbits in a miniature planetary system, the picture underlying the old quantum theory of Bohr and Sommerfeld. In fact, in May 1917, he would publish a paper in which he put his finger on some serious problems with this picture (Einstein 1917b, see Section 5.1.5 for discussion of this paper).

Einstein assumed a canonical distribution of the various states of the molecules. In a gas consisting of a large number of them, the relative frequency W_n of a molecule being in the state Z_n is given by

$$W_n = p_n e^{-\varepsilon_n/kT}, \tag{3.84}$$

with the usual Boltzmann factor $e^{-\varepsilon_n/kT}$ and the "statistical weight" p_n of the state Z_n (Einstein 1917a, p. 65).

In sec. 2, Einstein presented his assumptions about the exchange of energy and momentum between molecules and light quanta. He introduced three elementary processes through which a molecule can get from one quantum state to another, labeled (A), (B), and (B'). In modern terms they are: spontaneous emission (A), absorption (B), and stimulated or induced emission (B'). This last process was called negative absorption at the time. These three processes are illustrated in Fig. 3.4. The concrete visualization of the states Z_n and Z_m as energy levels of an electron in a molecule is ours and is completely arbitrary. Einstein did not offer any picture of the different energy states Z_n of the molecule.

The emission and absorption processes, as Einstein stressed, are the quantum analogues of absorption and emission of radiation by a classical Planck resonator (the term

---

[69] Einstein had anticipated this result in his Salzburg lecture: "According to the prevailing [wave] theory, an oscillating ion produces an outwardly propagated spherical wave. The opposite does not exist as an elementary process ... Thus, the elementary process of light radiation as such does not possess the character of reversibility ... [Experiments, however, indicate that] *the elementary process of radiation seems to be directed*" (Einstein 1909b, pp. 491–492; emphasis in the original).

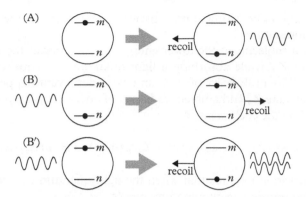

Figure 3.4 *The three basic processes in Einstein's quantum radiation theory: (A) spontaneous emission, (B) absorption, (B') stimulated emission/negative absorption.*

Einstein still used in this paper). The quantum analogue of the resonance frequency is some "frequency v characteristic for the index combination (m, n)" (Einstein 1917a, p. 66). At this point, Einstein left open how this frequency v depends on the energies ε_n and ε_m of the states Z_n and Z_m. Part of his derivation of the Planck law was that he could also derive the Bohr frequency condition $hv = \varepsilon_m - \varepsilon_n$ from the basic assumptions of his theory of radiation and the Wien displacement law.

Whereas classically emission and absorption are continuous and deterministic, the quantum counterparts of these processes are discontinuous and stochastic. This is a new element of Einstein's work in 1916. The stochastic nature of these processes is modeled after the law of radioactive decay. Einstein introduced coefficients A and B for the probability of his three processes. His paper thus became known as the "A and B coefficients" paper. We need to consider these three processes and the probability coefficients associated with them in more detail.

In process (A), a molecule in the higher energy state Z_m makes the transition to the lower energy state Z_n while spontaneously emitting a light quantum of frequency v_{nm}. The probability[70] of such a spontaneous emission event happening in a time interval Δt, given that the molecule is in state Z_m, is determined by the coefficient A_m^n, where the lower index refers to the initial state and the upper index refers to the final state of the molecule:

$$\text{Process (A):} \quad \text{Pr}(\text{spontaneous emission}|Z_m) = A_m^n \Delta t. \tag{3.85}$$

The relation between energy and momentum in special relativity tells us that a light quantum carrying off an amount of energy $\varepsilon_m - \varepsilon_n$ also carries off an amount of

---

[70] Einstein (1917a, pp. 66–67) considered the infinitesimal time interval dt and denoted the probabilities of all three processes (A), (B), and (B') by dW.

momentum $p = (\varepsilon_m - \varepsilon_n)/c$. Momentum conservation requires that the molecule recoils with minus that momentum.

In process (B), a molecule in the lower energy state Z_n makes the transition to the higher energy state Z_m while absorbing a light quantum of frequency ν_{nm} from the ambient radiation. The probability of such an absorption event happening during Δt, given a molecule in state Z_n and ambient radiation of energy density ρ at the frequency ν_{nm}, is determined by the coefficient B_n^m:

$$\text{Process (B):} \quad \Pr(\text{absorption}|Z_n, \rho(\nu_{nm})) = B_n^m \rho(\nu_{nm})\Delta t. \tag{3.86}$$

Momentum conservation requires that when the light quantum transfers its energy to the molecule, it also transfers its momentum, $p = (\varepsilon_m - \varepsilon_n)/c$.

Finally, in process (B'), a light quantum of frequency ν_{nm} triggers the emission of another light quantum of that same frequency by a molecule making a transition from the higher energy state Z_m to the lower energy state Z_n. The probability of such an induced or stimulated emission event happening during Δt, given a molecule in the state Z_m and ambient radiation of energy density ρ at the frequency ν_{nm}, is determined by the coefficient B_m^n:

$$\text{Process (B'):} \quad \Pr(\text{stimulated emission}|Z_m, \rho(\nu_{nm})) = B_m^n \rho(\nu_{nm})\Delta t. \tag{3.87}$$

As in the case of spontaneous emission, momentum conservation requires that the molecule recoils with momentum $p = -(\varepsilon_m - \varepsilon_n)/c$.

Einstein's quantum picture of energy and momentum exchange in emission and absorption is radically different from the classical picture. Here are two of the most striking differences. First, in classical theory, spherical waves are emitted and there is no recoil momentum. Quantum-theoretically, radiation is emitted in a single direction: the direction of the incoming radiation in the case of stimulated emission; a randomly selected direction in the case of spontaneous emission. Because of this, there will be recoil momentum. Second, in classical theory, the phase difference between the incoming wave and the oscillation of the resonator determines whether the resonator gains or loses energy when it interacts with incoming radiation. An energy gain is thus as likely as an energy loss. In the quantum theory, it is the initial state of the molecule that determines whether there will be an energy gain (absorption) or an energy loss (stimulated emission). Since under normal conditions the molecule is far more likely to be in the lower rather than in the higher energy state, an energy gain is far more likely than an energy loss. This alleviates the problem one ran into with classical theory that radiation tends to drain the energy from the matter it interacts with.

In sec. 3 of his paper, Einstein (1917b) presented the derivation of the Planck radiation law (up to a constant) and the Bohr frequency condition that he had already published before (Einstein 1916a). The two ingredients of this derivation are the Wien displacement law (cf. Section 2.2) and the condition that the elementary processes (A), (B), and (B') do not disturb thermal equilibrium. For this condition to hold, it has to be the case that in any time interval Δt and for any pair of states Z_n and Z_m, the probability of a transition from Z_n to Z_m is equal to the probability of a transition from Z_m to Z_n:[71]

$$\Pr(n \to m) = \Pr(m \to n). \tag{3.88}$$

The probability $\Pr(n \to m)$ is equal to the probability that a molecule is in the state Z_n times the probability that process (B) makes it jump from Z_n to Z_m:

$$\Pr(n \to m) = \Pr(Z_n)\,\Pr(n \to m | Z_n, \rho(\nu_{nm})). \tag{3.89}$$

Similarly, the probability $\Pr(m \to n)$ is equal to the probability that a molecule is in the state Z_m times the probability that either process (A) or process (B') makes it jump from Z_m to Z_n:

$$\Pr(m \to n) = \Pr(Z_m)\,\Pr(m \to n | Z_m, \rho(\nu_{nm})). \tag{3.90}$$

Using Eq. (3.84) and Eqs. (3.85)–(3.87) for the probabilities on the right-hand sides of Eqs. (3.89) and (3.90), we find:

$$\Pr(n \to m) = p_n e^{-\varepsilon_n/kT} B_n^m \rho \Delta t \tag{3.91}$$

and

$$\Pr(m \to n) = p_m e^{-\varepsilon_m/kT} \left(A_m^n + B_m^n \rho \right) \Delta t. \tag{3.92}$$

The condition (3.88) for equilibrium can thus be written as:

$$p_n B_n^m e^{-\varepsilon_n/kT} \rho = p_m B_m^n e^{-\varepsilon_m/kT} \left(\frac{A_m^n}{B_m^n} + \rho \right). \tag{3.93}$$

On the assumption that ρ goes to infinity when T goes to infinity, it follows from this condition that

$$p_n B_n^m = p_m B_m^n. \tag{3.94}$$

Eq. (3.94) thus simplifies to

$$e^{-\varepsilon_n/kT} \rho = e^{-\varepsilon_m/kT} \left(\frac{A_m^n}{B_m^n} + \rho \right). \tag{3.95}$$

---

[71] Since the condition must hold for any pair of states, we only need to consider one such pair, at least as long as there are no pairs of states with the same energy gap. Einstein (1916a, p. 320) explicitly made this simplifying assumption in the original version of his new derivation of the Planck law.

Multiplying both sides by $e^{\varepsilon_m/kT}$ and solving for ρ, we arrive at

$$\rho = \frac{A_m^n/B_m^n}{e^{(\varepsilon_m-\varepsilon_n)/kT}-1}. \tag{3.96}$$

Using the Wien displacement law in the form given by Planck, which says that ρ should be of the form $\nu^3 f(T/\nu)$ (cf. Eq. (2.42)), Einstein set

$$\varepsilon_m - \varepsilon_n \propto \nu \tag{3.97}$$

and

$$A_m^n/B_m^n \propto \nu^3. \tag{3.98}$$

Calling the proportionality constants h and α, respectively, we recover the Planck law (2.24) (up to a constant),

$$\rho = \frac{\alpha\nu^3}{e^{h\nu/kT}-1}, \tag{3.99}$$

and the Bohr frequency condition $\varepsilon_m - \varepsilon_n = h\nu$.[72] Einstein's derivation of the Planck law thus not only provided strong evidence for his new quantum theory of radiation but also for the Bohr frequency condition, perhaps the most dramatic departure from classical physics up to this point (see Sections 1.2.9 and 4.5.1).

3.6.2 Momentum fluctuations and the directed nature of radiation

The derivation of the Planck law in the preceding section does not depend on Einstein's assumptions about the exchange of momentum between molecules and light quanta. In the second part of his paper, however, Einstein (1917a, secs. 4–7) subjected these assumptions to a severe test. He analyzed the motion of molecules immersed in black-body radiation and the counterbalancing effects of random changes in their velocities due to emission and absorption of light quanta on the one hand and so-called "radiation

---

[72] If there were only absorption and spontaneous emission and no stimulated emission, we would have found the Wien law instead of the Planck law. Instead of Eq. (3.93), we would have

$$p_n B_n^m e^{-\varepsilon_n/kT}\rho = p_m A_m^n e^{-\varepsilon_m/kT}.$$

Using Eq. (3.94) to replace $(p_n/p_m)B_n^m$ by B_m^n, we can rewrite this equation as

$$\rho = (A_m^n/B_m^n)e^{-(\varepsilon_m-\varepsilon_n)/kT}.$$

Setting $A_m^n/B_m^n = \alpha\nu^3$ and $\varepsilon_m - \varepsilon_n = h\nu$ as before, we now arrive at Wien's law (2.25) (up to a constant):
$\rho = \alpha\nu^3 e^{-h\nu/kT}$.

friction" (*Strahlungsreibung*) on the other (Einstein 1909a, pp. 189–190; 1909b, p. 496). Through this analysis, in which he relied heavily on his assumptions about momentum exchange, he arrived at a differential equation for the energy density ρ of the radiation. Planck's radiation law turns out to be the solution of that equation. This would no longer be true, however, once the momentum assumptions are changed ever so slightly. This then provided strong support for those assumptions (Klein 1964, p. 20).

For this part of the paper, as he explicitly noted, Einstein (1917a, p. 69) drew heavily on "a method which is well known from the theory of Brownian motion." As we mentioned in Section 3.4.3, Einstein had used these same methods to analyze the motion of a mirror immersed in black-body radiation in 1909 and 1910 (Einstein 1909a, 1909b, Einstein and Hopf 1910).

Part of the condition that the elementary processes (A), (B), and (B′) do not disturb thermal equilibrium is that they preserve the Maxwell–Boltzmann velocity distribution of the molecules. Without loss of generality, Einstein could focus on the velocity's x-component. The three processes will affect the velocity v of a molecule in the x-direction—or, equivalently, its momentum Mv in the x-direction, where M is its mass—in two different ways. During any short time interval τ there will be radiation friction, proportional to minus the velocity v with some proportionality constant R (not to be confused with the ideal gas constant), and random fluctuations Δ in the molecule's momentum in the x-direction due to the stochastic nature of the absorption and emission processes. The condition that these two effects do not change the Maxwell–Boltzmann distribution is

$$\overline{(Mv - Rv\tau + \Delta)^2} = \overline{(Mv)^2}, \tag{3.100}$$

where the overbars denote time averages. Expanding the left-hand side, we find

$$M^2\overline{v^2} + R^2\tau^2\overline{v^2} + \overline{\Delta^2} - 2M\overline{v^2}R\tau + 2M\overline{v}\overline{\Delta} - 2R\overline{v}\tau\overline{\Delta} = M^2\overline{v^2}. \tag{3.101}$$

Choosing the time interval τ small enough so that τ^2 can be neglected and using that $\overline{v} = \overline{\Delta} = 0$, we see that only two terms in this relation survive (Einstein 1917a, p. 12):

$$\overline{\Delta^2} - 2M\overline{v^2}R\tau = 0 \tag{3.102}$$

Appealing to the equipartition theorem, Einstein set the molecule's average kinetic energy, $\frac{1}{2}M\overline{v^2}$, equal to $\frac{1}{2}kT$. Eq. (3.102) then turns into:[73]

$$\frac{\overline{\Delta^2}}{\tau} = 2RkT. \tag{3.103}$$

---

[73] This is an early instance of what is now known as the fluctuation-dissipation theorem.

Einstein now derived expressions for R and $\overline{\Delta^2}$ in terms of the radiation energy density ρ, using the assumptions about energy and momentum exchange between molecules and light quanta of his quantum radiation theory.[74] When these expressions are inserted into the equilibrium condition (3.103), this condition turns into a differential equation for ρ:

$$\frac{\rho h v}{3} = kT \left(\rho - \frac{v}{3} \frac{\partial \rho}{\partial v} \right) \left(1 - e^{-hv/kT} \right). \tag{3.104}$$

One readily verifies that the solution of this equation is the Planck law (3.99). Differentiating Eq. (3.99) with respect to v, we find

$$\frac{\partial \rho}{\partial v} = \frac{3 \alpha v^2}{e^{hv/kT} - 1} - \frac{\alpha v^3 (h/kT) e^{hv/kT}}{\left(e^{hv/kT} - 1 \right)^2}$$

$$= \frac{3\rho}{v} - \rho \frac{h/kT}{1 - e^{-hv/kT}}. \tag{3.105}$$

It follows that

$$\rho - \frac{v}{3} \frac{\partial \rho}{\partial v} = \frac{\rho}{3} \frac{hv/kT}{1 - e^{-hv/kT}}, \tag{3.106}$$

which is equivalent to Eq. (3.104). Einstein's quantum theory of radiation, especially its assumptions about the exchange of momentum between molecules and light quanta, thus passed a severe test and passed it with flying colors.

As we mentioned above, Einstein and Hopf (1910) went through a similar calculation on the basis of classical radiation theory.[75] The equation for ρ they arrived at can be found by letting h go to zero in Eq. (3.106):

$$\rho - \frac{v}{3} \frac{\partial \rho}{\partial v} = \frac{\rho}{3}. \tag{3.107}$$

Multiplying both sides by -3 and rearranging terms, we can rewrite this as

$$v \frac{\partial \rho}{\partial v} = 2\rho. \tag{3.108}$$

It follows that ρ must be proportional to v^2 as in the Rayleigh–Jeans law. This was the point of Einstein and Hopf's calculation. Their goal was to prove, without using any questionable assumptions, that classical theory leads to the Rayleigh–Jeans law, "which

---

[74] For a reconstruction of Einstein's derivations of expressions for R and $\overline{\Delta^2}$, see the web resource, *Momentum fluctuations.*

[75] For discussion of this paper by Einstein and Hopf see Pais (1982, p. 410) and Klein (1964, pp. 15–16, 20).

is in the most glaring contradiction with experiment" (p. 1114).[76] In their derivation of the Rayleigh–Jeans law, they noted,

> there is no question whatsoever of somehow arbitrarily transferring statistical considerations to radiation. The energy equipartition theorem was applied only to the translatory motion of oscillators [cf. Eqs. (3.102)–(3.103) in the quantum version of the calculation]. The successes of the kinetic theory of gases demonstrate that this law can be considered as thoroughly proved for translatory motion (p. 1115).

Einstein saw the quantum version of this analysis of momentum fluctuations as the most important part of the "*A* and *B* coefficients" paper. After reiterating that his theory yields both the Planck law and the Bohr frequency condition $\varepsilon_m - \varepsilon_n = h\nu$, he wrote:

> Most important, however, seems to me to be the result concerning the momentum transfer to the molecule due to the absorption and emission of radiation. If one of our assumptions about the momenta were to be changed, a violation of [the relation (3.103) expressing the condition for equilibrium between matter and radiation] would be produced; it seems hardly possible to maintain agreement with this relation, imposed by the theory of heat, other than on the basis of our assumptions (Einstein 1917a, pp. 75–76).

Given this result, Einstein confidently pronounced that

> outgoing radiation in the form of spherical waves does not exist. During the elementary process of radiative loss, the molecule suffers a recoil of magnitude $h\nu/c$ (p. 76).

This conclusion met with considerable resistance from his contemporaries. To give just one prominent example, Sommerfeld (1919, pp. 381–382), in the first edition of his influential book *Atombau und Spektrallinien* (see Chapter 6), explicitly denied that an atom would suffer a recoil upon emitting radiation: "*No momentum is transferred from the atom to the ether*" (Sommerfeld 1919, p. 382; emphasis in the original). In a footnote, Sommerfeld acknowledged Einstein's opposing view:

> We do not want to leave unmentioned, however, that Einstein has developed a very interesting theory about emission and absorption processes, in which every emission

---

[76] A letter written several decades later suggests that Einstein had already considered momentum fluctuations according to classical theory before 1905. On January 17, 1952, Einstein wrote to von Laue (the 'von' was added in 1913 when his father became a German nobleman) that "by 1905 I already knew for certain that [Maxwell's theory] led to the wrong fluctuations in radiation pressure and thereby to an incorrect Brownian motion of a mirror in a Planckian radiation cavity" (quoted in Rynasiewicz and Renn 2006, p. 25). One way to interpret this statement is that, by 1905, Einstein had somehow already found (a crude version of) Eq. (3.108) and noticed that it implied that $\rho \propto \nu^2$, which could not possibly be true for the correct law for the spectral distribution of black-body radiation. In the 1905 light quantum paper, Einstein used a much simpler argument, based on Planck's relation (3.26), to show that classical theory leads to what came to be known as the Rayleigh–Jeans law and the ultraviolet catastrophe (see Section 3.3.1).

process takes place in a certain direction and in which the change of momentum of an atom in the transition from [initial to final] configuration therefore does not vanish ... we, however, will treat the spherical wave as an element of experience in optics and will accordingly reject Einstein's theory (Sommerfeld 1919, p. 382, note).

In support of his position, Sommerfeld offered the argument that "there is no sufficient reason why an atom would prefer one direction for its [recoil] velocity over any other" (p. 382). Einstein (1917a, p. 76) had actually already conceded that the direction of the recoil velocity was a stochastic element in his theory. He had identified this as one of two shortcomings of his theory:

The weakness of the theory lies on the one hand in the fact that it does not get us any closer to making the connection with wave theory; on the other, that it leaves the duration and direction of the elementary processes to "chance" (p. 76).

Given Einstein's later attitude to quantum mechanics,[77] it is ironic that he himself consciously introduced a stochastic element into the theory.[78] Einstein thought this would be a temporary expedient. Noting that the recoil of a molecule absorbing or emitting a light quantum "is only determined by 'chance'," he hastened to add "according to the present state of the theory" (p. 76). "Nevertheless," he concluded, "I am fully confident that the approach chosen here is a reliable one" (p. 76).

As he wrote to Besso on September 6, 1916: "This all but clinches the reality of light quanta" (Einstein 1987–2018, Vol. 8, Doc. 254). Even though he did not make any further progress over the next few years, his confidence did not waver. In a letter to Besso of July 29, 1918, he wrote: "I have reflected during an incalculable number of hours on the question of the quanta, naturally without making any real progress. But I don't doubt anymore the *reality* of the quanta of radiation, although I am still entirely alone in having this conviction" (Bernstein 1991, p. 159). It would take another seven years and the discovery of the Compton effect for the rest of the physics community to follow suit (Stuewer 1975).

---

[77] For an overview of Einstein's later response to quantum mechanics, see Lehner (2014).

[78] Planck (1913, pp. 190–191) had already introduced such a stochastic element in his so-called "second theory". After having temporarily accepted that the energy of his oscillators can only take on discrete values (see Section 3.4.2), Planck now imagined that the energy of an oscillator of characteristic frequency v increases continuously until it reaches a value nhv (where n is an integer). Of all oscillators with energy nhv, a fraction η will then lose an amount of energy hv emitted as radiation, while a fraction $1 - \eta$ will continuously gain more energy until it reaches the value $(n + 1)hv$, at which point it once again becomes a matter of chance whether the oscillator will discontinuously lose or continuously gain energy.

4

The Birth of the Bohr Model

4.1 Introduction

In July 1913, Niels Bohr published Part One of his trilogy, "On the constitution of atoms and molecules" (Bohr 1913b). Part Two followed in September, Part Three in November (Bohr 1913c, 1913d). The first and most successful part contains Bohr's famous model for the hydrogen atom and hydrogen-like atoms, i.e., atoms with just one electron. With this model, which precariously combines classical mechanics and electrodynamics with quantum ideas, Bohr could account for some prominent features of the spectra of these atoms. In particular, he used the model to account for a formula for the wavelengths of a series of lines in the hydrogen spectrum first found by a Swiss school teacher, Johann Jakob Balmer (1885) (see Plate 17). Bohr could express a key factor in this Balmer formula—the Rydberg constant, named after the Swedish spectroscopist Johannes (Janne) Robert Rydberg (1890) (see Plate 16)—in terms of a combination of constants of nature including Planck's constant. This set the stage for concerted efforts during the next decade to study the effects of external electric and magnetic fields on the hydrogen spectrum, to take into account relativistic corrections, and to do the same for multi-electron atoms. These efforts will be the subject of Part Two of this volume (Chapters 5–7). In this chapter we cover the birth of Bohr's model.[1]

We follow Bohr during the two-year period from the dissertation that earned him a PhD from the University of Copenhagen in May 1911 to the completion of Part One of the Trilogy in April 1913. Bohr's (1911) dissertation shows that he was fully aware of the work, predominantly by German physicists, on black-body radiation and the specific heats of solids that, as we saw in Chapters 2 and 3, was driving the development of quantum theory in the preceding decade (Section 4.2). In September 1911, Bohr left Denmark to spend a year in England, first with J. J. Thomson in Cambridge, then with Ernest Rutherford in Manchester (Section 4.3). In Manchester, Bohr became familiar with Rutherford's idea of the nuclear atom according to which atoms consist of a small nucleus containing most of the atom's mass with a cloud of electrons swirling around it

---

[1] The most important secondary sources on this topic are Heilbron and Kuhn (1969), Heilbron (2013), and Kragh (2012, Chs. 1–2).

Constructing Quantum Mechanics. Anthony Duncan and Michel Janssen, Oxford University Press (2019).
© Anthony Duncan and Michel Janssen. DOI: 10.1093/oso/9780198845478.001.0001

(Section 4.3.3). Rutherford (1911) (see Plate 12) had introduced this qualitative model to account for experiments of Hans Geiger and Ernest Marsden (1909) on the scattering of α particles by different materials (Section 4.3.2). Bohr began building concrete quantitative models of Rutherford's nuclear atoms incorporating Planck's constant (Section 4.3.5). He recorded his results in what has become known as the Rutherford or Manchester Memorandum (Bohr 1912, Section 4.3.5). This document, written in July 1912, can be seen as a first draft of parts of the 1913 Trilogy, especially Parts Two and Three. In the Rutherford Memorandum, Bohr was not yet using his models to account for spectroscopic data. This only happened after his return to Denmark in September 1912 to take up a position as assistant to Martin Knudsen, the new professor of physics in Copenhagen. There, in February 1913, Hans Marius Hansen, a colleague recently returned from Göttingen, alerted Bohr to the Balmer formula (Section 4.4). Things then quickly fell into place and on March 6, 1913, Bohr mailed a version of Part One of the Trilogy to Rutherford with the request to communicate it to *Philosophical Magazine*. After a few rounds of revisions and a visit to Manchester to defend the length of the paper in person, Bohr submitted the final version on April 5, 1913. Parts Two and Three followed in the next few months.

Although Bohr is the central character of this chapter, other physicists helped pave the way for the Bohr model. We already mentioned Rutherford's contribution and Hansen's intervention. In Sections 4.3.1 and 4.3.4, we examine the classical and predominantly British tradition of atomic model building, starting with Thomson's (1904) plum-pudding model and the planetary ("Saturnian") model of the Japanese physicist Hantaro Nagaoka (1904).[2] George A. Schott (1907) (see Plate 13) subjected these classical models to a critical analysis and showed that they could never account for more than a handful of spectral lines. Schott's results were part of the reason that John W. Nicholson (1911b, 1911c, 1911d, 1912) (see Plate 14) developed a theory in which the familiar terrestrial elements are modeled as compounds of simpler hypothetical proto-elements, not to be found on Earth and existing in unbound form only in high-temperature environments such as the solar corona and stellar nebulae (Section 4.3.4; see also McCormmach 1966). This is reflected in the names of Nicholson's most important proto-elements: nebulium and coronium. In accordance with Schott's results, Nicholson assumed that his proto-elements had spectra with only a few lines. Unlike the theories of Thomson and Nagaoka, which are purely classical, Nicholson in the course of developing his theory suggested the quantization of both energy and angular momentum of the electrons circling in rings in his models.

In fact, neither Bohr nor Nicholson were the first to introduce Planck's constant in atomic modeling. That distinction goes to the Austrian physicist Arthur Erich Haas (1910a; 1910b; 1910c; see Section 4.3.4). Both Haas and Nicholson are cited in Part One of the Trilogy (Bohr 1913b, p. 6) but, while Haas's theory is only mentioned in passing, Nicholson's theory is compared to Bohr's own at several points in the paper.

---

[2] Although he had been working at Tokyo University since 1901, Nagaoka had been to Europe in the 1890s and he published his model in the British journal *Philosophical Magazine*.

Bohr probably first encountered Nicholson's theory in December 1912. Initially, he was both worried that his own theory was too similar to Nicholson's and puzzled that the two theories seemed to give conflicting results. Eventually, he found a way to reconcile the two (Section 4.4.1): Nicholson's results, he argued, apply to excited states, found only in stellar atmospheres, while his own apply to the "normal", or what we would now call the ground states of atoms. In addition to alerting him to the possibility of quantizing angular momentum rather than energy and inspiring him to include excited states in his theorizing, Nicholson's theory may also have primed Bohr for recognizing the potential of the models introduced in his Rutherford Memorandum to account for spectroscopic data.

In Section 4.5.1, we analyze Part One of the Trilogy, especially the derivation of the Balmer formula from Bohr's model of the hydrogen atom (Heilbron and Kuhn 1969, sec. 5). The published paper retains clear traces of earlier stages in Bohr's thinking, which allows us to reconstruct, at least to some extent, how he eventually arrived at the model that continues to be a staple of introductory textbooks to this day. In particular, we can reconstruct how Bohr arrived at what many at the time considered to be the most radical aspect of his model: the frequency of the radiation emitted or absorbed when an electron jumps from one orbit to another is not determined by the orbital frequency of either the initial or the final orbit but by the energy difference between them. Only for large quantum numbers does the radiation frequency merge with these two orbital frequencies (which in the limit of large quantum number also get closer and closer to one another). In Section 4.5.2, we discuss Parts Two and Three of the Trilogy. Finally, in Section 4.6, we examine some of the early evidence for and some of the early responses to Bohr's atomic model.

4.2 The dissertation: recognition of problems of classical theory

In 1903, Bohr enrolled at the University of Copenhagen, choosing physics as his major. This was the only university in Denmark in those days with a professor of physics on the faculty. This lone physics professor was Christian Christiansen, a colleague and friend of Bohr's father (Kragh 2012, p. 40).[3] At Christiansen's suggestion, Bohr chose the electron theory of metals of Drude (1900) and Lorentz (1905) as the topic for his Master's thesis (Bohr 1909). He decided to make it the subject of his dissertation as well (Bohr 1911).[4] We only cover the dissertation to the extent that it provides insight into what Bohr knew at the time about the problems facing classical theory and what part(s) of the theory he thought would have to be changed to overcome these problems.[5]

---

[3] See Aaserud (2013) for a fascinating portrait of the young Bohr and his courtship with Margrethe Nørlund, whom he married in 1912.

[4] For analysis of Bohr's dissertation, see Heilbron and Kuhn (1969, pp. 213–223) and Kragh (2012, pp. 44–45).

[5] See Staley (2005, 2008) for discussion of how the term "classical physics" acquired its modern meaning (cf. Section 1.2.3).

In the introduction, Bohr explained that "[t]he aim of [the dissertation] is to attempt to carry out the calculations for the various phenomena that are explained by the presence of free electrons in metals in as much generality as possible, while retaining the fundamental points of view underlying the theory of Lorentz" (pp. 299–300). He boiled down these "fundamental points of view" to two basic assumptions:

> [i] *[F]ree electrons are present in any piece of metal, their number depending on the nature and temperature of the metal, but their kind being the same in all metals* ... [ii] *[M]echanical heat equilibrium will exist between the free electrons and the atoms in a homogeneous piece of metal of uniform temperature and not subjected to external forces* (p. 300, Bohr's emphasis).

The specific-heat and radiation anomalies discussed in Chs. 2 and 3 posed a challenge to the second assumption:

> [E]xperiments on the specific heat of bodies show that thermal equilibrium ... cannot occur everywhere, in particular, not among the electrons supposed to be bound in great numbers inside the individual atoms.[6] Furthermore, it must be assumed that the Maxwell–Lorentz equations for the electro-magnetic phenomena are not strictly satisfied, for Lord Rayleigh's and Jeans' investigations on heat radiation have shown (see Chapter III) that, in an electromagnetic system for which these equations hold rigorously, statistical equilibrium in which the individual electrons possess kinetic energy cannot exist (p. 300).

To avoid these problems, Bohr assumed "that among free electrons (*contrary to those bound in atoms*) ... a statistical equilibrium as that envisaged above will occur" (p. 300; our emphasis). This suggests that Bohr's proposed escape from these difficulties was to deny the validity of ordinary mechanics for electrons bound in an atom. He drew similar conclusions in the final two chapters of the dissertation, one of which is already referred to in the quotation above.

Ch. 3 deals with the absorption and emission of thermal radiation (Bohr 1911, pp. 357–379). Bohr derived a formula for the spectral distribution of black-body radiation by using his elaboration of the electron theory of metals earlier in the dissertation to find the ratio of the emission and absorption coefficients of a metal.[7] Unsurprisingly, given that this theory was purely classical, his calculation reproduced the Rayleigh–Jeans law. Bohr had clearly kept up with the literature on black-body radiation, citing Rayleigh (1900b, 1905a, 1905b), Jeans (1905b, 1909), and Lorentz (1908c) (Bohr 1911, p. 378). As the passage below shows, he knew about the discussion concerning the Rayleigh–Jeans law and the Planck law triggered by Lorentz's (1908c) Rome lecture (see Section 3.4.2) and about the fluctuation arguments on the basis of which Einstein had

---

[6] Bohr followed Drude and Lorentz in assuming that "great numbers" of electrons are bound inside individual atoms. As we will see in Section 4.3.1, Thomson (1906) had already shown that this number is actually only on the order of the atomic weight of the atom (Kragh 2012, p. 32).

[7] As noted in Section 2.2, Kirchhoff had shown that this ratio is a universal function of frequency and temperature.

argued in 1909 that the Planck law requires that the Maxwell–Lorentz equations be modified (see Section 3.4.3):

> Jeans has expressed the opinion that we are not dealing with a real state of equilibrium in the experiments on heat radiation, but only with a state that changes very slowly, on account of the low power of absorption and emission of ordinary bodies for rays with very small periods of vibration. However, as appears from the discussion that has taken place on this question,[8] this is hardly sufficient to explain the phenomena found experimentally. Hence it seems impossible to explain the [Planck] law of heat radiation if one insists upon retaining the fundamental assumptions underlying the electromagnetic theory.[9] This is presumably due to the circumstance that the electromagnetic theory is not in accordance with the real conditions and can only give correct results when applied to a large number of electrons (as are present in ordinary bodies) or to determine the average motion of a single electron over comparatively long periods of time (such as in the calculation of the motion of cathode rays) *but cannot be used to examine the motion of a single electron within short intervals of time* (Bohr 1911, p. 378, our emphasis).

As in the passage from the introduction quoted above, Bohr thus suggested that there is something wrong with the laws governing the motion of individual electrons.

Ch. 4 of the dissertation is on magnetism (pp. 380–392). It starts with a short proof of what has since become known as the Bohr–van Leeuwen theorem.[10] Contrary to a suggestion by Thomson (1900–1901) in Paris in 1900, Bohr showed that a system of free electrons governed by classical theory and in thermal equilibrium cannot develop a magnetic moment as a result of the imposition of an external magnetic field. Thus, classical arguments purporting to explain the presence of either dia- or paramagnetism on the basis of a free electron gas are destined to fail. As Bohr (1911, p. 382) emphasized, "*a piece of metal in electric and thermal equilibrium will not possess any magnetic properties whatever due to the presence of free electrons.*" What about bound electrons? In a footnote, Bohr (1911, pp. 382–383) referred to the initial diamagnetism induced by a suddenly applied magnetic field due to the change in motion of bound electrons moving in closed orbits, as discussed by Paul Langevin (1905, p. 90). Langevin argued that the diamagnetism generated in this way would persist once the magnetic field settled down to a constant value, but Bohr pointed out that the reestablishment of statistical equilibrium—on the assumption that the bound electrons can exchange energy with one another and that their motion obeys the usual mechanical laws—would once again lead to a vanishing magnetic moment. Hence, Bohr (1911, p. 383) concluded, "it is not

---

8 At this point, Bohr cited another version of Lorentz's (1908b) Rome lecture, the response by Lummer and Pringsheim (1908) (in which they noted that, if the Rayleigh–Jeans law were correct, "steel at room temperature would have to emit a bright light"), and a further paper by Jeans (1908).

9 At this point, Bohr cited Einstein (1909a, 1909b) and Einstein and Hopf (1910) (see Section 3.4.3) as well as Planck (1910) (see Section 3.4.2).

10 Hendrika J. van Leeuwen (1919) discussed the same result in her dissertation. According to van Leeuwen (1921, p. 376), her PhD adviser Lorentz had independently found the result around the same time as Bohr (Heilbron and Kuhn 1969, p. 218, note 17). For an elegant proof of the theorem, see Van Vleck (1932, pp. 101–102).

possible to explain the magnetic properties of bodies on the basis of the electron theory." However, he qualified this conclusion by adding the following conditional clause: "if such effects as, e.g., the emission of energy are neglected, which might have the effect of preventing a statistical mechanical equilibrium among the bound electrons as that assumed for the free electrons." He then noted that "[t]he circumstances mentioned here seem to be intimately connected with those discussed at the end of the last chapter" (p. 383). In Bohr's estimation, in other words, both the inability to derive the Planck law for black-body radiation from classical theory (Ch. 3) and the inability to derive magnetic properties from such a theory (Ch. 4) were somehow connected to the breakdown of the usual electro-mechanical laws when applied to electron motion at subatomic dimensions, i.e., to the motion of bound rather than free electrons. In the introduction of the dissertation, as we saw above, Bohr (1911, p. 300) also referred to the special difficulties attached to the treatment of bound electrons in connection with the problem of specific heats.

In summary, Bohr's dissertation shows that he was primed for the necessity of abandoning some of the basic tenets of classical mechanics, electromagnetic theory, and statistical mechanics, prompted by the failures of purely classical theory in reproducing known features of absorption and emission of radiation by metals as well as their magnetic properties. Exactly which tenets to abandon, and what to replace them with, remained unclear, but the problems appeared to be closely connected to the behavior of bound as opposed to free electrons in the phenomena under study.[11]

4.3 The Rutherford Memorandum: atomic models and quantum theory

After obtaining his PhD in May 1911, Bohr won a scholarship from the Carlsberg Foundation to spend a year of postdoctoral studies in England. From September 1911 to March 1912, he was in Cambridge, where he tried in vain to interest Thomson, director of the Cavendish Laboratory, in the results of his dissertation. In December, Rutherford visited Cambridge and gave a talk. Rutherford had attended the Solvay conference (see Section 3.5) in Brussels in late October/early November 1911 and Bohr (1961, p. 1084) later recalled that this had been a topic of conversation when he and Rutherford first met. It is not clear, however, exactly when (November or December?) or where (Cambridge or Manchester?) this happened (Heilbron and Kuhn 1969, p. 233; Heilbron 2013, p. 151). In any event, Bohr was so taken with Rutherford that he arranged for a transfer to Manchester the following Spring. He arrived in Manchester in March 1912. This is where Bohr first encountered Rutherford's nuclear atom and began to develop detailed models for it. Before discussing the fruits of Bohr's labor (Section 4.3.3), we review the

---

[11] As Heilbron and Kuhn (1969, p. 222) conclude from their examination of Bohr's dissertation: "[T]he difficulty with magnetism strengthened and confirmed Bohr's conviction that the usual mechanical laws broke down when applied to rapidly moving electrons; and, even more than the radiation problem, it isolated the breakdown in the behavior of electrons bound in atoms."

prehistory of atomic models (Section 4.3.1) and the origin of Rutherford's nuclear atom (Section 4.3.2).

4.3.1   Prelude: classical atomic models (Thomson, Nagaoka, Schott)

By the first decade of the twentieth century, the basic features of the optical line spectra of a wide variety of elements had been identified, with whole sequences of lines grouped into series. In many cases an accurate quantitative representation of the frequencies/wavelengths of the series lines was available in the form of formulas of the Rydberg–Ritz type, which give the frequency of a spectral line as the difference of two terms (see Appendix B). These spectral lines were clearly emitted by electrons bound in atoms. A theoretical explanation of their frequencies therefore depended on the development of a calculable model of atomic structure. Until Haas's work in 1910 (see below), such models were entirely classical. Haas was the first to explicitly invoke Planck's constant in an attempt to fix basic features of atomic structure, such as the size of the atom (Hermann 1971, p. 90).

The difficulties of accounting for sharp line spectra, or for the fixed linear scale of atoms, in any atomic model in which mobile negative and positive charges were bound into atoms purely electrostatically were clearly exposed by Jeans (1901, pp. 425–427) in his paper, "The Mechanism of Radiation." Jeans went on to examine in detail the vibrational frequencies of "ideal atoms" in which stability is imposed by the introduction of a potential energy that is partly electrical (depending on the product of charges of two volume elements) and partly quasi-gravitational (depending on the product of the masses of two volume elements), with the positive and negative charges dispersed through the atom treated to a first approximation as continuous fluids. The aim of Jeans's calculations was to provide at least a qualitative justification (there were no actual quantitative confrontations with the spectral phenomenology) for the appearance of infinite spectral series. This was but one of a multitude of contorted attempts to connect the huge amount of information contained in the line spectra of the elements with specific physical models of atoms.

In his dissertation, Walter Ritz (see Plate 18), for instance, approached the problem starting with the spectral information encoded in the Balmer formula, and was led to an extremely baroque model of vibrating membranes with a very peculiar potential energy function in order to reproduce the Balmer formula (Ritz 1903).[12] A few years later, Ritz (1908a) abandoned this model in favor of an equally bizarre concoction of magnetic dipoles with electrons executing Larmor orbits on the axes of the dipoles, again in an attempt to explain Balmer's formula (and, with further exotic tweaks, the complex structure of spectra and the anomalous Zeeman effect!). These ambitious attempts to explain the spectral series phenomenology in a purely classical way received little or no

---

[12] See Appendix B, Section 6. For further historical analysis of Ritz's work in spectroscopy, see Hentschel (2012).

attention, although a paper by Ritz (1908b) later that year in the *Astrophysical Journal* is rightly remembered for the introduction, in a purely phenomenological fashion, of the combination principle. This principle says that the wavenumber or frequency of any spectral line can be written as the difference between two terms, both involving an integer. The combination principle survived the transition to quantum theory and was invaluable for the next two decades in deciphering the intricacies of atomic spectra.

A parallel development in the evolution of atomic structural theory was the development, beginning in 1899, of the "plum-pudding" model of J. J. Thomson (1904) and the planetary ("Saturnian") model of Nagaoka (1904).[13] In Thomson's model the positive charge was uniformly dispersed in a sphere, the surface of which was identified with the outer limits of the atom, while the negative electrons ("plums") moved around inside the positive "pudding". The Nagaoka model was a clear precursor of Rutherford's nuclear atom. In both models, radiative stability of the atom was a problem (moving bound electrons necessarily radiate in classical electrodynamics), as was mechanical stability, i.e., the stability of an orbit when subjected to a small perturbation. Radiative instability could be reduced on the assumption that the electrons are symmetrically located on rings, which eliminates the most dominant mode of energy loss, dipole radiation.[14] It could also be reduced if one assumed a large number of electrons (a continuous steady current ring, for instance, does not radiate at all). It should be kept in mind that, until about 1906, it was widely assumed that even the lighter atoms of the periodic table contained a large number of electrons. This was quite reasonable as long as one implicitly assumed a similar charge-to-mass ratio for the components of atoms carrying positive charge as the measured ratio for the negative electrons. An important paper by Thomson (1906), in which the number of electrons in atoms was estimated (on the basis of measurements of dispersion, X-ray scattering, and energy loss of beta rays passing through matter) to be *on the same order as the atomic weight*, undermined this assumption and thereby, ironically, Thomson's own model.

The problem of mechanical stability was worse for the Nagaoka model, where the electrostatic repulsion led to instability when the rotating ring of electrons was perturbed, due to the inter-electronic repulsion (in contrast to the case of Saturn's rings analyzed by Maxwell, which are stabilized by the attractive forces of gravitation). In this case the instability arose from perturbations in the plane of the orbit. In the Thomson model, at least for small numbers $n < 6$ in the ring, and given a sufficiently large angular velocity of rotation of the ring electrons, one had mechanical stability. Small perturbations would only lead to oscillatory wobbles of the electrons superimposed on the overall rotational motion in the ring. For n equal to 6 or larger, stability could be regained by inserting a negative charge (stationary electron) at the center of the atom (i.e., of the positive "pudding").

---

[13] For Thomson's model, see Kragh (2012, sec. 1.3, pp. 12–22) and Navarro (2012, p. 96), who places it in the context of Thomson's work on so-called Faraday tubes; for Nagaoka's model, see Yagi (1964), Conn and Turner (1965, pp. 111–119), and Kragh (2012, pp. 23–24).

[14] Thomson (1903) showed that for a ring of n nonrelativistic electrons with $\beta = v/c \ll 1$, the radiation rate decreases roughly like $(e\beta/2)^{2n}$. The radiation rate per electron for a ring of 6 electrons would therefore be about 10^{-17} the rate for a single electron for electrons with velocity $v/c \approx 1/100$, as is typical in atoms.

In either type of model, one could put aside the radiative instability issues and proceed to the calculation of the discrete frequencies associated with perturbations of a presumed ring of electrons rotating steadily with some angular velocity ω (in the normal state) of the atom. The discrete frequencies obtained in this way should obviously correspond to observed emission lines of excited atoms (all of this still in a purely classical context). Detailed calculations along these lines were performed by George A. Schott (see Plate 13) in early 1907. Schott was agnostic as to the details of the atomic model:

> There is no need at present to distinguish between the two possible alternatives: (1) the positive electricity constitutes a sphere of uniform electrification of atomic size, the negative electricity exists as corpuscles (negative electrons) moving inside the positive sphere (J.J. Thomson); (2) both positive and negative electricity exist as discrete charges—electrons—moving in closed orbits of atomic size [such as the negative charges in Nagaoka's model]. The following discussion applies to both types of theory (Schott 1907, p. 189).

The basic conclusion of Schott's investigations, stated already on the second page of his article, was negative: "It may be said at once that a single ring cannot be made to account for spectrum series or bands" (p. 190). The evidence of spectroscopy as regards the optical line spectra of elements clearly required three conditions to hold: (1) the emitted radiation had to be sufficiently intense to be observable; (2) the frequencies had to lie between the limits of the optical spectrum (generously interpreted by Schott as extending from the ultraviolet to the near infrared, 100 to 800 nm); and (3) the frequencies had to be given by a series formula of the Balmer–Rydberg–Kayser–Runge type. Schott showed that the imposition of just the first two conditions results in only a handful of lines of observable intensity arising from the perturbations of a ring of electrons. No such model, he noted, could thus be expected to explain the origin of infinite line series, with the lines accumulating to a limit point, as described by simple formulas of the Balmer–Rydberg type (p. 190).

We briefly examine the essential features of Schott's calculations, as their results are critical to the interpretation of Nicholson's atomic models discussed below. Without worrying about stability questions, Schott simply assumed that the n equally spaced electrons in the ring are rotating steadily (and stably) around the atomic center with angular velocity ω prior to being perturbed. He calculated the radiation emitted due to a small harmonic perturbation (of angular frequency q, in the rotating frame of the initial steady motion) superimposed on the circular motion of the electrons (either in the plane of the ring or perpendicular to it). He took the displacements of the electrons from their unperturbed locations to be a periodic sinusoidal function of the integer $i = 1, 2, \ldots, n$ giving the order of the n electrons in the ring. In other words, he assumed that the displacement of the ith electron in the ring would be proportional to $\sin(qt - k(2\pi i/n) + \alpha)$, where k is an integer called the "class" of the displacement. As can be read off from the argument of the sine function for an arbitrary fixed value of t, k is the number of "wavelengths" in a curve drawn through the displaced positions of the n electrons. Schott took the radius ρ of the ring to be on the order of 10^{-8} cm, which was known to be the order of magnitude of atomic sizes, and therefore much smaller (by three orders of magnitudes at least) than the relevant wavelengths λ of light.

The asymptotic (radiation regime) electric and magnetic fields generated by such displaced electron orbits were shown by Schott to contain harmonic series with frequencies $q + \omega(k + sn)$ for the sth harmonic. The radiation intensity falls off very rapidly with harmonic number s: for the lowest harmonic ($s = 0$) it is proportional to $(2\pi\rho/\lambda)^{2|k|}$, which again falls off rapidly with the class number k (as $2\pi\rho/\lambda$ is on the order of 10^{-3} in the optical). For example, Schott found that, for lines in the red, the radiation emitted by perturbations of class $k = 2$ was already 10^{-8} times weaker than that of class 0. The upshot was that a single ring of rotating electrons could only give rise to a few lines of observable intensity within the optical spectrum. The appearance of an infinite series of optical lines converging to a limit point was completely inexplicable in this framework.

More generally, the existence of a discrete spectrum emitted by atoms and molecules in an excited state was regarded early on (i.e., well before Bohr's trilogy) as an important clue to the underlying nature of atomic dynamics. In fact, Stark (1911, pp. 69–72)— in an extensive bibliography at the end of section 13, "Conclusions from the structure of series," of Ch. 2 of Vol. 2 of his *Prinzipien der Atomdynamik*—listed about 60 works by 32 different authors, bearing, as he put it, "on the dynamical interpretation of the structure of series spectra." The papers in this bibliography range from a 1871 paper by G. J. Stoney (in which "the conclusion is drawn that the lines in the spectra of gases are to be referred to periodic motions within the individual molecules, and not to the irregular journeys of the molecules amongst one another"), through papers by Nagaoka, Ritz, and Schott (but, interestingly, not Thomson, or the very recent work of Haas) up to the time of Stark's own book.

4.3.2   Scattering of α particles and Rutherford's nuclear atom

In the Spring of 1909, experiments performed in Manchester by two young assistants of Rutherford, Hans Geiger and Ernest Marsden (1909) (see Plate 12), established the existence of "diffuse reflexion" of α particles striking thin metal foils. This term was used to indicate the very unexpected deflection of α particles impinging on the metal surface by an angle greater than 90 degrees—in other words, particles reflected back from the surface, which could be detected by light flashes emitted when they impinged on a zinc sulfide screen placed on the same side of the foil as the radium-containing tube which provided the source of the α radiation. The existence of this diffuse reflection posed an immediate problem for the Thomson model of the atom. Presumably, the reflection was due to repulsive electrostatic forces originating in the positive charge $+Ne$ within the atom, which in the Thomson model is imagined uniformly dispersed throughout the volume of the atom. Given the estimates of the time, this would be a sphere with a radius of roughly $R = 10^{-8}$ cm. The maximum electric force exerted by this positive-charge "pudding" on the approaching α particle (of charge $+2e$) is attained at the surface of this sphere and is given by $F = 2Ne^2/R^2$. The average force experienced by the α particle as it traverses the atom will be substantially smaller. Not only will the force exerted by the positive-charge "pudding" decrease once the α particle enters the sphere of the atom, we are also neglecting the opposing forces exerted by the negative-charge "plums". Yet, even if we use $2Ne^2/R^2$ as our estimate of the force experienced by the α particle throughout its passage through the atom, we find that this force can never produce a change of momentum large enough for the α particle to be reflected back from the metal foil. If an α particle were to traverse the atom with average speed $u = 2R/\Delta t$ all the while subject to a repulsive force $F = 2Ne^2/R^2$, the impulse imparted to it would be $\Delta p = F\Delta t = 4Ne^2/uR$. If the particle is to be reflected back from the foil, which, according to Geiger and Marsden, occurred about one in 8000 times, the change in momentum Δp must be comparable to the initial momentum p. If M is the mass of the α particle, its momentum is on the order of $p = Mu$, where for u we can take the same velocity that we used in our estimate of Δp. In fact, for α particles with an energy of about 10 MeV (corresponding to a speed of about 2×10^9 cm/s, as stated by Rutherford) and a positive charge N on the order of 100,[15] passing through one of Thomson's plum-pudding atoms will barely slow down the α particle at all. We thus find that, even with our overly optimistic estimate of Δp, the ratio $\Delta p/p$ will only be of the order of 10^{-4}. Given the Thomson model, anything but a small deflection of the α particle from the initial direction of motion can thus be excluded. Reflections of the type seen by Geiger and Marsden cannot be reconciled with the Thomson scheme, unless we assume a positive charge millions of times greater than the electron charge. More precise experiments carried out by Geiger (1910) the following year allowed the determination of the full differential scattering cross section (i.e., the angular dependence), as well as the dependence on the nature of the metal used, the thickness of the foil, and even the dependence on the velocity of the α particles (which could be varied by interposing sheets of mica or aluminum). The quantitative results obtained in these experiments formed the empirical background for Rutherford's seminal paper, "The scattering of α and β particles by matter and the structure of the atom" (Rutherford 1911).[16]

After reviewing the main features of the Geiger and Marsden experiments, and in particular the inescapable conclusion from their measurements that "some of the α particles must suffer a deflexion of more than a right angle at a *single encounter*" (p. 670; our emphasis), something clearly impossible in an atomic model of the Thomson type, Rutherford states:

> The theory of J. J. Thomson is based on the assumption that the scattering due to a single atomic encounter is small, and the particular structure for the atom does not admit of a very large deflexion of an α particle in traversing a single atom, unless it be supposed that the diameter of the sphere of positive electricity is minute compared with the diameter of the sphere of influence of the atom.
>
> Since the α and β particles traverse the atom, it should be possible from a close study of the nature of the deflexion to form some idea of the constitution of the atom to produce the effects observed. In fact, the scattering of high-speed charged particles by the atoms of matter is one of the most promising methods of attack of this problem (p. 670).

---

[15] As mentioned above, Thomson (1906) had shown that the number of electrons was of the same order of magnitude as the atomic weight. Rutherford would deduce (see below) a value close to 100 for the nuclear charge of gold and platinum from the α-particle scattering data.

[16] For a fuller account of the genesis of the Rutherford model, see Heilbron (1968).

With these sentences, Rutherford can properly be said to have founded the enterprise of high-energy physics, which in the century subsequent to this paper has gone far beyond Rutherford's initial aim of understanding the structure of the atom in laying bare the catalog of elementary particles, with their interactions, down to distance scales 10^{-12} times smaller than an atomic dimension. The basic new hypothesis is also clear: the positive charge within an atom would have to be confined to a volume of dimension much smaller than that of the atom (more specifically, the assembly of negative electrons responsible for restoring electrical neutrality) in order for an approaching α particle to encounter electric fields of the size needed to reverse a 10 MeV α particle in its track.

Rutherford's paper is primarily concerned with exploring the consequences of a "nuclear model" of the atom,[17] and Thomson's model is only mentioned briefly, in sec. 5, where the relative importance of single and compound (i.e., multiple) scattering is examined. The quantitative analysis of the nuclear model begins in sec. 2, where, after explaining the negligible influence of the electrons on an α particle entering the atom, and approaching the central charge to a distance where a substantial deflection can occur, Rutherford calculates the relation between the angle of deflection and the impact parameter of the incoming particle (namely, the perpendicular distance from the nucleus to the straight line extrapolated from the initial trajectory of the incoming α particle). This result is then used in sec. 3 to derive a quantitative result for the differential cross section for (in modern terminology) Coulomb scattering,[18] essentially, the functional dependence of the probability of scattering of an α particle on (a) the deflection angle (inverse fourth power for small angles), (b) the central charge Ne (as N^2), and (c) the kinetic energy of the incoming particle (inverse square). The results of Geiger (1910) are then adduced as evidence that "the distribution for particles deflected between 30° and 150° from a thin gold-foil was in substantial agreement with the theory" (Rutherford 1911, p. 675).

In sec. 4, Rutherford considers the issue of loss of energy of the α particles in the case of lighter atomic targets. In sec. 5, the relative importance of single and multiple scattering in producing the final observed deflections is considered (with the conclusion that large deflections are in essentially all cases the result of a single encounter with an atom). In sec. 6, "Comparison of Theory with Experiments," he discusses various features of both α and β scattering from the point of view of the new model. In particular, Rutherford deduces a central charge value of approximately $N = 100$ for platinum, on the basis of the observation of Geiger and Marsden (1909) that roughly one in 8000 α particles are reflected back when falling on a thick plate of platinum.

[17] This term, or indeed the denotation "nucleus" does not appear in the paper, where the term "central charge" is used throughout for what would, by 1913, be universally dubbed the atomic nucleus. We will use the simpler term "nucleus" henceforth.
[18] Rutherford's formula for Coulomb scattering in fact survives the transition to modern quantum mechanics: Planck's constant is not relevant in the leading approximation.

In sec. 7, "General Considerations," the final section of the paper, Rutherford once again addresses the issue of the negligible influence of the electrons on the scattering, and closes a possible loophole, namely, the supposition that the nuclear charge Ne might actually consist of N separated unit charges (in a fashion reminiscent of Nicholson's atomic models, to be discussed below). The essential conclusion to be drawn from all this was clear:

> Considering the evidence as a whole, it seems simplest to suppose that the atom contains a central charge distributed through a very small volume, and that the large single deflections are due to the central charge as a whole, and not to its constituents (Rutherford 1911, p. 687).

As Rutherford famously put it later: "It was quite the most incredible event that has ever happened to me in my life. It was almost as incredible as if you fired a 15-inch shell at a piece of tissue paper and it came back and hit you" (quoted, for instance, in Andrade 1964, p. 111).

In the concluding paragraphs of his paper, Rutherford also mentions the "Saturnian" model proposed by Nagaoka in 1904, but only to point out that his assertions concerning the negligible effect of the electrons on the scattering would not be affected by their disposition in a ring-like configuration, rather than in a spherically symmetric cloud, as assumed earlier in the paper.

Additional experimental work by Geiger and Marsden (1913) would further explore—and confirm in much greater detail—the quantitative accuracy of Rutherford's cross-section formula for α-particle scattering. But by the late Spring of 1912, when Bohr was in residence in Manchester and himself vigorously engaged in research on α-particle scattering, the basic correctness of the new picture of the atom was already fully accepted by everyone in the Rutherford "orbit".

4.3.3   Bohr's first encounter with Rutherford's nuclear atom: energy loss of α particles traveling through matter

Upon arrival in Manchester in March 1912, Bohr initially divided his time between tinkering with some experiments on radioactivity and tinkering with his dissertation, which he tried to massage into a form suitable for publication (Heilbron and Kuhn 1969, pp. 234–236). In June, he gave up these pursuits and turned to a problem of immediate interest to Rutherford's group in the aftermath of the seminal experiments on α-particle scattering that had led to the development of the Rutherford nuclear-atom model (pp. 237–243). The problem was to understand the relation between the rate of energy loss of charged particles—in particular, α particles—passing through matter and the underlying atomic structure, i.e., the arrangement and motion of the charged electrons in the atoms (as the nuclei were much more massive, the energy transferred from the α particle to the atoms was almost entirely to the light electrons, the nuclei being regarded as static). This problem had been treated in a recent paper of Charles G. Darwin

(1912), grandson of the famous naturalist,[19] which Bohr read critically and decided to improve on, as he told his brother Harald in a letter of June 12, 1912 (Rosenfeld 1963, p. xvii–xviii).

A direct assault on this problem treating the electrons as bound orbiting particles in a Rutherford atom is sketched in a manuscript with calculations on "Dispersion [of light] and absorption of α-rays."[20] Bohr abandoned this direct approach because of the instability of the electron orbits under perturbation by the electric pulse generated by a passing charged particle. As it says on the cover sheet for this part of the manuscript:

> Given up for the time being since the calculation failed owing to the instability of the systems; cannot be continued without using one or another hypothesis (Bohr 1972–2008, Vol. 2, p. 46).

The hypothesis needed, as we will see in Section 4.3.5, is a quantum condition. Instead of Rutherford's model, Bohr used the Drude–Lorentz theory of dispersion, in which electrons in an atom are pictured as tiny harmonic oscillators.[21] In the spirit of this theory, he assumed that the effect of a perturbing electric field on the electrons in an atom could be equivalent to the excitation of harmonic oscillations of definite (angular) frequencies ω_s, with $s = 1, \ldots, r$, where r is the number of "dispersion electrons". This number r and the frequencies ω_s would be determined empirically from dispersion data. In the case of a passing charged particle, the sinusoidal electric field associated with an incident light wave in dispersion theory is replaced with an electric pulse generated at the location of the electron by the passing charged particle. This pulse results in the excitation of vibrations of the dispersion electrons, and the concomitant energy of excitation results in a loss of energy of the passing particle.

When Bohr left Manchester in late July 1912, he had started but not finished a paper on the absorption problem. He reported his struggles writing it in letters to his fiancé Margrethe Nørlund (Aaserud and Heilbron 2013, pp. 87–93, p. 166). The couple married in Copenhagen on August 1. They spent their honeymoon in England rather than in Norway, as originally planned, so that the groom could finalize the paper and deliver it in person to Rutherford (Aaserud 2013, p. 94). Rutherford communicated it to the *Philosophical Magazine*, where it appeared in January 1913 (Bohr 1913a). By comparison with the dispersion data for hydrogen gas, Bohr could at least roughly reproduce the α particle absorption data of T. S. Taylor (1909) for two different α particle energies, on the assumption that a hydrogen molecule, in its normal state, contains 2 electrons

---

[19] The title of Darwin's paper is "A theory of the absorption and scattering of the α rays." The term "scattering" refers to the deflections of α particles studied by Geiger and Marsden (see Section 4.3.2), whereas "absorption" is the term used at the time to refer to the energy loss of α particles moving through matter.

[20] This manuscript, consisting of about 40 sheets and divided into three parts with cover sheets in the hand of Mrs. Bohr, is discussed briefly by Heilbron and Kuhn (1969, p. 242). The cover sheets and six pages of calculations are included in Bohr (1972–2008, Vol. 2, pp. 45–55). On p. 624 of the same volume, a list of manuscripts in the Bohr archive with calculations related to this problem is given (items 3–7, the last item being the manuscript discussed by Heilbron and Kuhn).

[21] See Duncan and Janssen (2007, sec. 3.1), Jordi Taltavull (2017, Chs. 2–3), and the web resource *Classical dispersion theory*.

of (angular) dispersion frequency $\omega = 2.21 \times 10^{16} s^{-1}$. This picture of a hydrogen molecule followed from measurements of Clive and Maude Cuthbertson (1909, p. 166) interpreted in terms of the Lorentz–Drude dispersion theory. This particular frequency would come to play an important role in Bohr's first theory of quantized atomic systems (see Section 4.3.5). Bohr's comparison of dispersion and α-particle absorption data led him to conclude "with great certainty" that a "hydrogen atom contains only 1 electron outside the positively charged nucleus, and that a helium atom only contains 2 electrons" (Bohr 1913a, pp. 38–39). For heavier elements the deduced number of dispersion electrons was typically close to the nuclear charge expected on the basis of the Rutherford model (e.g., 18 instead of 16 for oxygen molecules, 14 instead of 13 for aluminum, but with considerably less agreement for even heavier elements like gold and lead).[22]

In the concluding sentence of his paper, Bohr intimated that "further information about the constitution of atoms" might be inferred from the absorption of α particles, which he promised would "be discussed in a later paper" (p. 39). On June 19, 1912, only a week after he had first mentioned the absorption problem to his brother, Bohr had written another letter to Harald enthusiastically telling him that he had "perhaps found out a little bit about the structure of atoms ... perhaps a little piece of reality" (Heilbron and Kuhn 1969, p. 238, p. 244). A month later, much of this optimism had evaporated. As he wrote to Margrethe on July 16, 1912:

> I have a paper that I must and will have written before I leave ... So many thoughts have rushed through my little head this year, and I have had so many visions; but in my hands there is but a dead treasure [presumably the unfinished paper on the absorption problem], for it is but indications of the most fleeting fantasy (Aaserud 2013, p. 92).

Bohr had recorded some of his "many visions" in notes he prepared for Rutherford. On July 6, 1912, Bohr had sent these handwritten notes to Rutherford with a one-line cover letter ("I send the remarks concerning the structure and stability of molecules for which you kindly asked"). This so-called Rutherford Memorandum was first published by Bohr's later collaborator Rosenfeld (1963, xxi–xxviii).[23] We return to this Memorandum in Section 4.3.5.

4.3.4 Interlude: Planck's constant enters atomic modeling (Haas, Nicholson)

Arthur Erich Haas was unwittingly injected into the early quantum theory of the atom by the requirement imposed by the evaluators of his habilitation thesis (on the history of

---

[22] See Hoyer (1973, pp. 178–186) for a detailed discussion of Bohr's calculations in this paper.

[23] The Memorandum can be found in transcription and in facsimile in Bohr (1972–2008, Vol. 2, pp. 135–158). On the envelope in which it has been preserved Bohr wrote: "First draft of the considerations contained in the paper 'On the constitution of atoms and molecules' [the title of the famous 1913 Trilogy] (written up to show these considerations to Prof. Rutherford) (June and July 1912)" (Rosenfeld 1963, xxi). Whereas the Memorandum itself is in English, the text on the envelope (with the exception of the title of the paper) is in Danish, which suggests that the text on the envelope was written only after Rutherford returned the Memorandum to Bohr.

the energy conservation principle) that he complete in addition an original investigation purely in physics before his habilitation and appointment as *Privatdozent* in the field of history of physics could be approved (Hermann 1971, pp. 90–91).[24] At the time there was considerable discussion—see, e.g., Einstein (1909a, p. 192)—that Planck's constant could be regarded as a derived quantity, to be expressed as a function of the electron charge and the speed of light, viewed as more fundamental. The ratio e^2/hc is dimensionless (later dubbed the fine structure constant), so perhaps there was some formula to be found of the form $h =$ dimensionless constant $\times\ e^2/c$. In his article on radiation theory in the *Enzyklopädie der mathematischen Wissenschaften*, Wien criticized this point of view and suggested that Planck's constant, "if it has any physical meaning, can probably be derived only from a universal property of atoms" (Wien 1909, p. 356). Haas apparently took Wien's comments very seriously, dove into the literature on thermal radiation, including Planck's (1906a) book, and by February 1910 had prepared an article entitled "On the electrodynamic meaning of the Planck radiation law and on a new determination of the elementary electrical quantum and the size of the hydrogen atom" (Haas 1910a).[25] Haas assumed a Thomson-type hydrogen atom, in which the electron executed circular orbits within or at the surface of a (singly charged) sphere of uniform charge density and radius a. His basic assumption—see, e.g., his *Jahrbuch* article (Haas 1910b, p. 263)—was that the maximum total energy (kinetic plus potential) of the atom should be equated to $h\nu$, where ν was the orbital frequency. The electrical potential energy, of course, is only determined up to an additive constant. Haas assumed that it is calculated with reference to the electron's electrical potential energy at the center of the positive sphere. With this choice, one finds that for an electron executing a circular orbit of radius r, the total energy is $(e^2/a^3)r^2$ for $r < a$, and $(3e^2/2a) - (e^2/2r)$ for $r > a$. In fact, Haas incorrectly found the total energy to be e^2/r for $r > a$, which would imply that the orbit of maximum energy corresponds to $r = a$. In fact, the energy continues to increase for external orbits, reaching a maximum $3e^2/2a$ for the orbit at infinity. In any case, Haas simply imposed the condition that for $r = a$, i.e., for the orbit right at the surface of the "plum pudding", which is mechanically indistinguishable from the orbit of the electron if the nucleus were a point particle,

$$|E_{\text{tot}}| = \frac{e^2}{a} = h\nu = h\frac{\omega}{2\pi}, \tag{4.1}$$

from which it follows that

$$\omega = \frac{2\pi e^2}{ah}. \tag{4.2}$$

---

[24] See Section 3.4.3 for an explanation of these features of the German university system in connection with Einstein's academic career.

[25] He reprised and extended these results in two subsequent papers (Haas 1910b, 1910c.) For discussion, see Hermann (1971, pp. 87–102).

Combining this result with Newton's second law,

$$m\omega^2 a = \frac{e^2}{a^2},$$
(4.3)

and solving for Planck's constant, one finds

$$h = 2e\pi \sqrt{am},$$
(4.4)

or, in his words, "the connection of the elementary quantum of action with the fundamental quantities of the electron theory" (p. 266). One could turn this around and regard his result as a determination of the radius a, and hence the size of the atom, in terms of Planck's constant and the electron charge and mass:

$$a = \frac{h^2}{4\pi^2 me^2},$$
(4.5)

which is exactly Bohr's formula for the radius of the ground state electron orbit in the hydrogen atom![26]

In another paper, Haas (1910c) tried to determine the Rydberg constant in terms of the fundamental quantities $m, e, h,$ and c. He almost succeeded! He related the limiting frequency ν^* in the Balmer formula to the total energy of his Thomson-type hydrogen atom:

$$h\nu^* = \frac{e^2}{a}.$$
(4.6)

In fact, this misses the mark in two ways: (i) the actual binding energy of the ground state, reckoning zero potential energy at infinity, is $e^2/2a$, and (ii) the limiting frequency in the Balmer formula refers to an excited end state of principal quantum number 2, not 1, hence with binding energy $e^2/8a$. Haas thus obtained a result for the Rydberg constant that is off by a factor of eight: $R = 16\pi^2 me^4/h^3 c$. Bohr's correct result is $R = 2\pi^2 me^4/h^3 c$. Despite these glaring problems, Haas's theory did attract the attention of leading physicists such as Lorentz and Sommerfeld (Hermann 1971, pp. 96–99; Kragh 2012, p. 21). Bohr (1913b, p. 6) dutifully cited Haas's (1910b) *Jahrbuch* article in Part One of the Trilogy, but Haas's work played no role in the genesis or further development of the Bohr model.

By contrast, the theory of John W. Nicholson (see Plate 14), to which we turn next, strongly influenced Bohr as he further developed the ideas about atomic structure in the Rutherford Memorandum in late 1912 and early 1913. At the annual meeting of the British Association for the Advancement of Science in Portsmouth in August–September

---

[26] The reason for the agreement is not hard to find. At the beginning of Part One of the Trilogy, Bohr (1913b, pp. 4–5) gave the binding energy in the ground state as Planck's constant times half the final frequency, $h\nu/2$. Hence $h\nu/2 = e^2/2a$, which is identical to Haas's condition Eq. (4.1).

1911, Nicholson, a survivor of the Cambridge mathematical tripos of 1904 and a lecturer at the Cavendish laboratory at the time,[27] presented an elaborate theory of atomic structure, aimed primarily at the fixing of atomic weights and the identification of spectral frequencies found in astrophysical contexts. The theory presented by Nicholson (1911b, 1911c, 1911d) later that year was purely classical but the following year he inferred from what turned out to be a mere numerological accident that the potential energy divided by the frequency of electrons circling in rings and the angular momentum of the ring had to be integral multiples of h and $h/2\pi$, respectively.

Nicholson's "structural theory of the chemical elements" (1911b) rested on three main assumptions:

1. Nicholson assumed that all mass is electromagnetic (1911b, p. 867). In Newtonian mechanics, mass is a primary property of all ponderable matter. As first pointed out by J. J. Thomson in 1881, the inertia of a moving charged sphere will appear to be greater than its Newtonian mass because the interaction of a moving charged sphere with its own electromagnetic field will result in additional inertia (Miller 1981, p. 43). In the early 1900s, the view became popular, especially among some German physicists, that the mass of a particle was entirely the result of the interaction of a particle's charge with its self-field (cf. Section 1.3.1). In this context, a relation between electromagnetic mass and electromagnetic energy was established, $m = \frac{4}{3}E/c^2$, which Einstein generalized to all energy while doing away with the factor $\frac{4}{3}$.[28] This assumption led Nicholson to a model that, while in the spirit of the Thomson model, was structurally more akin to planetary models of the Nagaoka (and more recently, Rutherford) variety. This is because the electromagnetic energy and hence the electromagnetic mass of a charge distribution depends on the volume it occupies and a mass on the order of that of an atom requires the positive charge(s) present in the atom to be confined to a volume much smaller than the size of the atom as a whole, say as given by the largest electron orbits. As Nicholson (1911b, p. 870) noted, "If the positive charges have a much smaller radius [than the negative charges], their inertia will greatly overweigh that of the electrons."

2. Nicholson took the existence of radioactivity, in particular of processes such as the emission of α particles (by now known to be ionized helium), to be strong evidence for the preexistence of a helium atom, as an independent entity, in the (for example) radium atom emitting it. Atoms of elements in the periodic table were thus to be regarded as compound entities, with several (moving) positive charges surrounded by a swarm of orbiting electrons. In modern terms, the atoms of the periodic table were really molecules!

---

[27] For biographical information on Nicholson, see Wilson (1956). From 1912 to 1921, Nicholson was professor of mathematics at King's College, London.

[28] For further discussion of this so-called electromagnetic worldview, see McCormmach (1970), Miller (1981), and Janssen and Mecklenburg (2007).

3. Nicholson assumed that the simplest possible systems, consisting of a positive charge orbited by a single ring of electrons, correspond to substances not present on Earth. He introduced four such "primary substances" or "protyles", which he called coronium, hydrogen, nebulium, and protofluorine with 2, 3, 4, and 5 electrons, respectively, in their electron ring (1911b, p. 867–869). These protyles are the true "atoms" in Nicholson's theory. The ordinary atoms of the periodic table correspond to "molecules" of bound protyles. The binding was presumed to be so great that the protyles could not be separated under terrestrial conditions, but could be in the much more violent environment of stellar or nebular atmospheres.

Nicholson's theory was an ingenious response to the problem emphasized by Schott, namely the extreme, and inadequate, simplicity of the line spectra arising classically from systems with a single electron ring. Stellar and nebular spectra were rife with strong lines that could not be assigned to the otherwise ubiquitous infinite Rydberg series. From the 1860s onward, several prominent lines had been found both in the spectra of nebulae and, during solar eclipses, in the spectrum of the solar corona that were nowhere to be found in the spectra of terrestrial elements, as studied in flame, gas discharge, or other laboratory sources of line spectra. The natural inference was to assign these lines to elements that did not exist on Earth. This is reflected in the names given to these elements in the astrophysical literature: nebulium and coronium (McCormmach 1966, p. 163, note 19).

Mendeleev had proposed the existence of additional light elements, e.g., newtonium, supposedly lighter than hydrogen. The case of helium, first found in the solar spectrum then later on Earth, gave added weight to such hypotheses. The work of Moseley (1913, 1914) finally established the completeness of the periodic table in its present form. However, the proper identification of the coronal and nebular lines, as arising from ionized iron and oxygen/nitrogen respectively, only came decades later.[29]

Nicholson associated single lines or pairs of lines with the spectra of his simpler protyles, each with only a single ring of electrons. Permanent radiative stability was no longer an issue in stars (or nebulae), as the protyles could be decaying and reforming constantly. However, to eliminate the dipole radiation coming from the circular motion, the strongest source of radiation instability, and to give the protyle atom a tolerably long lifetime (necessary given the sharpness of spectral lines), Nicholson assumed that the number of electrons in the ring had to be at least two. The baffling series spectra of normal atoms observed in terrestrial experiments could be put down to the much increased complexity of a system of several positive charges engaged in an intricate dance with orbiting electrons. In any case one could not, with the available theoretical machinery, calculate the expected spectrum for such complex systems.

In his first paper, Nicholson (1911b) restricted his attention to calculating the atomic weights of the protyles, and hence of the compound atoms occupying the periodic table, using his assumption that all mass is electromagnetic to determine the mass of each

---

[29] On the identification of the nebulium lines, see McCormmach (1966, p. 167, note 37).

primary atom as a function of its nuclear charge $+ne$, where in the neutral protyle there was a ring of n electrons to cancel the central charge. Here, Nicholson adopted Thomson's idea of a positive substance of fixed and universal charge density ρ, so a charge ne will take up a sphere of radius a such that $ne = \frac{4}{3}\pi a^3 \rho$. It follows that $a \propto n^{1/3}$. (This "classical electromagnetic radius of a nucleus" turns out to be on the order of 10^{-16} cm, far smaller than atomic sizes, or even than the sizes of nuclei as now known.) The electrostatic potential energy of such a sphere, identified with the inertial rest energy Mc^2, is proportional to $(ne)^2/a$, and hence to $n^{5/3}$ (p. 870). For the four lowest allowed cases then, $n = 2, 3, 4, 5$, the masses of the positive charge were in the ratio 3.1748, 6.2403, 10.079, 14.620 (p. 871). If oxygen is defined as having atomic weight 16.0, hydrogen works out as having atomic weight 1.008. Taking hydrogen to correspond to a protyle of $n = 3$ and rescaling the masses to the usual atomic weight scale, Nicholson found the atomic weights of his four protyles. An awkward feature, discussed only in passing by Nicholson (1911b, p. 868), was that one probably had to identify the normal hydrogen atom with (at least) two hydrogen protyles, as a single protyle would have too simple a spectrum—even for atomic hydrogen! One would then just double (for example) the occurrences of all other protyles while maintaining the right ratio of atomic weights. Nicholson then proceeded to concoct "molecules" of his basic protyles, finding atomic weights (by simply adding the masses of the constituent protyles) close to those of the observed elements of the periodic table.

Despite the numerological success of his theory in recovering the usual atomic weights, Nicholson was anxious to provide further confirmatory evidence, in the form of spectral line predictions. The spectrum could only be calculated (using the methods pioneered by Schott) for single-ring protyle atoms, hence the line frequencies obtained would be looked for in the alien environments of stars and nebulae where protyles were presumed to exist as independent entities. As mentioned above, some of the lines studied by Nicholson in astrophysical contexts are now known to be due to ionized oxygen and nitrogen, inaccessible at the time in terrestrial experiments. Since they could not be fitted into known Rydberg series, however, they were prime contenders for the line pairs or triples expected (à la Schott) from the perturbed oscillations of an electron ring with up to 5 electrons as in Nicholson's protyles. In the paper based on his presentation at the 1911 meeting of the British Association, which we have been considering so far, Nicholson (1911b, p. 865) simply asserted that the existence of his protyles had been confirmed by "a mathematical investigation of their spectra."

Around the time this paper was published (November 1911), another paper appeared in the *Monthly Notices of the Royal Astronomical Society* entitled "The Spectrum of Nebulium," in which Nicholson (1911c) applied the techniques developed earlier by J. J. Thomson and Schott to the calculation of the spectral frequencies to be expected from the perturbation of a rotating circular ring of four equally spaced electrons surrounding a positive nucleus of charge $+4e$, which, in Nicholson's scheme, was the neutral "nebulium" atom. For planetary models of Nagaoka type, it was already known that perturbations in the plane of the electron orbits would be unstable, but that the perturbations perpendicular to the orbit could be treated in the small-amplitude approximation and would yield a discrete set of perturbations of angular frequency

q (in the comoving frame of the unperturbed electrons, as in Schott's calculations). The instability in the plane of the orbit was simply ignored. The classical theory used throughout (Planck or his constant are nowhere mentioned) does not allow a determination of the radius of the initial orbit or the angular frequency ω of the electrons in the unperturbed orbit (cf. Section 4.3.5). As the derived frequencies were given as multiples of ω, Nicholson could only compute the ratios of spectral frequencies. The perturbation analysis gave only three strong lines, and the ratio of the first two agreed (to three significant figures) with the frequency ratio of two prominent lines seen in nebular spectra (now known to arise from ionized oxygen and nitrogen). From the actual frequencies, Nicholson could determine ω and then the radius of the circular ring as 3.453×10^{-8} cm, in the right ballpark for atomic sizes (in fact, quite a bit too large!). Nicholson went on to examine the spectra of ionized nebulium (either with less than 4 or even with surplus electrons, $n > 4$) and could correlate (with roughly the same level of accuracy) the frequencies obtained from perturbation calculations with observed nebular lines.

Nicholson then turned his attention to unexplained lines in the spectrum of the solar corona (now known to be from highly ionized iron). Following the same procedures as in his nebulium paper, Nicholson calculated frequency ratios for protofluorine ($n = 5$) and identified some of the coronal lines. He published these results in *Monthly Notices* about a month after the paper on nebulium (Nicholson 1911d). The remaining coronal lines were addressed in a second paper (Nicholson 1912), which appeared about six months later, in June 1912, around the same time that Bohr was preparing the Rutherford Memorandum. It is in this paper that Planck's constant made its first appearance in Nicholson's work, providing him with a way to relate the radius and frequency of the unperturbed orbits to a universal constant of Nature. Right at the beginning of the paper, in the second paragraph of the introduction, Nicholson introduced Planckian ideas in connection with spectra (in the present instance, of his protofluorine primary atom with five electrons):

> The constant of nature in terms of which these spectra can be expressed appears to be that of Planck in his recent quantum theory of energy.[30] It is evident that the model atoms with which we deal have many of the essential characteristics of Planck's "resonators" . . . It is not desirable, in this paper, to give any detailed account of the principles of Planck's theory, but we shall avail ourselves of the main result. This states that the energy possessed by a resonator is of the form $nh\nu$, where ν is the frequency of the resonator, n is an integral number, and h is an universal constant of nature [for which Nicholson quotes the value 6.548×10^{-27} erg·sec] (Nicholson 1912, p. 677).

Nicholson goes on to point out that, given the radius a and angular frequency ω of the electron ring for the protofluorine atom (fixed by matching to lines in the solar corona

---

[30] Nicholson cites two papers by Planck (1900a, 1901a) in *Annalen der Physik*. The first of these, submitted in November 1899, summarized Planck's derivation of the Wien law, which he would only replace by his new law for black-body radiation in October 1900 (see Sections 2.3–2.4).

in his previous paper), the magnitude of the potential energy of the ring of n (in this case 5) electrons, divided by the frequency $\omega/2\pi$, is

$$nma^2\omega^2 \cdot \frac{2\pi}{\omega} = 154.94 \times 10^{-27} \text{erg} \cdot \text{sec} = 23.66\,h, \tag{4.7}$$

where we used the virial theorem, according to which the total energy equals minus the kinetic energy, to set minus the potential energy equal to twice the kinetic energy $nmv^2 = nma^2\omega^2$. Nicholson suggested that, given the uncertainties in the contemporary values of the electron charge and mass, the 23.66 factor could just as well be 25. The reason for this choice came from the corresponding analysis for singly ionized ($n = 4$) and doubly ionized ($n = 3$) protofluorine, which gave energy/frequency multiples of 22 and 18 respectively: the "harmonic sequence" 25, 22, 18, 13, 7, 0 (for $n = 5, 4, 3, 2, 1, 0$ electrons) terminating as desired in zero energy for zero electrons.[31] Of more significance for Bohr's later development of atomic theory, Nicholson (1912, p. 679) noted that the quantization condition for energy divided by frequency

$$nma^2\omega^2 \cdot \frac{2\pi}{\omega} = 2\pi nma^2\omega = rh, \tag{4.8}$$

where r is an integer, is equivalent to a quantization condition for the total angular momentum of the n ring electrons

$$nma^2\omega = r\frac{h}{2\pi}. \tag{4.9}$$

He wrote:

> If therefore, the constant h of Planck has, as Sommerfeld[32] has suggested, an atomic significance, it may mean that the angular momentum of an atom can only rise or fall by discrete amounts when electrons leave or return. It is readily seen that this view presents less difficulty to the mind than the more usual interpretation, which is believed to involve an atomic [i.e. discrete] constitution of energy itself (Nicholson 1912, p. 679).[33]

Nicholson (1912, p. 681) realized that his new quantization condition did not provide a solution to the problem of explaining series formulas of Balmer–Rydberg–Kayser–Runge

---

[31] Usually, the term "harmonic sequence" is applied to a sequence of numbers whose reciprocals form an arithmetic sequence where successive numbers differ by a fixed amount. Here, the quantization numbers for the whole ring imply energy/frequency factors *per electron* of 5, 5.5, 6, 6.5, and 7, which is an arithmetic sequence with the angular momentum per electron equal to $\frac{1}{2}(15 - n)(h/2\pi)\omega$, for $n \neq 0$ electrons in the ring.

[32] There is no citation but Nicholson is probably referring to Sommerfeld (1911a). For discussion of this paper, see Section 1.3.1 and Eckert (2015b).

[33] In his obituary of Nicholson, William Wilson writes: "The discovery, by one of his colleagues, of the quantum conditions, $\int p_i\, dq_i = n_i h$, was partly due to Nicholson's angular momentum discovery and especially to his inspiring influence (*Phil. Mag.* June 1915)" (W. Wilson 1956, p. 211). The colleague (at King's College, London) was Wilson himself and the paper cited is W. Wilson (1915, see Section 5.1.2).

type. He reminded his readers that his theory was not to be applied to elements found on Earth (in the periodic table), which were complicated composites of his primary "protyles". But he emphasized that with the introduction of Planck's constant, in the form of a quantization condition, the absolute radii and orbital frequencies of his primary atoms could be fixed, without the need for empirical input in the form of frequencies of solar and nebular spectral lines.

4.3.5   Planck's constant enters Bohr's atomic modeling

We now return to the Memorandum that Bohr (1912) prepared for Rutherford in June/July 1912 (see Section 4.3.3). This memorandum provides a record of Bohr's first efforts to construct explicit models of atoms and molecules, imposing, inspired by Planck, a linear relation between energy and frequency. Bohr began the Memorandum by considering n equidistantly spaced electrons circling in a ring around the "kern" (nucleus) of the atom with a positive charge of ne. This two-dimensional arrangement is mathematically more tractable than three-dimensional ones. An error in his calculation of the total energy of the n electrons circling in such a ring made this arrangement look particularly promising to Bohr (1912, p. 136). He double-counted the positive electrostatic energy due to inter-electron repulsions and was thus led to believe that the total energy of the ring becomes positive for $n > 7$. This would mean that no more than seven electrons can be bound in an atom consisting of a single ring. Bohr saw this as an indication of a possible explanation of the periodicity of the table of elements, even though the period he thought he had found was seven rather than eight. Bohr's error, as pointed out by Heilbron and Kuhn (1969, p. 246, note 88), is particularly baffling given that, at the end of the Memorandum (Bohr 1912, pp. 142–143), he explicitly proved a special case of the virial theorem for the kind of arrangement shown in Fig. 4.1 below in which the repulsion between nuclei is stabilized by the attraction between nuclei and electrons circling in one or more rings around the line connecting these nuclei. Bohr correctly stated that for any such system the potential energy is equal to minus twice the kinetic energy. The energy of a single ring around one nucleus will thus always be negative and there is no limit to how many electrons such a ring can contain.

After getting off to an unfortunate start, Bohr switched to more fertile ground. He pointed out that purely mechanical equations for such ring systems cannot result in fixing the size of the rings and the periods of the orbits, which although connected through the equations of motion, can vary continuously.

The absolute identity of different atoms of the same substance had already puzzled Maxwell. In his article "Atom" in the 9th edition of the *Encyclopedia Britannica*, he wrote:

> Whether or not the conception of a multitude of beings existing from all eternity is in itself self-contradictory, the conception becomes palpably absurd when we attribute a relation of quantitative equality to all these beings. We are then forced to look beyond them to some common cause or common origin to explain why this singular relation of equality exists, rather than any one of the infinite number of possible relations of inequality (Maxwell 1952, Vol. 2, p. 482; quoted, e.g., in Stehle 1994, p. 224).

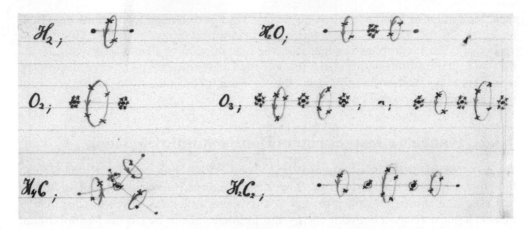

Figure 4.1 *Arrangement of electron rings in models for various molecules in the Rutherford Memorandum (Bohr 1912, p. 138 and p. 150). Image courtesy of the Niels Bohr Archive, Copenhagen.*

Bohr proposed to find (in Maxwell's words) a "common cause" by adopting the hypothesis that "for any stable ring [there] will be a definite ratio between the kinetic energy of an electron in the ring and the time of rotation." He stated that "no attempt of a mechanical foundation" of the hypothesis will be made, adding parenthetically "as it seems hopeless" (Bohr 1912, p. 137). Such a hypothesis, however, seemed to provide the only explanation of "a whole group of experimental results" supporting "conceptions of the mechanismus [sic] of the radiation as the ones proposed by Planck and Einstein" (ibid.). Bohr does not go into further detail but, as we have seen in Section 4.2, his dissertation shows that he was well acquainted with the work of Planck and Einstein. In a footnote added to the parenthetical clause ("as it seems hopeless"), Bohr briefly summarized another line of reasoning already found in his dissertation. Even though "it seems to be rigorously proved," he wrote, that classical mechanics fails for individual atoms, it can still be used *except* when dealing with "questions of stability (or of final statistical equilibriums [sic])" (ibid.).

Bohr then gave a list of four items that he claimed can be explained with his hypothesis:

1) the periodic law of the atomic volumes of the elements
2) Whiddington's law of the [proportionality] of the velocity of electrons needed to excite the characteristic [X-rays] of an element [to] the atomic weight [A] of that element.
3) (approximately) Bragg's law of the absorption of α-rays by different elements
4) ... the stability and heat of combination of some of the simple compounds (Bohr 1912, p. 137).

After drawing the "figurations" (arrangements) for the electron rings in a variety of molecules (H_2, H_2O, O_2, O_3, H_4C, H_2C_2; see Fig. 4.1), Bohr (1912, p. 140) applied his hypothesis to the hydrogen atom, the hydrogen molecule, the helium atom, and the (putative) helium molecule ("He$_2$"). Bohr modeled the hydrogen atom simply as a ring with a single electron rotating around the nucleus with charge $+e$. He modeled the hydrogen molecule as containing two electrons rotating on a ring bisecting a line connecting two such nuclei. Similarly, he represented the helium atom as a ring with two electrons rotating around a nucleus with charge $+2e$ and the helium "molecule" as a ring of four electrons bisecting and rotating around a line connecting two such nuclei. In all cases, Bohr was clearly considering the "normal" (i.e., stable, lowest energy, or ground) state of the system. Figure 4.2 gives our own representations of Bohr's models for the helium atom and the hydrogen molecule. Figure 4.3 shows the page of the Rutherford Memorandum with Bohr's own drawings and calculations for the hydrogen and helium atoms and molecules.

We examine the case of the hydrogen molecule in some detail (see Fig. 4.2b; see also Heilbron and Kuhn 1969, pp. 248–250). The geometry indicated in the figure is fixed by the condition of nuclear stability. The electrostatic repulsion of the nuclei is given by $e^2/4r^2$, where r is the distance from each nucleus to the center of the electron ring of radius a.[34] This should be balanced by the attraction of each nucleus to the pair of electrons. The component of the force of one of these electrons in the direction of the line connecting the two nuclei is equal to $e^2/(r^2 + a^2)$ times $r/\sqrt{r^2 + a^2}$. Nuclear stability thus requires that

$$\frac{e^2}{4r^2} = \frac{2e^2 r}{(a^2 + r^2)^{3/2}}. \tag{4.10}$$

Solving for r, we find that $r = a/\sqrt{3}$, as indicated in Fig. 4.2b.[35]

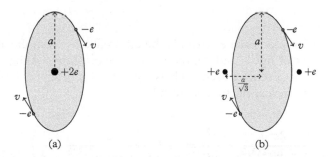

(a)   (b)

Figure 4.2 *Bohr's models of the helium atom (a) and the hydrogen molecule (b) in the Rutherford Memorandum (cf. Figs 4.1 and 4.3).*

---

[34] In Bohr's notation, a and r are reversed.
[35] From Eq. (4.10) it follows that $(a^2+r^2)^{3/2} = 8r^3$. Hence, $a^2 + r^2 = 8^{2/3}r^2 = 4r^2$ or $r^2 = a^2/3$.

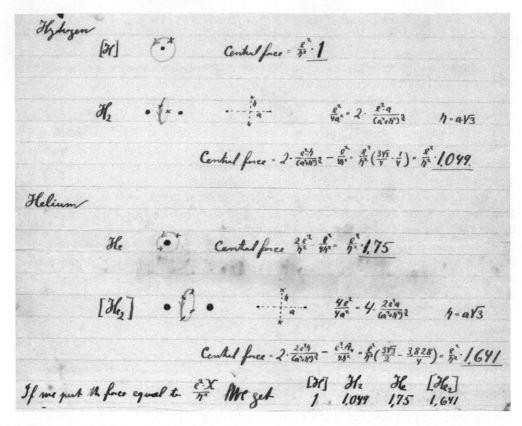

Figure 4.3 *Calculations for H, H_2, He, and He_2 in the Rutherford Memorandum (Bohr 1912, p. 140 and p. 154). Image courtesy of the Niels Bohr Archive, Copenhagen.*

The centripetal force on each electron is the difference between the attractive force coming from the pair of nuclei and the repulsive force $e^2/(2a)^2$ coming from the other electron. The former is given, for each nucleus, by $e^2/(r^2 + a^2)$ times $a/\sqrt{r^2 + a^2}$. So the net centripetal force on each electron is given by

$$\frac{2e^2 a}{(a^2 + r^2)^{3/2}} - \frac{e^2}{4a^2} = \frac{e^2}{a^2}\left(\frac{2}{(1 + (r/a)^2)^{3/2}} - \frac{1}{4}\right) \tag{4.11}$$

Inserting $(r/a)^2 = \frac{1}{3}$, we find that this reduces to

$$\frac{e^2}{a^2}\frac{3\sqrt{3} - 1}{4} = X\frac{e^2}{a^2}, \tag{4.12}$$

where $X \approx 1.049$ (see Fig. 4.3). Similar equations obviously hold for the hydrogen atom, with $X = 1$, the helium atom, with $X = 1.75$, and the helium "molecule", with $X \approx 1.641$ (see Fig. 4.3).[36]

Inserting Eq. (4.12) for any one of the models considered by Bohr (thus leaving X unspecified) into Newton's second law, we find $m\omega^2 a = X(e^2/a^2)$, where ω is the angular velocity of any one of the electrons in the model (so $\nu = \omega/2\pi$ is the cyclic frequency). We can rewrite this as

$$m\omega^2 a^3 = Xe^2. \tag{4.13}$$

Of course, we are still in the middle of Maxwell's quandary: although Eq. (4.13) shows that the value of one determines the value of the other, ω and a remain continuously variable quantities. At this point, however, Bohr (1912, p. 141) invoked the "*special Hypothesis $\underline{E = K\nu}$*", where E is the *kinetic* energy of *each* electron. Although Bohr did not set his mysterious proportionality constant K equal to h, he clearly realized, as we already saw, that K had to have something to do with Planck's constant.

Inserting $\frac{1}{2}m\omega^2 a^2$ for E_{kin} and $\omega/2\pi$ for ν, we can rewrite the relation

$$E_{\text{kin}} = K\nu \tag{4.14}$$

as $m\omega a^2 = K/\pi$. Squaring this relation, we find:

$$m\omega^2 a^4 = \frac{K^2}{m\pi^2} \tag{4.15}$$

Combining this last equation with Eq. (4.13), we arrive at

$$a = \frac{K^2}{\pi^2 Xme^2}. \tag{4.16}$$

For a given value of K the basic dimensions of the atom or molecule are now fixed, as are the angular and cyclic frequencies of the electrons:

$$\omega = \frac{K}{\pi ma^2} = \frac{\pi^3 me^4}{K^3}X^2, \quad \nu = \frac{\omega}{2\pi} = \frac{\pi^2 me^4}{2K^3}X^2, \tag{4.17}$$

where we used Eqs. (4.15) and (4.16) for ω and a, respectively.

---

[36] The expression for X in Eq. (4.12) for the hydrogen molecule, relabeled F, returns in Part Three of the Trilogy (Bohr 1913d, p. 863). In the case of the helium atom, the difference between attractive and repulsive forces on each electron is $(2e^2/a^2) - e^2/(4a^2) = (e^2/a^2)(2 - 1/4)$. Hence, for a helium atom, $X = 1.75$. The value for X for the helium "molecule" can be found in the same way as the value for the hydrogen molecule, except that the geometry gets a little more complicated because there are now four electrons in the ring between the two nuclei. In Part Two of the Trilogy, Bohr (1913c, p. 479) gives a formula for the factor F (his new notation for X) in the net centripetal force $F(e^2/a^2)$ on each electron in a ring of n electrons circling around a single nucleus. We derive this formula in Section 4.5.2 (see Eqs. (4.54)–(4.56)).

By the virial theorem Bohr proved for his models, $E_{\text{tot}} = -E_{\text{kin}}$, the total energy of n electrons is given by

$$E_{\text{tot}} = -nK\nu = -nX^2 \frac{\pi^2 m e^4}{2K^2}, \tag{4.18}$$

where we used Eq. (4.17) for ν. Bohr (1912, p. 141) wrote this relation as

$$E_{\text{tot}} = -nX^2 A, \tag{4.19}$$

omitting the simple algebra with which we established that $A = \pi^2 m e^4 / 2K^2$ (see Eqs. (4.13)–(4.18)).

The value of A is fixed once a value for K has been chosen. Bohr (1912, p. 141) reported in his Memorandum that $A \approx 1.3 \times 10^{-11}$ ergs. Given the values for e and m available to Bohr (off by a few percent), this value for A corresponds to a value for K of roughly $0.6h$, where h is Planck's constant.[37] As first suggested by Rosenfeld (1963, p. xxxi), Bohr appears to have fixed the value of K, and hence A, empirically, by substituting the (angular) dispersion frequency $\omega = 2.21 \times 10^{16}\text{s}^{-1}$ of molecular hydrogen into Eq. (4.17) and solving for K. As we saw in Section 4.3.3, he used this same dispersion frequency in his treatment of α-particle absorption. It can also be found on another page of the Memorandum—now in terms of the cyclic frequency "n [$= \omega/2\pi$] $= 0.3 \times 10^{+16}\text{s}^{-1}$"—in connection with an estimate of the temperature at which ultraviolet radiation would break up a hydrogen molecule (Bohr 1912, p. 140).[38] No indication is given in the Memorandum of how he arrived at the estimate for A (Rosenfeld surmises that this was discussed in a missing page of the Memorandum), so this remains speculative, but one can well imagine that Bohr realized that the identification of orbital frequencies with dispersion frequencies, valid for the forced harmonic oscillators of the Lorentz–Drude theory (where the frequency of the charged particle oscillation is independent of amplitude), could not possibly be valid for electrons in bound Keplerian orbits, and was therefore unwilling to be more explicit about his reasoning on this point.

With the value of K (and A) fixed, Bohr (1912, p. 141) could calculate the heat of formation of a hydrogen molecule from two hydrogen atoms. Since $X = 1$ for the two hydrogen atoms and $X \approx 1.049$ for a hydrogen molecule (cf. Eq. (4.12)), the energy released when two hydrogen atoms combine to form one hydrogen molecule is given by (see Eq. (4.19) with $n = 2$):

$$2(1.049^2 - 1.000^2)A. \tag{4.20}$$

---

[37] Note the similarity to the value posited for the binding energy of the ground state of the hydrogen atom in the first paper of the Trilogy, which corresponds to $\frac{1}{2}h\nu$, or $K = 0.5h$. This is also the appropriate proportionality factor for the harmonic oscillator, in its *first excited state* (if we ignore zero point energy), which has kinetic energy $\frac{1}{2}h\nu$.

[38] In the third paper of the Trilogy, Bohr (1913d, p. 864) once again used this frequency, now written as $\nu = 3.5 \times 10^{15}\text{s}^{-1}$ (Rosenfeld 1963, p. xxxi).

Bohr gave the value 1.9×10^5 cal. for NA, where N is the number of molecules in 1 cc. of the gas. The heat of formation for 1 cc. of the gas is thus 3.8×10^4 cal., a value similar to other known heats of formation (e.g., of water from hydrogen and oxygen). The corresponding calculation for helium (two helium atoms combining to form one helium "molecule") gives $4(1.641^2 - 1.75^2)A$ ($X = 1.75$ for a helium atom and $X \approx 1.641$ for a helium "molecule"). This heat of formation is negative, indicating that helium atoms do not combine to form molecules, as was well known.

These explicit calculations are the only ones in the Memorandum addressing one of the four topics (the fourth in the list given above) for which Bohr claimed to have found an explanation on the basis of his hypothesis that $E = K\nu$. The first two items, the periodic dependence of atomic size and the Whiddington (1911) law are mentioned explicitly in two letters written the following year, one to Rutherford of January 31, 1913 and one to the Hungarian chemist George de Hevesy of February 7, 1913 (Bohr 1972–2008, Vol. 2, pp. 579–580 and pp. 529–531, respectively).[39] In the former, Bohr emphasizes that his theory applies only to the ground state of atoms, and that it is supported by these two items:

> It seems therefore to me to be a reasonable hypothesis, to assume that the state of the systems considered in my calculations is to be identified with that of the atoms in their permanent (natural) state. (This hypothesis seems to be justified by the agreement between the theory and experiments on atom-volumes and Röntgen-rays, which I obtained *from the first moment* and which I have tried to trace still longer) (Vol. 2, p. 579, our emphasis).

Here, "from the first moment" presumably refers to the Memorandum of July 1912. The periodic variation of atomic size and the Whiddington formula are both addressed explicitly in Part Two of the Trilogy of 1913 (see Section 4.5.2). As Heilbron and Kuhn (1969, p. 283) note and as we shall see in Section 4.5.2, Bohr could easily have undertaken these investigations in late 1912, after his return to Copenhagen, but before he had taken the decisive steps leading to the deciperment of the Balmer formula in March 1913.

4.4 From the Rutherford Memorandum to the Trilogy

4.4.1 Bohr comparing his results to Nicholson's

Bohr had met Nicholson while at Cambridge in the Fall of 1911 and had been distinctly underwhelmed. Nicholson's (1911a) approach to the electron theory of metals, he thought, was "entirely preposterous."[40] He appears to have encountered Nicholson's

---

[39] These letters are also reproduced in Bohr (1963, pp. xxxvi–xxxvii and pp. xxxii–xxxiv (in facsimile), respectively).

[40] Bohr to C. W. Oseen, December 1, 1911 (Bohr 1972–2008, Vol. 1, p. 427).

papers on solar/nebular spectra for the first time in December 1912.[41] Despite his earlier poor impression of Nicholson, he clearly took them very seriously.[42]

At first Bohr was troubled by the very different results obtained by two theories apparently proceeding from identical premises: (*i*) the application of classical mechanics to rings of electrons electrostatically bound to a central positive charge, and (*ii*) the introduction—by Nicholson (1912, see Section 4.3.4) in a paper published in June and by Bohr (1912, see Section 4.3.5) in his Rutherford Memorandum of June/July— of a quantization condition that imposes a proportionality between ring energy and frequency, with a constant of proportionality determined by empirical input, which, in both cases, is on the order of Planck's constant.[43]

To see where the discrepancies are coming from, we compare Nicholson's (1911c) treatment of nebulium (his four-electron protyle) with the results we obtain when we use Bohr's methods in the Rutherford Memorandum for the corresponding system (a ring with four electrons circling around a single nucleus with charge $+4e$). With the help of Eq. (4.18), we calculate the binding energy per electron for this system. Following Bohr, we set $K = 0.6h$ (cf. our discussion at the end of Section 4.3.5). We also need a value for X. Both in the Rutherford Memorandum and the Trilogy, Bohr wrote the net centripetal force on each electron in a ring of n electrons circling around a single nucleus as $X(e^2/a^2)$ (where a is the radius of the ring). In Part Two of the Trilogy, Bohr (1913c, p. 479) gives a formula for X (renamed F) for arbitrary n, $F = N - s_n$, where N is the number of unit charges $+e$ on the nucleus and n is the number of electrons (cf. note 36 above and Eq. (4.56) below). Bohr (1913c, p. 482) provides a table of values of s_n for $n = 1, \ldots, 16$. From this table, we read off that $s_n = .957$ for $n = 4$. Hence, for the system under consideration, $X = F = 4 - .957 = 3.043$.[44] Dividing Eq. (4.18) for the system's total energy by $-n$ and using the values $K = 0.6h$ and $X = 3.043$, we find that the binding energy per electron is given by:

$$X^2 \frac{\pi^2 m e^4}{2K^2} = 1.4 \times 10^{-10} \text{ergs}. \tag{4.21}$$

Nicholson found a much lower value for the binding energy per electron in nebulium in his paper of November 1911. The value of this binding energy is not given in the paper but the values for the radius a of the electron ring and the angular frequency ω of

[41] Heilbron and Kuhn (1969, p. 260) surmise that Bohr might have seen (but dismissed) Nicholson's (1911b) paper on chemical elements earlier, as this paper was published in *Philosophical Magazine*, which Bohr probably read regularly. The spectral papers (Nicholson 1911c, 1911d) were published in *Monthly Notices of the Royal Astronomical Society*.

[42] The development of Bohr's thought discussed in this section is also covered in sec. IV (pp. 255–266) of Heilbron and Kuhn (1969).

[43] There is no evidence that Bohr was aware of the 1910 work of Haas prior to his development of the quantized hydrogen model of March 1913, even though it is mentioned in Part One of the Trilogy.

[44] In his paper on nebulium, Nicholson (1911c, p. 51) derived an equivalent formula.

electrons in this ring are (Nicholson 1911c, p. 56). Inserting the value for the velocity of an electron, given as $a\omega = 4.33 \times 10^{-3}c$ (where c is the velocity of light), into $\frac{1}{2}m(a\omega)^2$, which gives both the kinetic energy of the electron and, by the virial theorem, its binding energy, we find that the binding energy per electron in nebulium is only 7.7×10^{-12} ergs (where we used the modern value for the electron mass, 9.1×10^{-28} g). The radius of Nicholson's four-electron ring, given as 3.453×10^{-8} cm, is accordingly much *larger* than the radius of Bohr's four-electron ring, 2.6×10^{-9} cm.

Recall that Nicholson and Bohr determined the values of a and ω in completely different ways. As we saw in Section 4.3.5, classical theory does not fix the values of a and ω but only tells us that the value of one fixes the value of the other (see Eq. (4.13)). Bohr fixed the values of a and ω by imposing the condition $E_{\text{kin}} = K\nu$ in Eq. (4.14) with $K \approx .6h$ (a value, as we saw in Section 4.3.5, probably based on a dispersion frequency in molecular hydrogen). Nicholson, by contrast, fixed the values of a and ω for nebulium by matching radiation frequencies associated with transverse oscillations of the ring electrons (according to the purely classical analysis of Schott (1907)) to lines found in astrophysical spectra. It was not until the following June that he noted that the angular momentum of electrons in his protyles was always close to an integer multiple of Planck's constant (Nicholson 1912, see Section 4.3.4). Using the numerical values given above, we find that the angular momentum $m\omega a^2$ per electron in nebulium is about 3.9 times $h/2\pi$. Allowing for uncertainties in the measured values of the various quantities involved, nebulium should thus have 16 units of angular momentum $h/2\pi$. As we saw in Section 4.3.4, Nicholson (1912, p. 679) determined that protofluorine (with five electrons) has 25 such units.

At the time (i.e., in late 1912), Nicholson's success in matching numerous lines in astrophysical spectra to radiation frequencies of his protyles impressed Bohr (and others). Bohr does not seem to have entertained the possibility that Nicholson's results were simply wrong (or at least, an incorrect interpretation of the empirical situation). Instead, in a postscript to a Christmas card Bohr and his wife jointly sent to his brother Harald on December 23, 1912, Bohr wrote:

> Although it does not belong on a Christmas card, one of us would like to say that he thinks Nicholson's theory is not incompatible with his own. For the latter's [i.e., Bohr's] calculations should be valid for the final or classical state of the atoms, while Nicholson seems to be concerned with the atoms while they radiate, i.e., while the electrons are about to lose their energy, before they have occupied their final positions (Bohr 1972–2008, Vol. 2, p. 108).

In other words, Bohr assumed that Nicholson was dealing with atoms in a highly excited state, with the ring electrons much further out from the nucleus and with much smaller binding energy than they would have upon settling down in a presumed "final" lowest energy state, the normal, non-radiating permanent condition of a stable atom. That Nicholson was explicitly only considering light emitted in stars and nebulae, presumably from atoms in a very energetic environment, and therefore quite reasonably in an excited state, lent plausibility to this interpretation. Bohr reiterated the point more clearly in the letter to Rutherford of January 31, 1913 from which we already quoted in Section 4.3.5:

> The theory of Nicholson gives apparently results which are in striking disagreement with those I have obtained; and I therefore thought at first that the one or the other necessarily was altogether wrong. I have however now taken following point of view.
>
> In his calculations, Nicholson deals, as I, with systems of the same constitution as your atom-model; and in determining the dimensions and the energy of the systems, he, as I, seeks a basis in the relation between the energy and the frequency suggested by Planck's theory of radiation. The state of the systems considered in my calculations are [sic] however—between [i.e., among] states in conformity with the relation in question [i.e., the quantization condition $E = K\nu$ in Eq. (4.14) above]—characterized as the one in which the systems possess the smallest possible amount of energy, i.e. the one by the formation of which the greatest possible amount of energy is radiated away (Bohr 1972–2008, Vol. 2, p. 579).

The problem with this gloss on the discrepancy between the two theories is that, at the time of this letter, Bohr's theory had nothing to say about excited states. Bohr had only introduced a single quantization condition in his Rutherford Memorandum—the one stating that the kinetic energy per electron should be $K\nu$, with the same constant K for all atoms in their permanent "normal" state. So the plural in "between states in conformity with the relation" was purely speculative at this point. Bohr had yet to come up with a more general quantization condition that would apply to an atom's normal state as well as to its excited states. This deficiency would be remedied (at least for hydrogen) in Part One of the Trilogy.

One might think that Nicholson's theory would have impressed upon Bohr the need for dealing with excited as well as with normal states of atoms and that it would have inspired him to try to use the models he introduced in his Rutherford Memorandum to account for spectroscopic evidence. In fact, he explicitly told Rutherford in his letter of January 31, 1913, that—as McCormmach (1966, p. 177) put it—"he was not going to stray from his original purpose:"

> I must however remark that the considerations sketched here play no essential part in the investigation in my paper. I do not at all deal with the question of calculation of the frequencies corresponding to the lines in the visible spectrum. I have only tried, on the basis of the simple hypothesis, which I used from the beginning, to discuss the constitution of the atoms and molecules in their permanent state (Bohr 1972–2008, Vol. 2, p. 580).

Before long, however, Bohr would turn his attention to both excited states and spectroscopy.

Interest in Nicholson's theory slowly faded over the next few years (McCormmach 1966, pp. 179–184; Kragh 2012, pp. 111–116). Nicholson (1914a, 1914b, 1914c) published a number of papers criticizing his competitor's theory. Bohr was not overly worried by these criticisms. In a letter of May 20, 1914, he told Rutherford:

> I am very sorry that I have not yet found time to answer Nicholson's criticism . . . I do not however consider his objections as serious, and intend as soon as possible . . . to write a paper in which I shall try as clearly as possible to state my standpoint (Bohr 1972–2008, Vol. 2, p. 580).

Bohr had already drafted a short letter to *Nature* in response to Nicholson (Bohr 1913e), which he did not send, and would draft another response (Bohr 1914–1915), which he did not publish, before finally completing and publishing the paper he announced in this letter to Rutherford (Bohr 1915).[45] By this time it was becoming clear that Bohr's theory had superseded Nicholson's. In a report on radiation and quantum theory for London's Physical Society in 1914, Jeans concluded his discussion of Nicholson's theory with the consoling thought that it had at least paved "the way for the ultimate explanation of the phenomenon of the line spectrum" (Jeans 1914b, p. 50).

4.4.2 Enter the Balmer formula

Upon his return to Copenhagen in September 1912, Bohr was appointed assistant to Martin Knudsen. Knudsen replaced Bohr's PhD adviser, Christiansen, who had just retired. Fourteen years Bohr's senior, Knudsen had been a lecturer at the University of Copenhagen since 1901. Unlike Bohr, he had also attended the November 1911 Solvay conference, to which he had been invited even though he had not shown any interest in quantum theory (Kragh 2012, pp. 54–55). Knudsen and Bohr were the only two applicants for Christiansen's position. The University of Copenhagen clearly recognized Bohr's talent as a theoretician but was looking for an experimentalist. In a recommendation to the faculty we read that "[i]f a teaching position in mathematical physics were at issue—a position which our university unfortunately lacks—then there could hardly have been any doubt that Dr. Bohr would have been the right choice" (quoted in Aaserud 2013, pp. 88–89).

Bohr's new duties left him little time to turn the Rutherford Memorandum into a paper. As he told Rutherford in a letter of November 4, 1912:

> I am sorry not yet having been able to finish my paper on the atoms and send it to you, but till now I have had so much to do with some lecturing and laboratory work (assisting Prof. Knudsen) that I have had only very little time left (Bohr 1972–2008, Vol. 2, p. 577).

---

[45] One of the problems facing Nicholson's theory, which to our knowledge neither Nicholson nor his critics (including Bohr) ever addressed, is that it is at odds with Rutherford's (1911, p. 687) conclusion on the basis of Geiger and Marsden's (1909) experiments on the scattering of α particles that most of the mass of an atom is concentrated in a single nucleus (cf. Section 4.3.2). In Nicholson theory, after all, all atoms in the periodic table are compounds of his protyles and thus contain several nuclei.

What further slowed down Bohr's writing of the paper was that he kept finding new material to incorporate in it (Kragh 2012, p. 55).

In February 1913, Bohr came across a formula that caused him to change direction rather drastically in Part One of what would become the 1913 Trilogy. Sometime during that month—as Bohr recalled in a conversation with Rosenfeld in 1954 (Rosenfeld 1963, p. xl)—Hans Marius Hansen, a young Danish spectroscopist, asked Bohr how his approach to atomic structure could account for atomic spectra. Hansen had recently returned to Copenhagen from a two-year stint in Göttingen, where he had done work with Woldemar Voigt on the inverse Zeeman effect (the splitting of absorption lines by a magnetic field). This work would form the basis for his doctoral dissertation. Up to this point, Bohr had thought atomic spectra too complicated to be a useful signpost for further advances. Hansen pointed out to him that in some cases, the spectra seem to follow strikingly simple algebraic laws such as the Balmer formula. He directed Bohr to one of the standard references for series spectroscopy at the time, the second volume of *Prinzipien der Atomdynamik* (Stark 1911).

The second chapter of Stark's book, "The series spectrum," contains a clear exposition of the Balmer formula (sec. 9, pp. 43–45). The final formula in this section expresses the spectral frequency (actually, the inverse wavelength, differing from the frequency by a factor of the speed of light) as a difference of two terms:

$$\nu = \nu_0 \frac{m^2 - 4}{m^2} = \nu_0 - \frac{4\nu_0}{m^2} = \frac{N_0}{4} - \frac{N_0}{m^2}, \tag{4.22}$$

where $\nu_0 = 27\ 418.7\ \text{cm}^{-1}$, $N_0 = 4\nu_0$, and $m = 3, 4, 5, \ldots$. The constant N_0 is now known as the Rydberg constant and typically written as R. Eq. (4.22) gives the frequencies of the lines in the Balmer series in the visual range of the hydrogen spectrum.

According to Hermann (1971, p. 157), a slip of paper was found in Bohr's personal copy of Stark's book with a reference to a paper by H. A. Wilson, in which the more general, and more suggestive, formula for the spectral series of hydrogen appears (Wilson 1912, p. 662):

$$\nu = N \left(\frac{1}{n^2} - \frac{1}{m^2} \right), \tag{4.23}$$

where N is the Rydberg constant and n and m are integers such that $m > n$. This formula not only describes the Balmer series (for $n = 2$ and $m = 3, 4, 5, \ldots$) but also, for instance, the Lyman series in the ultraviolet (for $n = 1$ and $m = 2, 3, 4, \ldots$) and the Paschen series in the infrared (for $n = 3$ and $m = 4, 5, 6, \ldots$).

In his later years, Bohr told Rosenfeld on several occasions that "as soon as I saw Balmer's formula, the whole thing was immediately clear to me" (Rosenfeld 1963, pp. xxxxix). We should probably take this with a grain of salt, but at some point between first seeing the Balmer formula in February 1913 and sending a first version of Part One of the Trilogy to Rutherford on March 6, 1913, Bohr somehow made the necessary leap of identifying the individual terms of the right-hand sides of Eqs. (4.22) and (4.23) with quantized energy levels of the atom.

Setting $X = n = 1$ in Eq. (4.18) for the total energy of n electrons in the models that Bohr considered in the Rutherford Memorandum, we find that the total energy of the one electron in hydrogen is given by

$$E_{\text{tot}} = -\frac{\pi^2 m e^4}{2K^2},$$
(4.24)

where K is the proportionality constant in Bohr's quantization condition $E_{\text{kin}} = K\nu$ (see Eq. (4.14)). The kinetic energy of a Planck oscillator, which is half its total energy, can only take on the discrete values $E_{\text{kin}} = \frac{1}{2}\tau h\nu$, where τ is an integer. Treating Bohr's model for a hydrogen atom as somehow analogous to a Planck oscillator, as Bohr (1913c, p. 4) did in sec. 1 of Part One of the Trilogy, we are thus led to set

$$K = \frac{1}{2}\tau h.$$
(4.25)

Inserting this value into Eq. (4.24), we find that

$$E_{\text{tot}} = -\frac{2\pi^2 m e^4}{\tau^2 h^2}.$$
(4.26)

For the lowest-energy state, i.e., for $\tau = 1$, $K = .5h$. This is close to $K = .6h$, the value Bohr chose for K in the Rutherford Memorandum.

We arrive at Eq. (4.23), the Balmer formula in the form given by Wilson, if we assume that the energy the electron in a hydrogen atom loses when it goes from an initial energy state labeled by $\tau = \tau_1$ to a final energy state labeled by $\tau = \tau_2 < \tau_1$ is emitted as radiation with a frequency equal to this energy loss divided by Planck's constant. On this assumption and with the help of Eq. (4.26), we find that

$$\nu = \frac{E_{\text{tot}}^{(\tau_1)} - E_{\text{tot}}^{(\tau_2)}}{h} = \frac{2\pi^2 m e^4}{h^3}\left(\frac{1}{\tau_2^2} - \frac{1}{\tau_1^2}\right).$$
(4.27)

We recover Eq. (4.22), the Balmer formula as given in Stark's book, if we set $\tau_2 = 2$ and $\tau_1 = 3, 4, 5, \ldots$ and the Rydberg constant to:

$$R = \frac{2\pi^2 m e^4}{h^3}.$$
(4.28)

This is indeed the expression given in Part One of the Trilogy (p. 9). Inserting the values $e = 4.7 \times 10^{-10}$ esu, $e/m = 5.31 \times 10^{17}$ esu/g, and $h = 6.5 \times 10^{-27}$ erg ·sec., Bohr found the value

$$R = 3.1 \times 10^{15} \text{sec}^{-1}, \tag{4.29}$$

which, given the uncertainties in the values for the various constants in Eq. (4.28), is in agreement with the value for R determined on the basis of spectroscopic measurements,

$$R = 3.290 \times 10^{15} \text{sec}^{-1} \tag{4.30}$$

(Bohr 1913c, p. 9).

Bohr's route to this remarkable result was considerably more circuitous than the simple sequence of steps in Eqs. (4.24)–(4.28) suggests. That much is clear from the published version of Part One of the Trilogy (no letters or manuscripts survive that would allow us to reconstruct his reasoning in February–March 1913 in more detail). This is perfectly understandable. Equation (4.27) amounts to severing the relation between radiation frequencies and orbital frequencies. As Heilbron and Kuhn (1969, p. 263) point out, "[t]hat a spectral line of a given frequency must be produced by a charge vibrating at the same frequency was a consequence of electromagnetic theory which even Planck and Einstein had not thought to challenge." Bohr did not set out to challenge this standard result either. He found himself forced to do so in order to recover the Balmer formula.

4.5   The Trilogy: quantum atomic models and spectra

On March 6, 1913, Bohr sent Rutherford, as he wrote in a cover letter, "the first chapter of my paper on the constitution of atoms. I hope that the next chapters shall follow in a few weeks."[46] In the end, it took a few more *months* for Bohr to finish his Trilogy. After several rounds of revisions, Part One, consisting of 25 printed pages and dated April 5, 1913, appeared in the July issue of *Philosophical Magazine* (Bohr 1913b).

On March 20, 1913, Rutherford acknowledged receipt of Bohr's manuscript. He made it clear that he thought the content was very interesting. He also noted, however, that Bohr could have expressed himself more clearly and more concisely:

> There is one criticism of a minor character which I would make in the arrangement of the paper. I think in your endeavour to be clear you have a tendency to make your papers much too long, and a tendency to repeat your statements in different parts of the paper. I think that your paper really ought to be cut down, and I think this could be done without sacrificing anything to clearness. I do not know if you appreciate the fact that long papers have a way of frightening readers, who feel that they have not time to dip into them (Bohr 1972–2008, Vol. 2, p. 112–113).

---

[46] See Bohr (1972–2008, Vol. 2, p. 111–115, pp. 581–587) for the exchange between Bohr and Rutherford in March–June 1913 concerning (the publication of) the Trilogy. There are two letters from Rutherford and five (drafts of) letters from Bohr.

In a postscript, he added: "I suppose you have no objection to my using my judgement to cut out any matter I may consider unnecessary in your paper?" (p. 113). Rutherford's letter crossed the next letter from Bohr, dated March 21, with a revised and even longer version of Part One (in which he added, for instance, some comments on Nicholson's theory). Rutherford received this new version on March 25 and immediately sent another note to Bohr urging him to shorten the paper:

> I think the additions are excellent and appear quite reasonable; the difficulty is, however, that your paper is already rather full and long for a single paper. I really think it desirable that you should abbreviate some of the discussions to bring it within more reasonable compass. As you know, it is the custom in England to put things very shortly and tersely in contrast to the Germanic method, where it appears to be a virtue to be as long-winded as possible (p. 113).

This last observation is reminiscent of a comment Maxwell once made (in a letter of August 1873 to his friend Tait): "Boltzmann could not understand me on account of my shortness, and his length was and is an equal stumbling block to me" (quoted, for instance, in Klein 1970a, p. 97).

In his letter of March 20, Bohr had already announced that he was planning to come to Manchester to discuss the paper in person. This visit took place in early April. The result of this visit appears to have been that Bohr prevailed and Rutherford accepted that Part One of the Trilogy be published with only minor changes. "Years later," the editors of Bohr's *Collected Works* note, "Rutherford still remembered Bohr's inflexibility on this point" (Bohr 1972–2008, Vol. 2, p. 113).

Bohr enclosed Part Two of the Trilogy with a letter to Rutherford of June 10, 1913 (cf. note 46). Part Two (27 pages) appeared in September (Bohr 1913c). Bohr must have submitted Part Three sometime between June 22 and October 16, 1913. He sent letters to Rutherford on both dates. In the former, he wrote that "working on the third part is in great progress." In the latter, written shortly after having met with Rutherford in Birmingham, he did not mention the paper at all, suggesting he had meanwhile submitted it. Perhaps he handed it to Rutherford in person in Birmingham. Part Three (19 pages) appeared in November (Bohr 1913d).[47]

4.5.1   Part One: the hydrogen atom

Part One of Bohr's Trilogy begins with a short general introduction, in which Bohr recounts the α-particle scattering experiments that led Rutherford to propose his nuclear model of the atom (see Section 4.3.2) and argues, citing the proceedings of the Solvay conference (Langevin and de Broglie 1912), that quantum ideas are needed to account for its stability (Bohr 1913b, pp. 1–3).[48] In Part One, Bohr announced, he will present

---

[47] See sec. 10, "Omitted parts . . .," of the editorial note on the Trilogy for discussion of material that Bohr may have left out of these papers (1972–2008, Vol. 2, p. 127–130).
[48] Unless noted otherwise, all page references in this subsection are to Bohr (1913b).

"the mechanism of the binding of electrons by a positive nucleus . . . in relation to Planck's theory" and show how it can "account in a simple way for the line spectrum of hydrogen."

Getting down to business, Bohr considered an electron with mass m and charge $-e$ orbiting a nucleus with charge E (equal to e in the case of the hydrogen atom). Leaving aside the problem of radiative instability of this system for the moment, he examined the conditions for its mechanical stability. Initially allowed to be ellipses (with major axis $2a$), the orbits are assumed to be circles (of radius a) for most of the calculation. As this does not affect any of the results and simplifies the algebra, we will assume circular orbits right away. An electron in a circular orbit around a nucleus is the simplest of the configurations considered in the Rutherford Memorandum. The centripetal force needed for its mechanical stability is provided by the Coulomb force:

$$m(2\pi\omega)^2 a = \frac{eE}{a^2}, \tag{4.31}$$

where ω is Bohr's notation for the *cyclic* frequency of the electron orbit. The angular frequency, for which ω is typically used, is thus equal to $2\pi\omega$ in Bohr's notation.

The ionization energy W is equal to minus the total energy E_{tot} of the electron and hence, by the virial theorem, to E_{kin} (cf. Eq. (4.18)).[49] It is easy to verify this directly for this simple system. The potential energy of the electron in the Coulomb field of the nucleus is $-eE/a$. Its kinetic energy is

$$E_{\text{kin}} = \frac{1}{2}m(2\pi\omega a)^2 = \frac{eE}{2a}, \tag{4.32}$$

where in the last step we used Eq. (4.31). It follows that the ionization energy is given by

$$W = -(E_{\text{kin}} + E_{\text{pot}}) = \frac{eE}{2a}, \tag{4.33}$$

which is indeed just E_{kin}.

Solving for ω in Eq. (4.31) and eliminating a with the help of Eq. (4.33), we find

$$\omega^2 = \frac{eE}{4\pi^2 ma^3} = \frac{eE}{4\pi^2 m}\left(\frac{2W}{Ee}\right)^3 = \frac{2W^3}{m\pi^2 e^2 E^2}. \tag{4.34}$$

The (cyclic) orbital frequency is thus given by:

$$\omega = \sqrt{\frac{2}{m}}\frac{W^{3/2}}{\pi eE}. \tag{4.35}$$

---

[49] For elliptical orbits, $E_{\text{tot}} = -\overline{E_{\text{kin}}}$, where $\overline{E_{\text{kin}}}$ is the kinetic energy averaged over the orbit. For circular orbits, E_{kin} is constant.

With the help of Eq. (4.33), the radius of the orbit can likewise be expressed in terms of the ionization energy:

$$a = \frac{eE}{2W}.$$ (4.36)

Commenting on this last pair of equations, the first two given in his paper, Bohr noted that "if the value of W is not given, there will be no values of ω and a characteristic for the system in question" (Bohr 1913b, p. 3). As in the Rutherford Memorandum, Bohr would fix ω and a by imposing a quantum condition on W (cf. Eq. (4.13)–(4.17) in Section 4.3.5).

Bohr now addressed the system's radiative instability, reminding his readers that according to classical electrodynamics the accelerating electron should continuously emit radiation and spiral down into the nucleus. However, "actual atoms in their permanent state seem to have absolutely fixed dimensions and frequencies" and "after a certain amount of energy characteristic for the systems in question is radiated out, the systems will again settle down in a stable state of equilibrium, in which the distances apart of the particles are of the same order of magnitude as before" (p. 4). To account for these basic empirical facts, Bohr turned to Planck's so-called second theory (see note 78) in which an "atomic vibrator of [cyclic] frequency ν" absorbs radiation continuously but emits it only in discrete amounts of energy, $\tau h\nu$, where τ is an integer.[50] Rather than quantizing the ionization energy of his model of a hydrogen atom in analogy with the *kinetic* energy, $\frac{1}{2}\tau h\nu$, of a Planck oscillator, as we did at the end of Section 4.4.2 (see Eq. (4.25)), Bohr quantized it in analogy with the *total* energy of a Planck oscillator. As we saw—and as Bohr clearly realized—the factor $\frac{1}{2}$ is crucial to get the right value for the Rydberg constant. The rather unconvincing argument Bohr used to introduce this factor in sec. 1 of his paper is replaced in sec. 3 by an argument based on an embryonic version of what would become his correspondence principle.

At the beginning of sec. 1, Bohr assumed that, [*i*] when an electron is captured by a nucleus and settles down in a state with total energy

$$E_{\text{tot}}^{(\tau)} = -W_\tau = -\tau h\nu,$$ (4.37)

"homogeneous [i.e., monochromatic] radiation is emitted of a frequency ν, equal to [*ii*] half the frequency of revolution of the electron in its final orbit" (pp. 4–5). The feeble argument Bohr offered in support of the second assumption is that it "suggests itself, since the frequency of revolution of the electron at the beginning of the emission is 0" (p. 5). "The question, however, of the rigorous validity of both assumptions," he hastened to add in the next sentence, "will be more closely discussed in sec. 3" (p. 5).

Actually, by the end of sec. 1, Bohr had already changed his picture of the emission process to the one we now associate with the Bohr model. Emission of radiation no longer just occurs when electrons are captured by nuclei and atoms are formed but whenever

---

[50] Bohr (p. 4) cited Planck (1910, 1911a, 1912) at this point.

an electron jumps from one of the quantized energy states given by Eq. (4.37) to a lower one (Heilbron and Kuhn 1969, pp. 263–264). It is only with this new picture that Bohr could recover the Balmer formula. Bohr, however, did not bother to remove references to his initial inadequate picture of the emission process from his paper.[51] Leaving these traces, Bohr inadvertently justified Rutherford's criticism of the way he wrote the paper.

At the beginning of sec. 3, Bohr raised a strong objection to his original picture of the emission process. Assume that an electron captured by a nucleus and eventually settling down in some state labeled by τ emits τ quanta of energy $h\nu$ in the process, all in the form of electromagnetic waves of frequency ν. This is the assumption he made at the beginning of sec. 1. He now wanted to reject it. He wrote:

> Considering systems in which the frequency is a function of the energy, this assumption, however, may be regarded as improbable; for as soon as one quantum is sent out the frequency is altered (p. 12).

In the next sentence, he reassured the reader that "we can leave [i.e., give up] the assumption used and still retain" the quantization condition,

$$W_\tau = \tau h \frac{\omega}{2}, \tag{4.38}$$

arrived at on the basis of the dubious argument that Bohr had offered in sec. 1 to set $\nu = \omega/2$. Curiously, the objection Bohr raised at the beginning of sec. 3 against his own original theory repeats almost verbatim an objection he raised against Nicholson's theory toward the end of sec. 1:

> [S]ystems like those considered [by Nicholson] in which the frequency is a function of the energy, cannot emit a finite amount of homogeneous [i.e., monochromatic] radiation; for, as soon as the emission of radiation is started, the energy and also the frequency of the system are altered (p. 7).

We return now to sec. 1 and the quantization condition (4.38), postponing discussion of Bohr's new justification of it in sec. 3. Using Eq. (4.35) for ω in Eq. (4.38), we find an equation,

$$W_\tau = \frac{\tau h}{2} \left(\sqrt{\frac{2}{m}} \frac{W_\tau^{3/2}}{\pi eE} \right), \tag{4.39}$$

[51] In sec. 2, we find two such passages: "The condition for the appearance of the spectrum is, according to the above theory, that helium atoms are present in a state in which they have lost both their electrons" (p. 11); "Let us assume that the spectrum in question corresponds to the radiation emitted during the binding of an electron" (p. 12). In sec. 4, we find another one: "In analogy to the assumption used in this paper that the emission of line-spectra is due to the re-formation of atoms after one or more of the lightly bound electrons are removed ..." (p. 18). As we shall see at the end of Section 4.5.2, there is another reference to this old picture in the overall conclusion of the Trilogy at the end of Part Three (Bohr 1913d, p. 875).

in which W_τ occurs on both sides. This equation allows us to express W_τ as a function of the quantum number τ and various constants:

$$\sqrt{W_\tau} = \sqrt{\frac{m}{2}}\frac{2\pi eE}{\tau h}, \qquad W_\tau = \frac{2\pi^2 me^2 E^2}{\tau^2 h^2}. \tag{4.40}$$

Inserting these expressions into Eqs. (4.35) and (4.36), Bohr could fix the values of ω and a (p. 5):

$$\omega_\tau = \sqrt{\frac{2}{m}}\frac{\sqrt{W_\tau}W_\tau}{\pi eE} = \frac{2W_\tau}{\tau h} = \frac{4\pi^2 me^2 E^2}{\tau^3 h^3}, \qquad a_\tau = \frac{eE}{2W_\tau} = \frac{\tau^2 h^2}{4\pi^2 meE}. \tag{4.41}$$

Inserting the expression for W_τ in Eq. (4.40) into Eq. (4.37), we find the formula for the total energy of the state labeled τ that leads to the empirically confirmed expression for the Rydberg constant (cf. Eqs. (4.24)–(4.28)):

$$E_{tot}^{(\tau)} = -\frac{2\pi^2 me^2 E^2}{\tau^2 h^2}. \tag{4.42}$$

Before Bohr could derive the Balmer formula, however, he had to change his picture of the emission of radiation. Toward the end of sec. 1, he introduced his new picture. He now considered the "passing of the systems between different stationary states" (the quantized energy states labeled by τ) and introduced the assumption that this process "is followed by the emission of a homogeneous [i.e., monochromatic] radiation, for which the relation between the frequency and the amount of energy emitted is the one given by Planck's theory" (p. 7). This finally allowed Bohr to derive the Balmer formula at the beginning of sec. 2 (pp. 8–9). If a hydrogen atom (for which $E = e$) passes from a state labeled τ_1 to a state labeled τ_2, radiation will be emitted of the frequency:

$$\nu = \frac{E_{tot}^{(\tau_1)} - E_{tot}^{(\tau_2)}}{h} = \frac{2\pi^2 me^4}{h^3}\left(\frac{1}{\tau_2^2} - \frac{1}{\tau_1^2}\right), \tag{4.43}$$

which gives the Balmer formula and allowed Bohr to express the Rydberg constant in terms of the constants of nature e, m, and h (see Eq. (4.28)). Bohr had to pay a very high price for this success. He was forced to sever the relation between radiation and orbital frequencies.

Bohr next relaxed the quantum condition (4.38) to

$$W_\tau = f(\tau)h\omega, \tag{4.44}$$

where f is some yet to be determined function. Eqs. (4.39)–(4.43) remain valid as long as we substitute $2f(\tau)$ for τ everywhere. For hydrogen ($E = e$), the expression for the orbital frequency in Eq. (4.41) thus changes to:

$$\omega_\tau = \frac{\pi^2 m e^4}{2 f(\tau)^3 h^3},\tag{4.45}$$

while Eq. (4.43) for the radiation emitted when the electron goes from the state labeled τ_1 to the state labeled τ_2 changes to

$$\nu = \frac{\pi^2 m e^4}{2h^3}\left(\frac{1}{f(\tau_2)^2} - \frac{1}{f(\tau_1)^2}\right).\tag{4.46}$$

From the form of the Balmer formula, Bohr now inferred that $f(\tau)$ has to be proportional to τ:

$$f(\tau) = c\tau.\tag{4.47}$$

To determine the proportionality constant c, Bohr imposed the condition that for transitions between energy states characterized by large values of τ, the radiation frequency ν in Eq. (4.46) is equal to the orbital frequency in Eq. (4.45) (as the values of τ get larger and larger, the orbital frequencies of neighboring orbits gets closer and closer). For a transition from $\tau_1 = N$ to $\tau_2 = N - 1$, with $N \gg 1$, the latter becomes

$$\omega_N = \frac{\pi^2 m e^4}{2c^3 N^3 h^3} \approx \omega_{N-1},\tag{4.48}$$

while the former is given by:

$$\nu = \frac{\pi^2 m e^4}{2c^2 h^3}\left(\frac{1}{(N-1)^2} - \frac{1}{N^2}\right).\tag{4.49}$$

The expression in parenthesis can be written as

$$\frac{2N-1}{N^2(N-1)^2} \approx \frac{2}{N^3}.\tag{4.50}$$

Inserting this expression into Eq. (4.49) for ν and setting the result equal to Eq. (4.48) for ω_N, we find:

$$\frac{\pi^2 m e^4}{N^3 c^2 h^3} = \frac{\pi^2 m e^4}{2c^3 N^3 h^3},\tag{4.51}$$

from which we read off that $2c = 1$ or $c = \frac{1}{2}$. Bohr had thus found a much more convincing justification for the quantization condition (4.38) that he knew would lead to the correct expression for the Rydberg constant. But not only did he have to sever the link between radiation and orbital frequencies, except in the limit of high quantum numbers, he also had to *assume* rather than *derive* part of the Balmer formula.

At the end of sec. 3 of the paper, Bohr (1913b, p. 15) pointed out that the quantization of the ionization energy W in Eq. (4.38) can be rewritten in terms of the quantization of the angular momentum L. Bohr credits Nicholson (1912, p. 679) with having first suggested this reinterpretation (cf. Section 4.3.4, Eqs. (4.8)–(4.9)). For circular orbits with radius a, we already saw that $W = E_{\text{kin}}$ (see Eq. (4.33)) and that $E_{\text{kin}} = \frac{1}{2}m(2\pi\omega a)^2$ (see Eq. (4.32)). The angular momentum for such an orbit is thus related to the ionization energy via:

$$L = m(2\pi\omega a)a = \frac{W}{\pi\omega}. \tag{4.52}$$

It follows that the condition $W = \tau h\omega/2$ can be rewritten as

$$L = \tau \frac{h}{2\pi}. \tag{4.53}$$

Bohr used M instead of L and wrote $M = \tau M_0$ with $M_0 = h/2\pi = 1.04 \times 10^{-27}$ [erg·sec]. This is the form in which the quantization condition is consistently used in Parts Two and Three of the Trilogy.[52]

4.5.2 Parts Two and Three: multi-electron atoms and multi-atom molecules

In the Fall of 1913, Parts Two and Three of the Trilogy were published in the *Philosophical Magazine* (Bohr 1913c, 1913d).[53] They contain Bohr's speculations on the electronic structure of single atoms with more than one electron and of molecules. This is reflected in their titles: "Systems containing only a single nucleus" (Part Two) and "Systems containing several nuclei" (Part Three). As we mentioned at the end of Section 4.3, large parts of these papers were probably written well before Bohr's breakthrough in February and early March 1913 leading to the treatment of the hydrogen atom in Part One of the Trilogy. In fact, some of the results announced, but not derived, in the Rutherford Memorandum of June 1912—specifically, the periodic variation of atomic volumes and the Whiddington X-ray formula—are first addressed in detail in Part Two of the Trilogy.

---

[52] Bohr's readers, starting with Sommerfeld (1915a, p. 428, p. 431), can thus be forgiven for presenting the quantization of angular momentum in Eq. (4.53) as the quantization condition Bohr used all along. This reading of Bohr, however, drew the ire of two prominent commentators, who emphasized that it is a "recurrent myth" that "Bohr derived the Balmer formula by quantizing angular momentum" (Heilbron and Kuhn 1969, p. 280, note 157).

[53] These parts of the Trilogy are also discussed by Heilbron and Kuhn (1969, sec. VI), Kragh (1977; 2012, sec. 2.6) and Scerri (2007, pp. 188–197).

It is difficult to overstate the ambitious character of these two papers. Bohr was essentially attempting to lay the foundation for two enormous fields of modern physical science: atomic physics and quantum chemistry. In the absence of an adequate conceptual framework for this Herculean task, Bohr attempted to feel his way forward to an understanding of the electronic structure of multi-electron and multinuclear systems, extrapolating some of the ideas that had worked so well for hydrogen. The results obtained were, as we shall see, frequently in conflict with empirical information (scant as it was at the time), in which case Bohr either discounted the accuracy of the experiments or allowed phenomenological considerations to outweigh theoretical ones. His final assertions—e.g., on the number of electrons in the concentric rings supposed to exist in the normal states of larger atoms—often represent an amalgam of, and compromise between, theoretical argument and phenomenological input.

Nevertheless, in the opening of Part Two, Bohr identifies a number of important qualitative points concerning atomic structure that were indispensable for further progress, and, much as they may now appear common-place, were still matters of debate in 1913. In particular, he states three basic assumptions that have survived the transition to quantum mechanics and entered modern atomic physics essentially intact:

1. "On account of the small dimensions of the nucleus, its internal structure will not be of sensible influence on the constitution of the cluster of electrons, and consequently will have no effect on the ordinary physical and chemical properties of the atom. The latter properties on this theory will depend entirely on the total charge and mass of the nucleus" (Bohr 1913c, p. 477).

2. "[T]he internal structure of the nucleus will be of influence only on the phenomena of radioactivity" (ibid.)

3. "The total experimental evidence supports the hypothesis[54] that the actual number of electrons in a neutral atom with a few exceptions is equal to [the ordinal position of the corresponding element arranged by] atomic weight" (ibid).[55]

To make quantitative progress on atomic structure, Bohr (1913c) adopted three assumptions/hypotheses/conditions that have not survived the transition to modern quantum theory:

1. "[E]lectrons are arranged at equal angular intervals in coaxial rings rotating [in the same plane] around the nucleus" (Bohr 1913d, p. 477).

2. "[I]n the permanent [i.e., ground] state of an atom the angular momentum of every electron round the centre of its orbit is equal to the universal value $h/2\pi$" (ibid.).

---

[54] Here Bohr refers to a paper by the Dutch amateur physicist Antonius van den Broek (1913). For discussion of van den Broek's contributions, see Hirosige (1971) and Scerri (2007, pp. 165–169).
[55] The identity of atomic number and nuclear charge would soon be established by the work of Henry Moseley, which will be covered in more detail in Section 6.2.

3. A "condition of stability" (ibid.), according to which the total energy of the system in the ground state is less than in any neighboring configuration satisfying condition (2).

In treating atomic systems with more than one electron, Bohr immediately faced the enormous difficulties involved in the treatment of systems of three or more bodies subject to inverse square forces in classical Newtonian mechanics. Effectively, analytic solutions of the equations of motion in such cases were restricted to a very limited class of special configurations of enhanced symmetry, which in many cases are not even mechanically stable (i.e., small arbitrary displacements from the initial motion would lead to ever-increasing deviations).

The choice of a single ring of n equally spaced electrons at least allows Bohr to examine an analytically solvable system in which condition (2)—that each electron have constant angular momentum $h/2\pi$—can be maintained while upholding the laws of classical mechanics. Any one of these n electrons would experience a purely radial net centripetal force resulting from the electrostatic attraction of the nucleus and the repulsion of the $n-1$ other (symmetrically arranged) electrons (cf. Fig. 4.5). This net force can be written as $F(e^2/a^2)$, where a is the radius of the ring and F is Bohr's new notation for what he called X in the Rutherford Memorandum, a factor depending on n (cf. Eq. (4.12), note 36, and Eqs. (4.54)–(4.56) below). While these radial forces would leave the angular momentum of the electron unchanged, it was well known from the time of the Nagaoka model that, due to the inter-electron repulsion, this arrangement is not mechanically stable with respect to displacements of the electron motions in the plane of the ring. This is in sharp contrast to the case of purely attractive gravitational forces, where Maxwell, for instance, had been able to establish the stability of Saturn's rings.

Acknowledging the instability problem, Bohr (1913c, p. 480) reminds his readers that "we have assumed that the ordinary principles of mechanics cannot be used in the discussion of the problem in question." His way out of the problem was to invoke condition (2) above: the imposition of the fixed angular momentum for each electron would force, by fiat, the stability with respect to in-plane displacements. He produces a simple demonstration of this for a symmetrical displacement of all electrons, initially on a circular ring of radius a, to a ring of radius αa, where condition (2) ensures that the energy increases by an amount $(1-1/\alpha)^2$ times the initial kinetic energy. Since "the total energy of the new configuration is greater than in the original," Bohr (1913c, p. 480) concludes that, "[a]ccording to the condition of stability [i.e., condition (3) above] the system is consequently stable for the displacement concerned."

For displacements perpendicular to the ring, mechanical stability was obtained without the necessity to freeze the electronic angular momenta, provided the number of electrons in the ring does not exceed a given value for each value of the nuclear charge. Nuclei of charge 10 could stably accommodate a single ring of 8 electrons, nuclei of charge 20 a ring of 10 electrons, and so on (with the number of allowed electrons growing rather slowly with the nuclear charge). In fact, the stability with respect to perpendicular displacements was at the core of Nicholson's calculations of the oscillations of ring electrons (superimposed on the regular rotational motion of the ring). These oscillations were responsible for the radiation identified with coronal and nebular lines. Of course, in the new picture advanced by Bohr in Part One of the Trilogy, monochromatic spectral radiation emitted by atoms is explicitly quantal in nature, with a frequency determined by the Bohr condition, $E = h\nu$, and not by the frequencies present in the Fourier analysis of the mechanical motion, as in Nicholson's theory. Bohr still accepted the validity of Nicholson's results at this point, but reinterpreted the lines calculated by Nicholson as due to resonant scattering in coronal or nebular environments in which just those frequencies corresponding to stable transverse oscillations of ring electrons are preferentially absorbed and reemitted (cf. Bohr 1913b, p. 23; 1913c, p. 482).

Not included in the list of three hypotheses introduced by Bohr above is the implicit assumption,[56] certainly at the core of the procedure employed throughout Part One of the Trilogy, that the electrons in stationary states of the atom follow orbits determined by the classical equations of motion, i.e., that each electron is accelerated on the basis of the net electrostatic force it experiences due to the other charged particles in the system. It would hardly be possible otherwise to arrive at any quantitative statements concerning the binding energy of such states. Unfortunately, conditions (1)–(3) above are simply inconsistent with the validity of the classical equations once more than one electron ring is present.

The reason is not hard to uncover. The electrons in two different rings rotate with different frequencies (to satisfy condition (2)). This means that the net torque experienced by an electron in one ring due to the electrostatic repulsion of the electrons in another ring is both nonzero and time varying. As long as the usual laws of classical mechanics are strictly observed, it is thus impossible to maintain the angular momentum of each electron at the posited value $h/2\pi$. Also, the interaction of the rings will necessarily distort the motion of the electrons from a purely circular path, especially if the rings are close in radius.

Bohr was perfectly aware of these problems. His approach was simply to assume that the rings are sufficiently separated in radius and frequency that, *to a good enough approximation*, the electrons in a given ring "see" the electron charges on other rings as effectively continuous circular distributions of charge, so that the net electrostatic repulsion seen by an electron in a given ring is once again radial in direction (hence producing no net torque) and constant in time. With this assumption, Bohr was able to establish two important qualitative facts (once again assuming well-separated rings): (*a*) the effect of outer rings on inner electrons is small, and (*b*) the net effect of rings interior to a given electron is to effectively reduce the positive nuclear charge "seen" by this electron by the net negative charge carried by the inner rings (Bohr 1913c, pp. 482–486).

With these intuitions in place, Bohr proceeds, in sec. 3 of Part Two, to examine the normal (ground) state configuration and binding energy for the first four elements in the periodic table: hydrogen, helium, lithium, and beryllium. The concentric ring

---

[56] As we will see at the end of this section, however, this assumption is the second of five assumptions listed in the conclusion of the paper (Bohr 1913d, p. 874).

configuration assumed for the normal state of the atom of a certain element is indicated by the notation $N(n_1, n_2, ...)$, where N is the nuclear charge and $n_1, n_2, ...$ indicate the number of electrons in successive rings, starting with the innermost.

The treatment for hydrogen essentially reiterates the results of Part I, with the 1(1) configuration leading to a binding energy of $W_0 = 13$ eV (the modern value is 13.6 eV). Bohr compares this to the value of the ionization potential for hydrogen of about 11 volts found by J. J. Thomson (1912a), adding "[n]o other data, however, are available for hydrogen" (Bohr 1913c, p. 488). Bohr also finds a bound configuration 1(2), in other words a stable negative ion H^-, a hydrogen atom with an extra electron, with an ionization potential of $0.13W_0$. It was not until a decade and a half later that Hans Bethe (1929) established (through an elaborate quantum-mechanical variational calculation) that this hydrogen anion, as it is now called, actually exists as a bound system. Bethe found an ionization energy of about 0.75 eV, less than half the Bohr value of 1.69 eV. Beginning in the early 1940s, the existence of the hydrogen anion would eventually be confirmed on the basis of infrared absorption in the solar atmosphere.

The helium 2(2) configuration of two electrons at opposite positions on a single ring is exactly as proposed in the Rutherford Memorandum (see Fig. 4.2), except that the constant K is now set equal to $h/2$. The total binding energy for both electrons is found to be $6.13W_0$, where W_0 is the binding energy for the ground state of hydrogen. This gives an ionization potential (i.e., the energy required to remove a single electron) of $2.13W_0$, which Bohr (with the available values at the time) equates to 27 eV, quite a bit larger than the contemporary value of 20.5 eV due to James Franck and Gustav Hertz (1913) cited here by Bohr (1913c, p. 489). Both the theoretical and the experimental value were in error (in opposite directions). As we will see in Section 7.4, the correct understanding of the ground state of helium would prove to be one of the most critical difficulties of the old quantum theory and would eventually sound its death knell. Bohr does not comment on the discrepancy, apart from saying that "these values are of the same order of magnitude" (p. 489). He may not have trusted the experimental values, which were still in a state of flux. Bohr also shows that the configuration 2(3), the negative ion He^-, is higher in energy than the neutral atom so that, unlike the case of hydrogen, such an ion would not be stable.

With lithium we encounter the first case of Bohr mixing theoretical argument with phenomenological considerations in his approach to atomic theory. The chemical evidence clearly indicated that lithium was univalent, in other words, that one of the three electrons was less strongly bound and therefore more chemically active than the other two. On the other hand, direct application of the Bohr approach showed that the single ring configuration 3(3) was more tightly bound (by about 21 eV) than the two-ring configuration 3(2,1), in which the single electron in the outer ring was much less tightly bound than the inner two. In this case, the chemical evidence was dispositive for Bohr, who concludes that the 3(2,1) arrangement has to be accepted, whatever stability considerations might otherwise imply. The same quandary is encountered with beryllium (valence 2), where the more tightly bound 4(4) single-ring configuration is discarded in favor of the chemically more plausible 4(2,2) one.

1 (1)	9 (4, 4, 1)	17 (8, 4, 4, 1)
2 (2)	10 (8, 2)	18 (8, 8, 2)
3 (2, 1)	11 (8, 2, 1)	19 (8, 8, 2, 1)
4 (2, 2)	12 (8, 2, 2)	20 (8, 8, 2, 2)
5 (2, 3)	13 (8, 2, 3)	21 (8, 8, 2, 3)
6 (2, 4)	14 (8, 2, 4)	22 (8, 8, 2, 4)
7 (4, 3)	15 (8, 4, 3)	23 (8, 8, 4, 3)
8 (4, 2, 2)	16 (8, 4, 2, 2)	24 (8, 8, 4, 2, 2)

Figure 4.4 *Electron-ring configurations for the first 24 elements proposed in Part Two of the Trilogy (Bohr 1913c, p. 497).*

In sec. 4 ("Atoms containing greater numbers of electrons"), Bohr moves on to elements with larger atomic number, eventually producing a table (see Fig. 4.4) with hypothesized ring configurations for all the elements from hydrogen to chromium ($N = 24$) (p. 497). Phenomenological considerations often trump theoretical ones. As a result, Bohr's arguments are not particularly persuasive and we shall not cover them in detail. Through a curious argument involving the tendency for combination of two rings with the same number of electrons, Bohr infers that the inner rings of these atoms can only have 2, 4, or 8 electrons (p. 495). A new inner ring of 8 electrons appears once the screened nuclear charge (i.e., the nuclear charge minus the net charge of previous inner rings) reaches 10, as indicated by the transverse stability arguments. The outermost ring is simply forced to contain a number of electrons matching the chemical valence, as we saw earlier for lithium and beryllium. The basic idea seems to have been, presumably already at the time of the Rutherford Memorandum, that periodic properties of the elements arise by the filling of inner rings, at which point the properties of the element revert to at least qualitative similarity with the element(s) of lower atomic weight with the same number of electrons in the outermost ring.

The resulting configurations bear little or no resemblance to the shell structure that was to emerge much later, after the introduction of electron spin and the exclusion principle. In Volume Two of our book, we will examine Bohr's "second atomic theory", developed in the early 1920s in another attempt to decipher electron arrangements for the atoms of the periodic table. At that point, appeals to the correspondence principle replaced the stability considerations of 1913, but the results (again frequently guided by chemical information) were no more reliable.

In secs. 4 and 5 of Part Two of the Trilogy, Bohr returned to two problems (from the original list of four) which he had explicitly announced as successes of his theory of atomic configurations in the Rutherford Memorandum of the previous year (see Section 4.3.5).

First, there is the question of the periodic variation in atomic volume/size (Bohr 1913c, pp. 496–497). The atomic size was known to decrease in the periodic table going horizontally from the alkali metals across to the elements of valence 4 (in the earlier

The Trilogy: quantum atomic models and spectra   191

part of the table) and then increase again to the inert gases at the end of the row.[57] This empirical regularity provided Bohr with yet another argument for the electron-ring configurations for the different elements shown in Fig. 4.4.

Bohr sets the atomic volume equal to the cube of the radius of the largest of the concentric electron rings representing the arrangement of electrons in the ground state of the atom. He assumes for simplicity that the size of the outermost ring can be estimated by assuming the inner rings to screen the nuclear charge, so that the electrons in the outermost ring can be treated as circling around a single positive charge that neutralizes the charges of those electrons. Eq. (4.16) tells us that the radius of such a single electron ring is inversely proportional to the factor F (formerly X), which gives the strength of the net centripetal force $F(e^2/a^2)$ felt by each electron in the ring (see Eq. (4.12)).

Earlier in Part Two of the Trilogy, Bohr (1913c, p. 479) gives an expression for this factor F for the case of n electrons on a ring rotating around an equal but opposite charge at the center. We derive this formula with the help of Fig. 4.5. The centripetal force felt by each electron in the ring is given by the electrostatic attraction of the central effective charge, proportional to n, minus the net electrostatic repulsions of the $n - 1$

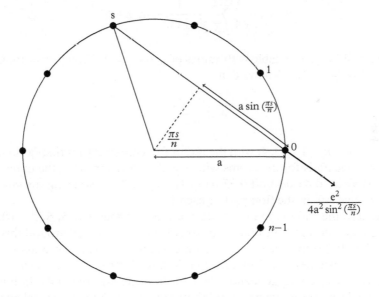

Figure 4.5 *Calculating the net inter-electron electrostatic repulsion for a ring with n electrons labeled 0 through n − 1.*

---

[57] What frequently complicates the determination of atomic volume/size is that elements are present as molecules (and in gaseous form) rather than in atomic form. Bohr (1913c, p. 497) acknowledges this: "It must, however, be borne in mind that the experimental determinations of atomic volumes in most cases are deduced from considerations of molecules rather than atoms."

other electrons in the ring. Referring to Fig. 4.5, we see that in a ring of radius a with n electrons (labeled by an index $s = 0, 1, 2, \ldots, n-1$) symmetrically arranged, the repulsive force on (say) the 0th electron due to the sth electron is of magnitude

$$\frac{e^2}{4a^2 \sin^2 (\pi s/n)}. \tag{4.54}$$

To find the component of this force in the (outward) radial direction we need to multiply by a further factor of $\sin (\pi s/n)$. We thus arrive at a radial centrifugal component

$$\frac{e^2}{4a^2 \sin (\pi s/n)}. \tag{4.55}$$

Combining these contributions for all electrons $s = 1, 2, \ldots, n-1$, and including the centripetal electrostatic attraction $n(e^2/a^2)$ of the nucleus, we find for the dimensionless factor F:

$$F = n - \frac{1}{4} \sum_{s=1}^{n-1} \frac{1}{\sin (\pi s/n)} \equiv n - s_n. \tag{4.56}$$

Bohr (1913c, p. 482) gives a table with values of s_n for $n = 1$ through $n = 16$. Given that $a \propto 1/F$ (see Eq. (4.16)), it follows that

$$a \propto \frac{1}{n - s_n}. \tag{4.57}$$

Inserting the values $n = 1, 2, 3, 4$, with the radius for a one-electron ring (as in hydrogen) denoted by a_0, one finds for the atoms with $n = 2, 3, 4$ electrons in the outermost ring, ring radii of $0.57\ a_0$, $0.41\ a_0$, and $0.33\ a_0$ (1913c, p. 478), displaying the characteristic decrease in atomic size in the first part of each row.

Unfortunately, the radii continue to decrease if one sets $n = 5, 6, \ldots$, whereas the atomic size should increase in the second part of each row. To get around this problem, Bohr (1913c, p. 497) transfers electrons to inner rings, starting a new outer one with the number of electrons determined by the (in this case electronegative) valence: thus $n = 3$ for nitrogen, $n = 2$ for oxygen, and $n = 1$ for fluorine (see Fig. 4.4). In this way, the atomic volume decreases as we go from lithium (element 3) to carbon (element 6) and increases again as we go from nitrogen (element 7) to fluorine (element 9), just as the empirical data suggest.

In sec. 5 of Part Two, Bohr (1913c, pp. 498–500) returns to the second item on the list of claimed accomplishments in the Rutherford Memorandum, Whiddington's (1911) law, asserting that the minimum velocity of incident electrons required to excite characteristic (K) X-ray radiation is proportional to the atomic weight of the target element. This proportionality also follows directly from the idea—recently advanced by Thomson (1912a, pp. 456–457) in a paper on ionization cited by Bohr (1913c,

p. 498)—that the characteristic X-rays are initiated by the ejection of a more tightly bound inner electron, followed by another electron occupying the vacancy left by the ejected one. The energy needed to remove the inner electron, which for a heavy atom was assumed to experience essentially unscreened the full nuclear charge, is just the binding energy. Rutherford's α-scattering experiments had by now made it clear that the nuclear charge corresponds to roughly half the atomic weight A. The binding energy is proportional to X^2 (see Eqs. (4.18)–(4.19)). The centripetal factor X (relabeled F in Part Two) for an inner electron is proportional to the (unscreened) nuclear charge and hence to A. To match the binding energy, the kinetic energy of the incident electron, $\frac{1}{2}mv^2$, must thus be proportional to A^2. It follows that $v \propto A$, which is Whiddington's law.

Part Three of the Trilogy (Bohr 1913d) is concerned with the electronic structure of bound systems with more than one nucleus, i.e., with molecules. Bohr maintains the molecular "figurations" adopted in the Rutherford Memorandum (see Fig. 4.1): the nuclei are arranged along a straight line and the electrons circulate in rings around the line connecting the nuclei. For systems with more than one ring, the analysis of the mechanical stability becomes quite intricate. Bohr therefore restricts his attention to diatomic molecules with a single ring containing only a few electrons circling around the inter-nuclear axis. For a system with two nuclei of charge Ne and a single ring of n electrons, Bohr performs (in sec. 2) a detailed transverse-stability analysis analogous to that performed in Part Two for a single atom. The resulting condition only allows the values ($N = 1$, $n = 2$) and ($N = 1$, $n = 3$). These values correspond to the hydrogen molecule H_2 and the hydrogen molecular anion H_2^-, respectively. In addition, he establishes the stability of these systems with respect to motion of the nuclei towards or away from each other. For larger atoms and molecules, as already shown in the figurations of the Memorandum (see Fig. 4.1), some of the electrons would be in rings in between the nuclei (thereby forming a "chemical bond"), while the rest would circulate with their respective nuclei at the center (sec. 5, p. 873).

The only detailed calculations performed in Part Three relate to the neutral hydrogen molecule (sec. 3). Bohr had already treated this system in the Memorandum, albeit with a slightly incorrect quantization condition. These calculations, accordingly, are a reprise and an extension of those we covered in Section 4.3.5 (see Fig. 4.2(b) and Eqs. (4.10)–(4.12)). With the correct angular-momentum quantization condition, Bohr now finds that it takes an energy of $0.2W_0$ (or about 2.7 eV) for two hydrogen atoms to form a hydrogen molecule. He acknowledges that the value measured by Irving Langmuir (1912) is more than twice as large (the modern experimental value is 4.75 eV). He attributes the discrepancy to the "indirect method employed" by Langmuir, who arrived at his value by measuring the heat conduction through hydrogen gas (Bohr 1913d, pp. 863–864).

Bohr (1913d, p. 866) cites experiments by Thomson (1912b, p. 253) in support of the existence of the other stable system he found, the molecular hydrogen anion, H_2^-, where the extra electron is bound to the neutral H_2 molecule with an energy of about 1.5 eV. In fact, this anion does not exist as a stably bound system, although a relatively long-lived (i.e., on the order of microseconds) resonance in this channel was finally observed in an accelerator mass spectrometer experiment in 2005 (Heber et al. 2006)!

In addition to the net binding energy of the neutral hydrogen molecule, Bohr (1913d, p. 864) calculates the frequency of oscillations obtained by displacing the electron ring parallel to the inter-nuclear axis (the stability of the system with respect to such displacements had already been established, with small displacements leading to harmonic oscillations around the equilibrium position). He finds a frequency of 3.8×10^{15} Hz, which is close to the value of 3.5×10^{15} Hz obtained from the dispersion measurements of Clive and Maude Cuthbertson (1909) that he had already used in his paper on the scattering of α particles (Bohr 1913a) and in the Rutherford Memorandum (see Sections 4.3.3 and 4.3.5). In those cases, however, he had interpreted it as an orbital frequency on the basis of the Drude–Lorentz dispersion theory. The dispersion measurements, analyzed on this basis, indicated the presence of *two* dispersion electrons of this frequency. In Part Three of the Trilogy, Bohr ingeniously manages to come up with another source of radiation of (roughly) the same frequency. Using his frequency condition, $E = h\nu$, he determines the frequency of the light emitted if one electron is removed from the hydrogen molecule, leaving a free electron, a hydrogen nucleus, and a neutral hydrogen atom, which he assumes to be in its ground state (Bohr had concluded earlier—and incorrectly—that the positive molecular hydrogen ion H_2^+ cannot exist). The total binding energy of this system is thus the binding energy W_0 of this one hydrogen atom. The difference between this energy and the energy of the original system, which Bohr sets equal to $2.2W_0$, corresponds to a frequency $\nu = 1.2W_0/h = 3.7 \times 10^{15}$ Hz. This frequency is close to the ring-oscillation frequency found earlier and this emission process thus provided Bohr with the desired second dispersion mode. It is hardly necessary to say that none of this ingenious argumentation would survive the transition to modern quantum mechanics.

Bohr was more successful (or, perhaps, lucky) in his discussion of the vibrational spectrum of the hydrogen molecule. Small displacements of the nuclei from their (stable) equilibrium points along the inter-nuclear axis would also lead to harmonic oscillations, the frequency of which could be calculated by simple mechanics. Bohr found the result 1.91×10^{14} Hz (the correct modern value is 1.32×10^{14} Hz), well into the infrared and of the same order of magnitude as "that calculated by Einstein's theory from the variation of the specific heat of hydrogen with temperature" (Bohr 1913d, p. 866) by Niels Bjerrum (1911, 1912), a lecturer in chemistry at the University of Copenhagen (Kragh 2012, p. 87, note 93). Bohr correctly interpreted the observed absence of any infrared absorption by hydrogen gas (including at this specific frequency) as due to the "symmetrical structure of the system" (Bohr 1913d, p. 866). In other words, the arrangement of the electrons centrally, and symmetrically, between the two nuclei implied that the molecule had no net permanent electric dipole to which the electric field of an incoming electromagnetic wave could couple:

> The complete absence of infrared absorption in hydrogen gas might be considered a strong argument in support of a constitution of a hydrogen molecule like that adopted here, compared with model-molecules in which the chemical bond is assumed to have its origin in an opposite charge of the entering atoms (Bohr 1913d, p. 866).

In the final section of Part Three (entitled "Concluding remarks") Bohr lists the five assumptions underlying his approach to atomic and molecular structure in the Trilogy:

1. That energy radiation is not emitted (or absorbed) in the continuous way assumed in the ordinary electrodynamics, but only during the passing of the systems between different "stationary" states.

2. That the dynamical equilibrium of the systems in the stationary states is governed by the ordinary laws of mechanics, while these laws do not hold for the passing of the systems between the different stationary states.

3. That the radiation emitted during the transition of a system between two stationary states is homogeneous [i.e., monochromatic], and that the relation between the frequency ν and the total amount of energy emitted E is given by $E = h\nu$, where h is Planck's constant.

4. That the different stationary states of a simple system consisting of an electron rotating around a positive nucleus are determined by the condition that the ratio between the total energy, emitted during the formation of the configuration, and the frequency of revolution of the electron is an entire multiple of $h/2$. Assuming that the orbit of the electron is circular, this assumption is equivalent with the assumption that the angular momentum of the electron round the nucleus is equal to an entire multiple of $h/2\pi$.

5. That the "permanent" state of an atomic system—i.e., the state in which the energy emitted is maximum—is determined by the condition that the angular momentum of every electron round the centre of its orbit is equal to $h/2\pi$ (Bohr 1913d, pp. 874–875).

We have already seen (and Bohr must surely have been aware) that conditions 2 and 5 are in fact incompatible for multi-electron atoms with more than one ring of electrons. In the following, we shall see that the old quantum theory continued to be plagued by inconsistencies, which were never (and could never be) resolved. In condition 4, we see Bohr reverting to his first (and least convincing) argument from Part One for his quantum condition, based on an analogy to Planck's second theory (see the discussion following Eq. (4.37)). Henceforth, he would avoid this way of stating the quantization condition, relying instead on correspondence arguments as providing the needed constraining principle.

If we step back for a moment and take a balance of the achievements of Bohr's remarkable efforts in 1913, laying the foundation for another twelve years of intense activity in atomic theory, it is difficult to avoid the conclusion that Bohr was fortunate to have developed his ideas in the *absence* of precise empirical data. With the exception of atomic spectroscopy, experimental atomic physics was in its infancy, and could only be expected to provide the crudest quantitative information about important parameters such as the binding energies of electrons in atoms (determined through successive ionization potentials) or even about the actual existence of hypothetical atomic systems (such as the molecular hydrogen anion, H_2^-, discussed above). Instead, the only area in which relatively precise empirical data existed, atomic spectroscopy, furnished Bohr with the Balmer formula, fortuitously adapted to the one type of system (hydrogen or hydrogen-like atoms) in which his quantization condition (in consequence of special features of Coulomb's law) actually yielded a result in consonance with modern quantum theory. In essentially all other cases, Bohr's calculations led to quantitatively (and in some cases, qualitatively) incorrect results, which could fortunately (for Bohr and the acceptance of his ideas) be argued away on the basis of uncertainties in the experimental situation. Within ten years the experimentalists would have eliminated many of the shortcomings of their earlier efforts, thereby precipitating several crises (see Chapter 7), which would necessitate the development of a very different theory.

4.6 Early evidence for the Bohr model: spectral lines in hydrogen and helium

Bohr's remarkable achievement of expressing the Rydberg constant appearing in the Balmer formula in terms of known fundamental constants might have been dismissed as a fortuitous accident, especially since the derivation involved some truly iconoclastic assumptions from the point of view of classical radiation theory. Already in Part One of the Trilogy, however, Bohr tremendously strengthened his case by using the new model to clear up a confusing set of spectroscopic observations going back to stellar spectra studied seventeen years earlier by the Harvard astronomer Edward Pickering (1896, 1897).[58] He studied the spectrum of the star ζ Puppis, a blue giant with an extremely high surface temperature on the order of 40 000 K, capable of producing substantial quantities of ionized helium. Pickering (1896, p. 369) detected a set of lines he called "very remarkable and unlike any other as yet obtained." He identified six lines in particular whose wavelengths fit a Balmer-like formula $\lambda = C_1 m^2/(m^2 - 4) - C_2$ (with $m = 5, 6, 7, 8, 9, 10$), but realized shortly thereafter (early 1897) that the lines in question could be described even more accurately by a formula of exactly the Balmer type:

$$\lambda = 3646.1 \frac{n^2}{n^2 - 16}, \quad n = (5, 7, 9), 11, 13, 15, \ldots \tag{4.58}$$

(the integers in parentheses correspond to lines too weak to detect in the stellar spectra). Remarkably, this formula also describes the original Balmer series wavelengths, if we take the even values $n = 6, 8, 10, \ldots$ for the running integer in Eq. (4.58). Moreover, the series limit (for n going to infinity) was evidently 3641.6 Å for both series (1 Ångström $= 10^{-10}$ m). This led Pickering (1897, p. 92) to assume that the series "is so closely allied to the hydrogen series, that it is probably due to that substance under conditions of temperature or pressure as yet unknown."

---

[58] For an excellent survey of the tangled history of the Pickering lines, see Robotti (1983).

The appearance of two series of lines exactly interspaced with each other, and with the same term limit, was already familiar from Rydberg's (1897) early investigations of alkali spectra (see Appendix B, section B.4). They correspond to the sharp and diffuse series in these elements (in later terms, lines arising from s to p and d to p transitions, with the coincident term limit corresponding to the binding energy of the same lowest p level). Rydberg naturally assumed that Pickering had uncovered the sharp series of lines for hydrogen, with the original Balmer series corresponding to the diffuse series. If one rewrites Eq. (4.58) to emphasize the relation to the Rydberg parametrization of series spectra (with the term limit $3646.1 = 4/R$), the "sharp" Pickering series takes the form of a Balmer series with half-integral running numbers

$$\frac{1}{\lambda} = R\left(\frac{1}{2^2} - \frac{1}{(n/2)^2}\right), \quad n = 5, 7, 9, \ldots \tag{4.59}$$

(with R the Rydberg constant for inverse wavelength). In fact, the spectral shift σ identified by Rydberg (see Eq. (B.5)) for the sharp lines for lithium was close to $1/2$, so the appearance of half-integral values $5/2$, $7/2$, $9/2$ etc. in Eq. (4.59) fit perfectly well in this picture. The analogy was completed by Rydberg's prediction, again in analogy to the alkali metals, of a new hydrogen series—the principal—in which the running and fixed terms were interchanged in comparison to the sharp series (see Eq. (B.7)), with the fixed term corresponding to a running number (in this case, half-integer) one less than the minimum allowed ($5/2$) in the sharp series. The wavelengths for this new series should thus follow the pattern

$$\frac{1}{\lambda} = R\left(\frac{1}{(3/2)^2} - \frac{1}{n^2}\right), \quad n = 2, 3, 4, \ldots, \tag{4.60}$$

which for $n = 2$ predicted a line at 4688 Å.

Rydberg's confidence in his attribution of the Pickering lines to a sharp hydrogen series and the predicted first principal series line at 4688 Å was solidified by the clear appearance of the latter (as a line even brighter than the known lines of hydrogen) in various stellar spectra published by Antonia C. Maury and Pickering (1897).[59] However, doubts about the association of this line with hydrogen would soon emerge. A coronal line at 4686 Å was detected in 1898 in an eclipse of the Sun, and by 1905 Norman Lockyer, an English astronomer long involved in the story of helium (Nath 2013), had found essentially coincident lines (at 4685.90 and 4685.97 Å respectively) in the solar spectrum and in the spectrum from a discharge tube containing only helium and oxygen.

Nevertheless, the identification with hydrogen continued to be stubbornly maintained right up to (and past) the appearance of the Bohr model in 1913, in which half-integral

---

[59] As Pickering notes in the preface to this publication: "This work was assigned to Miss Antonia C. Maury in 1888, and she alone is responsible for the classification contained in Part I. of this volume." See Sobel (2016) for a history of women working at the Harvard Observatory.

quantum numbers would have to be introduced to account for the Pickering lines *if* these lines were ascribed to hydrogen. The mathematical association of the Pickering lines with the Balmer series of hydrogen emphasized by Rydberg continued to exert a hypnotic hold on the interpretation of these lines. In a paper in 1912 reporting on the latest investigations of laboratory spectra of hydrogen, the astronomer Alfred Fowler continued to insist on the hydrogenic origin of both the Pickering lines and the 4686 "principal series" line, notwithstanding all evidence to the contrary:

> So far it has not been possible to obtain the lines in question from hydrogen alone under apparently identical conditions, although there is not the slightest difficulty in obtaining them when helium is also present. It is remarkable also that a very small proportion of hydrogen is sufficient to show 4686 brightly ... One hesitates to believe, however, that the presence of helium is an indispensable condition for the production of the Principal series of hydrogen, and further experiments on this point will be made. In the absence of strict experimental proof that 4686 and the associated lines are really due to hydrogen, Rydberg's theoretical investigations may be regarded as justifying such a conclusion (Fowler 1912, p. 65).

The confusion over the origin of the Pickering and Fowler lines was addressed head-on by Bohr in Part One of the Trilogy. Bohr (1913b, p. 10) observed that, for singly-ionized helium (a helium atom with a single electron but doubly charged nucleus), the Balmer formula Eq. (4.27) would acquire an extra factor of 4 in the numerator of the Rydberg constant, due to the square of the nuclear charge,

$$\nu = \frac{E_{\text{tot}}^{(\tau_1)} - E_{\text{tot}}^{(\tau_2)}}{h} = \frac{8\pi^2 m e^4}{h^3}\left(\frac{1}{\tau_2^2} - \frac{1}{\tau_1^2}\right). \tag{4.61}$$

Transferring a factor of 4 to the denominators, and writing the formula for inverse wavelengths $1/\lambda = \nu/c$ instead of frequencies (and in terms of the Rydberg constant $R = 2\pi^2 m e^4/h^3 c$) one obtains a formula exactly similar to the Balmer one for hydrogen, but with *halved* quantum numbers:

$$\frac{1}{\lambda} = R\left(\frac{1}{(\tau_2/2)^2} - \frac{1}{(\tau_1/2)^2}\right). \tag{4.62}$$

Comparing this with Eq. (4.60), one sees immediately that the 4688 line emerges as the consequence of a transition from a $\tau_1 = 4$ state to a $\tau_2 = 3$ state in ionized helium. The Pickering lines of Eq. (4.59) can similarly be attributed to transitions to level $\tau_2 = 4$ from odd levels $\tau_1 = 5, 7, 9, \ldots$. The transitions from even levels $\tau_1 = 6, 8, \ldots$ to $\tau_2 = 4$ would coincide in wavelength with the original Balmer series (corresponding in transitions from 3, 4, 5, ... to 2) and would therefore typically overlap with hydrogen lines, leading (if hydrogen were also present, as in stars) to "lines of a greater intensity than the rest of the lines in the series" (Bohr 1913b, p. 11), as observed.

In the letter to Rutherford of March 6, 1913 with which he enclosed the first version of Part One of the Trilogy, Bohr indicated his doubts that the Pickering–Rydberg–Fowler lines were really due to hydrogen, suggesting (in line with the observations of Lockyer) that helium was the culprit. The Copenhagen chemist Bjerrum, Bohr writes,

> suggested to me that if my point of view was right the lines might also appear in a tube filled with a mixture of helium and clorine [sic] . . . we have not in Copenhagen the opportunity to do such an experiment satisfactorily; I might therefore ask you, if you possibly would let it perform in your laboratory, or if you perhaps kindly would forward the suggestion to Mr. Fowler, which [sic] may have the arrangement used still standing (Bohr 1972–2008, Vol. 2, p. 582).

By September 1913, an associate of Rutherford's in Manchester, Evan J. Evans (1913), was able to establish the clear presence of the 4686 line in a helium tube discharge after scrupulous efforts had been made to eliminate all traces of hydrogen, with none of the conventional hydrogen lines visible. A few weeks later, Fowler (1913a) responded that the apparent absence of the usual hydrogen lines might simply be due to the intense nature of the discharges used (essential in creating the singly-ionized helium, as we now know), which somehow suppressed the usual hydrogen lines, even with some residual hydrogen present in the tube. More worrying was the small, but significant (i.e., outside observational errors) discrepancy between the observed wavelengths and those calculated by Bohr in Eq. (4.62).

This observation of Fowler's led, again within a few weeks, to Bohr's beautiful, and final, resolution of the issue, in a letter to *Nature* dated October 23, 1913 (Bohr 1913f). The discrepancies raised by Fowler were shown to be due to the neglect of nuclear motion effects: the electron mass m appearing in the Rydberg constant (see Eq. (4.61)) should really be replaced by the reduced mass $\mu \equiv mM/(M+m)$ in order to take into account the large but finite mass M of the nucleus, as in any treatment of bound two-body systems, even classically. This means that there is a slight difference (of the order of 4 parts in 10000) in the value of the Rydberg constant for hydrogen and helium, and once this is taken into account the discrepancies raised by Fowler disappear, as acknowledged by Fowler (1913b) himself in a note immediately following Bohr's letter.[60]

Bohr's resolution of the Pickering line mystery was clearly an important component in the early recognition of the importance of the new model for atomic spectroscopy.[61] Yet, Fowler was not the only spectroscopist who was so wedded to Rydberg's interpretation of the Pickering–Fowler lines that he had a hard time accepting Bohr's reinterpretation of

---

[60] Philosopher of science Imre Lakatos (1970, p. 149) used this episode as an illustration of what he called "monster-adjustment", defined as: "turning a counterexample, in light of some new theory, into an example."
[61] The famous experiments by Franck and Hertz (1913, 1914, 1916, 1919) only came to be recognized as evidence for the Bohr model several years later. Bohr cautiously referred to these experiments as supporting his theory in a paper published in the September 1915 issue of *Philosophical Magazine* (Bohr 1915). And it was not until four years later that Franck and Hertz (1919) themselves for the first time stated unequivocally and in print that their results confirmed Bohr's theory. See Kragh (2012, pp. 143–146) and Gearhart (2014) for historical discussions of these experiments and their connection to Bohr's theory.

them. At the Rydberg Centennial Conference in Lund, Sweden, in 1954, Bohr recalled that the Göttingen spectroscopist Carl Runge had warned "at a colloquium in Göttingen, against such apparently arbitrary use of spectral evidence by theoreticians who did not seem properly to appreciate the beauty and the harmony of the general pattern of series spectra, revealed [above] all by the ingenuity of Rydberg" (Kragh 2012, p. 123). This is just one example of the negative initial reaction to Bohr's atomic model, especially in Germany, which, unlike Great Britain, did not have a tradition of atomic model building (Kragh 2012, Ch. 3). Even Sommerfeld, whose elaboration of Bohr's model will be the focus of Part Two of this volume, initially had his reservations. On September 4, 1913, in a postcard acknowledging receipt of one or more parts of the Trilogy that Bohr had sent him, Sommerfeld wrote: "The problem of expressing the Rydberg–Ritz constant by Planck's h has for a long time been on my mind. Though for the present I am still rather sceptical about atomic models in general, calculating this constant is undoubtedly a great feat" (Kragh 2012, p. 92).

In the introduction (see Section 1.2.10), we already quoted a passage from Einstein's (1949, pp. 46–47) autobiographical notes in which he recalled how impressed he had been when he first learned of the Bohr model, despite its "insecure and contradictory foundation." In late September 1913, George de Hevesy wrote to Bohr from Vienna where he was attending the 85th annual *Naturforscherversammlung*. None of the talks at this conference mentioned Bohr's model. Einstein's talk was on gravity (Einstein 1913). Hevesy, however, had a conversation with Einstein about the Bohr model. Hevesy told Einstein that it had now been established that the Pickering–Fowler lines belong to helium. Einstein, Hevesy reported to Bohr, was "extremely astonished . . . and told me: 'Than the frequency of light does not depand at all on the frequency of the electron . . . this is an *enormous achievement*. The theory of Bohr must then be wright'" (Hevesy to Bohr, September 23, 1913; spelling errors in the original). A few weeks later, Hevesy reported this same conversation to Rutherford: "When I told him about the Fowler spectrum the big eyes of Einstein looked still bigger and he told me 'Then it is one of the greatest discoveries'" (Hevesy to Rutherford, October 14, 1913; both Hevesy letters are quoted in Kragh 2012, p. 95).

In early September 1913, a few weeks before the *Naturforscherversammlung*, the 83rd annual meeting of the British Association for the Advancement of Science took place in Birmingham.[62] Unlike their counterparts in Vienna, September 21–28, the scientists in attendance in Birmingham, September 10–17, heard about Bohr's new theory on several occasions. Bohr's (1913b) July paper in *Philosophical Magazine*, Part One of the Trilogy, was mentioned by Sir Oliver Lodge in his presidential address, entitled simply "Continuity." Lodge, an unrepentant ether theorist and spiritualist, conceded that the agreement between Bohr's theory and observation, "including a determination of series of spectrum lines," was "very remarkable" (Lodge 1914, p. 17), but made it clear that the theory and the discontinuity it entailed were not to his liking. That same

---

[62] For an account of the discussions about atomic models and radiation at this meeting, see the report, "Physics at the British Association," in the November 6 issue of *Nature*, pp. 304–309; the coverage in *The London Times*, September 12 and 13; and Jeans (1914a). See also Keller (1983, pp. 173–176), Jammer (1966, pp. 86–87), Mehra and Rechenberg (1982a, p. 193), and Kragh (2012, pp. 92–93).

day, both Thomson and Rutherford gave talks under the title "The structure of the atom." According to the report in *The Times* the next day, Rutherford talked about the nuclear atom and the alpha and beta scattering evidence for it, while Thomson used the occasion to criticize the quantum theory, arguing that "there were insuperable difficulties in supposing the energy to be 'done up' in units, and he illustrated this point by a simple analogy" (p. 10). The report in *Nature* also referred to this simple analogy (leaving it to the reader's imagination to fill in how it was supposed to work): "it will be long before his illustration of the quantum theory by pint-pots is forgotten" (p. 305).

The next day, September 12, a special session devoted to the problem of radiation was held. It drew an audience larger than the 350 people the large room at Mason College could hold, including two prominent foreign guests invited by the British Association, Lorentz and Mme. Curie. Jeans gave the opening lecture. As we saw in Chapter 3, he had closely followed the discussions about black-body radiation and specific heats in the preceding years. Jeans gently dismissed Thomson's qualms, saying he "would rather not adopt [Thomson's image,] as a pint pot could be conceived as sometimes only partially filled." Thomson got a laugh when he responded later in the session "that in using the image of the pint pot it had never occurred to him to imagine it as half-full. If Mr. Jeans could do that, his views of human nature fell short of the reality" (*The Times*, September 13, 1913, p. 10).

On a more serious note, Jeans mentioned Bohr's theory prominently and approvingly. He praised Bohr for his "most ingenious and suggestive, and I think we must add convincing, explanation of the laws of spectral series," though he also noted that "the only justification at present put forward for [the basic assumptions of Bohr's theory] is the very weighty one of success" (Jeans 1914a, p. 379). Other contributions to the session came from Lorentz, the experimentalist Pringsheim, who stressed the need for more accurate determinations of various constants, and Bohr himself, who "gave a short explanation of his atom" (*Nature*, November 6, 1913, pp. 305–306). More conservative opinions were voiced by Lodge, Larmor, Thomson, and Lord Rayleigh. The latter got a laugh when he prefaced his remarks by saying that "[w]hen he was a young man he had been of the opinion that no man over 70 ought to intervene in the discussion of rising theories. Now he had passed that limit, his conviction on that matter was no longer quite so strong" (*The Times*, September 13, 1913, p. 10).

All in all, the session was a great success for Bohr, who enthusiastically wrote to his wife two days later that Jeans had given "a very beautiful and kind presentation of my theory. I think that he is convinced that there is at least some reality behind my considerations" (Niels Bohr to Margrethe Bohr, September 14, 1913; quoted in Kragh 2012, p. 93). Decades later, in his 1958 Rutherford Memorial Lecture, he recalled that Jeans's "lucid exposition was, in fact, the first public expression of serious interest in considerations which outside [Rutherford's] Manchester group were generally received with much scepticism" (Bohr 1961, p. 1093; quoted in Kragh 2012, p. 132).[63]

---

[63] The following year, Jeans praised Bohr's theory again in the influential report on radiation and quantum theory that we already mentioned at the end of Section 4.4.1, calling it "very remarkable and intensely interesting" (Jeans 1914b, p. 51; quoted in Kragh 2012, p. 94).

Plate 1 *Wilhelm Wien.*

Plate 2 *Friedrich Paschen.*

Plate 3 *Heinrich Rubens.*

Plate 4 *Max Planck.*

Plate 5 *Hendrik Antoon Lorentz.*

Plate 6 *Boltzmann tombstone with*
$S = k \log W$.

Plate 7 *Planck tombstone with*
$h = 6.62 \cdot 10^{-34} W \cdot s^2$.

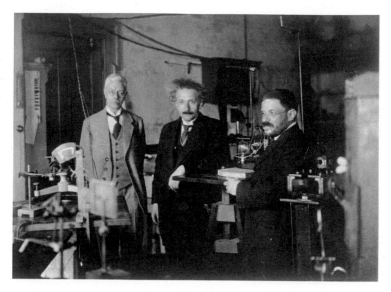

Plate 8 *Pieter Zeeman, Albert Einstein, and Paul Ehrenfest in Amsterdam.*

Plate 9 *The first Solvay Conference in Brussels in 1911.*

Plate 10 *Johannes Stark.*

Plate 11 *Peter Debye.*

Plate 12 *Hans Geiger and Ernest Rutherford in Schuster Laboratory in Manchester.*

Plate 13 *George A. Schott.*

Plate 14 *John Williams Nicholson.*

Plate 15 *Arnold Sommerfeld and Niels Bohr in Lund, 1919.*

Plate 16 *Johannes Rydberg.*

Plate 17 *Johann Jakob Balmer.*

Plate 18 *Walter Ritz.*

Plate 19 *Woldemar Voigt.*

Plate 20 *Max Laue (far left), Paul Sophus Epstein (to his left), Paul Ewald (far right), and other members of Sommerfeld's group in Munich in 1912.*

Plate 21 *Karl Schwarzschild.*

Plate 22 *Henry Moseley.*

Plate 24 *Paul Ewald, Alfred Landé, Mrs. Landé, Lise Meitner, Mr. Schmitz, and Ernst Back.*

Plate 23 *Wojciech (Adalbert) Rubinowicz.*

Plate 25 *Hans Kramers.*

Plate 26 *Seated: Max Born. Standing behind him (left to right): Carl Oseen, Niels Bohr, James Franck, and Oskar Klein.*

Part II

The Old Quantum Theory

5
Guiding Principles

The publication of Bohr's 1913 trilogy on the quantization of atomic systems and, in particular, the spectacular success of his ideas in accounting for the numerological magic of the Balmer formula for the spectra of hydrogen and ionized helium, clearly marks a watershed moment in the development of quantum theory. It comes at roughly the halfway point between Planck's introduction in 1900 of the quantum of action in the treatment of black-body radiation and the development of the new quantum mechanics of Heisenberg, Schrödinger and others in 1925/1926 and its formal completion by von Neumann in 1927. The possibility of a quantitative understanding of the rich supply of empirical data on the line spectra of atoms had predictable effects. First, there was a considerable increase in the number of physicists addressing quantum issues in atomic physics from 1913 on. Second, for the next twelve years, the analysis and interpretation of spectral data became the primary tool of quantum theory, always with the archetype of the Bohr–Rutherford "planetary model" of the atom at its core. After the appearance of modern quantum mechanics, the theory thus developed during the period between 1913 and 1925 became known as the old quantum theory.

This period perhaps represents the most difficult and complex part of the conceptual evolution of quantum mechanics. On the one hand, the theory was able to address, and indeed account quantitatively for, several important phenomena in atomic physics. Sommerfeld's derivation of the relativistic fine structure in hydrogenic (single-electron) atoms and in the X-ray spectra of a whole range of atoms, as well as the Schwarzschild–Epstein calculation of the splitting of the spectral lines of hydrogen by an external electric field, strongly suggested that physicists were on the right track. It hardly seemed possible that both of these highly nontrivial calculations could give the right answer by pure accident. On the other hand, the failure of the theory, especially from 1920 onwards, to deal convincingly with much of the spectroscopic data (the anomalous Zeeman effect and helium spectrum being the most prominent examples) increasingly led to the realization that minor tinkering with the basic principles employed to date would not meet the challenge of providing a truly comprehensive and reliable theory of atomic phenomena.

In the immediate aftermath of Bohr's 1913 papers, the task facing theorists working on quantum physics seemed clear: the ad hoc hypotheses introduced by Bohr, however successful in accounting for the spectrum of the hydrogen atom, needed to be supplanted

Constructing Quantum Mechanics. Anthony Duncan and Michel Janssen, Oxford University Press (2019).
© Anthony Duncan and Michel Janssen. DOI: 10.1093/oso/9780198845478.001.0001

by a more general set of instructions for the insertion of Planck's quantum, clearly the characteristic quantity of the new physics, into dynamical systems of greater generality than that of a single charged particle bound by a Coulomb potential. To deal with the flood of spectroscopic data being produced by the experimentalists, one needed to understand the extension of Bohr's results to atoms subject to external fields (electric or magnetic), as well as to atoms with more than one electron, and even to molecules, condensed systems (solids), etc. What was clearly needed was a guiding principle (or set of principles) that would extend Bohr's idea of a hybrid classical/quantum dynamics: classical dynamics holding for a selected set of "admissible" classical motions, quantum jumps between such motions leading, via the Bohr frequency condition, to a predictable line spectrum for the atomic system in question.

In this chapter, we discuss the three guiding principles that proved to be most important for the evolution of the old quantum theory from 1913 to 1925. First, we outline the development of the quantization conditions (or rules) introduced by Planck, Wilson, Ishiwara, Sommerfeld, Schwarzschild, and Epstein. These conditions formed the basis for essentially all explicit calculations in the old quantum theory up to its demise in 1925. Second, we give a brief account of the introduction and reception of the adiabatic principle, championed by Ehrenfest as an important criterion in determining precisely which classical quantities were the ones suitable for quantization. Finally, there is the correspondence principle of Bohr, introduced (without the name) already in the first paper of the 1913 trilogy, to justify the quantization of angular momentum of the electron in the hydrogen atom (see Section 4.5.1), but later undergoing considerable extension and deepening, right up to the point where it served as a critical component of the Kramers dispersion theory, which, in turn, triggered the development of matrix mechanics.

5.1   Quantization conditions

5.1.1   Planck

The extension of the early successes of quantum ideas in the areas of heat radiation, the photoelectric effect, and the specific heats of solids at low temperatures (to name three prominent examples prior to 1910) to more general problems involving both periodic (e.g., atomic theory) and aperiodic (e.g., β, γ, and X-ray emission processes) phenomena was the critical problem facing quantum theory in the first decade of its existence. During this period Planck appears to have held to the most conservative interpretation of energy quantization compatible with the need to introduce finite energy elements in the combinatorial analysis of the entropy of charged oscillators in the black-body problem (see Sections 2.6 and 3.4.2). For a considerable period of time (see Kuhn 1987, ch. V), Planck continued to maintain the possibility of continuously variable oscillator energies, with a quantized *exchange* of energy, first localized solely during periods of absorption of energy from an incident light wave, then later solely during periods of emission of electromagnetic energy by the oscillator.

For an oscillator of natural frequency ν, the equiprobable occupation of any point in energy intervals of width $h\nu$ was the only assumption needed for the combinatorial arguments employed in Planck's 1900 derivation of the black-body formula to hold (see Section 2.6).

In the first edition of his lectures on heat theory, Planck (1906a, p. 155) introduces a critical new idea in his reexamination of the entropy calculation for a system of (one-dimensional) simple harmonic oscillators. Recalling the essential role played by the phase space spanned by coordinates q and momenta p in classical statistical mechanics, where regions of equal volume are assigned equal a priori probability, Planck notes that, for a single oscillator, with mass m and natural cyclic frequency ν, the area in the (q, p)-plane enclosed by the constant-energy curve (cf. Fig. 5.1),

$$E = \frac{1}{2}m\omega^2 q^2 + \frac{1}{2m}p^2 \tag{5.1}$$

(with $\omega = 2\pi\nu$), which determines an ellipse with semi-major/minor axes $\sqrt{2E/m\omega^2}$ and $\sqrt{2mE}$,[1] is given by

$$\pi\sqrt{\frac{2E}{m\omega^2}} \cdot \sqrt{2mE} = E\frac{2\pi}{\omega} = \frac{E}{\nu} \tag{5.2}$$

(cf. Appendix A, Eqs. (A.169)–(A.171)). The area of the elliptical rings contained between two oscillator orbits of energy differing by a single quantum of energy $h\nu$ was therefore a universal constant, h, a quantity from which all reference to the specific

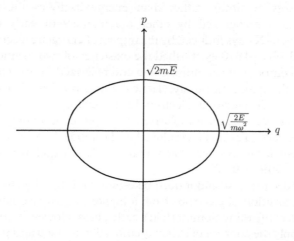

Figure 5.1 *Orbit in phase space (q, p) of a harmonic oscillator with energy E.*

---

[1] For $p = 0$, $q = \sqrt{2E/m\omega^2}$; for $q = 0$, $p = \sqrt{2mE}$.

mechanical (or electromagnetic) properties of the charged oscillator is absent. Planck was quite explicit about the fundamental change in attitude entailed by this observation. By this association of rings to elementary quanta of energy, he wrote,

> the quantum of action h is revealed to us with a new significance, namely, as the size of an elementary region in the phase space of an oscillator, *valid for oscillators of quite arbitrary frequency* (Planck 1906, p. 156, our emphasis).

In the second edition of his lectures on heat radiation, the quantization of the phase space and its relation to classical statistical mechanics is made explicit and generalized to systems with many degrees of freedom. The phase space must be regarded as sliced into regions where the quantity[2]

$$G = \int dq_1 \ldots dq_n dp_1 \ldots dp_n \qquad (5.3)$$

has a fixed value (namely, h^n), but where "the shape and position of the separate region elements are determined by the limits of the integral and must be determined anew in every separate case" (Planck 1913, p. 126). For one-dimensional systems such as the simple harmonic oscillator, the application of this idea was, of course, unproblematic, and it could even be extended trivially to new situations such as the anharmonic oscillator, with, for example, potential energy $V(q) = aq^2 + bq^3$, whose quantized energies were computed from the condition $\int dqdp = nh$ (in an expansion in the small parameter b) by Boguslawski (1914) in a study of pyroelectricity.

The significance of phase space area, with dimensions of action (distance × momentum = energy × time), rather than energy itself, as the proper focus of quantization was soon recognized by other practitioners of early quantum theory. Experimental studies of X-rays induced by the impact of energetic electrons on a metallic anode in the period 1905–1910 established the existence of two distinct components of the radiation. One component was anisotropic and polarized. It was correctly identified by Sommerfeld as the expected purely classical *Bremsstrahlung* ("braking radiation") emitted by a rapidly decelerating electron losing energy as it traverses the metal anode. In addition, there was an unpolarized isotropic component, with wavelengths characteristic of the particular metal employed in the anode.[3] For this latter component, Sommerfeld suggested, "it was very likely that the Planck quantum of action plays a role" (Sommerfeld 1909, p. 970).

Within the next decade, it would indeed be established that the characteristic X-rays resulted from the transition of electrons from a higher to a lower quantized orbit in the Bohr atomic model, confirming Sommerfeld's early guess. However, Sommerfeld's other early attempts to apply the concept of action quantization to the β and γ radiation emitted

---

[2] Planck uses φ_i, ψ_i for coordinates and momenta; here we revert to the more usual modern notation.

[3] Later, Sommerfeld (1911b) would retract his insistence on the purely classical character of the polarized component, applying his quantum action principle (see below) here as well.

by radioactive materials were not successful. As we saw in Section 1.3.1, Sommerfeld tried in vain to account for the energy of the γ rays frequently emitted together with β radiation on the basis his "*h*-hypothesis", according to which the total action of the emission process (energy emitted times the duration of the emission) be equal to a single Planck quantum h.[4] Sommerfeld's reluctance to accept Einstein's light-quantum proposal at face value (in which he was in line with the vast majority of physicists until the early 1920s) also led him to apply the same idea to the photoelectric effect: The continuous Maxwellian electromagnetic wave incident on a photoelectric cathode would be absorbed over a definite "accumulation time" again determined by an action quantization principle, at the conclusion of which the photoelectron would be emitted from the atom (Sommerfeld 1911a). As with Planck, it was crucial for Sommerfeld that the usual laws of classical physics (both mechanical and electromagnetic) be maintained at all times *except* during the sudden energy exchange processes (emission or absorption) characteristic of atomic phenomena.

Until the heyday of the old quantum theory, with the quantization rules of Wilson, Ishiwara, Sommerfeld, and Schwarzschild in the driving seat, Planck continued to insist that the introduction of discretization in black-body radiation was strictly localized at the moments of sudden transition of the oscillator energies, which, at some time between 1911 and 1912, he transferred from the absorption to the emission process. In his so called "second theory" (see Section 2.6), oscillators occupied evenly distributed points in phase space between the elliptical rings demarcating regions of area h, moving out continuously under the classically determined absorption of energy from the incident black-body radiation, until a phase boundary (one of the elliptical rings corresponding to energy nhv) was reached, at which point, *purely stochastically*, a constant fraction η emit all their energy nhv, while the remaining $1 - \eta$ move into the next higher ring with energy $nhv < E < (n + 1)hv$, continuing to gain energy by absorption. The statistical analysis of this theory gave results identical to that of Planck's original theory, with the exception of the addition of a constant zero-point energy $\frac{1}{2}hv$ to all average oscillator energies (quite reasonably, as the lowest energy oscillators were uniformly distributed in the innermost ellipse corresponding to energies $0 < E < hv$). At this point, most clearly in the early history of quantum physics, discretization and irreducible acausality become firmly linked in the structure of the theory (cf. note 78 at the end of Chapter 3). However, equally clearly, the discretization still did not apply to the allowed orbits of mechanical systems in phase space, but only to the interaction process between such systems and the (equally continuous) electromagnetic field.

As we saw previously in our discussion of the Bohr model, the first argument given by Bohr for the appearance of stationary states of quantized energy is, strangely enough, based precisely on this second theory of Planck in which mechanical motions of continuous energy are allowed! While Bohr takes over certain aspects of Planck's treatment of charged oscillators—specifically, the sudden emission of exactly n quanta of energy hv from an oscillator at the precise moment where it crosses a phase space boundary—he

---

[4] For a fuller account of this episode, see Eckert (2015a, pp. 11–12; 2015b).

quite clearly (in 1913) abandons the notion that the mechanical system can exist (and interact classically with the electromagnetic field) at continuous, nonquantized values of the energy (i.e., in the interior of the elliptical rings in the oscillator case). Instead, the usual classical laws of motion are now assumed to hold only for a special discrete set of mechanical motions. These are the electron orbits that are reached by an electron which begins with zero energy far from an isolated hydrogen nucleus and emits exactly $nh\nu$ of monochromatic electromagnetic radiation (of frequency ν), reaching a stationary orbit (obeying classical mechanics—i.e., following a Keplerian orbit) with binding energy exactly equal to $nh\nu$. As we saw in Section 4.5.1, the quantitative content of the theory is completed by making the fundamentally arbitrary assumption that the frequency ν is precisely one-half of the frequency (determined by Newtonian principles) of the final orbit. The interpretation of the energy quantization so achieved is only related later in the paper to the more familiar requirement that the angular momentum of the electron in the nth (by now circular) orbit should be exactly $n(h/2\pi)$.

It would have been perfectly possible for Bohr to have realized at this point that this quantization condition, equivalent to the requirement

$$\oint p_\vartheta \, d\vartheta = nh \tag{5.4}$$

(with ϑ and p_ϑ conjugate angular and angular momentum variables), is the exact analog for rotational motion of Planck's condition specifying the elliptical boundaries separating the regions of unit quantum of action in the vibrational motion of an oscillator,

$$\oint p \, dq = \text{area of boundary ellipse} = \frac{E}{\nu} = nh. \tag{5.5}$$

Bohr, however, does not seem to have realized this. Given the tectonic shift in the underlying conceptual framework (from continuously variable to discretely determined mechanical motions), this is perhaps understandable. In fact, the only part of Bohr's 1913 trilogy which can be considered the germ of an important guiding principle of general applicability to a wider range of quantum problems than the hydrogen atom is the "correspondence principle" (as it would later be called) argument discussed previously, to which we shall return later in this chapter. It represents an important conceptual strand in the development of the old quantum theory, especially from about 1918 on, right up to the dispersion theory which proved to be the transitional phase into the matrix mechanics of Heisenberg, Born, and Jordan.

The need for a more general approach to quantization than afforded either by Planck's treatment of oscillators or Bohr's approach to the hydrogen atom became increasingly pressing in the years following Bohr's trilogy. The enormous amount of spectroscopic data, especially in the optical regime, and from multi-electron atoms, required some sort of explanation on the basis of quantum principles. Quantization rules which could operate satisfactorily in the peculiar hybrid classical/quantum framework of the old quantum theory were finally independently formulated in 1915 by several workers: of primary importance for our account—by Planck and Sommerfeld, working in Berlin and Munich, respectively, and by William Wilson at King's College, London. The special significance of the quantized "phase integrals" (in Sommerfeld's terminology) employed by these authors as equivalent to action variables in the action-angle formalism was first pointed out by Karl Schwarzschild, first in a letter to Sommerfeld of March 1, 1916, and again shortly thereafter in his paper on the Stark effect (discussed below). We shall discuss the approaches of Planck and Wilson first, returning to Sommerfeld and the Munich school at the end of the section.

In late 1915, Planck returned to the problem of finding an appropriate generalization of the "slicing" of phase space into cells of finite volume he had initiated with his considerations of elliptical rings (for the one-dimensional oscillator) in the first edition of his lectures on heat radiation (Planck 1906a). Now the goal was to find a generalization to systems of many degrees of freedom (Planck 1916). In a system with n independent coordinates and momenta, the $2n$-dimensional phase space $(q_1, \ldots, q_n, p_1, \ldots, p_n)$ was to be sliced into cells of equal phase space volume by hypersurfaces corresponding to constant values of n independent functions[5]

$$g_1(q_i, p_i), \ldots, g_n(q_i, p_i). \tag{5.6}$$

The boundaries of the cells would be placed at quantized values for the g_i:

$$g_i(q_j, p_j) = n_i h, \quad n_i = 1, 2, 3, \ldots \tag{5.7}$$

Planck repeatedly refers back to the classical equivalence between phase space volume and a priori probability, an equivalence he insists on retaining in the quantum theory. To ensure the preservation of this relation, he therefore demands that

> [a]n orbit of the system in phase space can not, under any circumstances, cut through the boundaries specified by [Eq. (5.7)]; the system rather traverses its entire orbit either in the interior of an elementary region, or along the boundary of two elementary regions (Planck 1916, p. 387).

Here one is considering the purely classical motion presumed to obtain *between* quantum jumps occasioned, say, by absorption or emission of radiation. Certainly, the easiest way to ensure the above requirement is to insist that the quantities g_i represent *constants of the motion*. The enforcement of the independent quantization rules in Eq. (5.7) also requires a *factorization of the phase space measure*

---

[5] Planck used the more awkward notation $g, g', g'', \ldots$ for these functions, and φ_i, ψ_i for q_i, p_i. We have changed the notation for clarity. We are assuming here the non-degenerate case.

$$dG \equiv \int_{g_1}^{g_1+dg_1} \cdots \int_{g_n}^{g_n+dg_n} dq_1 dp_1 \cdots dq_n dp_n = dg_1 \cdots dg_n, \qquad (5.8)$$

ensuring that the volume of the elementary region of phase space corresponding to $n_1 h < g_1 < (n_1 + 1)h, \ldots, n_n h < g_n < (n_n + 1)h$ be precisely h^n. In the original example of the linear oscillator, with $n = 1$, g_1 could, indeed must, be chosen to be proportional to the energy (which is in fact the only conserved quantity).

In the remainder of his paper on phase space structure, Planck applies this idea to a number of case studies: rigid rotators in a plane or in three dimensions, a one-dimensional simple harmonic oscillator, the two-dimensional isotropic harmonic oscillator, and the Coulomb (e.g., hydrogen atom) problem, treated both nonrelativistically and with special relativistic mechanics, among others. Lacking an automatic procedure for finding suitable slicing functions g_i, Planck is forced to deduce the quantization conditions for each problem on a case-by-case basis. We shall illustrate his method with a single case, the two-dimensional oscillator (analyzed classically in detail in Appendix A, Section A.2). As our coordinates (q_1, q_2) we use polar coordinates (r, ϑ). The conjugate momenta (p_1, p_2) are (p_r, p_ϑ) (cf. Fig. 5.2). In these coordinates, the Hamiltonian for the system is given by

$$H = \frac{p_r^2}{2m} + \frac{p_\vartheta^2}{2mr^2} + \frac{1}{2}m\omega^2 r^2. \qquad (5.9)$$

H and p_ϑ (energy and angular momentum) are constants of the motion, so our slicing functions g_1 and g_2 must be chosen as functions of these. From Eq. (5.9) it follows that

$$H \geq \frac{p_\vartheta^2}{2mr^2} + \frac{1}{2}m\omega^2 r^2. \qquad (5.10)$$

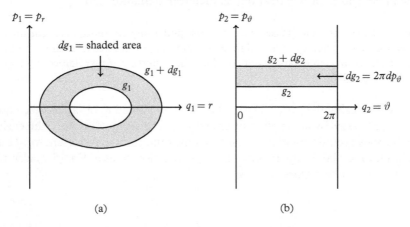

(a)   (b)

Figure 5.2 *Planck's "slicing" of (a) the radial phase space and (b) the angular phase space of a two-dimensional oscillator.*

Since

$$\left(\frac{p_\vartheta}{\sqrt{2mr}} - \sqrt{\frac{m}{2}}\omega r\right)^2 \geq 0, \tag{5.11}$$

or, equivalently,

$$\frac{p_\vartheta^2}{2mr^2} + \frac{1}{2}m\omega^2 r^2 \geq \omega p_\vartheta, \tag{5.12}$$

Eq. (5.10) implies that

$$H \geq \omega p_\vartheta. \tag{5.13}$$

The accessible phase space is thus bordered by the line $H = \omega p_\vartheta$. The (necessarily positive) functions g_1 and g_2, with a lower boundary at zero (ensuring that the appropriate quantization conditions are $g = nh$ and not $g = nh + $ constant) can thus be chosen as

$$g_1 = C_1(H - \omega p_\vartheta), \tag{5.14}$$

$$g_2 = C_2 p_\vartheta. \tag{5.15}$$

We now determine the normalization constants, starting with C_2. Eq. (5.8) with $(q_2, p_2) = (\vartheta, p_\vartheta)$ tells us that

$$dg_2 = \int_{g_2}^{g_2+dg_2} d\vartheta \, dp_\vartheta = 2\pi \, dp_\vartheta. \tag{5.16}$$

Comparison of this relation to Eq. (5.15) gives $C_2 = 2\pi$. To determine C_1, we consider an extremely eccentric elliptical orbit (almost the straight line through the force center), so that p_ϑ can be set to zero. Eq. (5.14) then reduces to $g_1 = C_1 H$, so g_1 must be proportional to H. For $p_\vartheta = 0$, the Hamiltonian in Eq. (5.9) is formally identical to the Hamiltonian for the one-dimensional oscillator. Using the quantization condition Eq. (5.5) for the one-dimensional oscillator and Eq. (5.8), with $(q_1, p_1) = (r, p_r)$, we find

$$dg_1 = \int_{g_1}^{g_1+dg_1} dr dp_r = \frac{1}{\nu} dH. \tag{5.17}$$

It follows that $C_1 = 1/\nu = 2\pi/\omega$.[6]

---

[6] Planck incorrectly identified the complete orbit in the straight-line limit as the portion where the particle leaves and returns to the force center once, rather than twice, as a comparison with the highly eccentric elliptical orbit case shows must be the case. This confusion indicates clearly the perils of the ad hoc case-by-case procedure employed by Planck to identify suitable quantization conditions.

To summarize, the appropriate quantization conditions for this model become

$$g_1 = \frac{2\pi}{\omega}H - 2\pi p_\vartheta = n_1 h, \tag{5.18}$$

$$g_2 = 2\pi p_\vartheta = n_2 h, \tag{5.19}$$

and, finally, for the energy

$$H = (n_1 + n_2)\frac{h}{2\pi}\omega = (n_1 + n_2)\hbar\omega. \tag{5.20}$$

Had Planck in late 1915 been au courant with the more sophisticated developments in analytical mechanics already familiar to theoretical astronomers working in celestial mechanics, he would have recognized immediately that the choice of action-angle variables (when they exist![7]) provides an automatic and immediate solution to the factorization requirement in Eq. (5.8) for the slicing functions g_i. Indeed (see Appendix A, Eq. (A.54)), the canonical transformation from general coordinates $(q_1, \ldots, q_n, p_1, \ldots, p_n)$ to action-angle variables $(w_1, \ldots, w_n, \mathcal{J}_1, \ldots, \mathcal{J}_n)$ has unit Jacobian,

$$\int dq_1 \ldots dq_n \, dp_1 \ldots dp_n = \int dw_1 \ldots dw_n \, d\mathcal{J}_1 \ldots d\mathcal{J}_n. \tag{5.21}$$

However, the angle variables w_i have unit period, so

$$\int dw_i = 1, \tag{5.22}$$

and Eq. (5.21) simply becomes

$$\int dq_1 \ldots dq_n \, dp_1 \ldots dp_n = \int d\mathcal{J}_1 \ldots d\mathcal{J}_n. \tag{5.23}$$

This allows us to identify directly the slicing functions with the action variables, $g_i = \mathcal{J}_i$. Moreover, by construction the action variables are constants of the motion, as desired. They would then, quite generally, be the appropriate instruments for quantization, at least in any problem which admits the introduction of action-angle variables (i.e., all separable conditionally periodic systems).

---

[7] The conditions for the applicability of the action-angle formalism are given in Appendix A, Section A.2.3.

The equivalence of Planck's procedure to the action-angle formalism is easily demonstrated in the two-dimensional oscillator case discussed above. From Eq. (A.146) we have that the integral defining the radial action variable $\mathcal{J}_r$ is[8]

$$\mathcal{J}_r = \oint p_r dr = 2 \int_{r_{\min}}^{r_{\max}} \sqrt{\left(2m\alpha_1 - \frac{\alpha_2^2}{r^2} - m^2 \omega^2 r^2 \right)} \, dr. \tag{5.24}$$

Here α_1 and α_2 are the Hamilton–Jacobi separation constants (see Section A.2.1) corresponding to energy (cf. Eqs. (A.143)–(A.144)) and angular momentum, respectively. More specifically, $\alpha_2 = p_\vartheta$ and $\mathcal{J}_\vartheta = \oint p_\vartheta d\vartheta = 2\pi\alpha_2$. Evaluating the integral in Eq. (5.24), one finds[9]

$$\mathcal{J}_r = \frac{2\pi}{\omega}\alpha_1 - 2\pi\alpha_2 = \frac{2\pi}{\omega}H - 2\pi p_\vartheta. \tag{5.25}$$

Comparison with Eq. (5.18) shows that the action variable $\mathcal{J}_r$ is just Planck's function g_1. Comparison of

$$\mathcal{J}_\vartheta = 2\pi\alpha_2 = 2\pi p_\vartheta \tag{5.26}$$

with Eq. (5.19) similarly shows that $\mathcal{J}_\vartheta$ is Planck's function g_2. The result of Planck's laborious procedure is thus completely equivalent to the resolution of the mechanical problem in action-angle variables. This was pointed out by Schwarzschild in a postcard to Sommerfeld of March 5, 1916 (Sommerfeld 2000, Doc. 241) and explicitly shown by Epstein (1918) (see Section 5.1.4).

5.1.2 Wilson and Ishiwara

In two important papers, submitted to the *Philosophical Magazine* in March and November 1915 respectively—and independently of the work of both Planck (described above) and Sommerfeld (described below)—William Wilson, working at King's College, London, introduced clearly and unambiguously the quantization rules for a multiply periodic system that would form the backbone of the old quantum theory for the next decade. He begins his first paper by quoting the remark by Jeans in his report on radiation and quantum theory (see Section 4.6) that Bohr's theory "is not inconsistent with the quantum-theory and is closely related to it" (Jeans 1914b, p. 51).[10] Given Jeans's

---

[8] Simply set $H = \alpha_1$ and $p_\vartheta = \alpha_2$ in Eq. (5.9) and solve for p_r.

[9] For a detailed version of the evaluation of an integral of this form, see, e.g., Duncan and Janssen (2014, p. 70).

[10] Wilson's paper was communicated to the *Philosophical Magazine* by Nicholson, who is also mentioned in the acknowledgments. Only much later did Wilson acknowledge the importance of Nicholson's (1912, p. 679) quantization of angular momentum for the quantum condition proposed in this paper (see Section 4.3.4, note 33).

assessment, Wilson expresses the goal of unifying Planck's treatment of the harmonic oscillator and Bohr's treatment of the hydrogen atom in terms of a "single form of quantum theory" (Wilson 1915, p. 795). However, Wilson immediately departs from a basic premise of Planck's second theory: namely, that the classical motion of the system (between the stochastic transitions characteristic of quantum behavior) could occur on orbits evenly distributed throughout phase space. Instead, Wilson insists that the only possible motions correspond to quantized values for the time integrals corresponding to the action[11] computed for each periodic coordinate separately.

More specifically, let us suppose that generalized coordinates $q_1, \ldots, q_n$ can be found such that the motion in coordinate q_i is periodic with frequency ν_i (hence period $1/\nu_i$). Let p_i be the momenta canonically conjugate to q_i. The total kinetic energy is supposed to take the form $\sum_i L_i$, with $L_i \equiv \frac{1}{2} A_i \dot{q}_i^2$, where the coefficients A_i may depend on the q_i. The conjugate momentum is then $p_i \equiv \partial L_i / \partial \dot{q}_i = A_i \dot{q}_i$. Wilson asserts that for each coordinate pair separately,

$$\int_0^{1/\nu_i} 2L_i dt = \int_0^{1/\nu_i} p_i \dot{q}_i dt = \int_0^{1/\nu_i} p_i dq_i = n_i h, \tag{5.27}$$

where the n_i are nonnegative integers. The quantities in Eq. (5.27) are in fact just the action variables $\mathcal{J}_i$ for such a system (see Section A.2), but, as with Planck, the terminology "action-angle variables" does not appear anywhere. Wilson's stated goal of a unifying principle encompassing both Planck's treatment of oscillators and Bohr's hydrogen atom model amounts to the observation that, for the oscillator, the requirement $\oint p \, dq = nh$ is exactly Planck's condition for the boundary orbits of the elliptical rings of area h, while for the hydrogen atom, an electron in a circular orbit, and therefore constant potential energy, must satisfy

$$2 \int_t^{t+\frac{1}{\nu}} L dt = 2L \int_t^{t+\frac{1}{\nu}} dt = \frac{2L}{\nu} = nh. \tag{5.28}$$

As the kinetic energy L is just one-half the angular momentum $\mathcal{J}$ times the angular speed ω (the unconventional notation for orbital angular momentum chosen to avoid confusion with the use of L for kinetic energy), the quantization condition $L = \frac{1}{2} nh\nu$ in Eq. (5.28) is equivalent to Bohr's quantization condition for angular momentum

$$\mathcal{J} = \frac{2}{\omega} L = n \frac{h}{2\pi}. \tag{5.29}$$

---

[11] The action integral considered by Wilson—the time integral of the kinetic energy, not the full Lagrangian—is sometimes called the "abbreviated" or "Maupertuis" action. The principle of least action in the Maupertuis version asserts that the physical path corresponds to the extremum of the abbreviated action over *energy conserving* paths.

In a second paper submitted to *Philosophical Magazine* in November 1915, Wilson (1916) addresses a rival proposal for quantization in terms of action integrals due to the Japanese physicist Jun Ishiwara. Ishiwara studied with Nagaoka in Tokyo and spent two years in Europe (working with Sommerfeld in Munich and Einstein in Zurich) before returning to Tohoku University in 1914. In his paper in the *Proceedings of the Tokyo Mathematico-Physical Society*, Ishiwara (1915) applies a single quantization condition to the average of the separate integrals in Eq. (5.27), with the integrals extended over the period of the whole system. This proposal therefore applies only to strictly periodic, not conditionally periodic, systems, and imposes a far less stringent restriction on the allowed classical motions than Wilson's proposed rule. Most of Wilson's paper is taken up by (in hindsight, correct) criticisms of the Ishiwara proposal. This is followed by a critical assessment of Planck's second theory, in which the mechanical system is allowed to occupy points uniformly in phase space, and not just at the boundaries identified by quantization conditions, where, as described earlier, Planck presumes quantum discontinuity appears in the guise of sudden stochastic emission processes (while absorption of electromagnetic energy occurs continuously and classically). Wilson criticized this peculiar asymmetry in Planck's treatment of the interaction of matter and radiation in the following terms:

> So long as no other way of accounting for the phenomena concerned can be found, such hypotheses may be justifiable; but no theory can be held to be satisfactory which applies one system of dynamics to the process of absorption and another to emission ... Moreover, there is at least one consequence of the continuous absorption hypothesis which is not in good accord with experimental results. It demands a photo-electric "accumulation period". The experiments of Marx and Lichtenecker [1913] seem to show, however, that this period (if it exists at all) must be much shorter than that calculated from the principles of Planck and Ishiwara (Wilson 1916, p. 159).

For our present purposes, in tracing the historical development of the quantization rules of the old quantum theory, the most intriguing part of Wilson's second paper appears in the final two pages. Wilson applies the quantization rule in Eq. (5.27) to the hydrogen atom, choosing polar variables (r, ϑ) as the generalized coordinates. In Wilson's notation, in which the coordinates and conjugate momenta $(r, \vartheta, p_r, p_\vartheta)$ are written as (q_1, q_2, p_1, p_2), the quantization conditions are

$$\int p_1 dq_1 = \rho h, \tag{5.30}$$

$$\int p_2 dq_2 = 2\pi p_2 = \sigma h, \tag{5.31}$$

where ρ and σ are integers. For a Kepler orbit of eccentricity ϵ and semi-major axis a, we have

$$q_1(= r) = \frac{K}{1 + \epsilon \cos \vartheta}, \tag{5.32}$$

where $K = a(1 - \epsilon^2)$. Hence,

$$p_1(= p_r) = m\dot{r}, \tag{5.33}$$

$$p_2(= p_\vartheta) = mr^2\dot{\vartheta} = \frac{mK^2\dot{\vartheta}}{(1 + \epsilon\cos\vartheta)^2}. \tag{5.34}$$

The radial integral in Eq. (5.30) can be turned into an angular integral. Taking the time derivative of Eq. (5.32),

$$\dot{q}_1(= \dot{r}) = \frac{K\epsilon\sin\vartheta}{(1 + \epsilon\cos\vartheta)^2}\dot{\vartheta}, \tag{5.35}$$

we find that

$$dq_1(= dr) = \frac{K\epsilon\sin\vartheta}{(1 + \epsilon\cos\vartheta)^2}d\vartheta. \tag{5.36}$$

Combining Eqs. (5.33)–(5.35), we can express $p_1 = p_r$ in terms of $p_2 = p_\vartheta$

$$p_1 = \frac{\epsilon\sin\vartheta}{K}p_2. \tag{5.37}$$

With the help of Eqs. (5.36) and (5.37), the radial integral in Eq. (5.30) can be converted into a straightforward angular integral and evaluated:

$$\int p_1\,dq_1 = p_2\epsilon^2 \int_0^{2\pi} \frac{\sin^2\vartheta}{(1 + \epsilon\cos\vartheta)^2}d\vartheta$$

$$= 2\pi p_2 \left(\frac{1}{\sqrt{1 - \epsilon^2}} - 1\right). \tag{5.38}$$

Substituting the quantization conditions (5.30) and (5.31) in Eq. (5.38), one finds

$$\rho = \sigma\left(\frac{1}{\sqrt{1 - \epsilon^2}} - 1\right), \tag{5.39}$$

or

$$\sqrt{1 - \epsilon^2} = \frac{\sigma}{\rho + \sigma}, \tag{5.40}$$

a quantization condition for the eccentricity of the allowed Keplerian orbits, which clearly include both circular and noncircular ($\rho \neq 0 \Rightarrow \epsilon \neq 0$) cases.

Wilson makes no further comment on this remarkable result (from which Sommerfeld was shortly to make so much hay). Introducing this result into the classical formula for the total energy of the bound electron, for example, Wilson would have immediately found, using Eqs. (5.31) and (5.40),

$$E = -\frac{me^4}{2p_2^2}(1 - \epsilon^2) = -\frac{2\pi^2 me^4}{h^2}\frac{1}{(\rho + \sigma)^2} \equiv -\frac{2\pi^2 me^4}{h^2}\frac{1}{N^2}, \tag{5.41}$$

which is just Bohr's formula, with the principal quantum number $N = \rho + \sigma$. This formula clearly indicates the degeneracy of the electron energy levels, which was soon to assume enormous importance in the elucidation of spectral structure by Sommerfeld and the Munich school.[12]

5.1.3 Sommerfeld

Independently of the work of Planck and Wilson discussed above, Sommerfeld, by the end of 1915,[13] had arrived at a quantization procedure entirely equivalent to that proposed by these authors. Sommerfeld, however, was less interested in finding formal principles on which quantum theory could be based than in finding a unified treatment of the rapidly increasing body of empirical knowledge on the optical spectra of atoms, both those with a single ("hydrogenic") and those with more than one electron. His report to the Royal Bavarian Academy of Sciences in Munich, presented on December 6, 1915 begins:

> The theory of the Balmer spectrum of hydrogen would seem at first sight to have been brought to a definitive conclusion by the wonderful investigations of N. Bohr.... Nevertheless, I would like to show that the theory of the Balmer series in a certain sense also displays a defect, as soon as one attempts to include non-circular (e.g., in the case of hydrogen, elliptical) orbits ... While other elements display an array of different (spectral) series (the principal series, associated series, and their combinations) and of different types (simple, doublet, triplet series), hydrogen has only the single Balmer series. According to the formulation to be presented here, this will be explained by the coalescence of an array of series into the Balmer series, namely, that each of the (Balmer) lines can arise in a definite number of distinct ways, not only through circular motions, but also from elliptical orbits of definite eccentricity (Sommerfeld 1915a, p. 425).

Sommerfeld then states the principal conclusion of the paper: the stationary states of the electron in a hydrogen atom are only fully identified if one introduces *two* independent quantum numbers, in Sommerfeld's notation n and n', instead of the single "principal" quantum number introduced by Bohr. With some slight alterations to bring

---

[12] In 1920, Sommerfeld would respond to Wilson's complaint that his work was inadequately cited in the first edition of *Atombau and Spektrallinien* (Sommerfeld 1919) by pointing to exactly this failure to draw the relevant phenomenological conclusions from his calculation. See Eckert (2013a, p. 44).

[13] For more details on this critical phase of Sommerfeld's work, see Eckert (2013a, 2014). For a brief summary of Sommerfeld's career up to this point, see Section 1.3.1.

Sommerfeld's notation into accordance with modern custom, this means that the formula for the reciprocal wavelength of spectral lines must now be written as

$$\frac{1}{\lambda} = R(\varphi(n, n') - \varphi(m, m')), \qquad (5.42)$$

where R is the Rydberg constant, and $\varphi(n, n')$ a function of the two quantum numbers. For hydrogen this function is $\varphi(n, n') = 1/(n+n')^2$, so Bohr's principal quantum number is just the sum $n + n'$. Evidently, different choices of n and n', holding their sum fixed, correspond to distinct orbits of equal energy: there is a *degeneracy* in the spectrum of the hydrogen atom. As Sommerfeld would soon show, this degeneracy is a special feature of the pure $1/r$ nuclear Coulomb potential experienced in a single-electron atom, as well as the use of nonrelativistic dynamics for the kinetic energy. In multi-electron atoms, the degeneracy is necessarily broken, leading to the appearance of several states of nearby energy. This, in turn, splits the single Balmer series into the array of principal, associated, etc. series found in multi-electron atoms.[14]

Sommerfeld's (1915a) introduction of quantization rules for periodic orbits in sec. 1 of his paper, "The quantum assumption for periodic orbits," begins with a nod to classical statistical mechanics, specifically the Liouville Principle associating equal a priori probability with regions of equal volume in (q, p) phase space. The additional feature of the quantum theory (as emphasized throughout Planck's work), compatible with this principle but going beyond it, is the requirement that this phase space, for each (q, p) canonical pair, be sliced into "elementary regions" with an area exactly equal to Planck's constant h. Sommerfeld recognizes the need to distinguish between the *librational* periodic motion (cf. Appendix A, Section A.2.3) in which both q and p are strictly periodic functions of time (as in the case of the linear harmonic oscillator) and *rotational* periodic motion, in which q is an angle variable and values of q differing by 2π are identified as physically equivalent. In the former case, Sommerfeld associates the allowed states with the successive elliptical curves which enclose annular regions, each of area h, with the lowest ("zeroth") state corresponding to the particle at rest (so the curve is simply the point at the origin, enclosing zero area). As each successive curve adds an area exactly equal to Planck's constant, the Nth curve must enclose area nh (cf. Fig. 5.2 (a)):

$$\int p_n(q)dq = nh. \qquad (5.43)$$

For the rotational case, Sommerfeld refers to a paper of Ehrenfest (1913b) on specific heats of diatomic molecules, in which the rotational motion of the molecule (thought of as a dumbbell with two point masses subject to uniform rotation around the central point) is quantized according to the requirement (derived from the adiabatic principle which we shall discuss in great detail in the following section) that twice the rotational kinetic energy times the period be an integral multiple of Planck's constant. Sommerfeld, following Ehrenfest, shows that the "orbits" in phase space corresponding to this

---

[14] See Appendix B for a review of spectroscopic terminology.

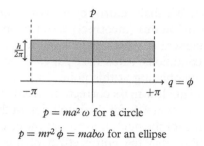

$$p = ma^2\,\omega \text{ for a circle}$$

$$p = mr^2\,\dot\phi = mab\omega \text{ for an ellipse}$$

Figure 5.3 *Quantization of angular phase space (Sommerfeld 1915a, cf. Fig. 5.2 (b)). Sommerfeld's caption for this figure reads: "$p = ma^2\omega$ for a circle; $p = mr^2\dot\varphi = mab\omega$ for an ellipse."*

requirement, plotted in the (q, p) plane, in which q is the angular orientation of a particle executing rotational motion and p the (constant) angular momentum, correspond to successive horizontal lines bounding regions of area $2\pi \times (h/2\pi) = h$. We have reproduced Sommerfeld's graphical representation of this slicing of rotational phase space in Fig. 5.3 (Sommerfeld uses φ rather than ϑ for the angular variable), where Sommerfeld's captions (translated here) clearly indicate that the quantization condition applies with equal validity to circular or elliptical motion (a point also emphasized by Wilson in his second paper). Once again, assuming the lowest orbit to correspond to zero angular momentum, the Nth allowed state must satisfy Eq. (5.43), or, as in this case p_n is constant, we find the Bohr quantization condition for angular momentum

$$p_n = n\frac{h}{2\pi}. \tag{5.44}$$

In sec. 3 of his paper, Sommerfeld derives the quantization condition in Eq. (5.40) for the eccentricity, using the notation n' (Wilson's ρ) for the radial quantum number and n (Wilson's σ) for the angular momentum quantum number. The result, as explained previously, is the new spectral formula (5.42), with $\varphi(n, n') = 1/(n + n')^2$, exhibiting the degeneracy alluded to by Sommerfeld in the introductory paragraphs. Writing $N = n + n'$ for the principal quantum number, and excluding the case $n = 0$, corresponding to a straight line path in which the electron would have to pass through the nucleus, Sommerfeld then concludes that for the energy level corresponding to principal quantum number N, there are precisely N possible orbits, the circular orbit found by Bohr, with ($n' = 0; n = N$), and $N - 1$ elliptical orbits with ($n' = 1, \ldots, N - 1; n = N - 1, \ldots, 1$). The single spectral line of Eq. (5.42) would therefore conceal a total of $N \cdot M$ (where $M = m + m'$) possible transitions. In an admittedly poorly motivated maneuver, Sommerfeld proposes a strange selection rule reducing this number:

$$m \geq n, \quad m' \geq n'. \tag{5.45}$$

In other words, the monotonic decrease of the total energy of the electron (and hence, of the principal quantum number) in any possible spontaneous transition is extended to the requirement that *neither* radial nor angular quantum number increase, even if the overall energy decreases. We shall return to this issue later, but mention here that Sommerfeld would soon abandon this (incorrect) assumption.

In addition to a clear statement of what would soon be commonly called the "Bohr–Sommerfeld quantization conditions", this first paper of Sommerfeld's contains two other developments worth mentioning at this point. In sec. 4, Sommerfeld generalizes Bohr's treatment of the finite nuclear mass corrections from circular to general elliptical orbits (crucial, as we saw in Section 4.6, for the resolution of the Pickering lines puzzle). Phase integrals are introduced for the nuclear as well as the electronic motion, and Bohr's conclusion for circular orbits—that the entire effect of the nuclear motion amounts simply to replacing the electron mass m by the reduced mass $\mu = mM/(M + m)$ of the electron and the nucleus—is extended to the more general set of orbits allowed by the new quantization rules.

In a final section, entitled "On the independence of the quantum condition from the choice of coordinates," Sommerfeld touches on a subtle issue that was later to expose one of the most critical conceptual flaws of the old quantum theory. It is certainly not obvious a priori (and indeed, it is false) that the results obtained for the allowed quantized states are independent of the particular set of generalized coordinates and momenta (q_i, p_i) employed in the solution of the classical dynamics of the system. Sommerfeld's results for the hydrogen atom depend crucially on the choice of polar coordinates (r, φ) in the plane of the orbit of the electron. A full specification of the electron position in three dimensions obviously requires a third coordinate, which we may choose to be z, the Cartesian coordinate perpendicular to the orbit plane.

Sommerfeld's argument for the privileged character of polar coordinates in the problem consists of two parts. First, he establishes the coordinate independence of the *sum* of all three phase integrals (including the z degree of freedom). Second, he argues that the phase integrals for the φ and z coordinates must be quantized. It then follows from the first assertion that the phase integral for the radial coordinate must also be quantized.

In more detail, the argument proceeds as follows. First, for any mechanical system in which the total kinetic energy takes the quadratic form $T = \sum_i A_i(q_1, q_2, \ldots)\dot{q}_i^2$ (this is clearly the case in the one-electron Coulomb problem), with the canonical momenta given by $p_i = \partial T/\partial \dot{q}_i = 2A_i \dot{q}_i$, we have

$$T = \frac{1}{2}\sum_i p_i \dot{q}_i. \tag{5.46}$$

We must now assume that the system is simply periodic, so that all of the coordinates have a common period τ.[15] Then twice the integral over time of the kinetic energy (the Maupertuis action), over a single period τ, is precisely the sum of the independent phase integrals subjected to quantization in the new approach:

---

[15] Essentially, this is the situation assumed by Ishiwara, as we saw in Section 5.1.2.

$$2 \int_0^\tau T dt = \sum_i \int_0^\tau p_i \dot{q}_i dt = \sum_i \oint p_i dq_i = \sum_i n_i h, \qquad (5.47)$$

where each phase integral extends over a single cycle. As the time integral on the left (essentially, the mean kinetic energy) has an invariant dynamical meaning, so must the sum of quantum numbers on the right.

The second part of the argument starts from the observation that φ and z are *cyclic* variables, i.e., they do not appear explicitly in either the kinetic or potential energy for the electron when written in (r, φ, z) coordinates. It follows that the conjugate momenta p_φ and p_z are constants of the motion and therefore must have an invariant mechanical meaning (in the case of the z-coordinate, one takes the orbit to lie in the xy-plane, with $z = 0$, so necessarily $p_z = 0$). One notes the similarity here to Planck's insistence that the g_i slicing functions correspond to constants of the motion. Having justified the quantization of the phase integrals for φ and z in this way, Sommerfeld concludes that the radial integral (as the difference of the sum in Eq. (5.47) and the sum of the φ and z phase integrals, the latter being zero) must also be quantized, even though the radial coordinate as such is clearly not cyclic. Sommerfeld then verifies explicitly that further point transformations of the radial coordinate, say by choosing a new coordinate $s = f(r)$, do not affect the radial quantization condition.

This at least superficially appealing argument conceals a multitude of serious difficulties—most critically, the appearance of ambiguities in the determination of the orbits—arising from the arbitrariness of coordinate choices in the old quantum theory, as we shall see in our discussion of the Stark effect in Section 6.3.

5.1.4 Schwarzschild, Epstein, and (once again) Sommerfeld

In a four-page letter to Sommerfeld written on March 1, 1916 while on military duty in Brussels, Karl Schwarzschild (see Plate 21) outlined the essential features of the quantization approach that would become the heart of the old quantum theory for the remaining period of its existence (Sommerfeld 2000, Doc. 240). Schwarzschild's training as an astronomer[16] had familiarized him with those aspects of analytical mechanics which were particularly fruitful in celestial mechanics (cf. Appendix, Section A.2), in particular the method of action-angle variables introduced by Delaunay and further developed by Jacobi, Poincaré, and Charlier.[17] A brief quote will indicate the essence of the information communicated by Schwarzschild in this important letter:

> I imagine the integration (of the equations of motion) carried out in the fashion of Jacobi, so that the a_k, w_k become a new system of canonical variables:

---

[16] For Schwarzschild's facility in astronomical matters, and Sommerfeld's lack thereof, see Eckert (2013b), pp. 132–133.
[17] For further discussion of the use of action-angle variables in the old quantum theory, see Nakane (2015).

due to the situation of the nucleus itself, for example due to a surrounding electron ring. For a field-free hydrogen nucleus the orientation of the orbit plane is physically undetermined due to a lack of a reference direction and is therefore also quantum-theoretically indeterminable. If we nevertheless introduce a quantum condition for the spatial orientation of the orbit in hydrogen, we mean this in the following sense: we imagine a direction in space to be distinguished by some (external or internal) physical cause, but let the strength of this cause decrease to zero, so as to obtain quantitative relations valid for field-free motion, but with the possibility of a definite orientation relative to a preferred direction (or plane) (Sommerfeld 1916a, pp. 28–29).

The reference to the "nucleus" with a "surrounding electron ring" here is highly suggestive of the core model which would be introduced five years later by Heisenberg in an attempt to model the complexities of multiplet structure and the Zeeman effect (see Sections 7.1.2 and 7.3.3). In any event, the reliance on "external or internal" origins for a distinguished direction clearly indicates that Sommerfeld is imagining circumstances in which one part of the atom may define a distinguished direction which affects electron orbits in another part.[18] From the point of view of the new approach introduced by Schwarzschild, in which quantization is to be applied to the action variables arising from the treatment of the mechanical problem in a so-called *separable* coordinate system,[19] the possibility of an additional quantization condition arises because the introduction of an external field (in the z-direction, say) leads to terms in the Hamiltonian that restrict the separability of the system in arbitrary coordinate systems. For example, if spherical coordinates are to be used, they must be oriented with the colatitude angle ϑ measured from the direction of (for example) an external magnetic field (say, in the z-direction), for the Hamilton–Jacobi procedure to go through. The orientation of the coordinate system (r, ϑ, ψ) itself is thus fixed by the physics of the problem, and for the single-electron (hydrogenic-atom) case we are led to a triplet of quantization conditions

$$\oint p_r dr = n'h \quad \text{(as before)}, \tag{5.49}$$

$$\oint p_\vartheta d\vartheta = n_2 h, \tag{5.50}$$

$$\oint p_\psi d\psi = n_1 h. \tag{5.51}$$

With a preferred direction selected by a putative external field in the z-direction, the field can be allowed to diminish in magnitude to the point where the electron orbits are again

[18] This point is made explicit in the first edition of *Atombau und Spektrallinien*, where Sommerfeld (1919, p. 408) states: "Such a distinguished direction can either be given through an external force field or an internal one."

[19] Roughly, this means that the equations for the mechanical problem can be split into a set of equations each of which involves only one of the system's coordinates. For a precise definition of a separable system in Hamilton–Jacobi theory, see Appendix A, Section A.1, Eq. (A.129).

Kepler ellipses, but now with the plane of the orbit occupying various orientations with respect to the z-axis. The radial quantization condition (5.49) operates as previously (see Eq. (5.38)), yielding

$$2\pi p \left(\frac{1}{\sqrt{1-\epsilon^2}} - 1 \right) = n'h, \tag{5.52}$$

where p is the magnitude of the total angular momentum vector, oriented orthogonal to the plane of the orbit, and by definition, at an angle α to the z-axis. The z-component of this vector, $p\cos\alpha$, is the canonical momentum conjugate to the azimuthal angle ψ, i.e., p_ψ, given by $2\pi p_\psi = n_1 h$ from Eq. (5.51).

The imposition of the second quantization condition is most easily arranged by a simple argument introduced by Sommerfeld (1919, p. 413) in *Atombau und Spektrallinien*, replacing the complicated geometrical integration of the 1916 *Annalen* paper. One recalls (see Eq. (5.27)) that for systems where the kinetic energy T is a quadratic function of the time derivatives of the coordinates q_i, one may write $T = \frac{1}{2} \sum_i p_i \dot{q}_i$, or, for the present case, using spherical coordinates

$$T = \frac{1}{2} \left(p_r \dot{r} + p_\vartheta \dot{\vartheta} + p_\psi \dot{\psi} \right) = \frac{1}{2} \left(p_r \dot{r} + p \dot{\varphi} \right). \tag{5.53}$$

In the second equality we have returned to polar coordinates (r, φ) in the orbit plane, so that the momentum p, as in Eq. (5.52), represents the total angular momentum, quantized previously as $p = n(h/2\pi)$. Evidently,

$$p\dot{\varphi} = p_\vartheta \dot{\vartheta} + p_\psi \dot{\psi}. \tag{5.54}$$

Integrating with respect to time over a single period (shared by all the coordinates, as the orbits are closed), we find

$$\int p \, d\varphi = 2\pi p = \int p_\vartheta \, d\vartheta + \int p_\psi \, d\psi = 2\pi (n_2 + n_1) h. \tag{5.55}$$

It follows that

$$p = (n_1 + n_2) \frac{h}{2\pi}, \tag{5.56}$$

so that the two new angular quantum numbers n_1 and n_2 are seen to sum to the previous azimuthal quantum number n. Moreover, the tilt angle α of the allowed orbits is now quantized, as

$$\cos\alpha = \frac{p_\psi}{p} = \frac{n_1}{n_1 + n_2}. \tag{5.57}$$

The formula for the energy in terms of eccentricity and angular momentum (see Eq. (5.41)) can now be exploited to give an updated version of the Bohr energy levels

$$E = -\frac{2\pi^2 m e^4}{h^2} \frac{1}{N^2}, \quad N = n' + n_1 + n_2, \tag{5.58}$$

revealing a new layer of degeneracy in the Bohr levels, corresponding to the various choices of n_1 and n_2 for a fixed value n of the total angular momentum quantum number. The new degree of freedom introduced by spatial quantization of electron orbits would turn out to play a central role, as we shall later see, in the analysis of external-field effects (Stark and Zeeman) in the old quantum theory, as well as in the modeling of multiplet structure in terms of intra-atomic fields emerging from a division of the atom into "core" and "valence" parts.

5.1.5   Einstein

We close our account of the quantization rules of the old quantum theory with a brief glance at a remarkable paper by Einstein (1917b), entitled "On the quantization rule (*Quantensatz*) of Sommerfeld and Epstein" (Schwarzschild is somewhat unkindly omitted), presented in May 1917. After a review of the classical Hamilton–Jacobi theory, which Einstein presents in a version strongly tinged with geometrical reasoning, the paper considers a classification of the classical dynamics of systems with several (n) degrees of freedom in terms of the existence of momentum vector fields on the n-dimensional coordinate space. Such fields are defined by following the orbits of the particle (or particles) for an infinite time as they traverse (by Hamiltonian motion) some compact subregion of coordinate space. A set of momenta p_i can then be assigned at each point $(q_1, \ldots, q_n)$ in coordinate space when the particle passes through this point. Einstein (1917b, p. 88) distinguishes two types of motion:

- Type (a): The p_i-systems repeat themselves, so that (in the region traversed) only a finite number of p_i-systems appear. In this case the momenta p_i can be represented as single or multiply valued functions of the coordinates q_i.

- Type (b): Infinitely many p_i-systems appear at any given point. In this case the p_i *cannot* be represented as functions of the q_i.

Einstein continues:

> One notices immediately, that Type (b) excludes the quantum condition formulated in [Eq. (5.48)]. On the other hand, classical statistical mechanics is essentially concerned *only* with Type (b), as only in this case is the microcanonical ensemble equivalent to the temporal evolution of a single system [i.e., the dynamics is ergodic].

Remarkably, Einstein is saying that the conditions required for the successes of classical statistical mechanics are in some sense incompatible with the proposed quantization rules of Sommerfeld and Epstein, successful or not as these may be for the very special systems under examination in atomic physics. Einstein's realization of the incompatibility of what we now refer to as "chaotic dynamics"—which is to say, but for a set of measure zero, *all* nonlinear classical dynamical problems—with the fundamental tenets of the old quantum theory does not seem to have met with any serious response, although it was cited in de Broglie's (1924) seminal thesis, and by Schrödinger (1926a) in the second paper of his wave mechanics trilogy.

Although Einstein's criticism of the Bohr–Sommerfeld approach was absolutely on the mark, it offered no way forward in treating the multiple pressing problems of atomic physics that spectroscopy was furnishing at the time, and was more or less forgotten until the field of semiclassical quantum chaos emerged in the early 1970s. At that point, Einstein's considerations (in particular, the classification of nonlinear dynamics in terms of invariant tori) became the focus of a new, and to this day extremely active, field.[20] One of the most important tools in formulating semiclassical treatments of general dynamical systems is the WKB approximation in wave mechanics.[21] We shall see later that this approximation, for the case of conditionally periodic systems, which are the ones amenable to the quantization rules of the old quantum theory, can be viewed as a half-way house between the old and new quantum theories. In particular, the WKB treatment of such systems leads precisely to the quantization rules (5.48), supplemented in certain cases by correction terms (the so-called "Maslov indices") which, as we shall see, explain, once included, some of the remarkable, if initially mysterious, successes of the old quantum theory in problems such as the relativistic fine structure and the Stark effect (Duncan and Janssen 2014).

5.2   The adiabatic principle

The quantization rules discussed in the previous section, now usually referred to as the "Bohr–Sommerfeld rules"—or, sometimes, the "Bohr–Sommerfeld–Wilson rules" or, more accurately, the "Planck–Wilson–Ishiwara–Sommerfeld–Schwarzschild–Epstein rules"—formed the basis for essentially all detailed quantitative studies of bound atomic systems in the decade 1916–1925 leading up to the new quantum mechanics of Heisenberg, Schrödinger, and others. Nevertheless, despite the natural expression of these rules in the language of action-angle variables (for those very special systems to which the action-angle formalism could at all be applied), a deep physical motivation for the choice of the canonical phase integrals as the natural objects for quantization was still lacking. Such a motivation was provided by Paul Ehrenfest (see Plate 8),[22] whose work on the extension of the adiabatic theorem of classical mechanics to quantum theory

---

[20] For enlightening discussions of the seminal role of Einstein's 1917 paper in the later development of quantum chaos, see Gutzwiller (1990) and Stone (2005, 2008).

[21] The WKB approximation is named after Gregor Wentzel (1926), Léon Brillouin (1926), and Kramers (1926), who developed it independently of one another shortly after the formulation of wave mechanics. It is sometimes called the WKBJ approximation because Harold Jeffreys (1924), had introduced it earlier in a different context.

[22] For a more extensive discussion of the history of the adiabatic principle than possible here, see Klein (1970a, Ch. 11) and the series of papers by Pérez and collaborators (Navarro and Pérez 2004, 2006; Pérez 2009).

became an influential strand (along with the correspondence principle to be discussed below) in the developments of the old quantum theory emanating from the Copenhagen school led by Bohr.[23]

5.2.1   Ehrenfest's early work on adiabatic invariants

In our review of classical mechanics in Appendix A (see Section A.1.4) we describe the mechanical adiabatic theorem of Boltzmann, which goes back to his early work (1866) on a mechanical foundation for the second law of thermodynamics, and is explained at length in the second volume of his *Lectures on the Principles of Mechanics* (Boltzmann 1904, sec. 48). The specialization of this theorem to the case of a slow variation in the parameters of the system (e.g., the volume of a container of gas) with no transfer of heat leads to the conclusion that for periodic systems the ratio of the average kinetic energy over a cycle to the frequency of the motion remains a constant in the limit that the transformation is carried out infinitely slowly. It should be emphasized that this result is *a purely mechanical one*: there is no appeal to statistical reasoning, and it can be applied to a single particle, or to a collection of particles each executing a bounded periodic motion.[24]

In a paper entitled "What features of the light quantum hypothesis play an essential role in the theory of heat radiation?," Ehrenfest (1911) had shown that the survival of the classical Wien displacement law (originally derived on the basis of thermodynamic arguments firmly based on the second law; cf. Section 2.2) in the quantum theory of Planck was critically dependent on the adiabatic invariance of the ratio of energy to frequency for each of the normal modes of radiation in a slowly contracting enclosure. Consider radiation with wavelengths in the interval $(\lambda, \lambda + \delta\lambda)$ in a container closed off by a reflecting piston. If the piston is moved slowly to reduce the volume available to the radiation, the increase in the radiation's energy E_λ due to the work done by the piston against the radiation pressure exerted on it is exactly proportional to the increase in the radiation's frequency $\nu = c/\lambda$ due to the Doppler shift produced by its reflection from the incoming piston. In fact, the adiabatic invariance of the ratio of energy to frequency holds separately for each of the normal modes of the radiation in the container.

By late 1912, Ehrenfest had begun to examine the extent to which the existence of adiabatic invariants of this sort, seemingly secure against the vagaries of quantum theory, could be used to provide a firm foothold in developing principles of quantization

---

[23] The conceptual importance of the adiabatic principle seems to have been recognized only belatedly, and with less than full enthusiasm, on the other hand, by Sommerfeld and his coworkers in Munich.

[24] Bohr seems to have suffered under the misapprehension that the adiabatic principle as promulgated by Ehrenfest contained an irreducibly thermodynamic component, which he wished to eliminate by renaming it "the principle of transformability of motion" (letter Bohr to Ehrenfest, May 1918). In fact, the theorem on which Ehrenfest relies is a purely mechanical one, as is apparent from the title of his paper of November 1913, "A mechanical theorem of Boltzmann and its relation to the theory of quanta" (Ehrenfest 1913c). Of course, the adiabatic principle has a profound importance in heat theory as well, as for example in Ehrenfest's earlier discussion of the Wien displacement law. The more general applicability of the principle in the case of black-body radiation is clear from the first line of Ehrenfest's November 1913 paper, discussed below: "When black *or also not black* radiation is compressed reversibly and adiabatically ..." (Ehrenfest 1913c, p. 591; our emphasis).

for a wider class of systems than those treated to date. In an extensive letter to Joffe of February 20, 1913, Ehrenfest refers to the Boltzmann theorem and indicates that the adiabatic principle selects the ratio of average *kinetic* energy to frequency (for any periodic mechanical system) as the suitable object of quantization. The essential intuition operative here is that, during an infinitely slow transformation of the parameters in the energy function of a quantum-mechanical system in an allowed quantum state, one expects the integer quantum number associated with any quantized aspect of the motion to be invariant during the transformation, as such an infinitely slow continuous variation in the conditions of the motion could hardly be expected to evoke a finite and discontinuous "quantum jump" to a different integral value.[25]

For a harmonic oscillator, the average kinetic energy $\overline{T}$ is, by the classical virial theorem, one half of the total energy, quantized in integer multiples of $h\nu$, so the adiabatic invariant $\overline{T}/\nu$ must be quantized in *half-integral quanta*:

$$\frac{\overline{T}}{\nu} = n\frac{h}{2}, \quad n = 0, 1, 2, \dots. \tag{5.59}$$

In May 1913, Ehrenfest (1913b) submitted a paper on the specific heat of diatomic gases, in response to a recent paper by Einstein and Stern (1913) on the same subject, in which an attempt was made to distinguish between the two versions of Planck's theory, the later version containing an additional zero-point energy $h\nu/2$, observable in specific heat measurements only if (as Einstein and Stern assumed) the frequency ν was temperature dependent. For our purposes,[26] the interesting aspect of Ehrenfest's paper is that it appears to be the first publication in which the proposed half-integral quantization rule (5.59), presumably motivated by adiabatic principle considerations, was explicitly implemented by Ehrenfest. For a rigid rotator (viewing the diatomic molecule as two point masses connected in a dumbbell configuration) of moment of inertia L and cyclic frequency of rotation ν, the rotational kinetic energy, subjected to the rule (5.59), becomes

$$\frac{1}{2}L(2\pi\nu)^2 = n\frac{h\nu}{2}, \tag{5.60}$$

where n is an integer (positive, negative, or, for a nonrotating molecule, zero). It follows that the angular momentum $p = L \cdot 2\pi\nu$ is quantized according to

---

[25] The adiabatic invariance of the entropy, as given by Boltzmann's famous formula in terms of the logarithm of the number of microstates of the system compatible with the given macroscopic variables, is clearly related to the non-redistribution (by quantum jumps) of occupied quantum states of the particles during an adiabatic change: as the occupation numbers n_i remain stable, so do combinatoric quantities such as $\sum_i n_i \ln n_i$. The connection between the adiabatic principle and Boltzmann's famous entropy formula seems first to have been enunciated explicitly by Einstein (1914b).

[26] For discussion of Ehrenfest's paper in the context of discussions about the specific heat of diatomic gases, see Gearhart (2010, sec. 4).

$$p = n\frac{h}{2\pi},$$ (5.61)

which is identical to the quantization rule Bohr introduced in the first paper of his 1913 trilogy (see Eq. (4.52)). Ehrenfest appears to have been unaware of Bohr's work at this time and should perhaps share the credit for the discovery of one of the most fruitful ingredients of the old quantum theory.

For the statistical application intended in this paper, one has to decide on the a priori probabilities assigned to the various allowed quantum states of the rotator. Ehrenfest puts it this way:

> If one denotes the orientation angle of the molecule by q, and the corresponding angular momentum ... by p, and considers the lines in the pq phase plane that lie between $q = -\pi$ and $q = +\pi$, then the allowed phase regions consist of the point
>
> $$q = p = 0$$
>
> and the line pairs
>
> $$p = \pm\frac{h}{2\pi}, \pm 2\frac{h}{2\pi}, \pm 3\frac{h}{2\pi}, \dots$$ (5.62)
>
> The point $p = q = 0$ and each of the line pairs ... should be regarded for statistical purposes as equiprobable regions of the phase space (Ehrenfest 1913b, p. 453, equation number added).

With this assumption, the canonical partition function of the system can be constructed by standard arguments, and the specific heat of a gas of such rotators evaluated, showing characteristic quantum deviations from classical equipartition at low temperatures. Strictly speaking, the stationary "point" $q = p = 0$ referred to by Ehrenfest should be replaced by a line ($p = 0$, $-\pi < q \leq +\pi$), as all orientation angles are equiprobable for a stationary rotator, but this pedantic observation has no effect on the subsequent argument. If one represents the quote given here pictorially, one arrives at precisely the figure displayed by Sommerfeld (with a reference to Ehrenfest's paper) and reproduced previously as Fig. 5.3 (see also Fig. 5.2 (b)).[27]

Later in 1913, in the paper mentioned previously, Ehrenfest explicitly introduced the adiabatic theorem (in the form originally given by Boltzmann) and argued for its use as a guide to quantization procedures in quantum theory. After defining the "adiabatic influencing" of a periodic mechanical system as one in which a set of external parameters $r_1, r_2, \dots$, on which both the kinetic and potential energy of the system may depend, are varied "infinitely slowly", the theorem is expressed as follows:

> With adiabatic influencing of a periodic system the quotient of the temporal mean of the kinetic energy $(\overline{T})$ and of the frequency (ν) remains unchanged (adiabatic relation). If δ' denotes an infinitesimal adiabatic change, P the original period, then:

---

[27] For rotational motion, it does not matter whether we choose $0 \leq q \leq 2\pi$ or $-\pi \leq q \leq +\pi$.

$$\delta' \left(\frac{\overline{T}}{\nu} \right) = \delta' \int_0^P T dt = 0.$$

(The action calculated over a period remains constant on adiabatic influencing) (Ehrenfest 1913c, p. 593).

The equality given here by Ehrenfest is to be interpreted as follows: in the limit where the external parameters are varied arbitrarily slowly, over a correspondingly increasing period of time, with the net effect of a *finite* change in the Hamiltonian of the system (with periodic motion maintained at all points in between for the instantaneous values of the Hamiltonian), then an initial motion with period P_A and mean kinetic energy $\overline{T}_A$ is transformed into a final motion with period P_B and mean kinetic energy $\overline{T}_B$, such that

$$P_A \overline{T}_A = P_B \overline{T}_B. \tag{5.63}$$

More succinctly, the theorem simply states that under adiabatic influencing, the integral of the kinetic energy over a single period remains fixed. Ehrenfest immediately makes three important points concerning this assertion:

1. In cases where the virial theorem holds, so that the average potential energy over a period stands in a fixed ratio to the average kinetic energy, the theorem can clearly be taken as applying to the total energy $E = T + V$, instead of just the kinetic energy. For example, for particles moving in one dimension in a potential $V(x) \propto |x|^n$, one has $\overline{V} = (2/n)\overline{T}$ and $E = (1 + 2/n)\overline{T}$. A quantization Ansatz applied to the kinetic energy would therefore immediately imply a corresponding quantization of the total energy.

2. The principle is not extendible, at least not in any obvious fashion, to nonperiodic motions.

3. The principle is subject to modification if singular motions are encountered, in particular, if the "periodic motion begins to detach itself into two or more separate motions" (Ehrenfest 1913c, p. 593).

Ehrenfest promptly gives an example of the final point (p. 594). A particle is allowed to bounce back and forth freely in a tube closed at both ends, while a repulsive field of force is established adiabatically in the interior of the tube, say in the middle. The initial motion is librational, with a full cycle corresponding to the particle moving from the left end of the tube (say at coordinate q_L) to the right end (at q_R) and then back again. At the instant when the maximum value of the repulsive potential energy in the tube exceeds the total energy of the particle, and depending on where the particle happens to be in the tube at that moment, either the left librational limit q_L or the right librational limit q_R jump suddenly and discontinuously to the value where the energy of the particle reaches this maximum value. The particle then executes a new vibrational periodic motion, with essentially the same average kinetic energy (if we assume that the region of non-vanishing potential is restricted to a very small region in the interior of the tube) as previously, but with two new possible periods corresponding to the location of the new internal barrier. Clearly, the adiabatic principle (5.63) fails in this case.

In the main application of the adiabatic principle examined by Ehrenfest (1913c) in sec. 3 of his paper, an attempt is made to reexamine the angular momentum quantization rule for a rigid rotator which Ehrenfest (1913b) had proposed in his paper on diatomic gases. Now the procedure was to connect adiabatically the quantization of rotational motion to the original Planck case of harmonic oscillatory (hence, librational) motion. The physical system employed by Ehrenfest is an ingenious reimagining of a classical pendulum, where one inserts a fixed electric dipole in a uniform electric field (giving a potential energy identical in mathematical form to the gravitational potential of a pendulum consisting of a massive particle attached to a rigid massless rod) and then causes the system to execute oscillations in a vertical plane containing the electric field vector. For small energy of the dipole, one has librational motion with the dipole oscillating around the equilibrium orientation where it is aligned with the electric field. However, by adiabatically decreasing the field, one eventually effects a crossover to rotational motion where the dipole, depending on the initial conditions, rotates clockwise or anti-clockwise through the vertical axis of the electric field (see Appendix A, Fig. A.3). If the electric field is made to vanish, the rotation becomes uniform, with a constant angular momentum. For the initial librational oscillations, one has the usual Planck quantization, giving the half-integral quanta discussed previously for the kinetic energy:

$$\frac{\overline{T}_{osc}}{\nu} = 0, \frac{h}{2}, 2\frac{h}{2}, 3\frac{h}{2}, \ldots \tag{5.64}$$

As Ehrenfest observes, after transition from libration to rotation, a complete cycle requires the negative end of the dipole (say) to go only once between $\vartheta = -\pi$ and $\vartheta = \pi$, where ϑ is the angle between the dipole moment and the electric field, whereas in the final oscillation (just before the transition to rotational motion) the librational character of the motion requires two passages between these termini. Ehrenfest therefore concludes that the appropriate period to use after rotational motion begins is actually *twice* $2\pi/\omega$, where ω is the angular speed of the dipole, and we can assume that the electric field is already negligibly small at the moment of transition (by adiabatically decreasing the dipole moment, say). One then obtains, applying Eq. (5.63), that

$$\frac{\overline{T}_{rot}}{\nu} = \frac{4\pi}{\omega}\overline{T}_{rot} = 0, \frac{h}{2}, 2\frac{h}{2}, 3\frac{h}{2}, \ldots \tag{5.65}$$

and, as the angular momentum (p in Ehrenfest's notation) is related to the rotational kinetic energy by $T_{rot} = p\omega/2$, we now have the quantization rule for angular momentum

$$p = 0, \pm\frac{h}{4\pi}, \pm2\frac{h}{4\pi}, \pm3\frac{h}{4\pi}, \ldots \tag{5.66}$$

exactly half the values claimed previously in his study of diatomic gases (see Eq. 5.62).[28]

---

[28] The analysis of the puzzle of half-integral quanta given here closely follows Duncan and Pérez (2016).

Ehrenfest (1913c, p. 596) apologizes in a footnote for having gotten the wrong result in the earlier paper, but points out that the factor of two could be absorbed in the (anyway unknown) moment of inertia of the gas molecules. At first it would seem that Ehrenfest seems to have ignored his own cautionary admonition (point 3 enumerated above), as the transition from libration to rotation in this system obviously involves a bifurcation of the motion. In the case of the dipole (or pendulum), the transition from libration to rotation involves a singular boundary motion where the period actually goes to infinity: if the energy of the dipole is exactly that required to reach alignment with the field, the time required to do so goes to infinity, and the requirement that adiabatic changes occur slowly *over a single period of the motion* is clearly violated. Nevertheless, this objection is a red herring[29] and does not explain the peculiar (and incorrect!) result involving half-integral quanta obtained here by Ehrenfest, as we shall now show.

The divergence of the period of motion is related to the fact that the potential energy approaches its maximum quadratically: $V(\vartheta) \propto -\cos\vartheta$, which is approximately equal to $1 - \frac{1}{2}(\vartheta - \pi)^2$ for $\vartheta \to \pi$. We can easily avoid this divergence by using a potential that approaches its maximum less rapidly, for example, the "saw-tooth" potential shown in Fig. 5.4,

$$V(\vartheta) = \frac{a}{\pi}|\vartheta|, \qquad -\pi < \vartheta \le +\pi. \tag{5.67}$$

The linear rise of the potential is analogous to a uniform gravitational field[30] where it is clear that when given sufficient energy, the particle only needs a finite amount of time to climb and roll over the hill, even in the boundary case where the energy exactly matches the maximum potential energy value a. Evidently, the adiabatic principle ought to be

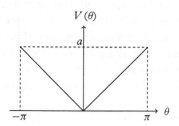

Figure 5.4 *A "saw-tooth" potential allowing a transition from libration to rotation without infinite periods. Based on Fig. 3 in Duncan and Pérez (2016, p. 6).*

[29] In particular, even for the case of the dipole/pendulum, with a cosine dependence of the potential energy, the set of initial conditions for which the particle actually arrives for the first time at the top of the potential hill with exactly zero momentum forms a set of measure zero. In any actual case, the particle will encounter the top of the hill for the first time with nonzero momentum, and pass smoothly over into the rotational phase of the motion. Explicit calculation shows that here as well the integral over the kinetic energy for a single pass around the circuit is perfectly continuous through the transition. Below we will present the results for a saw-tooth potential simply to emphasize that the problem of Ehrenfest's half-quanta occurs even in a case where there is no question of a motion of infinite period, however unlikely, occurring.

[30] One could imagine, for example, a point particle rolling on a circular track with elevation given by Eq. (5.67).

applicable in some form in this case, as the transition from libration to rotation occurs completely continuously (in position and momentum, hence in potential and kinetic energy). The adiabatic influencing of the system is achieved by setting $a(t) = a_0(1 - \epsilon t)$, where $\epsilon\tau \ll 1$ for the largest period τ encountered over the duration of the adiabatic change. We begin the process at $t = 0$ with the potential energy at the peak of the hill a_0 much greater than the kinetic energy of the particle, which therefore executes oscillations around $\vartheta = 0$, with the usual Planck quantization in effect:

$$\frac{\overline{T}_{\text{osc}}}{\nu} = \int_{\vartheta_-}^{\vartheta_+} p \, d\vartheta = n\frac{h}{2}. \tag{5.68}$$

where we have used $T dt = \frac{1}{2} p \dot{\vartheta} \, dt = \frac{1}{2} p d\vartheta$, and the fact that during libration a complete period corresponds to two transits between the minimum ϑ_- and maximum ϑ_+ angles of the motion. As the height of the hill is gradually lowered, we reach a point in which $\vartheta_- \to -\pi$ and $\vartheta_+ \to +\pi$, with the quantity on the left-hand side of Eq. (5.68) having maintained (as an adiabatic invariant) its initial value. As the particle passes from oscillation to rotation the potential energy changes only infinitesimally, so the mean kinetic energy for the final complete oscillation matches the mean kinetic energy for the first complete rotation, as does the time required to pass from $-\pi$ to $+\pi$, so we now have

$$\int_{-\pi}^{+\pi} p \, d\vartheta = \frac{\overline{T}_{\text{osc}}}{\nu} = n\frac{h}{2}. \tag{5.69}$$

This is clearly seen in Fig. 5.5, obtained by a straightforward numerical integration of Hamilton's equations for the system. Note that the flat red line for the total energy, at three times the value of the kinetic energy, confirms the virial theorem for the system for the period of time when it undergoes bounded oscillatory motion. At no point does the period of the motion diverge, as the green line shows. The flat blue curve indicates the constancy of the kinetic energy integral $\int p d\vartheta$ over the entire adiabatic change, through the transition from oscillation to rotation occurring at $t \simeq 400$, *provided we interpret the period ν throughout as the time required for a single pass between the librational limits, or for a single rotation.*

Continuing the adiabatic process, we can further decrease the height of the potential hill to zero, at which time the particle is executing uniform rotation, with a now constant angular momentum p. From the quantum condition

$$\int_{-\pi}^{+\pi} p \, d\vartheta = 2\pi p = n\frac{h}{2}, \tag{5.70}$$

Ehrenfest concluded that this angular momentum is quantized in units of $h/4\pi = \hbar/2$:

$$p = n\frac{h}{4\pi}. \tag{5.71}$$

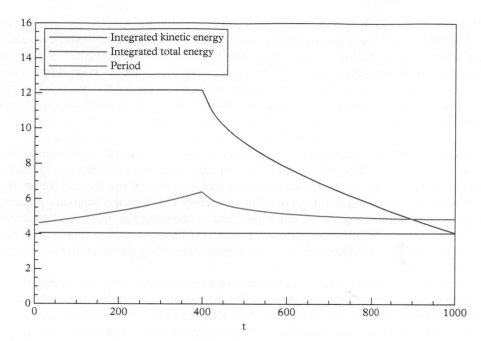

Figure 5.5 *Period integral of kinetic energy, total energy, and period of motion in potential of Fig. 5.4. Based on Fig. 4 in Duncan and Pérez (2016, p. 7).*

We cannot take refuge here in any singular occurrences at the transition point from libration to rotation: as the bottom line in Fig. 5.5 shows clearly, the adiabatic invariant in Eq. (5.71) remains constant right throughout the transition, and the period of the motion (center line) remains finite throughout. The final result, of course, is incorrect: we can hardly suppose that Ehrenfest discovered half-integral spin twelve years ahead of Goudsmit and Uhlenbeck! It is true that a bifurcation of orbits in phase space occurs at the transition, as the particle, depending on initial conditions, can end up rotating either clockwise or anti-clockwise in the rotational phase of the motion, but this cannot be regarded as affecting the adiabatic invariance of $\overline{T}/\nu$, as is confirmed by explicit calculation.

It should be emphasized that the present situation differs essentially from the other example of bifurcation given by Ehrenfest and discussed briefly above, the particle bouncing back and forth in a tube in which a repulsive force field is gradually established. There, at the moment when the particle is no longer able to overcome the potential barrier, one of the limits of the libration (q_L or q_R) changes discontinuously and infinitely rapidly, leading to a discontinuity in the period (and hence the value of the time integral of the kinetic energy over a single oscillation). As we have just seen, no such discontinuity occurs in the dipole (or pendulum) problem.

The resolution of the apparent failure of the adiabatic theorem to reproduce integral quantization of rotational angular momentum (in units of $\hbar$) cannot really be understood within the context of the conceptual framework of the old quantum theory: instead, we need to look ahead for a few moments to the wave mechanics that would eventually supplant it. For the period of time where the particle classically executes oscillations around $\vartheta = 0$, the system is described instantaneously by a wave function $\psi(\vartheta)$ which must be continuous over the entire angular range available to the particle (as the potential is finite everywhere, the wave function's first derivative must be continuous as well). In particular, we must have $\lim_{\vartheta \to -\pi} \psi(\vartheta) = \lim_{\vartheta \to +\pi} \psi(\vartheta)$, so only wave functions that are even under reflection $\vartheta \to -\vartheta$ are allowed.[31] This means that only states corresponding to *even* values of the quantum number n are physically admissible: $n = 0, 2, 4, 6, \ldots$ etc. The odd values of n, corresponding to wave functions satisfying $\psi(\vartheta) = -\psi(-\vartheta)$ are excluded: an exclusion which is simply incomprehensible in a picture where a localized point particle (in the earlier phase of the adiabatic process) executes oscillations which never reach the peak of the potential hill.[32] Consequently, the allowed values of angular momentum on the right-hand side of Eq. (5.69) do in fact correspond to *integral* multiples of $h/2\pi$ as originally claimed by Ehrenfest in his paper on diatomic gases.

After this experience, it is hard to avoid the conclusion that the very slow acceptance of the adiabatic theorem as a guiding principle for the old quantum theory (almost six years between Ehrenfest's first ruminations of late 1912 to Bohr's full-throated acceptance in the 1918 papers on the quantum theory of line spectra) was not unconnected to the frequent appearance of distracting subtleties of this sort which really could not be resolved within the framework of the old theory. It appears that Ehrenfest was never fully satisfied that the puzzle raised by half-integral quanta in the transition from oscillation to rotation had been satisfactorily resolved. As his student Jan Burgers, who contributed some important papers on the adiabatic principle himself (Burgers 1917a, 1917b, 1917c), recalled decades later:

> One of the problems which always was before Ehrenfest...is the question of the transition from the oscillator to the rotating system: the pendulum which turns over and the changing quantum rule you get in such a case. It made the impression there was some deep mystery in it which perhaps might explain things, but I did not get any hold on it. Ehrenfest was turning around with it, but it remained as far as I know at that time an open problem.[33]

---

[31] The wave functions in question are Airy functions for the potential of Eq. (5.67), Mathieu functions if we stick to the original dipole/pendulum case, $V(\vartheta) = \frac{1}{2}a(1 - \cos(\vartheta))$.

[32] The exclusion of odd quantum numbers is reconcilable as follows with the bifurcation phenomenon emphasized by Ehrenfest: The real even bound state wave functions appropriate for the oscillatory motion each connect to a pair of possible rotating complex wave functions corresponding to clockwise or anti-clockwise motion of the particle in the rotational phase, with equal and opposite angular momentum ... Explicit integration of the time-dependent Schrödinger equation for the adiabatic process leads to a final wave function which is an even linear combination of these two rotational wave functions.

[33] AHQP, p. 62 of transcript of an interview by Kuhn of Burgers, September 6, 1962.

5.2.2 Ehrenfest's 1916 paper on the adiabatic principle

In the Spring of 1916, after an exchange of letters in which Ehrenfest pointed out to Sommerfeld the relation of his previous work on the adiabatic theorem to the new quantization rules introduced by Sommerfeld in his paper on the Balmer series and the subsequent more comprehensive papers on the quantum theory of spectral lines,[34] Ehrenfest prepared a new paper on the adiabatic theorem. This paper, entitled "On adiabatic changes of a system in connection with the quantum theory" (Ehrenfest 1916a), was communicated by Lorentz to the Amsterdam Academy on June 24, 1916. A German translation (Ehrenfest 1916b) was received by *Annalen der Physik* on July 22, 1916. In this paper Ehrenfest adopts Einstein's (1914b) terminology: the "mechanical theorem of Boltzmann" or the "adiabatic relation" has now become the *adiabatic hypothesis*, which, according to Einstein, is the assertion that

> If a system is exposed to adiabatic influences, the "admissible" motions are transformed into "admissible" (Ehrenfest 1916a, p. 577).[35]

Here, the "admissible" motions clearly refer to the discrete set of allowed classical motions (for bounded motion) selected by the quantization rules of the theory, whatever they might be. With his usual clarity, Ehrenfest states his objectives for the present paper:

1. To formulate the adiabatic law as sharply as possible ...

2. To indicate what great significance must be ascribed to the "adiabatic invariants" in the quantum theory [and to link the adiabatic hypothesis] to the quantum hypothesis of Planck, Debye, Bohr, Sommerfeld ...

3. To point out difficulties, which [arise if the adiabatic changes pass] through singular motions.

4. To show at least how the adiabatic problems are connected with the statistical mechanical bases of the second law of thermodynamics. The statistical mechanical explanation Boltzmann gave of it rests on statistical foundations which are destroyed by the introduction of the quanta (p. 578).

Also typical of Ehrenfest he adds, "Hoping that others may succeed in removing the difficulties I was not able to surmount, I will publish my considerations" (ibid.).

Ehrenfest's new "sharp" formulation of the adiabatic law hardly differs at all from the previous version (Ehrenfest 1913c, p. 593, quoted above). Imagine subjecting a system undergoing periodic motion (at all stages of the adiabatic influencing) to the very slow

---

[34] See Ehrenfest to Sommerfeld, April 1916 (Sommerfeld 2000, Doc. 254, pp. 555–557), discussed in Eckert (2013b, p.213). We return to Sommerfeld's attitude toward the adiabatic principle at the end of this section.

[35] A literal translation of Einstein's (1914b, p. 826) statement reads: "In a reversible adiabatic change each quantum-theoretically possible state goes over into another such state."

variation of a set of external parameters $a_1, a_2, \ldots$ The adiabatic law now says that, given that

$$\delta' \int_0^P 2T dt = 0, \tag{5.72}$$

where P the period of the motion and δ' denotes the net change induced by the adiabatic process, then

$$\frac{2\overline{T}}{\nu} \text{ is an adiabatic invariant.} \tag{5.73}$$

The new factor of two simply removes the peculiar half-quanta of the previous papers, as one sees immediately in a system of a single degree of freedom:

$$\int_0^P 2T dt = \int_0^P p\dot{q} dt = \int_0^P p dq = nh. \tag{5.74}$$

The adiabatic invariance of the left-hand side here means that the equality to nh in the case of a "Planck resonator" (i.e., one-dimensional simple harmonic oscillator) implies that the phase integral $\int_0^P p dq$ must also equal nh for the admissible motions of any anharmonic oscillator adiabatically connected to the simple harmonic one, e.g., with potential function $V(q)$ containing cubic, quartic, etc. powers of q, in addition to the normal quadratic dependence. Ehrenfest points out that just such an assumption had previously been used for the quantization of non-harmonic motion by Debye (1914) in his 1913 Wolfskehl lectures in Göttingen and Boguslawski (1914) in his work on pyroelectricity.

Apart from the result in Eq. (5.72) applicable to systems of one degree of freedom, Ehrenfest emphasizes three other important matters for the more general case of a system of n degrees of freedom (1916a, sec. 5, p. 582, remarks A, B, and D).[36] Assuming the usual quadratic structure of the kinetic energy, $T = \sum_{i=1}^n A_i(q_1, q_2, \ldots)\dot{q}_i^2 = \frac{1}{2}\sum_i p_i\dot{q}_i$, one has immediately the adiabatic invariance of

$$\int_0^P 2T dt = \int_0^P \sum_{i=1}^n p_i\dot{q}_i dt = \sum_i \iint dq_i dp_i. \tag{5.75}$$

Moreover, the sum on the right-hand side, as pointed out originally by Poincaré, is unchanged under coordinate transformations: it has, gratifyingly, a coordinate-invariant meaning (remark A). As an adiabatic invariant, the *sum* of the phase integrals on the right-hand side qualifies as a suitable object for quantization in integral multiples of Planck's

---

[36] Remark C is that, in the one-dimensional case, the quantity $2\overline{T}/\nu = \iint dq dp$ is the only adiabatic invariant (ibid.).

constant.[37] This does not, of course, imply that the individual phase integrals are so quantized. Ehrenfest then points out that systems (of which two examples will follow later in the paper) can be found in which the individual phase integrals can indeed be shown to be adiabatic invariants, yielding consequently additional, independent, quantization conditions (remark B).[38] Finally, Ehrenfest quotes a theorem of Paul Hertz, to the effect that the volume enclosed by any fixed energy surface in the full $2n$-dimensional phase space is invariant under an adiabatic influencing of the system (remark D). No further use of this observation is made in the paper.

Ehrenfest's final example of the application of the adiabatic hypothesis to a system with a single degree of freedom is the troublesome case of the dipole—now converted to a simple pendulum undergoing ever larger oscillations—discussed above. Here, the doubling of the frequency at the transition from libration to rotation is simply ignored, and Ehrenfest now obtains—effectively, by *violating* the adiabatic continuity of the temporal integral of the kinetic energy!—the result that the angular momentum in the rotating case is quantized integrally according to $p = n(h/2\pi)$ (with $n = 0, 1, 2, \ldots$), for which, by now, in the aftermath of the Bohr atom and Sommerfeld's work on line spectra, there is simply overwhelming evidence. There is a clear sense of unease with this result, as the following remark, appearing right after it, makes clear:

> The considerations given above must still be completed, especially with view to the difficulty that during the adiabatic change the singular, *nonperiodic* motion is passed, which forms the limit between the pendulum motions and the rotations. It must therefore be investigated, how the invariants of both kinds of motion are connected (pp. 584–585).

As we saw, no such singular motion is actually responsible for the incorrect quantization obtained if one ignores that only half of the normally allowed Planck resonator states are actually *quantum-theoretically admissible* during the pendulum (i.e., oscillatory) part of the motion, a restriction comprehensible only in the light of later developments.

In the first of the two examples of systems of more than a single degree of freedom, Ehrenfest analyses the Sommerfeld treatment of central force problems from the point of the adiabatic hypothesis. Imagine a particle of mass m moving under the influence of an adiabatically variable central potential $\chi(r, a_i(t))$, where $\{a_1(t), a_2(t), \ldots\}$ is a set of slowly varying adiabatic parameters. The instantaneous equation of motion for the radial component in polar coordinates $(q_1, q_2) = (r, \varphi)$ is

$$m\ddot{r} = mr\dot{\varphi}^2 - \frac{d\chi}{dr}. \tag{5.76}$$

---

[37] Recall Sommerfeld's argument for coordinate independence of quantization conditions, based on exactly this sum (cf. Eq. (5.45)).

[38] In a note added in proof, Ehrenfest (1916a, pp. 596–597) raised the related question of whether the parts of Hamilton's principal function that get quantized separately in Schwarzschild and Epstein's treatment of the Stark effect are each individually adiabatic invariants. For a brief discussion and further references, see Duncan and Janssen (2015, pp. 236–237).

The first term on the right-hand side is just the centrifugal force. The central nature of the force implies that the angular momentum $mr^2\dot\varphi$, which we here denote by p_2 (with $q_2 = \varphi$ the conjugate coordinate), is constant:

$$\frac{dp_2}{dt} = \frac{d}{dt}(mr^2\dot\varphi) = 0. \tag{5.77}$$

Note that the validity of Eq. (5.77) is quite independent of the rate at which the adiabatic parameters $a_i(t)$ are varied, provided only that the potential remain central at all times. Thus the angular momentum p_2 is an absolute invariant of the motion and thus, a fortiori, an adiabatic invariant, and a suitable candidate for quantization by the standard Ansatz

$$\iint dq_2 dp_2 = 2\pi p_2 = n'h. \tag{5.78}$$

Following Sommerfeld's notation, Ehrenfest uses n and n' instead of the more natural n_1 and n_2 for the radial and angular quantum numbers, respectively.[39] Eliminating $\dot\varphi$ from Eq. (5.76), using $p_2 = mr^2\dot\varphi$, we find[40]

$$m\ddot r = \frac{p_2^2}{mr^3} - \frac{d\chi}{dr} = -\frac{d\Phi}{dr}, \tag{5.79}$$

where the effective potential

$$\Phi \equiv \frac{p_2^2}{2mr^2} + \chi(r, a_i) \tag{5.80}$$

is the sum of a repulsive centrifugal term and the potential χ, which we now restrict to giving us an attractive force, be it a Coulomb force or the restoring force of a harmonic oscillator. In either case, the resulting motion leads to oscillatory libration between two extreme points r_A and r_B in the $(q_1, p_1) = (r, p_r)$ phase plane. By the usual procedure (as by suitable variation of the adiabatic parameters the system can be brought back to a one-dimensional Planck oscillator), this motion is subject to the quantization rule

$$\iint dq_1 dp_1 = 2\int_{r_A}^{r_B} p_1(r)dr = nh, \tag{5.81}$$

where p_1 is the radial momentum conjugate to $q_1 = r$. This then is how Ehrenfest could justify the Sommerfeld rules on the basis of the adiabatic hypothesis. One sees here also the special role played by cyclic variables (such as the angle φ), which do not appear in

---

[39] Actually, the radial and angular quantum numbers are interchanged from Sommerfeld's usage in the discussion of central potentials in sec. 7; in sec. 9, in the discussion of the two-dimensional harmonic oscillator, Ehrenfest returns to Sommerfeld's usage, n for the angular momentum quantum number, n' for the radial quantum number (Ehrenfest 1916a, p. 586, p. 589).

[40] Ehrenfest (1916a, p. 585, Eq. (21)) erroneously has mr^3/p_2^2 instead of p_2^2/mr^3 in this equation.

the Hamiltonian, and are therefore canonically conjugate to automatic constants of the motion ("super-adiabatic invariants" as it were). As previously emphasized by Planck in his choice of slicing functions, or Sommerfeld in his search for suitable phase integrals, such quantities are natural candidates for quantization under the adiabatic philosophy as well. In sec. 9 of his paper, entitled "Difficulties which occur, if the adiabatic reversible change gives rise to a singular motion," Ehrenfest (1916a, pp. 589–591) considers the subtleties that arise when the adiabatic hypothesis is applied in cases where the transition is effected through, or into, a degenerate system, in which the energy of the state does not uniquely determine the underlying quantum numbers. The example used is the isotropic two-dimensional oscillator which we have discussed before on a number of occasions. The isotropic Hamiltonian (our notation)

$$H_{\text{iso}} = \frac{p_1^2}{2m} + \frac{p_2^2}{2m} + \frac{1}{2}m\omega^2(q_1^2 + q_2^2), \tag{5.82}$$

where $(q_1, q_2) = (x, y)$ are just the usual Cartesian coordinates, is treated as arising from an adiabatic influencing of an initially anisotropic system with Hamiltonian,

$$H_{\text{aniso}} = \frac{p_1^2}{2m} + \frac{p_2^2}{2m} + \frac{1}{2}m\omega_1^2 q_1^2 + \frac{1}{2}m\omega_2^2 q_2^2. \tag{5.83}$$

We set

$$\omega_1 = \omega + \frac{1}{2}\delta(t) \quad \text{and} \quad \omega_2 = \omega - \frac{1}{2}\delta(t) \tag{5.84}$$

and allow $\delta(t)$ to go to zero slowly (i.e., $\dot{\delta} \ll \omega\delta$). In this case, unlike the situation for the dipole/pendulum transition from oscillation to rotation, a violation of the requirement that the adiabatic change be performed slowly over a period of variation of the adiabatic invariant of interest is unavoidable. Initially, with the anisotropic potential, the angular momentum is not conserved: the particle classically executes Lissajous figures of varying orientation and eccentricity. If we assume the motion initially to correspond to oscillations in q_1 and q_2 quantized with energy $n_1\hbar\omega_1$ and $n_2\hbar\omega_2$, respectively, we can begin the adiabatic process with

$$q_1 = A_1 \cos(\omega_1 t) \quad \text{and} \quad q_2 = A_2 \cos(\omega_2 t + \vartheta_0), \tag{5.85}$$

where the amplitudes are quantized,[41]

$$A_i = \sqrt{\frac{2\hbar n_i}{m\omega_i}}, \quad \text{with } i = 1, 2. \tag{5.86}$$

---

[41] Inserting Eqs. (5.85) for $q_i(t)$ into the expression for the energy of the oscillatory motion, $\frac{1}{2}m\dot{q}_i^2 + \frac{1}{2}mq_i^2$, we find that this energy is equal to $\frac{1}{2}m\omega_i^2 A_i^2$. Setting this equal to $n_i\hbar\omega_i$, we arrive at Eq. (5.86) for A_i.

The angular momentum is easily calculated, if we restrict ourself to the evolution over a period in which the frequencies change only a little,[42]

$$L = m(q_1 \dot{q}_2 - \dot{q}_1 q_2) = 2\hbar \sqrt{\frac{n_1 n_2}{\omega_1 \omega_2}} (\omega \sin(\delta t - \vartheta_0) + \tfrac{1}{2}\delta \sin(2\omega t + \vartheta_0)). \qquad (5.87)$$

As isotropy is approached ($\delta \rightarrow 0$), we see that the surviving part of the angular momentum oscillates with a period $2\pi/\delta(t)$ that is going to infinity. This slow oscillation results from the beat frequency of the ω_1 and ω_2 oscillations approaching degeneracy, and is a characteristic feature in any system where the parameters are varied to produce a new system with greater degeneracy than that at the start of the adiabatic process. Moreover, the limiting value of the angular momentum obtained if one actually carries out the variation, say by numerical calculation, does not agree with the quantized value $n\hbar\omega$ given by the Sommerfeld quantization of the isotropic system with $H = (p_1^2 + p_2^2)/2m + \tfrac{1}{2}m\omega^2 r^2$, where one obtains

$$E = (n + n')\hbar\omega, \qquad (5.88)$$

$$L = p_\varphi = n\hbar\omega. \qquad (5.89)$$

Here n' and n are the radial and the angular quantum number, respectively. Superficially, this seems to agree with the initial energy quantization carried out in Cartesian coordinates,

$$E = n_1 \hbar\omega_1 + n_2 \hbar\omega_2 \rightarrow (n_1 + n_2)\hbar\omega. \qquad (5.90)$$

However, as is apparent from Eq. (5.87), the limiting value of the angular momentum is dependent on the initial phase ϑ_0. As Ehrenfest comments

> On the other hand, it is not evident why in this way only one of the discrete values in [Eq. (5.89)] for the moment of [the angular] momentum would be obtained (Ehrenfest 1916a, p. 590).

Moreover, as Ehrenfest points out (in a remark attributed to Epstein in a footnote), the circular orbits in the polar quantization approach, where $n' = 0$ but n is an arbitrary integer, do not seem to be the same set as those in the Cartesian approach, where a

---

[42] In more detail, the calculation proceeds as follows. Using Eqs. (5.85)–(5.86), we find

$$q_1 \dot{q}_2 = -A_1 A_2 \omega_2 \cos(\omega_1 t) \sin(\omega_2 t + \vartheta_0), \quad -\dot{q}_1 q_2 = A_1 A_2 \omega_1 \sin(\omega_1 t) \cos(\omega_2 t + \vartheta_0),$$

where $A_1 A_2 = (2\hbar/m)\sqrt{n_1 n_2/\omega_1 \omega_2}$. Substituting $\omega + \tfrac{1}{2}\delta$ and $\omega - \tfrac{1}{2}\delta$ (we neglect the time dependence of δ in differentiating q_1 and q_2) for the factors ω_1 and ω_2 multiplying the sines and cosines in these two expressions, we note that the sum of these expressions consists of a term with ω multiplying the sine of the difference of $\omega_1 t$ and $\omega_2 t + \vartheta_0$ and a term with $\tfrac{1}{2}\delta$ multiplying the sine of the sum of $\omega_1 t$ and $\omega_2 t + \vartheta_0$. Using that $\omega_1 t - \omega_2 t - \vartheta_0 = \delta t - \vartheta_0$ and that $\omega_1 t + \omega_2 t + \vartheta_0 = 2\omega t + \vartheta_0$, we arrive at Eq. (5.87).

circular orbit must have $n_1 = n_2$, hence $E = N\hbar\omega$, with N *even. Different quantization methods in degenerate systems give different sets of quantum-theoretically admissible orbits!* This observation of Epstein's, as we shall argue later in our discussion of the Stark effect, indicates a profound inadequacy of the old quantum theory which could only finally be resolved with the advent of the new quantum mechanics of Heisenberg and Schrödinger (Duncan and Janssen 2014, 2015).[43] There are two appendices to Ehrenfest's (1916a, pp. 591–596) paper which merit a brief mention here. In the first, Ehrenfest gives a proof of the adiabatic invariance of $2\overline{T}/\nu$ for periodic systems of several degrees of freedom. A proof for the more application-rich case of general conditionally periodic motions would be provided soon[44] by Ehrenfest's student Burgers (a simplified version of which is discussed in Appendix A, Section A.1.4). In the second appendix, Ehrenfest gives an account of the Sommerfeld quantization of the isotropic two-dimensional oscillator in polar coordinates, in the course of which he comments on, and corrects, the previously noted error of a factor of 2 in Planck's treatment of the radial quantization of this system (cf. footnote 6 above).

Finally, in a note added in proof, Ehrenfest (1916a, pp. 596–597) references two recent papers by Epstein (1916b, 1916c) in which the whole apparatus of separation of variables, Hamilton–Jacobi theory, and action-angle variables is at last brought to bear on the problem of the Stark effect, to which we shall return in Section 6.2.

5.2.3   The adiabatic principle in Bohr's 1918 paper

In the Spring of 1916, Bohr submitted to the *Philosophical Magazine*, and then withdrew, a paper "On the application of the quantum theory to periodic systems" (Bohr 1916), in which he makes considerable use of the adiabatic invariance of $\overline{T}/\nu$, as discussed in the 1913 papers of Ehrenfest. His withdrawal of the paper was occasioned by his reception of Sommerfeld's paper on the quantum theory of line spectra, with its extensive discussion of the new quantization rules and applications to relativistic fine structure (discussed in detail below). Bohr comments on the peculiar half-quanta observed by Ehrenfest in the transition from oscillation to rotation in the dipole problem discussed above, and draws the same incorrect conclusion, that the problem must somehow lie in a singular motion of infinite period (Duncan and Pérez 2016, p. 5):

---

[43] The adiabatic process treated here, when carried out in wave mechanics, by explicit solution of the time-dependent Schrödinger equation, exhibits absolutely no discontinuity in the wave function of the particle. For example, if we begin the process with the particle in the $(n_1 = 1, n_2 = 1)$ state of the anisotropic oscillator, with the angular dependence of the wave function $\sin\varphi\cos\varphi$, this angular dependence is preserved to the point of isotropy, where the same wave function may be reexpressed as a linear combination of $n = +2$ and $n = -2$ wave functions, as $\sin\varphi\cos\varphi = \frac{1}{2}\sin 2\varphi = \frac{1}{4}(e^{+2i\varphi} + e^{-2i\varphi})$.

[44] It appears that the problem of establishing an adiabatic theorem applicable to conditionally periodic systems was assigned to Burgers at about the time of publication of the paper presently under discussion, i.e., the Summer of 1916. The result was a trilogy of papers presented to the Amsterdam Academy from November 1916 through January 1917 (Burgers 1917a, 1917b, 1917c), in which both nondegenerate and partially (or completely) degenerate systems were considered.

It does not seem possible, however, in the way indicated to effect a complete transformation from the vibrating system to a freely rotating doublet. In the critical state the frequency of vibration will be infinitely small and the state cannot be reached in a finite time (Bohr 1916, p. 438).

As we have seen, the correct explanation lies in an incorrect identification of the initially admissible quantum states, not in any need to pass through a singular motion.

Two years later, in the first part of his massive paper "On the quantum theory of line spectra," Bohr (1918) placed the adiabatic hypothesis, now renamed "the principle of mechanical transformability", at the center of his attempts to build a systematic framework for quantization of atomic systems. By that time, the successful attack on the Stark effect through quantizing action variables by Schwarzschild (1916) and Epstein (1916b) (see Section 6.3) had provided the old quantum theory with perhaps its most impressive quantitative validation after the elucidation of the Balmer series in the first installment of Bohr's 1913 trilogy. Moreover, Burgers' 1917 trilogy had put the mechanical adiabatic theorem on a rigorous mathematical foundation for the important class of conditionally periodic systems, with the action variables of such systems playing the central role as distinguished adiabatic invariants.

A brief quote will indicate the central role which Bohr attaches to this principle:

> It seems natural to assume that ... the motion of an atomic system in the stationary states can be calculated by direct application of classical mechanics, not only under constant external conditions, but in general also during a slow and uniform variation of these conditions. This assumption, which may be denoted as the principle of the "*mechanical transformability*" of the stationary states, has been introduced in the quantum theory by Ehrenfest and is, as it will be seen in the following sections, of great importance in the discussion of the conditions to be used to fix the stationary states of an atomic system among the continuous multitude of mechanically possible motions (Bohr 1918, pp.8–9).

After giving a demonstration of the invariance of the kinetic energy integral under adiabatic changes for periodic systems (essentially, a reprise of the proof in the appendix of Ehrenfest's (1916a) paper), Bohr generalizes the adiabatic hypothesis for one-dimensional systems, in the form $I = \int pdq = nh$, to the special class of systems of s degrees of freedom (q_k, p_k), with $k = 1, 2, \ldots, s$, in which the energy can be written as a sum of terms E_k each involving only the coordinate pair (q_k, p_k). As the motion separates trivially into independent motions for each coordinate, the quantization condition in the one-dimensional case has an obvious replacement, with the adiabatically invariant action integral for each coordinate (for adiabatic influencings of the system which maintain the coordinate independence) undergoing an independent quantization:

$$I_k = \int p_k dq_k = n_k h, \quad (k = 1, \ldots, s). \tag{5.91}$$

It is a simple exercise in applying the chain rule of differentiation to verify that under arbitrary coordinate transformations maintaining the independence of coordinates, i.e., $q_k \to \tilde{q}_k(q_k)$ (for each k separately),

$$\tilde{p}_k = p_k \frac{dq_k}{d\tilde{q}_k}, \quad d\tilde{q}_k = \frac{d\tilde{q}_k}{dq_k} dq_k, \tag{5.92}$$

so the I_k have a coordinate invariant meaning (within this restricted class of coordinate transformations).

The generalization from this very restricted class of systems (in Bohr's terminology, systems in which the different coordinates are "dynamical independent" of each other) to a general s-dimensional conditionally periodic system is then explained, making full use of the rich lore of analytical mechanical technology (Hamilton–Jacobi theory, action-angle variables, etc) which had by now been transmitted to the quantum physics community from the celestial astronomers. The result is formally identical to Eq. (5.91), but written more explicitly (Bohr 1918, Eqs. 21 and 22):

$$I_k = \int p_k(q_k, \alpha_1, \dots \alpha_s) dq_k \quad (k = 1, \dots, s), \tag{5.93}$$

$$I_k = n_k h \quad (k = 1, \dots, s), \tag{5.94}$$

where $\alpha_1, \alpha_2, \dots, \alpha_s$ are the separation constants which arise naturally in the solution of the Hamilton–Jacobi equation for a separable s-dimensional system.[45]

Apart from serving as the underlying motivation for the quantization conditions, the "mechanical transformability" principle is primarily used by Bohr as an important ingredient in determining the "a priori probability" of quantum states, i.e., in determining what we would nowadays call the degree of degeneracy of a state specified by some set of quantum numbers. For a one-dimensional system, specification of the quantum number n determines the orbit apart from an arbitrary choice of initial phase (location of the particle on the orbit at time zero), and it is therefore reasonable to suppose that such a classical motion can be associated with exactly one and only one quantum state.

For *nondegenerate* multidimensional systems in which the s coordinates are dynamically independent, one expects that a similar argument ensures unique specification of a quantum state once the n_k (with $k = 1, 2, \dots, s$) are specified. All such states must then be assigned equal a priori probability when statistical considerations à la Boltzmann come into play. The adiabatic argument then ensures that the same equality holds for any other systems obtainable from such a system by adiabatic influencing ("mechanical transformation") provided no degenerate systems (with the usual problems of singular motions of infinite period) are encountered along the way.

The final generalization, to all conditionally periodic systems, is argued heuristically by falling back on the Planck slicing of phase space which we discussed in Section 5.1, in an interesting application of the correspondence principle. First, Bohr observes that for such conditionally periodic systems the volume of phase space corresponding to an

---

[45] We shall shortly be seeing explicitly how all of this works for the Stark effect.

increment of unity in each of the quantum numbers n_k is equal to h^s at all points $(n_1, n_2, \ldots, n_s)$ (cf. Eqs. (5.8) and (5.23)). In the limit of large quantum numbers, the states vary relatively little over a change of unity in each of these large quantum numbers, and one therefore gets agreement with the results of classical Boltzmannian statistical mechanics (assigning phase space points uniformly through the phase space volume) by assigning exactly equal a priori probability to each of the discretely enumerated states identified by a unique choice of $(n_1, n_2, \ldots, n_s)$. Bohr concludes:

> The fact that the last considerations hold for every nondegenerate conditionally periodic system suggests the assumption that in general *the a priori probability will be the same for all the states determined by* [Eq. (5.91)], even if it should not be possible to transform the given system into a system of independent degrees of freedom without passing through degenerate systems. . . . When we consider a degenerate system, however, we cannot assume that the different stationary states are a priori equally probable. In such a case the stationary states will be characterized by a number of conditions less than the number of degrees of freedom, and the probability of a given state must be determined from the number of different stationary states of some nondegenerate system which will coincide in the given state, if the latter system is continuously transformed into the degenerate system in question (Bohr 1918, p. 26).

This quote states (unusually clearly for Bohr) the strategy to be followed to uncover the underlying degeneracies which allow distinct quantum states to masquerade as the same (for example, by leading to coinciding spectral lines): One imagines a nearby, nondegenerate system, perhaps induced by the application of a weak external field, as in the Stark effect, in which the degeneracy is lifted, and the individual underlying quantum states exposed as distinct. The need to elucidate the underlying "deep structure" of a quantum system by identifying an adequate number of independent quantum conditions would persist all the way through to von Neumann's axiomatization of the new quantum mechanics in 1927, where it takes the form of the specification of a complete set of commuting observables (von Neumann 1927, p. 260, cf. Duncan and Janssen 2013, pp. 251–252).

5.2.4   Sommerfeld's attitude to the adiabatic principle

It has been suggested, and the historical evidence would seem to support this, that Ehrenfest's tentative, probing style of enquiry evident particularly in his papers on the adiabatic principle was not to Sommerfeld's liking, incompatible as it was with the latter's straight-ahead, calculational methodology (Eckert 2013b, p. 256). Ehrenfest's letters to Sommerfeld emphasizing the conceptual importance of the adiabatic principle as providing an indispensable underpinning of the Bohr–Sommerfeld–Wilson quantization principles received at most noncommittal responses. In the first edition of Sommerfeld's (1919) *Atombau und Spektrallinien*, the "Bible" of atomic theory, the adiabatic hypothesis is mentioned only once (though with full credit given to Ehrenfest for its introduction into the quantum theory), as a justification for the application of Larmor's theorem (see Section 7.2) to the normal Zeeman effect, in which a magnetic field is turned on adiabatically (i.e., the fractional change of the field in each orbit of the electron being very small). By the third edition, the adiabatic hypothesis has been promoted to a full section (Sommerfeld 1922a, Ch. 5, sec. 7), with Einstein given credit for its introduction (on the basis of a short comment made at the Solvay 1911 conference), and Ehrenfest granted credit for its "name, and systematic development."[46]

Once again, the principle is applied primarily to the examples of the normal Zeeman effect, and in this case also to the Stark effect, in an attempt to justify the veto of states of zero magnetic quantum number in the magnetic field case by an adiabatic transition from a system with only an electric field present.[47] After giving a few simple mechanical examples (the ubiquitous pendulum pops up here again), Sommerfeld makes an explicit reference to the adiabatic invariance of the phase integrals.[48] Then follows the strange assertion:

> Inasmuch as we provide a proof of this theorem [i.e., adiabatic invariance of the phase integrals of a conditionally periodic system], we provide a justification of our general quantization Ansatz [i.e., quantization of the phase integrals] from the standpoint of the adiabatic hypothesis; on the other hand, if we regard this quantum Ansatz as certain (*gesichert*), we have proven the adiabatic hypothesis for the entire class of conditionally periodic systems (Sommerfeld 1922a, p. 378).

It would appear that even at this date, Sommerfeld is unable, or unwilling, to accord conceptual primacy to the adiabatic principle, or to see it as in any way more fundamental as the quantization principles in Eq. (5.93), which, as it would appear, *define* the content of the old quantum theory for Sommerfeld.

5.3 The correspondence principle

Last, but certainly not least, among the guiding principles with which physicists attempted to pick a path through the mysterious and confusing terrain formed by the mass of empirical data confronting them in the decade or so which we associate with the old quantum theory is the *Correspondence Principle* of Bohr.

It was, of course, known before Bohr that in certain limiting regimes the new laws containing Planck's constant, and describing phenomena where nonclassical behavior was clearly present, would approach the classical form for these phenomena. The earliest such case was black-body radiation itself, where the cross-over to the Rayleigh–Jeans law indicated that the classical equipartition theorem appeared to be valid, but only, in

---

[46] The usage "adiabatic hypothesis" is actually due to Einstein (see note 34).

[47] The veto of states with zero component of angular momentum along the direction of an external electric field will be explained below, in our treatment of the Stark effect in Section 6.3.

[48] As shown by Burgers. Referencing Burgers, Sommerfeld (1922a, Appendix 12) provides a version of Burgers' proof (see also Appendix A, Section A.1.4).

Rayleigh's words, for the "graver modes" (i.e., the low frequency end of the spectrum; see Sections 2.4 and 3.3.1). Another example was the specific heat of solids (cf. Section 3.5), which obeyed the classical Dulong–Petit law (except at very low temperatures) provided the natural frequencies of the constituent oscillators was low (in the infrared).

As we saw in Section 4.5.1, in our discussion of the introduction of the Bohr model of atomic structure, Bohr used precisely such a requirement (without however applying the terminology "correspondence principle") to justify his quantization rule for the energy of the stationary states, "by comparing calculations of the energy radiation in the region of slow vibrations ...with calculations based on the ordinary mechanics" (Bohr 1913b, p. 12). The notion that quantum mechanics must be consistent at some level with classical mechanics can scarcely be considered a product of genius, but, as with the adiabatic principle, the careful application of this superficially obvious requirement frequently turned out to be both highly nontrivial and extremely fruitful. To briefly review Bohr's argument of 1913 (see Section 4.5.1, Eqs. (4.47)–(4.51)): if we assume that for very high values of the principal quantum number N (such states are now called "Rydberg states"), where successive energy levels are very closely spaced (and therefore, almost continuous, as in classical theory), the usual laws of electromagnetic radiation from an oscillating charge apply, then one can match the frequency ν of the emitted radiation in a transition from level N to level $N-1$ to the cyclic frequency of the electron in the initial N^{th} orbit (which, as N is large, differs very little on a relative basis from the frequency in the final orbit). As we saw previously, this matching precisely agrees with the constants appearing in the Balmer formula, and in particular with Bohr's adoption of the quantization rule $L = N(h/2\pi)$ for the angular momentum in the Nth stationary state (at this point, with circular orbits only).

The requirement of consistency, or correspondence, between quantum and classical behavior in appropriate limits is used here to provide a post facto justification for the quantum Ansatz employed to select the distinguished quantum states. The correspondence principle, as it would later be called, would be used repeatedly in this way up to 1925, most critically in the development of the Kramers–Heisenberg dispersion theory which would provide the point of transition to the matrix mechanics of 1925 (Duncan and Janssen 2007). The correspondence principle also allowed Bohr and his collaborators in Copenhagen, first and foremost his Dutch assistant Hendrik "Hans" Kramers (see Plate 25), to determine the polarization of radiation emitted in quantum transitions and to give quantitative, if approximate, estimates of the intensities of spectral lines. If the intensity was found to be zero, such estimates amounted to selection rules declaring the associated transitions to be forbidden.

In the first paper of the 1913 trilogy, Bohr mentions elliptical orbits, and points out that the appearance of higher harmonics of the fundamental frequency emitted in high Rydberg states would be unavoidable once the orbits developed a nonzero eccentricity. For a circular orbit, the Cartesian coordinates, however oriented, contain purely sinusoidal variations, $r_i \propto \cos(\omega t + \vartheta_i)$, where $r_i = (x, y, z)$ with $i = 1, 2, 3$, and the period T is $2\pi/\omega$. For an elliptical orbit of period T, we have periodic but not harmonic behavior, $r_i(t + T) = r_i(t)$, whence the Fourier expansion

$$r_i(t) = \sum_\tau C_{i\tau} \cos\left(\tau\omega t + \vartheta_{i\tau}\right), \tag{5.95}$$

with τ running over positive integers or zero. For elliptical orbits, at least some of the higher coefficients $C_{i\tau}$, with $\tau = 2, 3, \ldots$, would necessarily be nonzero. Classically, such a motion of a charged particle would result in radiation at the fundamental (angular) frequency ω, as well as at higher harmonics $\tau\omega$ (with $\tau = 2, 3, \ldots$) of this frequency. Bohr suggests that such higher harmonics would naturally be associated, in the high quantum number regime, with transitions of principal quantum number N to $N - \tau$, as follows clearly from the Bohr frequency condition, together with the approximately uniform spacing of the energies for $\tau \ll N$.

By the time of writing of his comprehensive 1918 paper "On the quantum theory of line spectra" (Bohr 1918), Bohr had acquainted himself with the (by now indispensable) tools of post-Hamilton–Jacobi analytical mechanics. In revisiting the issue of "the relation between the theory of spectra of atomic systems . . . and the ordinary theory of radiation" (Bohr 1918, p. 14), he accordingly helps himself to what had become the standard formulation of the quantization principles in terms of action variables.[49] In Part I of his paper ("On the general theory"), Bohr first considers one-dimensional systems and then moves on quickly to the consideration of conditionally periodic systems with s degrees of freedom. In such systems (see Appendix A, section A.2.3) the fixed frequencies associated with each of the angle coordinates are given as partial derivatives of the energy with respect to the action variables (cf. Eq. (A.163)):

$$\nu_i(\mathcal{J}_1, \ldots, \mathcal{J}_s) = \frac{\partial H(\mathcal{J}_1, \ldots, \mathcal{J}_s)}{\partial \mathcal{J}_i}, \tag{5.96}$$

where the action variables are, of course, quantized, $\mathcal{J}_i = n_i h$. If one now considers a transition between two states specified by quantum numbers n_i' (higher energy) and n_i'' (lower energy), where all the quantum numbers are taken to be very large, with $n_i' - n_i'' \ll n_i'$, we may assume the variation of the energy function $H(\mathcal{J}_i)$ to be so gradual between the initial and final state as to allow the approximation

$$E' - E'' \approx \sum_i \frac{\partial H(\mathcal{J}_1, \ldots, \mathcal{J}_s)}{\partial \mathcal{J}_i} \Delta\mathcal{J}_i = \sum_i \nu_i h(n_i' - n_i''), \tag{5.97}$$

so that the frequency of radiation emitted in a transition between the two states is given by

$$\nu = \frac{1}{h}(E' - E'') = \sum_i \nu_i(n_i' - n_i''). \tag{5.98}$$

---

[49] We shall modify various confusing aspects of Bohr's notation in our discussion to assist the modern reader: for example, ν (instead of ω) is used throughout for cyclic frequencies, and action variables are termed $\mathcal{J}$ rather than I.

In terms of action-angle variables, the generalized coordinates describing a conditionally periodic system can be expressed in a generalized Fourier expansion (see Appendix A, Eq. (A.167)). Following Bohr's (1918, p. 33) notation (with cosines rather than complex exponentials), we can write this expansion as

$$q_i(\mathcal{J}_1, \mathcal{J}_2, \ldots, \mathcal{J}_n; t) = \sum_{\vec{\tau}} C_{i\vec{\tau}}(\mathcal{J}_1, \ldots, \mathcal{J}_n) \cos\left(2\pi(\vec{\tau} \cdot \vec{w} + c_{i\vec{\tau}})\right), \qquad (5.99)$$

where $C_{i\vec{\tau}}$ is a real amplitude, $c_{i\vec{\tau}}$ a real phase, and

$$\vec{w} \equiv (w_1, \ldots, w_n), \quad \vec{\tau} \equiv (\tau_1, \ldots, \tau_n), \qquad (5.100)$$

with $w_i = \nu_i t$ and τ_i taking on all integer values (positive, negative, or zero). According to classical electrodynamics, the frequency of the emitted radiation therefore corresponds to the values

$$\sum_i \tau_i \nu_i, \qquad (5.101)$$

i.e., integer linear combinations of the fundamental frequencies ν_i. This result "corresponds" exactly to the quantum relation (5.98), with

$$\tau_i = n_i' - n_i''. \qquad (5.102)$$

The term "correspondence principle" as such does not occur in the main body of Bohr's 1918 paper (Parts I and II). It does appear, however, in an appendix to Part III, "On the spectra of elements of higher atomic number," which only saw the light of day in November 1922, although a manuscript existed already in 1918 as the first two parts went to press. In the first paragraph of this appendix, Bohr explains the motivation for his choice of terminology:

> The problem treated in [sec. 1 of Part III, on the "General structure of series spectra"] offers a simple application of the point of view developed in Part I and denoted there as a formal connection, or analogy, between the quantum theory and the classical electromagnetic theory of radiation. In order to prevent the possible misunderstanding that it is here a question of a direct connection between the description of the phenomena according to the quantum theory and according to classical electrodynamics, in later papers of the author the law in which this analogy appears is designated as the "correspondence principle". It is a question indeed of a purely *quantum theory theorem*, which joins itself directly to the formulation of the fundamental principles of the quantum theory, and which affirms the existence of a connection between the possibility of any transition between two stationary states accompanied by radiation, and the appearance in the (classical) motion of a certain harmonic oscillation component which may be denoted as the oscillation "corresponding" to the transition (Bohr 1922a, p. 178; emphasis added).

We shall henceforth use the term "correspondence principle" to denote arguments of the type described in this quote.

One of the most significant uses of the correspondence principle as a predictive tool in the old quantum theory was undoubtedly in the elucidation of selection rules for atomic transitions. In classical electromagnetic theory, one determines the emission of radiation from a charged particle by subjecting the coordinates of the motion to Fourier analysis and examining the amplitude of the Fourier component corresponding to each mode of definite frequency. The intensity of the emission at each frequency is then given by a well known formula and is proportional to the square of the corresponding Fourier amplitude. If the coordinates q_i in Eq. (5.99) are chosen to be Cartesian (e.g., x, y, z for $i = 1, 2, 3$), the classical intensity of radiation corresponding to an electric dipole oriented in the i'th direction is proportional to $C_{i\vec{\tau}}^2$. In the quantum theory, this classical intensity has to be interpreted in terms of a probability per unit time of a spontaneous transition from a state characterized by the quantum numbers n_i' to a state characterized by the quantum numbers $n_i'' = n_i' - \tau_i$.

Of course, just as the frequency of emitted radiation in transitions between states at low quantum numbers bears no simple relation to the orbital frequency of the electron either before or after a quantum jump, one could not expect this association between Fourier amplitudes and intensities/transition probabilities to hold except in the limit of very large quantum numbers, i.e., for Rydberg states (in modern terminology), which at the time were completely inaccessible to experimental investigation. The use of the correspondence principle as a predictive tool in the accessible region of low quantum numbers therefore required some type of extension. The "deepening" of the correspondence principle undertaken by Bohr in 1918 that provided the desired extension consisted of two related but basically separate assumptions:

1. In cases in which for all classical motions of the emitting particle some Fourier amplitude, or set of amplitudes, $C_{i\vec{\tau}}$ were exactly zero, the corresponding quantum transition could be assumed to have vanishing probability—effectively, to be forbidden. This would then mean that certain lines which would otherwise be expected would be absent from the emission spectrum. Clearly, this is an assumption: strictly speaking, all that could really be demanded by correspondence principle arguments is that the probability of such transitions should go to zero in the limit of large quantum numbers. Nevertheless, "selection rules" (the term, *Auswahlprinzip* in German, is due to Sommerfeld, and does not appear in Bohr's 1918 paper) of this type were widely accepted and regarded as essentially certain consequences of correspondence reasoning.

2. In general, for non-vanishing Fourier amplitudes, Bohr assumed that even for low quantum numbers, the relative intensities of transitions could at least approximately be calculated in terms of the $C_{i\vec{\tau}}^2$. In particular, the associated classical interpretation would provide specific information concerning the polarization of the emitted light for each allowed transition. The insights provided by the correspondence principle in such cases were admittedly of a more qualitative nature.

The example given by Bohr to illustrate these points (Bohr 1918, pp. 32–34) is extremely illuminating, and will be reviewed here in some detail. The system considered is that of a particle moving in three dimensions, with the dynamics possessing an exact axial symmetry with respect to the z-direction. Choosing cylindrical coordinates $q_1 = \rho = \sqrt{x^2 + y^2}, q_2 = z, q_3 = \varphi = \arctan(y/x)$,[50] this implies that the Hamiltonian $H(q_1, q_2, p_1, p_2, p_3)$ is not a function of q_3, and the conjugate variable p_3, the angular momentum component in the z-direction, is consequently a constant of the motion. The dynamics is presumed to be such that the motion is conditionally periodic in the remaining two variables q_1 and q_2 (see Appendix A, Eq. (A.167)). In other words, with appropriate choice of angle variables $w_i(t) = v_i t$ $(i = 1, 2)$,

$$q_1(t) = \rho(t) = F_1(w_1, w_2), \tag{5.103}$$

$$q_2(t) = z(t) = F_2(w_1, w_2), \tag{5.104}$$

where the F_i $(i = 1, 2)$ are multiply periodic functions, so that for any integers n_1 and n_2 (positive, negative or zero),

$$F_i(w_1 + n_1, w_2 + n_2) = F_i(w_1, w_2). \tag{5.105}$$

Such functions, under mild regularity assumptions, always possess a double Fourier expansion, either of complex exponential form (cf. Appendix A, Eq. (A.167)), or, equivalently, in terms of real sines and cosines:

$$q_i(t) = \sum_{\vec{\tau}} \left(A_{i\vec{\tau}} \cos(2\pi \vec{\tau} \cdot \vec{w}) + B_{i\vec{\tau}} \sin(2\pi \vec{\tau} \cdot \vec{w}) \right)$$

$$= \sum_{\vec{\tau}} C_{i\vec{\tau}} \cos(2\pi(\vec{\tau} \cdot \vec{w} + c_{i\vec{\tau}})), \tag{5.106}$$

with $i = 1, 2$ and $\vec{\tau} = (\tau_1, \tau_2)$. The dependence of the amplitudes $C_{i\vec{\tau}}$ and phases $c_{i\vec{\tau}}$ on the action variables and initial conditions will be suppressed throughout our discussion.

The time dependence of the angular variable $q_3 = \varphi$ is a bit more complicated in the general case, as the derivative of the Hamiltonian with respect to p_3 may contain q_1 and q_2 (if the particle is in the presence of a magnetic field, for example), so we must set

$$\dot{q}_3 = \dot{\varphi} = \frac{\partial H(q_1, q_2, p_1, p_2, p_3)}{\partial p_3} = \sum_{\vec{\tau}} C_{3\vec{\tau}} \cos(2\pi(\vec{\tau} \cdot \vec{w} + c_{3\vec{\tau}})). \tag{5.107}$$

In fact, the derivative may in general contain any or all of the coordinates q_1, q_2, p_1, p_2, which all possess multiply periodic behavior with respect to the angle variables w_1

---

[50] Following standard modern notation, we choose φ rather than Bohr's ϑ for the azimuthal angle. We will change Bohr's notation in other inessential ways as well to maximize clarity of exposition.

and w_2. Separating out the constant term with $\vec{\tau} = 0$ in the sum in Eq. (5.107), and defining $C_{3,\vec{\tau}=0} \cos{(2\pi c_{3,\vec{\tau}=0})} \equiv 2\pi \nu_3$, we may integrate Eq. (5.107) with respect to time, obtaining

$$q_3(t) = \varphi(t) = 2\pi \nu_3 t + \sum_{\vec{\tau}} \tilde{C}_{3\vec{\tau}} \cos{(2\pi(\vec{\tau} \cdot \vec{w} + \tilde{c}_{3\vec{\tau}}))}. \tag{5.108}$$

The multiply periodic cosines and sines present in the sum in Eq. (5.107) for non-zero $\vec{\tau}$ simply integrate to another multiply periodic sum with altered coefficients and phases, as is reflected in the tildes added to $C_{3\vec{\tau}}$ and $c_{3\vec{\tau}}$ in Eq. (5.108).

In order to apply the classical electromagnetic theory most directly, we should go back to Cartesian coordinates ($x = \rho \cos{\varphi}, y = \rho \sin{\varphi}, z$), as the polarization properties of the emitted radiation are most easily read off in terms of these coordinates. The particle's motion in the x-direction is the product of ρ, given by Eq. (5.106) for $i = 1$, and the cosine of φ, given by Eq. (5.108). We can rewrite this product using standard trigonometric identities and the fact that the cosine of a multiply periodic function is itself a multiply-periodic function. In doing so, we encounter expressions such as

$$\cos{(2\pi\nu_3 t)} \cos{\left(\sum_{\vec{\tau}} \tilde{C}_{3\vec{\tau}} \cos{(2\pi(\vec{\tau} \cdot \vec{w} + \tilde{c}_{3\vec{\tau}}))} \right)} = \tag{5.109}$$

$$\frac{1}{2} \sum_{\vec{\tau}} \tilde{\tilde{C}}_{3\vec{\tau}} \left(\cos{\left(2\pi(\nu_3 t + \vec{\tau} \cdot \vec{w} + \tilde{\tilde{c}}_{3\vec{\tau}}) \right)} + \cos{\left(2\pi(\nu_3 t - \vec{\tau} \cdot \vec{w} - \tilde{\tilde{c}}_{3\vec{\tau}}) \right)} \right).$$

By contrast, the component of the particle's motion in the z-direction (given by Eq. (5.106) for $i = 2$) only contains Fourier modes with time-dependence of the form

$$z(t) = \sum_{\vec{\tau}} C_{2\vec{\tau}} \cos{(2\pi(\vec{\tau} \cdot \vec{w} + c_{2\vec{\tau}}))}$$

$$= \sum_{\vec{\tau}} C_{2\vec{\tau}} \cos{(2\pi((\tau_1 \nu_1 + \tau_2 \nu_2)t + c_{2\vec{\tau}}))}, \tag{5.110}$$

without the $2\pi\nu_3 t$ term. The upshot of arguments of this type can be summarized as follows:

1. The particle's oscillation in the x- and y-directions perpendicular to the axis of symmetry can be decomposed into Fourier modes of the form

$$\cos{(2\pi(\tau_1 \nu_1 + \tau_2 \nu_2 \pm \nu_3)t + \delta)},$$

where τ_1 and τ_2 are positive, negative, or zero, and δ is an uninteresting constant phase. The corresponding oscillating dipole moment $\vec{p} = e(x(t), y(t))$, where e is the particle's electric charge, will result (provided the corresponding amplitude coefficient is non-zero) in the emission of light with frequency $|\tau_1 \nu_1 + \tau_2 \nu_2 \pm \nu_3|$, which will be right or left circularly polarized when viewed along the axis of symmetry (the z-axis here) and linearly polarized when viewed along a line perpendicular to that axis.[51] The relation $\tau_i = n_i' - n_i''$ (see Eq. (5.102)) tells us that this radiation is produced by a transition between quantum states in which n_1 and n_2 change by arbitrary integer values and n_3 changes by ± 1.

2. The particle's oscillation in the z-direction results in the emission of z-polarized light of frequency $|\tau_1 \nu_1 + \tau_2 \nu_2|$ in all directions with the exception of the z-direction. Note that for this component, the ν_3 term is absent. This radiation is thus produced by transitions in which only n_1 and/or n_2 change while n_3 stays the same.

3. As Bohr points out, the existence of Fourier components in which the integers τ_i are negative as well as positive implies the possibility of transitions in which the overall energy decreases but in which some of the quantum numbers *increase*. This is contrary to a hypothetical selection rule initially advanced by Sommerfeld (cf. Section 5.1.3), according to which not only the principal quantum number, but also the separate subsidiary quantum numbers would each have to decrease in a transition corresponding to emission of radiation.

With very little in the way of detailed assumptions concerning the dynamics, we see that the correspondence principle now leads to some impressively detailed statements concerning the nature of the radiation emitted in possible quantum transitions. In Bohr's words:

> On the present theory we shall consequently expect that in this case only two kinds of transitions between stationary states given by [Eq. (5.91)] will be possible. In both of these n_1 and n_2 may vary by an arbitrary number of units, but in the first kind of transition, which will give rise to a radiation polarised parallel to the axis of the system, n_3 will remain unchanged, while in the second kind or transition n_3 will decrease or increase by one unit and the emitted radiation will be circularly polarised round the axis in the same direction as or the opposite of that of the rotation of the particle respectively.

As a simple and explicit example of an axially symmetric system of the type just discussed, we need look no further than the case of the electron in the hydrogen atom subject to the imposition of uniform electric and/or magnetic fields in the z-direction.[52]

---

[51] Recall that the direction of polarization at any distant point is obtained by projecting the dipole moment onto the plane containing the z-axis and perpendicular to the radius vector of that point.

[52] The "crossed-field" situation, in which the electric and magnetic fields pointed in a different direction, would present a much thornier problem for the old quantum theory. Bohr initially thought this problem to be incompatible with a conditionally periodic formulation (mistakenly), but as Pauli would describe in his *Handbuch* article (Pauli 1926, p. 159), the real problem with crossed fields is that one could construct an adiabatic transformation from allowed electron orbits, with nonzero angular momentum, to the long forbidden nucleus-crossing orbits with zero angular momentum, by using such fields.

The Hamiltonian for this system is just (using ρ, z, φ instead of q_1, q_2, q_3 for physical transparency, and neglecting terms of second order in the magnetic field)

$$H = \frac{p_\rho^2}{2m} + \frac{p_z^2}{2m} + \frac{p_\varphi^2}{2m\rho^2} - \frac{e^2}{\sqrt{\rho^2 + z^2}} + e\mathcal{E}z + \frac{e\mathcal{H}p_\varphi}{2mc}, \qquad (5.111)$$

where both electric $\vec{\mathcal{E}} = \mathcal{E}\hat{z}$ and magnetic $\vec{\mathcal{H}} = \mathcal{H}\hat{z}$ fields point in the z direction. The azimuthal angle φ is cyclic, and the Hamiltonian is precisely of the form assumed above. In particular, the selection and polarization rules deduced above must apply to the Stark effect ($\mathcal{E} \neq 0, \mathcal{H} = 0$), or to the Zeeman effect ($\mathcal{H} \neq 0, \mathcal{E} = 0$) in hydrogen, both of which we shall be discussing in great detail later (see Sections 6.3 and 7.2).

The second use of the deeper version of the correspondence principle described above—the calculation of relative intensities of spectral lines—required two distinct pieces of information to be combined. The amount of light of a specific frequency detected from a gas discharge clearly depends on (a) the number of atoms present in the gas at any given time with an electron in the specific initial stationary state giving rise to the transition in question, and (b) the probability per unit time that an electron in such a state will make a transition to the final state corresponding to the spectral line being observed. Only the second of these quantities has a direct connection to classical electromagnetic theory and the Fourier decomposition of the motion of charged particles. Assuming the existence of thermal equilibrium for the gas (rarely the case in most gas discharges, unfortunately), the occupation number for levels in gas atoms is determined by a Boltzmann factor. In practice, Bohr assumed occupation numbers to be roughly equal for states of equal a priori probability and, at least approximately, equal energy.

In the Stark effect, for example, the level splittings induced by available electric fields were sufficiently small that one could assume that the initial and final state energy levels had to a good approximation equal Boltzmann factors, reducing the calculation of the relative intensity of spectral lines in this case to the evaluation of the Fourier components of the motion (including the perturbing effects of the electric field). Bohr gave this highly nontrivial assignment to the aforementioned Hans Kramers, his brilliant student and assistant, of whom we will hear much more in the following chapters. In his doctoral thesis, in a tour-de-force of mathematical analysis, Kramers (1919) succeeded in a complete analytic evaluation of the Fourier components of the perturbed motion, both for the Stark effect and for the case of the relativistic corrections to the motion of the electron in a hydrogen atom. The comparison with empirical evidence, as we shall see, was initially (mildly) encouraging, but hardly conclusive, especially not when compared to the remarkable success of the calculations of the energy levels themselves, which were to provide, in Sommerfeld's (1919, p. 458) words, "a striking achievement of our field and a beautiful capstone on the edifice of atomic theory."

The correspondence principle continued to serve as a guiding principle of the old quantum theory right up to the point at which this theory, via the theory of dispersion developed by Kramers, and transmuted alchemically by Heisenberg into matrix mechanics, was finally supplanted by modern quantum mechanics. As we shall see later in our study of this remarkable conceptual transition, the most important ingredients of the Kramers dispersion theory, in addition to the technical apparatus imported from celestial mechanics, were (*i*) the quantum theory of radiation developed in Einstein's famous "A and B coefficients" paper (see Section 3.6), and (*ii*) the correspondence principle used to guide the construction of a dispersion formula for the index of refraction of a gas in terms of the quantized energy levels of its component atoms or molecules. At this point, the correspondence principle became a precise mathematical prescription, wherein derivatives in the classical theory (as in Eq. (5.96)) were replaced in a systematic way by finite difference quotients. In order to maintain a modicum of chronological order in our presentation, however, we shall defer discussion of this final transmutation of the correspondence principle in 1923/1924 to Volume 2.

6

Successes

In the years following the introduction of the Bohr model in 1913, quantum theory increasingly became the focus of attention for essentially all physicists concerned with constructing and analyzing mechanical models of atomic structure, with optical and X-ray spectroscopy providing by far the richest empirical testbed for assessing the validity of these models. We are fortunate to have, from 1919 through 1924, the series of four editions of Sommerfeld's *Atombau und Spektallinien* (Atomic Structure and Spectral Lines) in which the intense interaction between the empirical material and theoretical developments in the old quantum theory can be followed step by step, as described by a participant who was in every respect "at the center of the action". A complete account of all the twists and turns in this story would require far more space than available in this book and would only submerge the essential features of the development in a mass of unenlightening details.

We have therefore decided to begin the story of the development of quantum theory in the twelve years between the emergence of the Bohr model and the appearance of the quantum mechanics of Heisenberg and Schrödinger by highlighting, in this chapter, three cases in which the old quantum theory achieved spectacular successes in the interpretation and explanation of experimental results—the derivation of the relativistic fine-structure of hydrogen and ionized helium by Sommerfeld, the interpretation of X-ray multiplets in terms of relativistic fine-structure (also by Sommerfeld), and the elucidation of the Stark effect by Schwarzschild and Epstein. Our topics are chosen from the period following the introduction of the Bohr-Sommerfeld quantization rules in 1916, as we have already discussed the triumphs of the Bohr model in its first phase, in explaining the spectrum of hydrogen and ionized helium (see Section 4.6).

A short comment is needed here on our characterization of the topics described here as "successes of the old quantum theory". In hindsight, one knows, of course, that this theory was incorrect in fundamental ways, and, as we shall see, the "successes" to be described were in many respects lucky accidents. At the time, however, the most prominent practioners of the old quantum theory regarded these developments as striking successes of the theory. Sommerfeld's extension of Bohr's theory greatly impressed his colleagues (Eckert 2014, pp. 153–154). Einstein, for instance, told Sommerfeld in a letter of August 3, 1916: "Your spectral analyses number among my finest experiences in physics. Only through them, Bohr's idea becomes entirely convincing" (Einstein

Constructing Quantum Mechanics. Anthony Duncan and Michel Janssen, Oxford University Press (2019).
© Anthony Duncan and Michel Janssen. DOI: 10.1093/oso/9780198845478.001.0001

1987–2018, Vol. 8, Doc. 246). Though Einstein also had his reservations about the old quantum theory, many physicists were convinced, up to the early 1920s, that the theoretical basis for a complete description of atomic structure was already in hand, with only some troublesome details to be resolved by a more elaborate version of the same ideas. The appearance of a series of striking *failures* of the theory, which would completely undermine this optimism by the middle of the third decade of the twentieth century, will be the subject of the next chapter.

6.1   Fine structure

In the Spring and early Summer of 1916, Sommerfeld prepared and submitted for publication a long paper on "The Quantum Theory of Spectral Lines," which appeared in two parts in *Annalen der Physik* (Sommerfeld 1916a, 1916b). After reviewing his previously proposed quantization of elliptical orbits in the nonrelativistic approximation (see Section 5.1.3 of the previous chapter) and introducing the idea of space quantization in the course of extending the treatment of the quantum Kepler problem from a two dimensional to a fully three dimensional one (see Section 5.1.4), Sommerfeld turns to the problem of including special-relativistic effects in the quantization of the electron energy levels in hydrogen. Before explaining what Sommerfeld achieved in this direction, we should briefly review the discussion of the non-relativistic situation in the first part of the paper (entitled "Theory of the Balmer series"), as the treatment here is considerably more extensive than in his December 1915 report discussed previously.

The degeneracy of the energy levels in the Balmer lines was illustrated graphically by drawings of the actual circular and elliptical orbits for the lowest energy levels. We reproduce these depictions here (see Fig. 6.1), albeit in a somewhat different format, with the nucleus always at the origin (i.e. one focus of the corresponding ellipse), and the orbits drawn to scale and coplanar for convenience, although the orientation of the plane of the orbits and the semi-major axis is of course arbitrary. Formally, orbits exist

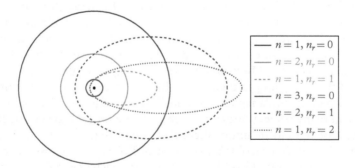

Figure 6.1 *Quantized orbits of the electron in hydrogen, drawn to scale, for principal quantum number $n_r + n = 1, 2, 3$. The black dot denotes the nucleus.*

with $n = 0, n_r \neq 0$:[1] they are the straight-line paths in which the electron falls directly into and through the nucleus, reaching infinite velocity at the center, assuming a point-like nuclear charge. Such orbits are excluded out of hand by Sommerfeld:

> $n = 0$ means $p = 0$ (angular momentum), in other words the degeneration of the elliptical plane to a double straight-line. Such a path is geometrically impossible, as it would so-to-speak pass right through the nucleus, and dynamically impermissible, because the velocity would become infinitely large. Here one would already have to include a relativistic treatment, which we shall give in Part II.

In fact, Sommerfeld goes considerably further in his discussion of the relative probabilities of the states, assigning a relative probability (*Häufigkeitsmass*) $n/(n + n_r)$ (l/n in modern notation, with l the angular momentum and n the principal quantum number) to the state with quantum numbers n, n_r (and principal quantum number $n + n_r$). This appears to be a pure guess, based on the simplest interpolation between the already excluded straight-line orbits with $n = 0$ and the assumed most probable circular orbits with $n_r = 0$. The formula actually follows approximately the degeneracy factor $2l + 1$ of modern quantum mechanics, but Sommerfeld uses it to estimate the intensity of spectral lines, without taking into account that this depends also on the intrinsic probability of transition, requiring in the old theory (later managed via the correspondence principle; cf. Section 5.3) the calculation of Fourier amplitudes of the motion (and in modern theory the calculation of a nontrivial matrix element). In addition, he provisionally appends the selection rule (later criticized by Bohr, as we saw previously) that in any allowed transition $(m, m_r) \to (n, n_r)$, both quantum numbers, and not just their sum, the principal quantum number, must remain constant or decrease: $m \geq n$, $m_r \geq n_r$.

As Sommerfeld realized, relativistic effects would become relevant not only for the forbidden straight-line orbits, but would be especially prominent for the highly eccentric orbits corresponding to smaller values of n, where the electron would pass closer to the nucleus and speed up. In particular, the relativistic effects would break the degeneracy of the states in the Balmer formula, by altering the kinetic energy from the Newtonian value, with greater alteration for the states with smaller n (the angular momentum). This would in turn result in a splitting of the Balmer lines, as $\varphi(n, n')$ (with Sommerfeld notation n' for n_r) in the Balmer formula Eq. (5.42) would now depend on n and n' individually, and not just the sum $n + n'$, i.e., Bohr's principal quantum number.

In the language preferred by Sommerfeld (but considered inartful by modern physicists; see, e.g., Okun (1989)), relativistic effects were to be included by replacing the constant mass of Newtonian mechanics by a variable relativistic mass m, related to the rest mass m_0 by

$$m = \frac{m_0}{\sqrt{1 - v^2/c^2}} = \frac{m_0}{\sqrt{1 - \beta^2}}. \tag{6.1}$$

---

[1] Sommerfeld uses the notation n' for the radial quantum number, rather than n_r used here.

The first task is simply to repeat the classical calculation of the Kepler orbits, now including the effects of such a variable mass. The conservation of angular momentum still holds, of course, so the motion is restricted to a plane, which we may take to be the x-y plane (with $z = 0$), where the relevant polar coordinates are (r, φ), with $x = r \cos \varphi, y = r \sin \varphi$. For an electron of charge $-e$ bound to a nucleus of charge $+Ze$ at the origin,[2] Newton's second law for the x coordinate reads

$$\frac{d}{dt}(m\dot{x}) = -\frac{Ze^2}{r^2} \cos \varphi. \tag{6.2}$$

The angular momentum $p_\varphi = mr^2\dot{\varphi}$ is still a constant of the motion[3] and we may use it to replace time derivatives with angular derivatives,

$$\frac{d\varphi}{dt} = \frac{p_\varphi}{mr^2} \Rightarrow \frac{d}{dt} = \frac{p_\varphi}{mr^2}\frac{d}{d\varphi}. \tag{6.3}$$

The momentum in the x-direction may therefore be reexpressed, introducing the notation $\sigma = 1/r$,

$$m\dot{x} = \frac{p_\varphi}{r^2}\frac{d}{d\varphi}(r \cos \varphi) = p_\varphi \left(-\frac{1}{r} \sin \varphi + \frac{1}{r^2}\frac{dr}{d\varphi} \cos \varphi \right)$$

$$= -p_\varphi \left(\sigma \sin \varphi + \frac{d\sigma}{d\varphi} \cos \varphi \right). \tag{6.4}$$

Using Eq. (6.3) to evaluate the left-hand side of Eq. (6.2), we find:

$$\frac{d}{dt}(m\dot{x}) = -\frac{p_\varphi^2}{mr^2}\frac{d}{d\varphi}\left(\sigma \sin \varphi + \frac{d\sigma}{d\varphi} \cos \varphi \right) = -\frac{p_\varphi^2}{mr^2}\left(\sigma + \frac{d^2\sigma}{d\varphi^2} \right) \cos \varphi. \tag{6.5}$$

Equating this result to the Coulomb force on the right-hand side of Eq. (6.2), we arrive at

$$\frac{d^2\sigma}{d\varphi^2} + \sigma = \frac{Ze^2 m}{p_\varphi^2}, \tag{6.6}$$

where the right-hand side is still varying through the orbit as a consequence of the variable relativistic mass m. As the conserved total energy (i.e., relativistic kinetic energy plus Coulomb potential energy) is $E = (m - m_0)c^2 - Ze^2/r$, we can rewrite the relativistic mass as

---

[2] We use the modern notation for the nuclear charge here: Sommerfeld writes $Ze = E$.
[3] Sommerfeld uses the notation p for the *angular* momentum: in order to distinguish it from the radial momentum p_r, shortly to be introduced in the quantization procedure, we have added the subscript.

$$m = m_0 + \frac{E}{c^2} + \frac{Ze^2}{c^2 r} = m_0 + \frac{E}{c^2} + \frac{Ze^2}{c^2} \sigma. \tag{6.7}$$

Inserting this on the right-hand side of Eq. (6.6) and rearranging, we find a second-order differential equation (of simple harmonic type) for σ as a function of φ:

$$\frac{d^2\sigma}{d\varphi^2} + \left(1 - \left(\frac{Ze^2}{p_\varphi c}\right)^2\right)\sigma = \frac{Ze^2 m_0}{p_\varphi^2}\left(1 + \frac{E}{m_0 c^2}\right), \tag{6.8}$$

where now the quantity on the right-hand side is strictly constant during the motion. Introducing

$$\gamma^2 \equiv 1 - \left(\frac{Ze^2}{p_\varphi c}\right)^2, \tag{6.9}$$

we can write the general solution of this forced simple harmonic equation as

$$\sigma(\varphi) = \frac{1}{r(\varphi)} = A\cos(\gamma\varphi) + B\sin(\gamma\varphi) + C, \tag{6.10}$$

where C is the right-hand side of Eq. (6.8) divided by γ^2:

$$C \equiv \frac{Ze^2 m_0}{\gamma^2 p_\varphi^2}\left(1 + \frac{E}{m_0 c^2}\right). \tag{6.11}$$

For convenience, one may orient the x-axis so that $B = 0$ and write Eq. (6.10) as

$$\frac{1}{r} = C\left(1 + \frac{A}{C}\cos(\gamma\varphi)\right). \tag{6.12}$$

Comparing this expression with the general equation for an ellipse, Eq. (5.32),

$$\frac{1}{r} = \frac{1 + \epsilon \cos \vartheta}{K}, \tag{6.13}$$

we set A/C equal to the eccentricity ϵ and rewrite Eq. (6.12) as:

$$\frac{1}{r} = C(1 + \epsilon \cos(\gamma\varphi)). \tag{6.14}$$

It follows from this that the aphelion and perihelion distances are given by

$$r_{\max} = \frac{1}{C(1 - \epsilon)}, \quad r_{\min} = \frac{1}{C(1 + \epsilon)}. \tag{6.15}$$

The semi-major axis is given by

$$a = \frac{1}{2}(r_{max} + r_{min}) = \frac{1}{C(1 - \epsilon^2)}. \tag{6.16}$$

We can thus replace C by $(a(1-\epsilon^2))^{-1}$ in the equation of orbit, Eq. (6.14):

$$\frac{1}{r} = \frac{1}{a(1 - \epsilon^2)}(1 + \epsilon \cos(\gamma\varphi)). \tag{6.17}$$

As γ is less than unity, the particle must move through an angle φ greater than 2π to return to perihelion at r_{min}. The perihelion thus precesses, advancing by an angle

$$2\pi \left(\frac{1}{\gamma} - 1\right) \tag{6.18}$$

per revolution. For the hydrogen atom $(Z = 1)$, with the electron in the lowest elliptical orbit, $p_\varphi = \hbar$, Eq. (6.9) gives

$$\gamma = \sqrt{1 - \alpha^2}, \tag{6.19}$$

where

$$\alpha \equiv \frac{e^2}{\hbar c} \approx \frac{1}{137} \tag{6.20}$$

is Sommerfeld's *fine-structure constant*, so named for reasons that will become clear shortly.

Evidently in this case $\gamma = 0.9999734$ is very close to unity, and the precession angle per revolution amounts to 0.000167 radians.[4] Choosing a more relativistic case, with $\gamma = 0.91$, allows us to display the precessional motion more clearly, as in Fig. 6.2.

In the classical solution given above (see Eqs. (6.10)–(6.11)), the energy E and angular momentum p_φ, as well as the constants A and B, are continuously variable quantities. The quantization of this system of two degrees of freedom (r and φ) now follows precisely the path Sommerfeld followed in the non-relativistic case.

---

[4] As Sommerfeld (1919, p. 329) points out in the first edition of *Atombau und Spektrallinien*, the precession given by the formula $2\pi(\frac{1}{\gamma} - 1)$ amounts to only $7''$ per century for Mercury, instead of the observed anomalous (i.e., not due to other planets) precession of $43''$ per century. The explanation of this discrepancy was, of course, the first great triumph of Einstein's general theory of relativity (Earman and Janssen 1993).

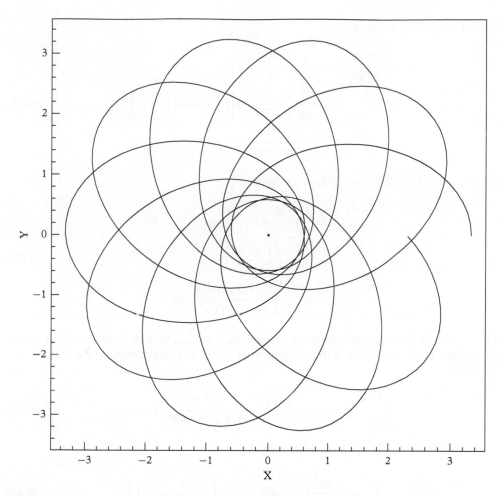

Figure 6.2 *Relativistic Kepler precession for a particle with* $\gamma = 0.91$ *(the red dot indicates the center of the force).*

The radial momentum p_r is given in the relativistic case by (recall that m is the variable relativistic mass)

$$p_r = m\dot{r}. \tag{6.21}$$

Using $p_\varphi = mr^2\dot{\varphi}$, one can write

$$p_r = m\frac{dr}{d\varphi}\dot{\varphi} = \frac{p_\varphi}{r^2}\frac{dr}{d\varphi}. \tag{6.22}$$

Using Eq. (6.17), we have

$$
\begin{aligned}
\frac{1}{r}\frac{dr}{d\varphi} &= -r\frac{d}{d\varphi}\left(\frac{1}{r}\right) \\
&= -r\frac{d}{d\varphi}\left(\frac{1}{a(1-\epsilon^2)}(1+\epsilon\cos(\gamma\varphi))\right) \\
&= \frac{\epsilon\gamma\sin(\gamma\varphi)}{1+\epsilon\cos(\gamma\varphi)}.
\end{aligned}
\tag{6.23}
$$

Finally, using Eqs. (6.22)–(6.23), we obtain

$$
\begin{aligned}
p_r dr &= p_r\frac{dr}{d\varphi}d\varphi \\
&= p_\varphi\left(\frac{1}{r}\frac{dr}{d\varphi}\right)^2 d\varphi \\
&= p_\varphi\frac{\epsilon^2\gamma^2\sin^2(\gamma\varphi)}{(1+\epsilon\cos(\gamma\varphi))^2}d\varphi.
\end{aligned}
\tag{6.24}
$$

A complete cycle in r means we must go from $r_{\min}$ to $r_{\max}$ and back again, which means a complete cycle (0 to 2π) in the variable $\psi \equiv \gamma\varphi$. The radial quantization condition is therefore

$$
\int_{\psi=0}^{\psi=2\pi} p_r dr = p_\varphi\epsilon^2\gamma\int_0^{2\pi}\frac{\sin^2\psi}{(1+\epsilon\cos\psi)^2}d\psi = n_r h.
\tag{6.25}
$$

Evaluating the integral (cf. Eq. (5.38)), we find

$$
\int_0^{2\pi}\frac{\sin^2\psi}{(1+\epsilon\cos\psi)^2}d\psi = \frac{2\pi}{\epsilon^2}\left(\frac{1}{\sqrt{1-\epsilon^2}}-1\right).
\tag{6.26}
$$

The quantization condition thus becomes

$$
2\pi p_\varphi\gamma\left(\frac{1}{\sqrt{1-\epsilon^2}}-1\right) = n_r h.
\tag{6.27}
$$

The angular quantization condition works precisely as before in the non-relativistic case:

$$
\int_0^{2\pi} p_\varphi d\varphi = 2\pi p_\varphi = nh.
\tag{6.28}
$$

Here we follow Sommerfeld and use the notation n (instead of the more explicit n_φ) for the azimuthal (angular momentum) quantum number.

Inserting nh for $2\pi p_\varphi$ in Eq. (6.27), we arrive at a formula for the eccentricity in terms of the radial and azimuthal quantum numbers:

$$\frac{1}{\sqrt{1-\epsilon^2}} - 1 = \frac{n_r}{\gamma n} \Rightarrow \sqrt{1-\epsilon^2} = \frac{1}{1 + \dfrac{n_r}{\gamma n}}. \tag{6.29}$$

At this point it is convenient to perform a check on the results by going to the nonrelativistic limit. In the limit that $c \to \infty$, we see that $\gamma \to 1$ and we get

$$\sqrt{1-\epsilon^2} = \left(1 + \frac{n_r}{n}\right)^{-1} = \frac{n}{n_r + n}, \tag{6.30}$$

which is Eq. (5.40) (in Wilson's notation), the quantization condition for eccentricity found independently by Wilson and Sommerfeld.

Inserting $p_\varphi = n\hbar$ in Eq. (6.9) for γ, we find

$$\gamma = \sqrt{1 - \frac{Z^2 e^4}{n^2 \hbar^2 c^2}} = \sqrt{1 - \frac{Z^2 \alpha^2}{n^2}}, \tag{6.31}$$

where, in the last step, we used Eq. (6.20) for α. Eq. (6.31) shows that Sommerfeld's fine-structure constant controls the size of departures from nonrelativistic behavior.[5] The parameter γ differs most from unity (i.e., the particle is most relativistic) for low values of n (these are the most eccentric orbits, with the particle approaching most closely to the nucleus) or for high Z, for the heavy elements, where we shall soon see that the results remain useful for X-ray spectra despite the presence of many electrons in the atom. Note that the formula clearly breaks down completely for the straight-line paths through the nucleus with $n = 0$, previously excluded by Sommerfeld, where we obtain $\sqrt{-\infty}$.

The determination of the energy of the quantized orbits follows once we express the eccentricity in terms of perihelion and aphelion distances. We shall follow a slightly less painful approach than Sommerfeld, beginning with the relativistic expression for the Hamiltonian (defined as the full relativistic energy, minus the here uninteresting rest energy):

$$H = \sqrt{m_0^2 c^4 + c^2 \left(p_r^2 + \frac{p_\varphi^2}{r^2}\right)} - \frac{Ze^2}{r} - m_0 c^2, \tag{6.32}$$

---

[5] Not surprisingly, the value of v/c for the electron in the ground state of the Bohr hydrogen atom is just α—see below.

where we have written p^2, the momentum squared, in terms of the radial momentum p_r and the angular momentum p_φ. Setting $H = E$ and $\sigma = 1/r$, we find, after some rearranging and squaring

$$m_0^2 c^4 + c^2 p_r^2 + c^2 p_\varphi^2 \sigma^2 = \left(E + Ze^2\sigma + m_0 c^2\right)^2. \tag{6.33}$$

The radial momentum p_r is thus given by an expression quadratic in σ:

$$c^2 p_r^2 = (Z^2 e^4 - c^2 p_\varphi^2)\sigma^2 + 2Ze^2(E + m_0 c^2)\sigma + E(E + 2m_0 c^2). \tag{6.34}$$

Calling the roots of the corresponding quadratic equation σ_1 and σ_2, we can write Eq. (6.34) as

$$c^2 p_r^2 = -(Z^2 e^4 - c^2 p_\varphi^2)(\sigma - \sigma_1)(\sigma_2 - \sigma). \tag{6.35}$$

The roots correspond to the turning points of the radial motion: $\sigma_1 = 1/r_{max}$ and $\sigma_2 = 1/r_{min}$. Using the definition of γ in Eq. (6.9) to substitute $-c^2 p_\varphi^2 \gamma^2$ for $Z^2 e^4 - c^2 p_\varphi^2$, we can rewrite Eq. (6.35) as

$$c^2 p_r^2 = -c^2 p_\varphi^2 \gamma^2 (\sigma^2 - (\sigma_1 + \sigma_2)\sigma + \sigma_1 \sigma_2). \tag{6.36}$$

Comparing this equation with Eq. (6.34), we read off the following expressions for the product and the sum of the two roots, respectively:

$$\sigma_1 \sigma_2 = -\frac{E^2 + 2m_0 c^2 E}{c^2 p_\varphi^2 \gamma^2} \tag{6.37}$$

$$\sigma_1 + \sigma_2 = \frac{2Ze^2(E + m_0 c^2)}{c^2 p_\varphi^2 \gamma^2}. \tag{6.38}$$

We can use the quantization condition (6.25) for p_φ and the expression (6.20) for α to write

$$c p_\varphi = c n \hbar = \frac{ne^2}{\alpha}. \tag{6.39}$$

Substituting this expression into the denominators of Eqs. (6.37)–(6.38) and rewriting the numerator of the former, we find:

$$\sigma_1 \sigma_2 = \frac{\alpha^2 \left(m_0^2 c^4 - (E + m_0 c^2)^2\right)}{n^2 e^4 \gamma^2} \tag{6.40}$$

$$\sigma_1 + \sigma_2 = \frac{2\alpha^2 Z(E + m_0 c^2)}{n^2 e^2 \gamma^2}. \tag{6.41}$$

The eccentricity can also be expressed in terms of the roots σ_1 and σ_2:

$$\epsilon = \frac{r_{\text{max}} - r_{\text{min}}}{r_{\text{max}} + r_{\text{min}}} = \frac{\left(\dfrac{1}{\sigma_1} - \dfrac{1}{\sigma_2}\right)}{\left(\dfrac{1}{\sigma_1} + \dfrac{1}{\sigma_2}\right)} = \frac{\sigma_2 - \sigma_1}{\sigma_2 + \sigma_1}. \tag{6.42}$$

It follows that

$$\sqrt{1 - \epsilon^2} = \frac{2\sqrt{\sigma_1 \sigma_2}}{\sigma_1 + \sigma_2}. \tag{6.43}$$

With the help of Eqs. (6.40)–(6.41), this can be rewritten as

$$\sqrt{1 - \epsilon^2} = \frac{n\gamma}{Z\alpha(E + m_0 c^2)} \sqrt{m_0^2 c^4 - (E + m_0 c^2)^2}. \tag{6.44}$$

Comparing this expression to the quantized expression for $\sqrt{1 - \epsilon^2}$ in Eq. (6.29), we find

$$\frac{n\gamma}{n\gamma + n_r} = \frac{n\gamma}{Z\alpha(E + m_0 c^2)} \sqrt{m_0^2 c^4 - (E + m_0 c^2)^2}, \tag{6.45}$$

or, equivalently,

$$\frac{E + m_0 c^2}{\sqrt{m_0^2 c^4 - (E + m_0 c^2)^2}} = \frac{n_r + \gamma n}{Z\alpha}. \tag{6.46}$$

Squaring and solving for $(E + m_0 c^2)^2$, we arrive at

$$(E + m_0 c^2)^2 = m_0^2 c^4 \frac{1}{1 + \left(\dfrac{Z\alpha}{n_r + \gamma n}\right)^2}. \tag{6.47}$$

So the orbit energy (with the rest energy removed) is

$$E = m_0 c^2 \left(\frac{1}{\sqrt{1 + \left(\dfrac{Z\alpha}{n_r + \gamma n} \right)^2}} - 1 \right). \qquad (6.48)$$

This remarkably compact and mysterious formula contains not only the Bohr energy levels, but also, as we shall now see, the small relativistic fine-structure corrections to the Balmer formula. These were to prove one of the most impressive successes of the old quantum theory. The importance of relativistic corrections can be read off directly from Eq. (6.31), where we see that the figure of merit for such corrections is the quantity $Z\alpha/n$, which in fact turns out to be precisely the value of v/c for the electron in the circular orbit with principal quantum number n. For atoms at the lower end of the periodic table (with $Z < 10$, say), $Z\alpha/n < 0.1$ and an expansion in powers of this quantity converges rapidly. In the case of X-ray spectra, discussed later in this chapter, Z may be large, in which case it may be necessary to include higher terms in the expansion, or, more directly, one may employ the exact formula Eq. (6.48).

Assuming the nuclear charge Z is not too large, we may expand Eq. (6.48) in powers of the fine-structure constant α. Following Sommerfeld (1916a, p. 54), we find:

$$E = m_0 c^2 \left(-\frac{1}{2} \frac{Z^2 \alpha^2}{(n_r + \gamma n)^2} + \frac{3}{8} \frac{Z^4 \alpha^4}{(n_r + n)^4} + O(\alpha^6) \right). \qquad (6.49)$$

Using the expression for γ to write:

$$\frac{1}{(n_r + \gamma n)^2} = \frac{1}{\left(n_r + n\sqrt{1 - Z^2\alpha^2/n^2} \right)^2}$$

$$= \frac{1}{(n_r + n)^2} \left(1 + \frac{Z^2 \alpha^2}{n(n_r + n)} + O(\alpha^4) \right) \qquad (6.50)$$

we can further expand Eq. (6.49) as

$$E = m_0 c^2 \left(-\frac{1}{2} \frac{Z^2 \alpha^2}{(n_r + n)^2} - \frac{1}{2} \frac{Z^4 \alpha^4}{n(n_r + n)^3} + \frac{3}{8} \frac{Z^4 \alpha^4}{(n_r + n)^4} + \dots \right) \qquad (6.51)$$

$$= m_0 c^2 \left(-\frac{1}{2} \frac{Z^2 \alpha^2}{(n_r + n)^2} - \frac{1}{2} \frac{Z^4 \alpha^4}{(n_r + n)^4} \left(\frac{n_r}{n} + \frac{1}{4} \right) + \dots \right). \qquad (6.52)$$

The first term is just the Balmer–Bohr energy, the second gives the famous Sommerfeld fine structure.

To translate this expression into modern notation, we replace $n_r + n$ by n, the principal quantum number, and n by l, the angular-momentum quantum number:

$$E = -\frac{Z^2 e^4 m_0}{2\hbar^2 n^2}\left(1 + Z^2 \alpha^2 \left(\frac{1}{nl} - \frac{3}{4n^2}\right) + O(\alpha^4)\right).$$
(6.53)

Note that, for fixed principal quantum number n, the relativistic correction contains a term which goes inversely as the angular momentum quantum number l, hence is largest for the most eccentric orbits with lowest angular momentum. The energy decreases (i.e. the binding energy *increases*) with increasing eccentricity as the expansion of the relativistic energy

$$\sqrt{m_0^2 c^4 + p^2 c^2} = m_0 c^2 + \frac{p^2}{2m_0} - \frac{p^4}{8m_0^3 c^2} + \cdots$$
(6.54)

has a negative sign in the third term.

The spectroscopic consequences of Eq. (6.52) for the Balmer lines in the spectrum of hydrogen, or more generally, of any atom with a single electron and a nucleus of charge Ze, are immediate. Identifying the individual stationary states by the integer pair (n, n_r) (Sommerfeld's (n, n')), with $N = n + n_r$ the principal quantum number, we find for example for the energy difference between the circular orbit $(N, 0)$ and the least eccentric elliptical orbit $(N - 1, 1)$ with principal quantum number N

$$E_{N-1,1} - E_{N,0} = -\frac{Z^2 e^4 m_0}{2\hbar^2 N^2} Z^2 \alpha^2 \frac{1}{N}\left(\frac{1}{N-1} - \frac{1}{N}\right)$$
$$= -\frac{Rhc\alpha^2}{N^4}\frac{Z^4}{N-1}$$
(6.55)

where

$$R \equiv \frac{e^4 m_0}{2h\hbar^2 c}$$
(6.56)

is the Rydberg constant. For the next pair (if present), one finds similarly

$$E_{N-2,2} - E_{N-1,1} = -\frac{Rhc\alpha^2}{N^4}\frac{Z^4}{N - 3 + 2/N}$$
(6.57)

and so on. From these results follow immediately two properties of enormous phenomenological consequence:

1. The fine structure splittings fall off rapidly with increasing principal quantum number, for large N, with the total splitting $E_{1,N-1} - E_{N,0}$ falling like $1/N^3$. This

means that, given the spectroscopic resolutions available at the time, the splitting of the levels would only be detectable for the lowest principal quantum numbers. Indeed, as we shall see, accurate measurement of the fine structure in hydrogen already proved extremely difficult even for $N = 2$.

2. The energy splittings increase rapidly (for one-electron atoms) with increasing atomic number, like Z^4. Again, given the technological resources of experimenters at this time, the only hydrogenic atom available for spectroscopic analysis, other than hydrogen itself, was ionized helium, He$^+$, with $Z = 2$. However, the Z^4 dependence would play a critical role in the elucidation of the fine structure of X-ray spectra, as we shall see in the next section, and even (as we shall see in Volume 2) in the discovery of electron spin.

An immediate consequence of Sommerfeld's relativistic treatment is the conclusion that the first excited state of hydrogen—in the language of the times, the "constant term $1/2^2$ of the Balmer series"—is actually a doublet. The theoretical value for the energy difference of these two states, expressed conventionally as an inverse wavelength (but using, confusingly, the symbol v, now associated with frequency, for inverse wavelength), is given by Eq. (6.55) divided by hc for $N = 2, Z = 1$:

$$\Delta v_H = \frac{R\alpha^2}{2^4} = 0.3652\,\mathrm{cm}^{-1}. \tag{6.58}$$

Actual measurements of this splitting for the H$_\alpha$ line ($N = 3$ to $N = 2$ transition) with atomic hydrogen in 1916, as quoted by Sommerfeld (1916a, p. 68), varied between 0.31 and 0.42 cm^{-1}. The primary difficulty was the smearing out of the spectral lines due to Doppler broadening: being the lightest of the elements, the thermal motion of hydrogen atoms in a gas discharge at normal temperatures was larger than for any other gas. The problem could be reduced somewhat by going to lower temperatures, but the first precision measurements were obtained by Friedrich Paschen (1916) (see Plate 2 and note 4 in Chapter 2) in helium, where, in virtue of the higher mass of the helium atom, the Doppler broadening is considerably reduced (Kragh 2003). Furthermore, the energy splittings are 16 times larger (Z^4, with $Z = 2$) in helium as compared to hydrogen, although this advantage is somewhat reduced as one needs to consider the $N = 4$ to $N = 3$ transitions in helium which are in the mid-optical range, rather than the now ultraviolet Balmer lines. Thus, from Eq. (6.53), one finds $E_{3,0} - E_{1,2} = (512/81)\Delta v_H$. The net effect is that the overall splitting *in wavelength*, which is the relevant spectroscopic parameter, of the $N = 3$ levels in ionized helium is about three times larger than that of the $N = 2$ level in hydrogen, which, taken together with the improved Doppler situation, allowed Paschen to make (for the standards of the time) a precision determination of Δv_H. His result (Paschen 1916, p. 910),

$$\Delta v_H = 0.3645 \pm 0.0045\,\mathrm{cm}^{-1}, \tag{6.59}$$

was in complete agreement with Sommerfeld's theoretical result Eq. (6.58).

In his spectroscopic work, the Tübingen experimentalist relied heavily on the work of his theorist colleague in Munich. Paschen was initially stymied by contradictions between his data and Bohr's original theory, but was able to resolve them with the help of Sommerfeld's elaboration of the theory. As he wrote in the acknowledgments of his paper, Sommerfeld's "indefatigable efforts made it possible to recognize in the imperfect experimental results the wonderful laws of his [i.e., Sommerfeld's] beautiful theory" (Paschen 1916, p. 940). In a letter of May 21, 1916, Paschen had told Sommerfeld that "without your theory these results would not have been found" (Eckert 2014, p. 153). Following Paschen's suggestion in this letter, the two men coordinated the publication of their results. Paschen's (1916) experimental paper was received by *Annalen der Physik* on July 1, 1916, Sommerfeld's (1916a, 1916b) two-part theoretical paper on July 5 (Eckert 2014, p. 153).

In addition to providing confirmation of the relativistic energy displacements implied by Sommerfeld's fine-structure formula, Paschen's spectroscopic data supplied the relative intensities of the array of lines originating from a single original line in Bohr's 1913 theory. In particular, the lines at 4686 Å in ionized helium corresponding to the $N = 4$ to $N = 3$ transitions should correspond to a total of $4 \times 3 = 12$ possible transitions (recall that $N = n + n_r$ with $n = 1, 2, 3, \ldots$ and $n_r = 0, 1, 2, \ldots$). Some of these might be blended together for lack of sufficient resolution, but the clear absence of a line would imply a selection rule forbidding the corresponding transition.

Sommerfeld had initially proposed the selection rule Eq. (5.45), according to which *both* the radial and azimuthal (i.e., angular-momentum) quantum numbers would be required to decrease or remain constant in a transition involving the emission of light. He was soon forced to relax this rule by Paschen's empirical data, which showed the clear presence of transitions in which the azimuthal quantum number increased. Sommerfeld thereupon only excluded transitions $(m, m') \to (n, n')$ in which $m' < n'$. This rules out three of the twelve possible transitions from $N = 4$ to $N = 3$: $(4, 0) \to (2, 1)$, $(4, 0) \to (1, 2)$, and $(3, 1) \to (1, 2)$. Given the rather poor resolution available at the time (1916), Sommerfeld was able to convince himself of the existence of the remaining nine lines. By the time he published the first edition of *Atombau und Spektrallinien*, Sommerfeld (1919), following work of his assistant Wojciech Rubinowicz (1918a, 1918b) (see Plate 23), had reexamined the issue of selection rules in radiative transitions and come to the conclusion, on the basis of energy and angular-momentum conservation arguments, that the radial quantum number is irrelevant, and that the appropriate selection rule involves only the angular momentum. He argued that, for allowed transitions,

$$\Delta n = 0, \pm 1 \tag{6.60}$$

(p. 393). By this time, Sommerfeld had access to Bohr's (1918) monograph on the quantum theory of line spectra, and in particular to his correspondence-principle derivation of selection rules, according to which the option $\Delta n = 0$ in Eq. (6.60) is excluded, leaving only transitions with $\Delta n = \pm 1$.

The argument for this restriction is as follows. For systems where the azimuthal angle φ is cyclic, we already saw in Section 5.3 that the motion in the x-y plane has a Fourier expansion given by Eq. (5.109), while the motion in the z direction has the expansion Eq. (5.110). We now specialize to a hydrogenic atom and choose the orbit of the electron to lie in the x-y plane. The angular momentum must then point in the z-direction, and the quantum number associated with $q_3 = \varphi$ is just Sommerfeld's n in the relativistic Coulomb problem.[6] Moreover, by choice of coordinates, the z motion vanishes (i.e., the coefficients $C_{2\overline{7}}$ in Eq. (5.110) vanish), while the factor $\cos(2\pi v_3 t) = \frac{1}{2}(e^{2\pi i v_3 t} + e^{-2\pi i v_3 t})$ in Eq. (5.109) implies that, by Bohr's extended correspondence principle, only transitions by which the associated quantum number increases or decreases by unity can occur. Whence the more stringent selection rule $\Delta n = \pm 1$. This reduces the number of allowed transitions from twelve down to five, which, *mirabile dictu*, actually agrees with the result obtained by applying the correct modern selection rules.

That electron spin is nowhere in sight in any of the preceding discussion makes the success of Sommerfeld's treatment of fine structure quite astonishing, given the essential role played by spin in modern derivations of fine structure.[7] In fact, several fortuitous successes of the old quantum theory, first among them Sommerfeld's calculation of the relativistic fine structure, have an illuminating theoretical explanation in terms of the semi-classical limit of the modern theory. It suffices to say here that the Dirac formula for the relativistic fine structure of hydrogen, incorporating both electron spin and relativistic kinematics in a wave-mechanical context, takes *precisely* the same form as Eq. (6.48), with the replacement $n \rightarrow j + \frac{1}{2}$, where j is the total angular momentum quantum number (orbital plus spin) of the electron, taking half-integral values $j = \frac{1}{2}, \frac{3}{2}, \ldots$. Accordingly, the quantity $j + \frac{1}{2}$ appearing in the Dirac formula assumes exactly the set of values as the azimuthal quantum number n, namely strictly positive integer values 1, 2, 3, etc. The energy levels predicted by Sommerfeld's formula are thus precisely the correct ones, although with, of course, a completely incorrect identification and interpretation of the quantum numbers appearing in the formula! Moreover, the radiative transitions allowed by the modern dipole selection rule $\Delta l = \pm 1$, $\Delta s = 0$ (orbital angular momentum changes by one unit, with no spin change) lead to just the same set of spectral lines (for hydrogenic atoms) as expected by Sommerfeld on the basis of the correspondence-principle-inspired rule, which also required $\Delta n = \pm 1$. At the heart of this apparent miracle are several serendipitous cancellations which result in the formal identity of the Sommerfeld and Dirac formulas stated above.[8]

---

[6] External fields are assumed to be zero here: thus, $\mathcal{E} = \mathcal{H} = 0$ in Eq. (5.111).

[7] For a brief discussion of this issue, see, e.g., Eckert (2013a, pp. 58–60). Yourgrau and Mandelstam (1979) argued that the neglect of wave-mechanical effects and the neglect of spin canceled each other out. Biedenharn (1983) took issue with this assessment.

[8] For a more detailed account of the development of selection rules by Rubinowicz and Sommerfeld in the old quantum theory, and a discussion of the surprising success of the Sommerfeld rules in accounting for the observed lines, even in the absence of an understanding of electron spin, see the web resource *Conservation Laws and Selection Rules*.

6.2 X-ray spectra

The development of modern quantum theory was assisted at several critical points by the appearance and ingenious utilization of new technologies opening up wholly new areas of empirical investigation. The infrared bolometers of Langley in the late nineteenth century played a crucial role in the study of the black-body spectrum leading to Planck's great discovery; the cathode ray tubes of Geissler and Crookes, and their later gas discharge descendants, were essential in the discovery of the electron and the study of optical spectra; and Röntgen's discovery in 1895 of X-rays, as we shall now see, was the starting point of a new empirical tool, X-ray spectroscopy, which provided insights of enormous importance throughout the development of a quantum theory of atomic structure.

Let us briefly review the essential qualitative features of X-ray physics, to the extent relevant for present purposes. When energetic electrons (accelerated through voltages ranging from hundreds to tens of thousands of volts) are allowed to strike a metal plate, two types of electromagnetic radiation are emitted, with wavelengths in the range from tenths of an Angstrom (very "hard" X-rays) to a hundred Angstroms (very "soft"). The radiation is partly continuous, partly a superposition of discrete homogeneous (i.e., monochromatic, of well-defined wavelength/frequency) components, with the particular discrete wavelengths appearing characteristic of the particular metal employed.

The *electromagnetic* nature of this penetrating radiation was initially unknown, but by the early years of the twentieth century, experiments of Hermanus Haga, Cornelis Wind, Bernhard Walter, Robert Pohl, and others had established the presence of diffraction when the radiation was passed through very narrow slits, and the wave (whence, with high probability, electromagnetic) character of the radiation, with a wavelength much smaller than optical, was thereby confirmed. In the first decade of the twentieth century, detailed investigations by Charles Barkla, in which the absorption coefficient of X-rays (percentage decrease in intensity on passage through a standard thickness of aluminum) was used as a marker for the as yet unmeasurable wavelength of the radiation, showed that the characteristic part (i.e., the discrete part dependent on the specific metal used as the anode) of the radiation typically consisted of two components, a shorter wavelength "*K*" component and a longer wavelength "*L*" component.[9]

By 1911, Whiddington, at the Cavendish Laboratory in Cambridge, had established the empirical rule that the minimum velocity of the electrons in an X-ray tube producing primary X radiation which was then used to induce characteristic *K* radiation in a second metal plate composed of an element of atomic weight A was proportional to A (see Section 4.5.2 for Bohr's explanation of this rule). The continuous component found by the early researchers is the bremsstrahlung (braking radiation) emitted by the electron as it decelerates on entering the metal, as a result of multiple collisions with the metal atoms. The discrete components of the radiation, with the advent of the Rutherford–Bohr planetary atomic model, were soon interpreted (e.g. Kossel (1914)) as the result of

---

[9] The original notation was "*A*" and "*B*": the letters were later shifted to allow for the possible discovery, never to come, of an even harder component than *K*.

ejection of an inner electron of a metal atom of energy E, followed shortly thereafter by the replacement of the missing electron in the atom by another of higher energy E', with the energy difference emitted as the outer electron changes state appearing in the form of X-ray radiation, via the Bohr frequency condition $h\nu = E' - E$. The fact that both E and E' were restricted, according to the Bohr model, to quantized values (depending of course on the particular metal in use) then accounted for the discrete frequencies found in the characteristic radiation.

The electromagnetic nature of X-rays made it clear that an accurate measurement of the wavelength of the radiation, comparable to that available in optical spectra, was an obvious precondition for the quantitative usefulness of X-rays as a window into the structure of the atom. Diffraction gratings of the kind used in optical spectroscopy were useless for this purpose, as the technology of the time allowed line spacings of no less than about 10^{-5} cm, much larger than the wavelengths to be measured. The solution to the problem appeared more or less simultaneously in England and Germany, with the work of the Braggs (father and son) in the former, and Max von Laue in the latter country. One simply had to take advantage of diffraction gratings provided free of charge by Nature—crystal lattices—with a "line separation", the distance, or lattice spacing, between successive planes of atoms in the crystal, far smaller (on the order of 10^{-8} cm) than those achievable in ordinary gratings by human technology. Constructive interference of monochromatic X-rays of wavelength λ, impinging on planes of atoms separated by lattice spacing d at an angle θ to the crystal surface, is achieved if the *Bragg condition* holds:

$$n\lambda = 2d\sin\theta, \quad n = 1, 2, 3, \ldots \tag{6.61}$$

An absolute measurement of wavelength required therefore the determination of the lattice spacing d, which was achieved by 1913 with a measurement $d = 2.814 \times 10^{-8}$ cm for rock salt, a value used by Henry Moseley (see Plate 22) in his seminal investigations to be described shortly. The salient point to note here is that crystal diffraction allows the determination of wavelengths over a range of several hundred (eight to nine octaves!), from a tenth of an Angstrom to tens of Angstroms, in other words over a far greater range than available with optical spectra, which range from 3000 to 7000 Angstrom (just over a single octave).

In practice, one is limited by the inequality $\lambda = 2d\sin\theta/n < 2d$, following from Eq. (6.61), and by the available crystals, to wavelengths less than about 10 Angstroms. There were also difficulties with strong absorption of the softer (longer wavelength) X-rays in air, requiring the development of vacuum X-ray spectrometers. Nevertheless, by one of those (in retrospect) remarkable coincidences, the technology of X-ray spectroscopy became available for use in atomic physics at precisely the moment when the Bohr model, and the understanding of atomic structure more generally, could profit most from it. Looking back on this moment from the vantage point of the vast store of spectroscopic knowledge accumulated in the years up to the publication of the first edition of *Atombau und Spektrallinien*, Sommerfeld states

In fact, the spectroscopy of Röntgen rays (X-rays) has provided results that are in many respects simpler and more satisfying than the spectroscopy of the optical regime, which stretches into unsurveyable realms ...The Röntgen rays come from the interior of the atom, where the organization of the electrons obeys simple laws under the influence of the unweakened nuclear charge; the visible spectrum arises at the periphery of the atom, where the electrons are more numerous and the nuclear charge loses its regulating influence ...A further reason (for the difference) can be adduced: Röntgen spectroscopy had from its beginning the new atomic theory of Bohr at its side as a guide and leader, while optical spectroscopy was without theoretical leadership for decades, and had to develop on its own the essential features from which the atomic theory could be built up (Sommerfeld 1919, p. 152).

The immediate utility of the Bohr model as an interpretational tool for the results of X-ray spectroscopy was apparent in the remarkable investigations of Moseley in 1913. Henry G. J. Moseley[10] received his MA in mathematics and natural science from Oxford in 1910, moving from there to join Rutherford's laboratory in Manchester in September of that year, at precisely the point when Rutherford was on the verge of formulating and publishing his nuclear model of the atom (which appeared in print in the *Philosophical Magazine* in April 1911).

A year later, after some work on radioactivity suggested by Rutherford had been completed, Moseley became interested in the new field of X-ray crystal spectrometry, which had been demonstrated experimentally by Friedrich and Knipping in Munich in the Summer of 1912, following suggestions of von Laue. As no one in Manchester was an expert in X-ray physics at the time, Moseley made a short visit to W. H. Bragg in Leeds, an expert in the field, to acquire the essential information needed to begin an experimental program. This program, carried out initially in collaboration with C. G. Darwin, was soon able to establish the existence of sharp peaks of the radiation reflected from a crystal surface, in accordance with the relation Eq. (6.61) proposed in late 1912 by the younger Bragg (W.L.), at the time a student at Cambridge. The result of this work was a measurement of the wavelengths of five X-ray lines in the spectrum of platinum (the metal used in the X-ray tube), with the primary uncertainty a geometrical factor arising from the uncertain value of the lattice spacing d for the various diffracting crystals used (potassium ferrocyanide, selenite, and rock salt), which affected the absolute, but not relative, values of the wavelengths obtained. This uncertainty would soon be resolved by the determination in Summer 1913 by W. L. Bragg of the crystal structure and lattice spacing of rock salt, mentioned previously.

By the Autumn of 1913, Moseley and Darwin had parted company (amicably), and Moseley continued on his own, with the intention of systematically surveying the high frequency (i.e., X-ray) spectra of a range of elements. The primary aim seems to have been to resolve the still noisome issue of the physical relevance of atomic number Z versus atomic weight A in determining atomic structure. A Dutch lawyer turned amateur

---

[10] For the life and work of Moseley, see Heilbron (1974, 1966).

physicist Antonius van den Broek[11] had recently published a paper, in which the atomic number Z, giving the serial order of the element in the periodic table,[12] rather than the atomic weight A, was identified as the determining factor not just for chemical properties, but also for the optical and X-ray spectra of the atom (van den Broek 1913). In particular, the atomic number is explicitly identified with the nuclear charge (in units of the electron charge). The frequency of the most prominent K radiation, the K_α line, should therefore bear a simple relation to Z, not A. This could be easily tested by checking the X-ray spectra for the occasional adjacent pairs of elements in the periodic table in which an increase in atomic number is accompanied by a *decrease* in atomic weight (e.g. cobalt and nickel at places 27 and 28 respectively).

Moseley employed an apparatus in which the elements to be targeted by cathode rays were mounted sequentially on a truck inside the X-ray tube, and could be conveniently brought under fire seriatim without disassembling the whole apparatus. Measurement of the quantity Q_K for the K_α line, defined in modern notation as

$$Q_K \equiv \sqrt{\frac{\nu}{\frac{3}{4}R}}, \tag{6.62}$$

then gave the remarkably simple result (Moseley 1913), apparent from inspection of a table of both quantities, that $Q_K \approx Z - 1$, to an accuracy usually better than 1% over a selection of accessible elements in the range of atomic numbers from 13 (aluminum) to 47 (silver). We remind the reader here that ν refers to inverse wavelength, not frequency. The Q_K values for the anomalous cobalt–nickel pair are in accordance with their atomic numbers, as Moseley expected, and not their atomic weight.[13] Rephrasing Eq. (6.62), one has for the frequency of the strongest K line in the spectrum of element of atomic number Z,

$$\nu = \frac{3}{4}R(Z - 1)^2, \tag{6.63}$$

a relation which has now become known as "Moseley's Law".

The dramatic simplicity of Moseley's Law was displayed even more clearly in a second paper in 1914 (Moseley 1914), where a plot is given of $\sqrt{\nu}$ as a function of atomic number Z over a range stretching from $Z = 13$ (Al) to $Z = 79$ (Au). The plot is reproduced here as Fig. 6.3, and shows six distinct X-ray series, with the linear dependence on Z apparent from straight line fits to the data (the lowest two lines

---

[11]   For discussion of van den Broek's contributions, see Hirosige (1971), Pais (1986), and Scerri (2007, pp. 165–169).

[12]   At this point, the notation N was more common for atomic number, and was used by Moseley.

[13]   The historical importance of Moseley's work in resolving several puzzles in the ordering of elements, and in particular in the discovery of "missing elements", is well described by Heilbron (1966), who also wrote a biography of Moseley (Heilbron 1974).

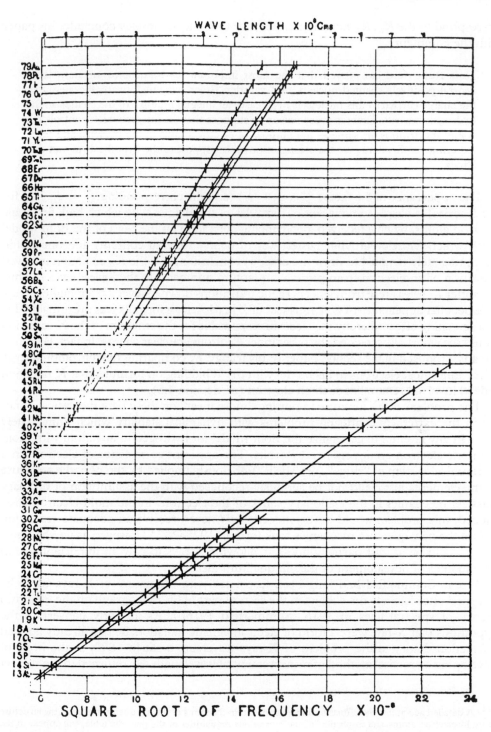

Figure 6.3 *Moseley's (1914) X-ray data.*

correspond to the K_α, K_β lines, the upper four to L lines). Moseley concludes his paper with the comment

> Now Rutherford has proved that the most important constituent of an atom is its central positively charged nucleus, and van den Broek has put forward the view that the charge carried by this nucleus is in all cases an integral multiple of the charge on the hydrogen nucleus. There is every reason to suppose that the integer which controls the X-ray spectrum is the same as the number of electrical units in the nucleus, and these experiments therefore give the strongest possible support to the hypothesis of van den Broek.

The appearance of the factor $\frac{3}{4} (= \frac{1}{1^2} - \frac{1}{2^2})$ (and the corresponding $\frac{5}{36} = \frac{1}{2^2} - \frac{1}{3^2}$ for the L series) is clearly reminiscent of formulas of the Balmer type which had, for single electron atoms, by now received a convincing derivation on the basis of the Bohr model (with which Moseley, at Manchester, was of course fully au courant) and one would be immediately tempted to draw the obvious conclusion, that the K_α radiation involved transition of an electron from principal quantum number 2 to the lowest level 1, while the softer L series arose from less energetic transitions down to level 2 from higher levels. The peculiar appearance of $Z - 1$ instead of Z in Moseley's formula was presumably due to a screening effect whereby the electron responsible for the emission of the X-radiation would see a partially screened (by other inner electrons) nuclear charge. In fact, Bohr had already proposed in the second paper of his trilogy, by comparing the ionization energy of the innermost electrons in his model with the Whiddington formula, that K radiation indeed involved removal of such an electron (cf. Section 4.5).

The problem at this juncture in putting "more meat" on these bare theoretical bones lay in the extremely confusing problems with dynamical stability which immediately arose in any attempt to generalize the Bohr model for hydrogen to any atom with more than one electron occupying nearby orbits. The apparent coexistence of electrons in a hierarchical system of "rings" (or "shells") obviously implied by the periodic structure of the elements just could not be seriously reconciled with the assumption of ordinary classical mechanics in the stationary states of the electrons in the multi-electron case. Indeed, this problem would never, indeed *could* never, receive an adequate resolution before the new quantum mechanics took form in 1925–26.[14] However, as we shall now see, it could be at least partially finessed, allowing X-rays to play an important role in certain lines of development of the old quantum theory.

Arnold Sommerfeld had long been fascinated with the nature of X-rays when he returned once again to the subject in the second part of his monograph on the quantum theory of spectral lines in 1916 (Sommerfeld 1916b). As an early convert to the hypothesis that X-rays were a form of electromagnetic radiation, he had already applied diffraction theory to the problem of X-rays traversing a slit in 1899, in the hope of developing a practical method of measurement of wavelength. He returned to the subject

---

[14] A detailed account of the efforts to understand, within the terms of the old quantum theory, the structure of multi-electron atoms, and in particular, the mysterious regularities of the periodic table, will appear in the second volume of this work.

about a decade later, showing that the anisotropic (and polarized) component of the primary X radiation emitted from a target struck by energetic electrons had precisely the angular distribution expected from classical bremsstrahlung (the radiation emitted by a rapidly decelerating charge). The characteristic radiation component (isotropic, unpolarized, and discrete) on the other hand would be due to some internal rearrangement process in the atom where the "memory" of the directional properties of the impinging electron had been lost. For this part, Sommerfeld strongly suspected the involvement of the Planck quantum. There followed papers in 1911 in which Sommerfeld introduced his ill-fated "*h*-theory" (discussed briefly in Section 1.3.1), an attempt to introduce the Planck quantum into the emission process for γ rays, interpreted as the electromagnetic radiation accompanying the sudden emission of β-electrons in a radioactive process, analogous to the production of X-rays by cathode rays in a X-ray tube.

The work of Bohr in 1913 and Moseley in 1914 resulted in a complete reorientation of Sommerfeld's thinking about X-rays. In the section of the 1916 *Annalen* article (Sommerfeld 1916b) entitled "Theory of Röntgen Spectra," he begins by reviewing Moseley's results, and the result, Eq. (6.63), appears immediately. The two-fold objective of his discussion is stated directly: (a) criticism of the peculiar $Z - 1$ factor, and (b) improvement of the result by inclusion of relativistic effects. In addition to new theoretical considerations, the discussion would have the advantage of considerably improved experimental data, obtained in the two years since Moseley's final paper on X-ray spectra by Malmer, Siegbahn, Stenström, Friman, and others.

In accordance with the new insights provided by the Bohr atomic model, the focus turns to the individual "terms" in the spectral formula, which directly correspond to allowed energy levels for individual electrons, each now viewed as associated with a definite orbit. The identification of atomic number Z (exactly this notation is introduced here by Sommerfeld, as he reserves N, commonly used at the time for atomic number, for the Rydberg constant, for which we employ the modern notation R) with the nuclear charge is now a given. As regards the *effective charge* to be used in the term formulas,

> In the following, by "effective nuclear charge", we shall mean that charge with which the nucleus actually acts on the electron orbits of the corresponding term. If there are z electrons inside of the orbit in question, in the immediate neighborhood of the nucleus, then the effective nuclear charge is the remaining $E = (Z - z)e$. One would naturally assume that z is a positive whole number, but we should not restrict ourself to this assumption a priori.

Thus, Moseley's $Z - 1$ factor will be replaced by a more general factor, $Z - k$ for the K terms (corresponding to the screening effect of electrons in the innermost orbits, with principal quantum number 1), $Z - l$ for the L terms (principal quantum number 2), etc. On the other hand, it will be assumed that the screening effect of the inner electrons on electrons, at least in K- and L-orbits, consists *only* of a reduction of the effective nuclear charge, leaving intact the Coulombic and spherically symmetric $1/r$ form of the electrostatic potential seen by the emitting electron. By the intuition advanced by Sommerfeld in the quote at the beginning of this section, the overwhelming influence of the nucleus on the innermost electrons of a large atom suggests that these electrons behave in ways that are "hydrogen-like". Accordingly, the characteristic rational factors $\frac{1}{1^2}, \frac{1}{2^2}$ in the K- and L-term formulas will not be altered. The actual values of the nuclear charge defects $k, l, \ldots$ are to be determined initially by an empirical fit to the data.

Sommerfeld next turns to the issue of the relativistic fine structure, as revealed by the appearance of doublets in both the K_α and L_α lines. While the K term for an electron in the lowest orbit (principal quantum number) corresponds to a single circular orbit (see Fig. 6.1), electrons at principal quantum number 2 have the option of a circular $(2,0)$ (in the (n, n_r) notation) or elliptical $(1,1)$ orbit, with energies split by the relativistic corrections, as in Eq. (6.55). In fact, the doublet structure of both cases has the same origin, in the energy splitting $E_{2,0} - E_{1,1}$. For the K_α line, a doublet naturally appears due to the energy difference in the initial L-term electron orbits, with the final K orbit being a singlet. For the L_α line, which corresponds to a transition from an electron with principal quantum number $N = 3$ to either of the final state L-orbits, [15] we again have a doublet, with the *same* splitting as for the K_α line, as had already been established experimentally by Kossel. In fact, Sommerfeld indicates (Sommerfeld 1916b, p. 129) that the equality of the splittings provided the initial impetus for his application of the relativistic fine structure results to X-ray spectra. The inverse-wavelength splitting $\Delta \nu$ (recall, $\nu = 1/\lambda$ here), to lowest order in powers of the fine-structure constant α, is from Eq. (6.55), setting $N = 2$,

$$\Delta \nu = \frac{E_{2,0} - E_{1,1}}{hc} = R\frac{(Z-l)^4}{2^4}\alpha^2 = (Z-l)^4 \Delta \nu_H, \qquad (6.64)$$

where $\Delta \nu_H$ is the corresponding (much diminished) splitting for hydrogen, known by extrapolation from ionized helium measurements to be $0.365 \ \mathrm{cm}^{-1}$, as we saw earlier. This result can be rearranged simply to give

$$\Delta \nu_H = \frac{1}{\tilde{Z}^4}\Delta \nu, \quad \tilde{Z} \equiv Z - l \qquad (6.65)$$

and viewed as a method of extrapolating measurements of the doublet splitting in the X-ray spectra of a wide range of atoms back to the archetypal, but hard to measure directly, hydrogen value. To the extent that the theoretical explanation of the doublet splitting is valid, one then expects to find a single value for the quantity on the right-hand-side of Eq. (6.65), independent of the atomic number of the element used. In effect, one is using the fine structure formulas to allow an extraction of the value of the fine structure constant—to the extent that the results so obtained for different atoms are consistent, the theory is validated.

Sommerfeld provides a graphical illustration of the success of this procedure in Figure 1 of Sommerfeld (1916b), part of which we reproduce here (Fig. 6.4). The measurement of the L doublet turns out to be more accurate than for the K doublet, as the longer

---

[15] Recall that the relativistic splittings fall off as $1/N^4$, so that the $N = 3$ terms could be regarded as forming a single degenerate level, to a first approximation.

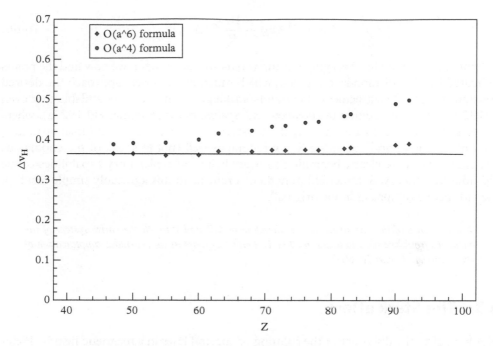

Figure 6.4 *Extrapolation of L doublet splittings to hydrogen value.*

wavelengths of the L lines implies, as $\Delta\lambda = \lambda^2\Delta\nu$, a larger wavelength splitting for the same value of $\Delta\nu$. A plot of $\Delta\nu_H$ as defined in Eq. (6.65) versus atomic number Z for the L doublet, using data of Siegbahn and Moseley, is shown in Fig. 6.4. The red circles indicate the results obtained from the lowest order formula Eq. (6.65), while the horizontal line indicates the previously obtained hydrogen value $\Delta\nu_H$. The nuclear charge defect l has by now been changed from Moseley's value of 1 to 3.5 to improve agreement with theory. As Sommerfeld comments, the (red circular) points lie on a curve which is impressively close to the hydrogen value 0.365 over a wide range of atomic numbers, from Z=47 (silver) to Z=92 (uranium),[16] and, in his usual optimistic fashion, he concludes "our rule is therefore exactly confirmed"!

Sommerfeld would presumably have been deliriously happy if he had taken the trouble to improve the approximation Eq. (6.64) further, by including the order α^6 terms in the relativistic energy (i.e., the next terms in the post-Newtonian approximation), which he had already calculated in equation (22) of (Sommerfeld 1916a).[17] Including these terms, Eq. (6.65) is modified to

---

[16] The less accurately measured K doublet values, available for the lighter elements from Z=24 to Z=47, showed considerably more scatter around the hydrogen value.

[17] By the time this topic is reviewed in *Atombau und Spektrallinien* some years later, the role of the higher terms in causing the curvature of the uncorrected data is understood.

$$\Delta v = \tilde{Z}^4 \Delta v_H + \frac{10}{R} \tilde{Z}^6 (\Delta v_H)^2. \tag{6.66}$$

Solving this equation for Δv_H and plotting versus atomic number, we now find the points indicated by blue diamonds in Fig. 6.4, which much more closely approach the desired constant value. The agreement is certainly striking, and it is understandable that even in 1922, in the third edition of *Atombau und Spektrallinien* (Sommerfeld 1922a), when, as we shall see in the next chapter, the mysteries of the complex multiplet structure, not only in the optical but in the X-ray regime as well (not to speak of the anomalous Zeeman effect), are clearly beginning to overwhelm the explanatory capabilities of the old quantum theory, Sommerfeld nonetheless returns to this agreeably simple result of the old theory (emphasis in the original):

> *The existence of the fine structure can therefore be followed through the entire system of the elements, from hydrogen to uranium. The L-doublet appears as an enormous amplification of the hydrogen doublet* (p. 609).

6.3 The Stark effect

Not long after the discovery of the splitting of spectral lines in a magnetic field by Pieter Zeeman (1896), the question naturally arose as to whether a similar effect was to be expected in the case of electric fields. In accordance with the prevalent view at the turn of the century that spectral lines were associated with harmonic vibrations of elastically bound electrons, an early attempt was made by Woldemar Voigt (1901) to estimate theoretically (and purely classically) the expected alteration in oscillation frequencies. Voigt obtained a result that was far too small to allow detection with then current spectroscopic technology. In 1913, however, undeterred by this pessimistic prediction, Johannes Stark (1913a, 1913b) (see Plate 10), and essentially simultaneously, the Italian physicist Antonino Lo Surdo (1913), succeeded in measuring an unambiguous splitting in the presence of an applied electric field of the Balmer lines of the visible hydrogen spectrum.[18] The almost simultaneous appearance of the Bohr model led very soon to attempts to understand the Stark–Lo Surdo results (much larger than those predicted by Voigt) in terms of the Bohr planetary model of the atom. A paper by Emil Warburg (1913) which appears to be one of the first attempts along these lines concluded however

---

[18] For the discussion of the Stark effect in this section, we drew on Duncan and Janssen (2014, 2015). The six-part paper in *Annalen der Physik*, entitled "Observations of the effect of the electric field on spectral lines," in which Stark and two collaborators, Georg Wendt and Heinrich Kirschbaum, reported their findings (Stark 1914; Stark and Wendt 1914; Stark and Kirschbaum 1914; Stark 1915) (the first one a reprint with minor corrections of Stark 1913) is reprinted in Hermann (1965b), preceded by a historical introduction about the discovery of the Stark effect (Hermann 1965a). For discussion of Lo Surdo's work, see Leone, Paoletti and Robotti (2004), who also discuss Lo Surdo's politics. Just as Stark became a strong supporter of the Nazis in Germany, Lo Surdo became a strong supporter of fascism in Italy. As noted by Kox (2013b, p. A66), both men "lived to see the demise of their respective ideologies and ended their lives as pariahs in the scientific world of their home countries."

that, rather than a splitting of the Balmer lines into five monochromatic components, the application of the electric field should actually lead to a smearing out of the original line, although the theory did give the correct order of magnitude for the total wavelength splitting.

This result was soon contested by Bohr (1914a). The Bohr argument is hardly convincing in retrospect, as he considers the energetic shift obtained classically by averaging the additional electrostatic potential energy (due to a weak applied electric field) of an electron in an elliptical orbit of semi-major axis a and eccentricity ϵ, inclined at an angle φ to the electric field E, and finds, for the *extreme cases* of eccentricity $\epsilon = 1$ and $\varphi = 0, \pi$

$$\Delta E = \pm \frac{3}{2} a e E. \tag{6.67}$$

If one then inserts the quantized Bohr model result (obtained of course for *circular* orbits!), $a = (h^2/4\pi^2 m e^2) n^2$, one obtains, for the extreme values of the possible shift in energy levels

$$\Delta = \pm \frac{3h^2}{8\pi^2 m e} E n^2, \quad n = 1, 2, 3, \dots \tag{6.68}$$

This formula turns out remarkably to be very similar to the correct result, and certainly accounts for the order of magnitude of the splittings found by Stark, although it is very hard to see why the estimate of an upper bound which Bohr claims to be estimating can be considered evidence for a set of sharp spectral lines instead of a continuously smeared line as claimed by Warburg. We should also recall that at this point Sommerfeld's quantization of eccentricity ϵ, as well as the notion of spatial quantization of orbits (which would select discrete values for the orientation angle φ), had yet to be introduced.

The correct treatment of the Stark effect, to the extent possible within the boundaries of the old quantum theory, was to emerge with surprising rapidity once the quantization procedure of Schwarzschild described in Section 5.1.4 above took hold (Eckert 2013a, sec. 4.2). In early March 1916, as soon as he had received the letter of Schwarzschild discussed previously, Sommerfeld suggested to his assistant Paul Epstein that a treatment of the Stark effect along these lines would be a suitable subject for Epstein's Habilitation thesis (for the University of Zurich). Epstein had already been immersed in the intricacies of analytical classical dynamics, in particular, in the appropriate treatment of the problem of a particle moving under the influence of two inverse square gravitationally attracting force centers, which, as Schwarzschild had pointed out two years earlier, became in the limit in which one force center was removed to infinity (with a corresponding increase in the mass of the receding object) the problem of a uniform force field superimposed on a single gravitational force center. The appropriate orthogonal coordinate system in which the classical Hamilton–Jacobi equations become separable was already known from the time of Jacobi (1884, Ch. 29): elliptical coordinates for two centers separated by a finite distance, parabolic coordinates once the aforesaid limit was taken.

Within the next few weeks, Sommerfeld received notice from Schwarzschild that he was well on the way to a resolution of the Stark problem, and Epstein redoubled his efforts, arriving at the now famous energy level formula by the morning of March 21. That same afternoon, a letter from Schwarzschild to Sommerfeld arrived in Munich, with almost identical results (there was a discrepancy in the assignment of quantum numbers in the formula, which we shall discuss below). In the end, quite fittingly, Schwarzschild and Epstein submitted their results on adjacent days. Epstein (1916a) submitted a preliminary note to *Physikalische Zeitschrift* on March 29 (it was published April 15); Schwarzschild (1916) submitted his paper to the Berlin academy on March 30. On the date of actual publication of Schwarzschild's paper, May 11, 1916, he succumbed to the autoimmune skin disease pemphigus. He was only 42 years old. Remarkably, Schwarzschild had carried out his ground-breaking work on quantization conditions and the Stark effect during his extraordinarily debilitating illness, an achievement on par perhaps with Schubert's completion of his remarkable last piano sonatas in the throes of his final illness (Schubert died at age 31). In his obituary, Sommerfeld (1916e) emphasized the importance of Schwarzschild's contributions to quantum theory.

Epstein's (1916b) full paper on the Stark effect was received by *Annalen der Physik* on May 9, two days before Schwarzschild's death. In our discussion of the treatment of the Stark effect by Schwarzschild and Epstein, we will mainly be following this paper by Epstein. The parabolic coordinates we employ are those adopted by Kramers (1919, p. 301) in his doctoral thesis, which slightly simplify the algebra.[19] We begin with the Hamiltonian for an electron (reduced mass μ, charge $-e$) in a hydrogen atom in an external electric field $\mathcal{E}$ in the z-direction, given in Gaussian units by:

$$H = \frac{\vec{p}^2}{2\mu} - \frac{e^2}{r} + e\mathcal{E}z, \tag{6.69}$$

where $\vec{p}^2 \equiv p_x^2 + p_y^2 + p_z^2$, with (p_x, p_y, p_z) the momenta conjugate to the coordinates (x, y, z).

We switch to parabolic coordinates (ξ, η, φ), related to Cartesian coordinates (x, y, z) via

$$z = \frac{\xi - \eta}{2}, \quad x + iy = \sqrt{\xi\eta}\, e^{i\varphi}. \tag{6.70}$$

This coordinate transformation is illustrated in Fig. 6.5. In parabolic coordinates the Hamiltonian in Eq. (6.69) is given by:

---

[19]   In terms of Cartesian coordinates (x, y, z), $r = \sqrt{x^2 + y^2 + z^2}$, Schwarzschild uses coordinates $\lambda = (r+z)/2, \mu = (r-z)/2$, and $\varphi = \arctan(y/x)$, if we reorient the electric field to point in the z, rather than the x direction. With a similar reorientation, Epstein's coordinates are $\xi_E = \sqrt{r+z} = \sqrt{2\lambda}, \eta_E = \sqrt{r-z} = \sqrt{2\mu}$ and φ. Finally, the coordinates used by Kramers, which we adopt here, are $\xi = \xi_E^2, \eta = \eta_E^2$ and φ. These three systems of coordinates are related by a point transformation of the form $q_i \rightarrow q_i'(q_i)$, which leaves the phase integrals (action variables) invariant, as we have discussed previously.

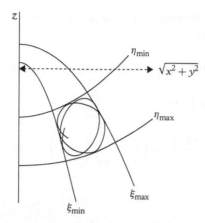

Figure 6.5 *Parabolic coordinates. This figure is taken from Epstein (1916b, p. 498) but the labeling has been changed to reflect the definition of the coordinate transformation $(x, y, z) \rightarrow (\xi, \eta, \varphi)$ as given in Eq. (6.70), which follows Kramers (1919, p. 301, Eq. (43)) rather than Epstein (1916b, p. 495, Eqs. (19)–(20)). The figure shows what Epstein calls a "meridian plane" (ibid.), a plane through the z-axis and the position of the electron. This plane rotates around the z-axis as the electron orbits the nucleus. Within this plane, the electron stays between ξ_{min} and ξ_{max} and between η_{min} and η_{max} (Duncan and Janssen 2014, p. 69, Fig. 1).*

$$H = \frac{1}{2\mu} \left(\frac{4\xi}{\xi + \eta} p_\xi^2 + \frac{4\eta}{\xi + \eta} p_\eta^2 + \frac{1}{\xi\eta} p_\varphi^2 \right) - \frac{2e^2}{\xi + \eta} + \frac{1}{2} e\mathcal{E}(\xi - \eta), \qquad (6.71)$$

where $(p_\xi, p_\eta, p_\varphi)$ are the momenta conjugate to (ξ, η, φ).[20] We now follow the procedure outlined in Section A.2.1 of Appendix A to obtain the Hamilton–Jacobi equation for this system: namely, one introduces the Hamilton–Jacobi function $S(\xi, \eta, \varphi)$ and replaces the momentum variables by the appropriate partial derivatives of S, $p_\xi \rightarrow \partial S/\partial\xi$, etc. in the total energy $E = H(\xi, \eta, p_\xi, p_\eta, p_\varphi)$, which is then set equal to the constant α_1. After multiplying both sides by $2\mu(\xi + \eta)$, one finds

$$4\xi \left(\frac{\partial S}{\partial\xi} \right)^2 + 4\eta \left(\frac{\partial S}{\partial\eta} \right)^2 + \left(\frac{1}{\xi} + \frac{1}{\eta} \right) \left(\frac{\partial S}{\partial\varphi} \right)^2 - 4\mu e^2 + \mu e\mathcal{E}(\xi^2 - \eta^2)$$

$$= 2\mu(\xi + \eta)\alpha_1. \qquad (6.72)$$

---

[20]   This result follows from the translation of the kinetic energy $K = \frac{1}{2}\mu(\dot{x}^2 + \dot{y}^2 + \dot{z}^2)$ into parabolic coordinates. One finds after a short calculation

$$K = \frac{\mu}{8} \left((1 + \eta/\xi)\dot{\xi}^2 + (1 + \xi/\eta)\dot{\eta}^2 \right) + \frac{\mu}{2} \xi\eta\dot{\varphi}^2.$$

Whence $p_\xi = \partial K/\partial\dot{\xi} = (\mu/4)(1 + \eta/\xi)\dot{\xi}$, etc. The Hamiltonian is then obtained in the usual way from the Lagrangian $\mathcal{L} = K + (e^2/r) - e\mathcal{E}z$ by computing $H = p_\xi\dot{\xi} + p_\eta\dot{\eta} + p_\varphi\dot{\varphi} - \mathcal{L}$ and eliminating time derivatives in favor of canonical momenta.

The separability of this equation hinges on the possibility of introducing a separated form for the Hamilton–Jacobi function,

$$S(\xi, \eta, \varphi) = S_\xi(\xi) + S_\eta(\eta) + S_\varphi(\varphi), \tag{6.73}$$

whence Eq. (6.72) takes the form

$$4\xi \left(\frac{\partial S_\xi}{\partial \xi}\right)^2 + 4\eta \left(\frac{\partial S_\eta}{\partial \eta}\right)^2 + \left(\frac{1}{\xi} + \frac{1}{\eta}\right) \left(\frac{\partial S_\varphi}{\partial \varphi}\right)^2 - 4\mu e^2 + \mu e \mathcal{E}(\xi^2 - \eta^2)$$

$$= 2\mu(\xi + \eta)\alpha_1. \tag{6.74}$$

If one solves this equation for $\partial S_\varphi / \partial \varphi$, one finds

$$\frac{\partial S_\varphi(\varphi)}{\partial \varphi} = \text{function only of } \xi, \eta. \tag{6.75}$$

Since the left-hand side can only contain φ and the right-hand side only ξ, η, both must be constant, and we can set

$$\frac{\partial S_\varphi}{\partial \varphi} = \alpha_3. \tag{6.76}$$

Replacing $\partial S_\varphi / \partial \varphi$ by α_3 in Eq. (6.72) and putting all terms with η on the left, all terms with ξ on the right side,

$$4\eta \left(\frac{dS_\eta}{d\eta}\right)^2 + \frac{\alpha_3^2}{\eta} - 2\mu e^2 - \mu e \mathcal{E} \eta^2 - 2\mu \alpha_1 \eta$$

$$= -\left(4\xi \left(\frac{dS_\xi}{d\xi}\right)^2 + \frac{\alpha_3^2}{\xi} - 2\mu e^2 + \mu e \mathcal{E} \xi^2 - 2\mu \alpha_1 \xi\right). \tag{6.77}$$

Since the left-hand side only depends on η while the right-hand side only depends on ξ, both must be equal to a constant, which we henceforth dub $2\alpha_2$. We thus arrive at a pair of equations, each an ordinary differential equation in a single variable

$$4\xi \left(\frac{dS_\xi}{d\xi}\right)^2 + \frac{\alpha_3^2}{\xi} - 2\mu e^2 + \mu e \mathcal{E} \xi^2 - 2\mu \alpha_1 \xi = -2\alpha_2, \tag{6.78}$$

$$4\eta \left(\frac{dS_\eta}{d\eta}\right)^2 + \frac{\alpha_3^2}{\eta} - 2\mu e^2 - \mu e \mathcal{E} \eta^2 - 2\mu \alpha_1 \eta = +2\alpha_2. \tag{6.79}$$

The constants α_1, α_2, and α_3 are called the *separation constants* (cf. Section A.2.1). From Eqs. (6.78)–(6.79) we may solve for the derivatives representing the conjugate momenta p_ξ and p_η in the Hamilton–Jacobi formalism, which, together with the previously determined angular momentum p_φ in Eq. (6.76), gives:

$$p_\xi = \frac{dS_\xi}{d\xi} = \frac{1}{2\xi}\sqrt{-\alpha_3^2 + 2(\mu e^2 - \alpha_2)\xi + 2\mu\alpha_1\xi^2 - \mu e \mathcal{E}\xi^3}, \qquad (6.80)$$

$$p_\eta = \frac{dS_\eta}{d\eta} = \frac{1}{2\eta}\sqrt{-\alpha_3^2 + 2(\mu e^2 + \alpha_2)\eta + 2\mu\alpha_1\eta^2 + \mu e \mathcal{E}\eta^3}, \qquad (6.81)$$

$$p_\varphi = \frac{dS_\varphi}{d\varphi} = \alpha_3. \qquad (6.82)$$

The next step, as indicated already by Schwarzschild in his letter to Sommerfeld (see Section 5.1.4), is to choose a very particular set of new canonical momenta, namely, the action variables defined by phase integrals for each of the coordinates allowing a separation of the motion. Specifically, we define

$$I_\xi(\alpha_1, \alpha_2, \alpha_3) = \oint p_\xi \, d\xi, \qquad (6.83)$$

$$I_\eta(\alpha_1, \alpha_2, \alpha_3) = \oint p_\eta \, d\eta, \qquad (6.84)$$

$$I_\varphi(\alpha_3) = \oint p_\varphi \, d\varphi. \qquad (6.85)$$

From Fig. 6.5, it is apparent that the motion is *librational* in the coordinates ξ, η, and *rotational* in the azimuthal angle φ (cf. Appendix, Section A.2.3). Specifically, this means that $\oint$ means $2\int_{\xi_{\min}}^{\xi_{\max}}$ in Eq. (6.83), $2\int_{\eta_{\min}}^{\eta_{\max}}$ in Eq. (6.84), and $\int_0^{2\pi}$ in Eq. (6.85). By convention, one chooses the direction in which the circuit is traversed to give a positive result for the action variables. We have indicated in Eqs. (6.83)–(6.85) the explicit dependence of the action variables on the separation constants α_1, α_2, α_3: I_ξ and I_η evidently depend on all three, while $I_\varphi = 2\pi\alpha_3$ only depends on the third (the magnitude of the component of the angular momentum in the z-direction, which is the only component of angular momentum conserved in the presence of an external electric field in the z-direction, which exerts a torque on the electron perpendicular to the z-direction).

The action variables may be chosen classically to have any (positive) continuous values we please, and it is apparent that for general choices of these quantities, the three equations above may be inverted to give the separation constants, and in particular the conserved energy $E = \alpha_1$, as a function of I_ξ, I_η, and I_φ. It is also clear that in general the frequencies $\nu_\xi = \partial E/\partial I_\xi$, etc. of the motion in the three coordinates take on independent real values and are, save for a set of choices of measure zero, incommensurate with one another. We have a classic case of *nondegenerate* conditionally periodic motion: the orbit of the electron fills out densely[21] the rotation solid obtained by rotating the curved quadrilateral in Fig. 6.5 around the vertical axis. The turning points of the libration for the type of motion we are interested in here (where the electron remains bound to the nucleus) consist of the two roots of the cubic functions under the square roots in Eq. (6.80) or Eq. (6.81) which remain finite when the electric field is taken to zero. For example, if we define the expression under the square root sign in Eq. (6.80) as

$$F(\xi) = -\alpha_3^2 + 2(\mu e^2 - \alpha_2)\xi + 2\mu\alpha_1\xi^2 - \mu e\mathcal{E}\xi^3, \qquad (6.86)$$

a plot of $F(\xi)$ for some conveniently chosen parameters (note: α_1 is negative for a bound electron) shows the generic structure displayed in Fig. 6.6. For $\xi_{min} < \xi < \xi_{max}$, the particle simply librates between the turning points ξ_{min} and ξ_{max}, in the region where $F(\xi) > 0$, as shown in Fig. 6.5.[22] Similar considerations hold for the motion in the η coordinate.

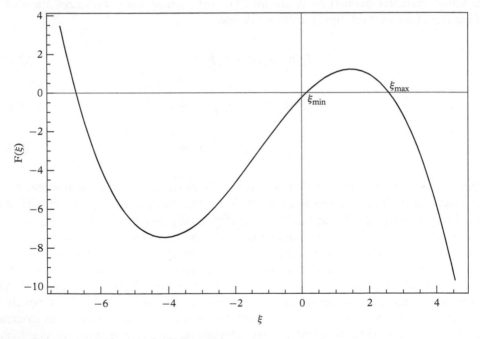

Figure 6.6 *Cubic function determining libration in ξ.*

[21] i.e., comes arbitrarily close to any point in
[22] It may help the reader's intuition to think of the function plotted in Fig. 6.6 as the kinetic energy of ξ motion of the electron (roughly speaking, $E - V$ for the ξ coordinate), or as the inverted potential energy. The electric field is taken much stronger than in realistic cases for reasons of scale, so that the third root (on the left in Fig. 6.6, unphysical as $\xi > 0$) can be easily seen. For the analogous $F(\eta)$, the third root appears at large positive η. For positive $\mathcal{E}$, the electron would be removed from the atom if initially placed beyond this point.

It is not possible to solve analytically for the separation constants in terms of the action variables for general values of the field, but the calculation can be carried out as an expansion in powers of the weak electric field $\mathcal{E}$, which turns out to be perfectly appropriate given the electric fields at the disposal of Stark et al. At the lowest order, we simply set $\mathcal{E} = 0$, and the cubic function $F(\xi)$ in Eq. (6.86) becomes a quadratic, which we may write

$$F(\xi) = (-2\mu\alpha_1)(\xi - a)(b - \xi), \quad a+b = \frac{\alpha_2 - \mu e^2}{\mu\alpha_1}, \quad ab = -\frac{\alpha_3^2}{2\mu\alpha_1}. \tag{6.87}$$

With the aid of the elementary integral

$$\int_a^b \frac{1}{\xi}\sqrt{(\xi - a)(b - \xi)}d\xi = \frac{\pi}{2}\left(a + b - 2\sqrt{ab}\right), \tag{6.88}$$

we find, for vanishing electric field,

$$I_\xi = \sqrt{-2\mu\alpha_1}\int_a^b \frac{1}{\xi}\sqrt{(\xi - a)(b - \xi)}\ d\xi$$

$$= \pi\left(\frac{\mu e^2 - \alpha_2}{\sqrt{-2\mu\alpha_1}} - \alpha_3\right). \tag{6.89}$$

The result for I_η is easily obtained by changing the sign of α_2 (cf. Eqs. (6.80)–(6.81), with $\mathcal{E} = 0$),

$$I_\eta = \pi\left(\frac{\mu e^2 + \alpha_2}{\sqrt{-2\mu\alpha_1}} - \alpha_3\right), \tag{6.90}$$

while the third action variable is given simply by $I_\varphi = 2\pi\alpha_3$. Adding all three action variables gives

$$I_\xi + I_\eta + I_\varphi = \frac{2\pi\mu e^2}{\sqrt{-2\mu\alpha_1}}, \tag{6.91}$$

from which we may immediately solve for the energy (to lowest order, i.e., for vanishing electric field)

$$\alpha_1 \equiv E^{(0)} = -\frac{2\pi^2\mu e^4}{(I_\xi + I_\eta + I_\varphi)^2}. \tag{6.92}$$

The frequencies of the motion in each of the three coordinates now coincide, as $\nu_\xi = \partial E/\partial I_\xi = \partial E/\partial I_\eta = \nu_\eta$, etc. We have *simple periodic motion* with a single frequency, and the orbit degenerates to a one-dimensional manifold, the Kepler ellipses of the pure Coulomb problem.

Once the electric field is switched on, the motion becomes nondegenerate, and we must content ourself with an expansion of the separation constants, and in particular the energy α_1 in powers of the electric field strength $\mathcal{E}$, where we shall write $E^{(n)}$ for the contribution to the energy of order $\mathcal{E}^n$.

$$E = E^{(0)} + E^{(1)} + E^{(2)} + O(\mathcal{E}^3). \tag{6.93}$$

Carried through to second order in $\mathcal{E}$, the result is

$$E^{(0)} = -\frac{2\pi^2 \mu e^4}{(I_\xi + I_\eta + I_\varphi)^2},$$

$$E^{(1)} = \frac{3\mathcal{E}}{8\pi^2 \mu e}(I_\xi + I_\eta + I_\varphi)(I_\xi - I_\eta),$$

$$E^{(2)} = -\frac{\mathcal{E}^2}{1024\pi^6 \mu^3 e^6}(I_\xi + I_\eta + I_\varphi)^4$$

$$\times \left(17(I_\xi + I_\eta + I_\varphi)^2 - 3(I_\xi - I_\eta)^2 - 9I_\varphi^2\right). \tag{6.94}$$

The first-order Stark effect computed by Schwarzschild and Epstein in 1916 corresponds to the second line, linear in the applied field. The calculation involves some mildly unpleasant algebra, which we have reproduced in full elsewhere (Duncan and Janssen 2014). The quadratic term (corresponding to the third line) is the result of considerably more painful calculations by Epstein (1916d) and Ali Moustafa Mosharrafa (1922), and shows clearly that the motion is fully nondegenerate once the electric field is present. Including only terms up to first order in $\mathcal{E}$, there is a remnant degeneracy, as the energy is a function of only two independent combinations of the action variables. This would lead Schwarzschild to an incorrect assignment of quantum numbers, as we shall see below.

So far the procedure has been entirely classical. Quantization, *à la* Schwarzschild–Epstein, is a matter of moments. One simply restricts the allowed values for the action variables in Eq. (6.94) to integer multiples of Planck's constant:[23]

$$I_\xi = n_\xi h, \quad I_\eta = n_\eta h, \quad I_\varphi = n_\varphi h, \tag{6.95}$$

---

[23] Epstein uses n_1, n_2, and n_3 for the quantum numbers; we adopt the notation used here as it clarifies the association with the different coordinates. Schwarzschild uses m_1 and m_2 for two independent quantum numbers which appear in the energy formula through first order in the electric field (see below).

where the quantum numbers n_ξ, n_η, and n_φ are positive integers or zero. The first two terms in the expansion of the energy then become

$$E_{n_\xi,n_\eta,n_\varphi} = -\frac{2\pi^2 \mu e^4}{h^2(n_\xi + n_\eta + n_\varphi)^2} + \frac{3\mathcal{E}h^2}{8\pi^2\mu e}(n_\xi + n_\eta + n_\varphi)(n_\xi - n_\eta). \qquad (6.96)$$

If we identify $n \equiv n_\xi + n_\eta + n_\varphi$ with Bohr's principal quantum number, the first term turns into the Balmer–Bohr formula for the bound stationary states of hydrogen, and the second yields the much desired first-order Stark shifts in the quantum theory. The circular orbits of the first paper of Bohr's 1913 trilogy correspond to the subset of orbits (for vanishing field) with $n_\xi = n_\eta = 0$, $n_\varphi = 1, 2, 3, \ldots$, for which the first-order Stark shift is zero. But just as with Sommerfeld's treatment of the Coulomb problem in polar coordinates, we see that the theory now yields a plethora of other possible orbits, fixed elliptical ones for zero electric field, but precessing slowly as in Fig. 6.5 once we turn on a weak electric field (i.e., one for which the second term in Eq. (6.96) is much less than the first). These orbits will be displaced in energy (to first order in the electric field) from the circular ones as long as $n_\xi \neq n_\eta$.

A remarkable feature of the analysis above of perturbed Kepler orbits in the Stark effect is the dependence of the actual geometry of the orbits obtained on the coordinate system used: in particular, the closed elliptical orbits obtained using parabolic coordinates, in the limit of vanishing electric field, are geometrically *different* (with, for example, different eccentricities) from those obtained in Sommerfeld's initial analysis using polar coordinates (giving the orbits depicted in Fig. 6.1). This orbit ambiguity problem[24] represents a clear conceptual flaw in the old quantum theory, which could only be resolved once orbits had been replaced by wave-functions, and the concept of superposition of states had entered the theory. In contrast, and fortunately for the success of the old quantum theory, the empirically accessible *energies* associated with the quantized states were immune from this ambiguity problem (for degenerate systems).

As in the case of Sommerfeld's formula Eq. (6.48) for the relativistic fine-structure, the old-quantum-theory formula (6.96) for the energy shifts induced by an applied electric field, to first order in the field, agrees exactly *in form* with the result obtained later by Epstein (1926) and Schrödinger (1926b) in wave mechanics (Duncan and Janssen 2014, sec. 4). The physical content and interpretation of the quantities appearing in the formula, however, are completely different in the two theories. In particular, they agree in their phenomenological consequences *despite* the incorrect assignment of quantum numbers in the earlier theory. Nevertheless, what at first sight appears to be almost miraculous agreement can be seen, in both the Stark effect as well as the fine structure, to be the result of a fortuitous numerological accident. We shall return to this point later once the elements of the wave mechanical theory necessary for the semiclassical derivation of the quantization rules of the old quantum theory have been introduced.

---

[24] For detailed discussion, see Duncan and Janssen (2014, sec. 3; 2015, sec. 2.3) and the Web Resource on *Orbits*.

From Eq. (6.96) follows immediately an expression for the shift $\Delta \nu$ (with ν, as usual at the time, meaning the inverse wavelength, $1/\lambda$) observed in the spectral line arising from the transition from the $(m_\xi, m_\eta, m_\varphi)$ to the $(n_\xi, n_\eta, n_\varphi)$ orbit:

$$\Delta \nu = \frac{C\mathcal{E}}{hc} \left(m(m_\xi - m_\eta) - n(n_\xi - n_\eta) \right), \quad \text{with} \quad C \equiv \frac{3h^2}{8\pi^2 \mu e}, \tag{6.97}$$

where we have introduced the simplified notation $m = m_\xi + m_\eta + m_\varphi$, $n = n_\xi + n_\eta + n_\varphi$ for the principal quantum numbers associated with the initial and final orbit. Epstein (1916b, p. 509) points out that the result confirms two important features of the empirically observed splittings: (*i*) they are to the available accuracy proportional to the field strength, and (*ii*) the split spectral lines are disposed symmetrically above and below a central unshifted line.

In addition, it was known that the different components were polarized when viewed from the direction orthogonal to the field: some were polarized in the z (or field) direction (the so-called "parallel" components), others perpendicular to the field, as would be expected for circularly polarized radiation when viewed *in the direction orthogonal to the field*. Epstein frankly admitted that his calculations could not account for the relative intensity of the lines, nor for the empirically observed rule that parallel polarized components corresponded to the change in the azimuthal quantum number $m_\varphi - n_\varphi$ being an even number, the perpendicular components to $m_\varphi - n_\varphi$ odd. This association would shortly be explained by the correspondence-principle-based selection rules of Bohr (cf. Section 5.3).

In his discussion of the classical structure of the orbits, Epstein had earlier in his paper noted that the orbits with $n_\varphi = 0$ would correspond to the electron orbit remaining in a meridional plane containing the z-axis (direction of electric field). If either n_ξ or n_η were in addition to be zero, the orbit would degenerate to a straight-line path passing through the nucleus, which would result in a singular motion for which even the classical analysis would presumably break down (for relativistic reasons, at the very least). Epstein allowed for $n_\varphi = 0$ provided both n_ξ and n_η were nonzero: in this case, the precessing orbit of the electron would eventually pass arbitrarily close to the nucleus but not through it. Orbits of this latter type would shortly be excluded by Bohr (1918, p. 27 and p. 75). In any event, given that the energy levels with $n_\xi \neq 0$, $n_\eta \neq 0$, $n_\varphi = 0$ are degenerate with others with $n_\varphi \neq 0$, we may assume the latter restriction in the comparison with data. With $n_\varphi \neq 0$, we have $n_\xi + n_\eta \leq n - 1$. We therefore have $n_\xi - n_\eta$ taking on the $2n - 1$ possible values from $-(n-1)$ to $n-1$: the initially degenerate Bohr levels at principal quantum number n are now split into $2n - 1$ levels, with the central level unshifted.

Fig. 6.7 shows the Stark effect for the first Balmer H_α line in the hydrogen spectrum, corresponding to transitions from $n = 3$ to $n = 2$. For $\mathcal{E} \neq 0$, as indicated by the arrows in Fig. 6.7, the energy loss in a quantum jump from $n = 3$ to $n = 2$ can take on fifteen different values, resulting in frequency shifts $\Delta \nu$ ranging from $-8C\mathcal{E}/hc$ to $+8C\mathcal{E}/hc$. This means that the frequency of the Balmer line H_α, emitted in the transition from $n = 3$ to $n = 2$, splits into fifteen different frequencies. Illustrating Epstein's general observation

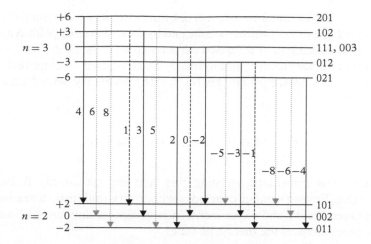

Figure 6.7 *Stark effect for the Balmer line H_α in the hydrogen spectrum: splittings ΔE (in units of $C\mathcal{E}$ with $\mathcal{E}$ the strength of the electric field and $C \equiv 3h^2/8\pi^2 e\mu$) of energy levels for $n = 2$ and $n = 3$ [horizontal lines with values of ΔE to the left and values of quantum numbers $(n_\xi, n_\eta, n_\varphi)$ to the right]; splittings Δv (in units of $C\mathcal{E}/hc$) of the frequency of the radiation emitted in transitions from $n = 3$ to $n = 2$ [arrows with values of Δv to the left—solid arrows: parallel polarization; dashed arrows: perpendicular polarization; dotted arrows: violation of selection rule]. The figure is not drawn to scale: the energy gap between the $n = 2$ and $n = 3$ levels is much greater than the level splittings (Duncan and Janssen 2015, p. 240, Fig. 1).*

noted above, the fourteen shifted frequencies lie symmetrically on opposite sides of the unshifted one.

Epstein eliminated six of these fourteen shifted frequencies, three on each side of the unshifted one. He adopted, at least initially, the selection rule initially proposed by Sommerfeld (cf. Eq. (5.45)), requiring each individual quantum number (and not just the principal quantum number) to decrease, or remain fixed, in an allowed transition (Epstein 1916b, p. 511),

$$m_\xi \geq n_\xi, m_\eta \geq n_\eta, m_\varphi \geq n_\varphi. \tag{6.98}$$

According to this selection rule, the six transitions in which any one of the three quantum numbers increases are forbidden. These are the transitions represented by dotted arrows in Fig. 6.7.[25] The corresponding lines were either absent or exceedingly faint in Stark's spectroscopic data, which supported Sommerfeld's selection rule. The nine remaining transitions all matched lines clearly present in Stark's data: the six transitions indicated

---

[25] Under this selection rule, the transitions '(003) → (011)' and '(003) → (101)' are also forbidden but $E_{003} = E_{111}$ even if $\mathcal{E} \neq 0$ and the transitions '(111) → (011)' and '(111) → (101)' are allowed, so this does not affect the number of lines (cf. Fig. 6.7).

by solid arrows (with $\Delta \nu$ equal to $\pm 2, \pm 3, \pm 4$ times $C\mathcal{E}/h$) producing light polarized parallel to the field; the three transitions indicated by dashed arrows (with $\Delta \nu$ equal to 0, ± 1 times $C\mathcal{E}/h$) producing light polarized perpendicular to the field.

The splittings of other Balmer lines found by Stark violated Sommerfeld's selection rule. To match Stark's data, Epstein (1916b, p. 516) eventually settled on a modified version of the rule,

$$n_\xi \leq m_\xi, \quad n_\eta \leq m_\eta, \quad n_\varphi \leq m_\varphi + 1, \tag{6.99}$$

and emphasized that transitions violating this rule are not strictly forbidden, just highly improbable. Epstein's paper concludes with a very optimistic assessment of the agreement between theory and experiment. In language reminiscent of Sommerfeld's most effusive promotions of the theory, he states

> We believe that the results communicated here provide a new proof of striking cogency of the correctness of the Bohr model. In our knowledge of the mechanism of the atom we are standing at the beginning of a new development, and it would seem that there is still much to expect from the quantum theory in the area of atomic physics (Epstein 1916b, p. 520).

Schwarzschild's (1916) paper, "On the Quantum Hypothesis," devotes only eight of its twenty-one pages to the Stark effect. There is an extensive introduction outlining the general quantization procedure in terms of action-angle variables, and the discussion of the Stark effect is followed by a second application of the method to the quantization of rigid body motion. The formula Schwarzschild obtains for the energy shift to first order in the electric field is (notwithstanding Epstein's later assertion in a 1962 interview with Heilbron that "he had the wrong formula") precisely equivalent to Epstein's. Schwarzschild treats the problem as one with a remaining degeneracy, as, *if we work only to first order in the field*, the energy depends only on two independent action variables, which we may take to be $I_1' = 2I_\xi + I_\varphi, I_2' = 2I_\eta + I_\varphi$ (cf. the first line of Eq. (6.94)). If this degeneracy were indeed a feature of the exact problem (to all orders of the electric field) one would have to choose a new set of action-angle variables, as Schwarzschild explains in detail in the first part of his paper, involving only the independent action variables appearing in the energy, and chosen so that the kinematic boundary of each variable is zero (so that no constant terms would appear on the right side of the quantization conditions Eq. (6.95)). These new variables would then be subject to quantization: in particular, Schwarzschild then chooses

$$I_1' = n_1 h, \quad I_2' = n_2 h, \tag{6.100}$$

and finds for the shift in frequency in the transition $(m_1, m_2) \rightarrow (n_1, n_2)$ to first order in $\mathcal{E}$,

$$\Delta v = \frac{C\mathcal{E}}{4hc}(m_1^2 - m_2^2 - (n_1^2 - n_2^2)), \quad \text{with} \quad C \equiv \frac{3h^2}{8\pi^2\mu e}. \tag{6.101}$$

This is exactly equivalent to Epstein's result Eq. (6.97), with the identification $n_1 = 2n_\xi + n_\varphi$, $n_2 = 2n_\eta + n_\varphi$. The problem with Schwarzschild's subsequent analysis, as compared with Epstein, is that the geometrical significance of various choices for the quantum numbers has been submerged, and that when Schwarzschild examines the possible states with principal quantum number 2, for example, and therefore $n_1 + n_2 = 4$, he allows all possible values $(4, 0), (3, 1), (2, 2), (1, 3), (0, 4)$ for (n_1, n_2), including the orbits $(4, 0), (0, 4)$ corresponding to $(n_\xi, n_\eta, n_\varphi) = (2, 0, 0), (0, 2, 0)$, which are the straight-line paths through the nucleus excluded by Epstein. In particular, for the integer quantity $Z = m(m_\xi - m_\eta) - n(n_\xi - n_\eta)$ in Eq. (6.97), which for Epstein can only take the values up to 8 (ignoring selection rules) for the $H_\alpha(3 \rightarrow 2)$ transition, Schwarzschild finds values up to 13. When Sommerfeld's selection rule is applied in the form $m_1 \geq n_1$, $m_2 \geq n_2$, this is reduced to allowed values $Z = 0, 1, 2, 3, 4, 5$, one more than Epstein's allowed transitions, with $Z = 0, 1, 2, 3, 4$ (cf. Fig. 6.7). As Schwarzschild points out, the values of Z inferred from the measured Δv do indeed come close to integer values (for example, he quotes 0.0, 0.9, 2.1, 3.0, 3.9 for the H_α line). Despite these minor differences,[26] both Schwarzschild and Epstein deserve credit for the first quantitatively successful energy quantization of an atomic system in an external field in the old quantum theory.

A much more systematic and thorough analysis of the line-intensity question and selection rules operative in the Stark effect, based on the extended correspondence principle of Bohr discussed in Section 5.3, would follow within two years in the doctoral thesis Kramers wrote in Copenhagen under Bohr's supervision. We recall here briefly the essence of Bohr's ideas. A transition between two orbits will be accompanied by radiation with a certain polarization and a certain intensity whenever the relevant coefficient in the Fourier expansion of the classically calculated orbit is non-vanishing. For very high quantum numbers it does not matter whether we consider the Fourier expansion of the initial or of the final orbit. For low quantum numbers, however, this clearly does matter, rendering Bohr's prescription ambiguous. Should we consider the Fourier expansion of the initial or of the final orbit? Some average of the two perhaps? Or a "suitably weighted" average over initial and final orbit and all orbits in between?

In a tour-de-force of classical analysis, Kramers was able to compute analytical expressions for the Fourier coefficients $C_{i\bar\tau}$ in the expansion Eq. (5.106), using the parabolic coordinates adapted to the theory of the Stark effect. With this information, Kramers could explicitly carry through Bohr's correspondence-principle approach, as discussed above in Section 5.3 (see, in particular the discussion from Eq. (5.103) to Eq. (5.110)), to account for the polarizations and intensities of the various components into which the Balmer lines split in the Stark effect. By distinguishing Fourier expansions

---

[26] Schwarzschild (1916, p. 564) acknowledges in a footnote added during proofs that Epstein has carried through the calculation "in an associated, less degenerate system" and obtained a "almost complete agreement with experiment, and also found a rule for the polarization." The flaw in Schwarzschild's degeneracy analysis is clearly exposed by Epstein (1916d, p. 183).

of the motion in the direction of the field (z) and in the plane perpendicular to the field ($x + iy$), Kramers could account for the polarizations found by Stark. The exponent for the Fourier expansion of z, Eq. (5.110), both for initial and final orbit, does not contain τ_3, which according to Bohr's correspondence principle should be set equal to $\Delta n_\varphi \equiv m_\varphi - n_\varphi$. This suggests that $\Delta n_\varphi = 0$ for all transitions in which radiation polarized parallel to the field is emitted. Similarly, only terms with $\tau_3 = \pm 1$ are present in the exponent of the Fourier expansion of x or y (as in Eq. (5.109)). This suggests that $\Delta n_\varphi = \pm 1$ for all transitions in which radiation polarized perpendicular to the field is emitted. Fig. 6.7 shows that these conclusions based on the correspondence principle are supported by Stark's findings. The solid-arrow transitions (parallel polarization) all have $\Delta n_\varphi = 0$; the dashed-arrow transitions (perpendicular polarization) all have $\Delta n_\varphi = \pm 1$.

Kramers could also account, at least qualitatively, for the intensities of the various components Stark had found. In principle, Kramers used the average of the squares of coefficients of the relevant Fourier components of the initial and the final orbits to estimate the intensity of the corresponding line. However, even in cases where a certain frequency was completely absent from the Fourier expansion of both the initial and the final orbit, Kramers left open the possibility that the corresponding line might appear in the spectrum, albeit only faintly, as its frequency might be present in the Fourier expansion of some intermediate orbit. Kramers thus allowed several lines that are forbidden by the selection rules of Sommerfeld and Epstein (see Eqs. (6.98) and (6.99)). As we saw above, Epstein had ruled out six possibilities for the transition $n = 3 \rightarrow n = 2$ (see the dotted arrows in Fig. 6.7 with values $\pm 5, \pm 6, \pm 8$ times $C\mathcal{E}/hc$ for $\Delta \nu$). Kramers predicted (correctly as it turned out) that these components of the Stark splitting of H_α had just escaped notice so far because of their low intensity (Kramers 1919, Appendix, Fig. 1). Measurements of line intensities in the early 1920s were sufficiently rough that good agreement with Kramers' theoretical estimates could be claimed (indeed, in his Nobel lecture in 1922, Bohr was to claim that "the theory reproduces completely the main feature of the experimental results"). Later, more accurate measurements by John Stuart Foster and Laura Chalk (1929) found definite discrepancies with Kramers' results, but by that time the correct wave-mechanical calculations were available and essential harmony between theory and experiment had been reestablished (Duncan and Janssen 2015, pp. 246–250).

Despite this convoluted history, it is clear from the published literature of the time that the agreement between the measured Stark effect line wavelengths and those predicted by the old quantum theory (which, up to first order, exactly coincide with those of wave mechanics) was regarded for the nine years of dominance of the Bohr–Sommerfeld theory as a spectacular success. A brief extract from the first (1919) edition of Sommerfeld's "bible" suffices to make the point:

> The theory of the Zeeman effect and quite especially that of the Stark effect belongs among the most impressive achievements of our field, and forms a beautiful capstone in the edifice of atomic physics (Sommerfeld 1919, p. 457).

As we shall soon see, Sommerfeld's optimism concerning the Zeeman effect, which (unlike the Stark effect) depends critically on an understanding of the as yet unknown intrinsic spin angular momentum of the electron, was certainly premature. Difficulties with the anomalous Zeeman effect, and more generally with the complex multiplet spectra encountered especially in the optical regime, along with the problems one ran into using the Bohr–Sommerfeld methodology to untangle the dynamics of multi-electron atoms (even for helium, the smallest), would within a few years ring the death knell of the old quantum theory.

7

Failures

In the previous chapter, we identified three areas in which the old quantum theory, in the period following the initial successes of the Bohr model with explaining the spectra of hydrogenic (single electron) atoms, was considered to have furnished convincing quantitative explanations of atomic phenomena: namely, the relativistic fine structure as given by the Bohr–Sommerfeld quantization procedure, the application of this fine structure to X-ray spectra, and finally, the Stark effect splitting of spectral lines in an electric field. In this chapter, we trace the growing unease and dissatisfaction with the conceptual framework of the old quantum theory. This arose from the inability of this theory to provide accounts which were both physically convincing and quantitatively accurate of several important aspects of atomic behavior which emerged from the constantly growing and increasingly accurate spectroscopic material provided by experiment in the period from 1916 to 1925.

As in the previous chapter, we have selected three topics for close examination which were at the center of developments in quantum theory in this period. All were, after tremendous effort, eventually assessed by contemporary physicists as failures of the old theory. The problems in question were (a) the interpretation of the complex multiplet structure of multi-electron atoms, in which the single levels of the first Bohr model were seen to dissolve into several separate energy levels which could *not* be interpreted simply in terms of orbits of varying eccentricity, (b) the peculiar and ultimately inexplicable (within the old quantum theory) patterns of spectral lines which emerged in the Zeeman effect once magnetic fields were imposed on atomic systems, and (c) the problem of the mechanics of the simplest multi-electron atom, helium, and, in particular, the inability of the theory to correctly predict the ionization energy of the helium atom in its ground state. By 1924, the inadequacies of the old quantum theory in all these areas had been fully exposed, and the expectation was widespread that further progress would require a complete reevaluation of the theoretical assumptions on which the quantum theory of atoms had previously been based.

As we saw in Chapter 6, the successes of the old quantum theory were not unqualified successes. Some of them, such as the explanation of the fine structure of various spectra, were largely accidental; others, such as the explanation of the Stark effect, turned out to be incomplete. Similarly, not all failures discussed in this chapter were complete failures. In the case of the Zeeman effect, for instance, Sommerfeld and others eventually

Constructing Quantum Mechanics. Anthony Duncan and Michel Janssen, Oxford University Press (2019).
© Anthony Duncan and Michel Janssen. DOI: 10.1093/oso/9780198845478.001.0001

found empirical rules accounting for the line splittings in impressive quantitative detail. That the Zeeman effect was nonetheless seen as a failure was because of persistent dynamical inconsistencies in a succession of mechanical models constructed to capture these empirical regularities. In hindsight, we can see that the inability to develop such models was due, in large part, to the absence of the notion of spin. The extent to which physicists were able to account for the Zeeman effect without what in hindsight looks like an absolutely critical ingredient is certainly remarkable but, as they themselves eventually recognized, their efforts came up short.

7.1   The complex structure of spectral multiplets

The Bohr model was faced from the very first moments of its existence with the problem of accounting not just for the simple spectra of hydrogen and, a little later, ionized helium, but with the plethora of increasingly accurate spectroscopic data for atoms with more than one electron which had been accumulating steadily from the time of Fraunhofer.[1] The periodic structure of the elements, with chemical properties recurring at regular intervals, suggested the existence of ordered "rings" or "shells" of electronic orbits of equal principal quantum number which would be added to sequentially as one increased the atomic number (after the work of Moseley, clearly identified with the nuclear charge and hence the number of orbiting electrons in the neutral atom). Once the rare gases were reached, the next electron would then be forced, for some as yet unknown reason of dynamical stability, into the next higher orbit. In particular, the alkali metals in group I of the periodic table would in this view have a single "valence" electron circulating outside the orbits of the inner electrons, more loosely attached to the nucleus and primarily responsible therefore for the chemical and, in particular, optical properties (hence the term *Leuchtelektron* in the German literature) of the atom.

Gratifyingly, of the non-hydrogenic atoms, the alkalis had long been known to display the simplest optical spectra, falling into the now well classified series (principal, sharp, diffuse, etc.), discussed in detail in Appendix B. Moreover, formulas of Rydberg–Ritz type obviously cried out for a quantitative explanation on the basis of the new Bohr (and soon, Bohr–Sommerfeld) methodology. The algebraic affinity of these formulas to the marvelously successful Balmer formula encouraged the hope that a plausible explanation on the basis of the new atomic model would be possible, especially after the success of the theory in accounting for subtle features like the relativistic fine structure in the hydrogen (and ionized helium) case in 1916, as we saw in the preceding chapter.

And it fell once again to Sommerfeld to answer the call.

---

[1] For a fairly extensive review of the history of spectroscopy from the time of Fraunhofer to the emergence of the Bohr atomic model, see Appendix B. We shall frequently rely in the forthcoming material on some of the nomenclature and phenomenology described there.

7.1.1 Sommerfeld on multiplets

In a paper of late 1916 (Sommerfeld 1916c) and in the first edition of *Atombau und Spektrallinien* (Sommerfeld 1919, p. 503), he examined the effect of an inner ring of electrons on a single outer electron whose orbit is coplanar with the ring. The inner circular ring (of radius a) is presumed to be occupied by $Z - k$ equally spaced electrons, with Z the atomic number, so that for radial coordinate (distance from nucleus) of the outer electron $r \gg a$, it sees effectively a screened nuclear charge $(Z - (Z - k))e = ke$, and asymptotically, an electrostatic potential $- ke^2/r$. For the neutral alkali elements, as we henceforth assume, $k = 1$. For more general values of r, Sommerfeld assumed that the speed with which the electrons in the rings circulated was large enough in comparison to that of the outer electron that the electrostatic potential energy due to the former could be modeled by an uniform charge distribution on the (circular) ring. The result was that the net potential energy seen by the electron was (a) a function only of r (provided of course that the orbits of the valence and inner electrons were coplanar), and (b) possessed an expansion for large r of the form [2]

$$U(r) = -\frac{e^2}{r} + \frac{b_1}{r^2} + \frac{c_1}{r^3} + \frac{b_2}{r^4} + \frac{c_2}{r^5} + \dots \qquad (7.1)$$

Assuming that the inner circular ring had its center at the position of the nucleus,[3] one can show that the coefficients $b_1, b_2, \dots$ vanish, leaving only odd inverse powers of r. For each bound quantized orbit, let us define a quantity ϵ through the binding energy $W = -E$ by

$$W = -E = hc\epsilon. \qquad (7.2)$$

Differences of ϵ between two states involved in a radiative transition then give directly the wavenumber (i.e., inverse wavelength, $1/\lambda$) of the transition: we may therefore associate ϵ with the spectral terms of Rydberg and Ritz. Allowed values of ϵ are to be determined by writing the energy of the system (here in polar coordinates),

$$E = -hc\epsilon = \frac{p_r^2}{2m} + \frac{p_\varphi^2}{2mr^2} + U(r), \qquad (7.3)$$

---

[2] The simple form $-e^2/r$ of the leading term follows from the fact that at very large r a test charge will "see" only the net interior charge of $-e$, effectively concentrated at the origin.

[3] Sommerfeld admitted that the effect of the outer electron might well be to displace the nucleus from the center of the ring, but proceeded with his simplifying assumption nonetheless. Landé (1919a, 1920) would later calculate the effects of such a displacement, in his theory of helium, discussed below in Section 7.3.

solving for the radial momentum p_r (with the angular momentum quantized as usual, $p_\varphi = n(h/2\pi) = n\hbar$),

$$p_r = \sqrt{-A + \frac{2B}{r} - \frac{C}{r^2} - \frac{D_1}{r^3} - \frac{D_2}{r^5} + \dots}, \tag{7.4}$$

with

$$A = 2mhc\epsilon, \quad B = me^2, \quad C = n^2\hbar^2, \quad D_1 = 2mc_1, \quad D_2 = 2mc_2, \dots \tag{7.5}$$

(Sommerfeld 1916c, p. 352) and then enforcing the radial quantization condition,[4]

$$\oint p_r dr = n_r h, \tag{7.6}$$

where the integral is taken once up and once down between the roots r_- and r_+ of the quantity under the square root, which reduce to the roots of the quadratic which remains when the non-Coulombic terms $D_1, D_2, \dots$ are neglected. In the latter case, one has

$$\oint \sqrt{\left(-A + \frac{2B}{r} - \frac{C}{r^2}\right)} dr = 2\pi \left(\frac{B}{\sqrt{A}} - \sqrt{C}\right)$$

$$= \frac{2\pi me^2}{\sqrt{2mhc\epsilon}} - 2\pi n\hbar = n_r h. \tag{7.7}$$

Rewriting this last equality as

$$\frac{\sqrt{2m}\pi e^2}{\sqrt{\epsilon ch}} = (n + n_r)h, \tag{7.8}$$

we find that

$$\epsilon = \frac{N}{(n + n_r)^2} \quad \text{with } N \equiv \frac{2m\pi^2 e^4}{ch^3}. \tag{7.9}$$

In this "zeroth order" approximation, one thus just recovers the Balmer expression for the term values, with N the Rydberg constant. Expanding the square root in Eq. (7.4) to include corrections of first order in D_1, Sommerfeld (1916c, p. 354) found that the right-hand side of Eq. (7.8) acquires a correction depending on B and C (but not on A, and therefore not, through ϵ, on n_r, the radial quantum number):

---

[4] For clarity, we use, as previously, the notation n_r, rather than Sommerfeld's n', for the radial quantum number.

$$n + n_r \rightarrow n + n_r + a(n) \text{ with } a(n) = \frac{\pi m e^2 D_1}{h \hbar^3 n^3} = \frac{m^2 e^2 c_1}{\hbar^4 n^3}. \qquad (7.10)$$

In other words, he found a constant shift (or "spectral defect") in the effective principal quantum number of exactly the type proposed by Rydberg, *provided we associate all the running terms of a given series with orbits of varying radial quantum number n_r but with a given, fixed angular momentum quantum number n.*[5] If one carries the expansion further, including terms of second order in D_1 and first order in D_2, the shift $a(n)$ is modified (by terms falling faster by four powers of n) and a new term appears on the right-hand side of Eq. (7.8), proportional to the coefficient A, i.e., to the term ϵ itself, multiplied again by a function of the angular momentum n (1916c, p. 357):

$$n + n_r \rightarrow n + n_r + a(n) + \alpha(n)\epsilon. \qquad (7.11)$$

Consequently, at this order, Eq. (7.9) changes to the following relation implicitly determining ϵ (which appears on both the left- and right-hand sides):

$$\epsilon = \frac{N}{(n + n_r + a(n) + \alpha(n)\epsilon)^2}. \qquad (7.12)$$

This result is identical to Ritz's Ansatz Eq. (B.21), taking $\epsilon = (m, \alpha, \beta)$, with the innocuous changes in notation

$$m, \alpha, \beta \text{ (Ritz)} \rightarrow n + n_r, a(n), \alpha(n) \text{ (Sommerfeld)} \qquad (7.13)$$

Sommerfeld was fully aware of the shortcomings of this argument, but frankly admitted it was impossible to make progress in the problem of atoms with more than one electron without drastic simplifications:

> As the many-body problem is too difficult, we must proceed in an approximate fashion. We will therefore treat the motion of the inner electrons as given and treat . . . the motion of the outer electrons in the field determined by the nucleus and the inner electrons. We thereby neglect the back-reaction of the outer electrons on the inner ones (Sommerfeld 1916c, p. 349).

The result obtained would also seem to depend on the very specific ring geometry for the inner electrons adopted by Sommerfeld.[6] A few years later, Gregor Wentzel (1923), one

---

[5] A few years later, Bohr (1921, pp. 141–144) proposed an argument based on the correspondence principle giving the form of the Rydberg spectral shifts, in agreement with Sommerfeld's lowest-order result (7.10). The argument does not rely on the specific configuration of the core electrons. For discussion of Bohr's argument, see Darrigol (1992, pp. 152–153).

[6] For example, the displacement of the inner electron(s) due to the outer was explicitly taken into account in the treatment of helium a few years later by Landé (1920), which led to corrections to Sommerfeld's results.

of his assistants,[7] showed that the only critical assumption is that the effective potential seen by the outer electron(s) is central, i.e., a function $U(r)$ only of the distance r to the nucleus. In modern quantum mechanics, according to the standard method used in atomic theory for calculating electronic levels, the Hartree–Fock approximation (again, an approximation, but frequently a very good one!), the effective potential due to a closed shell of electrons can be shown rigorously to be central, so this assumption is (fortunately for the success of the old quantum theory in this area) actually on very solid footing for the alkali metals, where the single valence electron is moving under the influence of the nucleus together with an assembly of closed shells formed by the "inner electrons".[8]

Wentzel's argument, in the notation used above, goes as follows. If we assume that the valence electron orbit is determined by an effective central potential $U(r)$, the radial quantization condition takes the form

$$\oint p_r dr = \oint \sqrt{-2m(hc\epsilon + U(r)) - \frac{\hbar^2 n^2}{r^2}}\, dr = n_r h. \tag{7.14}$$

Define a "spectral defect" Δ by

$$h\Delta \equiv \oint p_r dr - \oint \sqrt{-2m(hc\epsilon - \frac{e^2}{r}) - \frac{\hbar^2 n^2}{r^2}}\, dr, \tag{7.15}$$

which is basically the difference between the radial integral with the "correct" potential $U(r)$ and the same integral where only the asymptotic, fully screened remnant potential (for the valence electron in the alkali case, as previously) is used. In the latter case the integral can be performed analytically (use Eq. (7.7) with $A \equiv 2mhc\epsilon$, $B \equiv me^2$, and $C \equiv \hbar^2 n^2$), so we have

$$h\Delta = n_r h - \left(\frac{\sqrt{2m}\pi e^2}{\sqrt{hc\epsilon}} - nh \right), \tag{7.16}$$

from which it follows that

$$\epsilon = \frac{N}{(n + n_r - \Delta)^2}. \tag{7.17}$$

---

[7] In 1921, Gregor Wentzel and Adolf Kratzer became Sommerfeld's assistants, succeeding Wilhelm Lenz and Paul Ewald (see Plates 20 and 24), who were called to chairs in theoretical physics in Hamburg and Stuttgart, respectively (Eckert 2013b, p. 254).

[8] Even for the alkaline earths in group 2, the central approximation is quite good, as the excitation of one of the two valence electrons leaves a core consisting of closed shells plus a single electron, with the properties of the effective potential determined mainly by the former. See Bethe and Jackiw (1986), pp. 65–72.

Inspection of Eq. (7.15) shows that the quantum defect is only a function of ϵ and the angular momentum quantum number n (apart from fundamental constants, e, m, c, etc.), with the former decreasing like an inverse quadratic for large n. It may therefore be expanded in a power series in ϵ, and keeping just the first two terms,

$$\Delta = -(a(n) + \alpha(n)\epsilon + O(\epsilon^2)),\tag{7.18}$$

we see that we recover precisely Sommerfeld's result Eq. (7.12), regardless of the detailed mathematical form of $U(r)$. Since the correct effective potential (in Hartree–Fock approximation) behaves like the screened potential $-e^2/r$ plus *exponentially decreasing* (rather than inverse power, as in Sommerfeld) terms, the more general form of the argument is essential if we are to understand (in retrospect) the success of the old quantum theory in accounting dynamically for the Rydberg–Ritz formula.

In Ch. 4, section 2 of the first edition of *Atombau und Spektrallinien*, Sommerfeld (1919) reviews the current empirical situation with respect to optical series spectra. The phenomenological basis of the series terminology[9]—the association of lines into principal, diffuse, and sharp series—is explained ("structure and multiplicity of the lines, prominence, blurred or sharp, behavior in the Zeeman effect," etc.). As usual in the German literature, following Kayser and Runge, the terminology principal (*Hauptserie*), first (diffuse), and second (sharp) associated series (*I, II Nebenserie*) is used, with the obvious abbreviations H.S., I. N.S., and II. N.S. Sommerfeld then introduces the successive representations of the term values introduced by Rydberg and Ritz, adopting the notation of the latter, where a generic term is written (m, α, β), satisfying Eq. (B.21).[10] The running terms of the sharp, principal, and diffuse series are then distinguished by the notation $(\alpha, \beta) \to (s, \sigma), (p, \pi)$, and (d, δ) respectively. By this time, a new series in the infrared had been recognized, and dubbed the "Bergmann" series (B.S.), with the running term in this series denoted (m, b, β). For simplicity, the Ritz constants $\sigma, \pi, \delta, \beta$ were usually suppressed, and Sommerfeld summarizes the four most well studied series in the following table (p. 234):

$$H.S. \quad v\left(=\frac{1}{\lambda}\right) = (\tfrac{3}{2}, s) - (m, p), \quad m = 2, 3, 4, \ldots$$

$$I.N.S. \quad v\left(=\frac{1}{\lambda}\right) = (2, p) - (m, d), \quad m = 3, 4, 5, \ldots$$

$$II.N.S. \quad v\left(=\frac{1}{\lambda}\right) = (2, p) - (m + \tfrac{1}{2}, s), \quad m = 2, 3, 4, \ldots$$

$$B.S. \quad v\left(=\frac{1}{\lambda}\right) = (3, d) - (m, b), \quad m = 4, 5, 6, \ldots\tag{7.19}$$

[9] See Appendix B.
[10] Sommerfeld uses (m, a, α).

The half-integral values chosen for the sharp terms are here a hangover of the confusion initiated by the discovery of the Pickering lines (see Section 4.6), and the fact that the spectral defect for the lighter alkalis is close to one-half, as explained earlier.[11]

After a review of the term structure from the "bird's eye" point of view, in which the multiplet (doublet or triplet for alkalis and alkaline earths) structure of the lines is ignored, Sommerfeld then explains the extension of the Rydberg–Ritz schema to the "component series" (*Teilserien*) whereby the component lines in a given series are identified by an additional index, e.g., (m, p_i), with $i = 1, 2$ for the doublet series, and $i = 1, 2, 3$ for triplet series. The empirical regularities concerning the splittings within each series (discussed in Appendix B in our review of atomic spectroscopy) are then summarized.

The association of series terms with specific orbits in the Bohr–Sommerfeld theory follows later in the chapter after the Bohr theory, and its elaboration by Sommerfeld to include elliptical orbits, is presented in sections 3–5, where the theory is developed for "hydrogen-like" atoms (atoms with a single electron). Sommerfeld then returns in section 6 to the "hydrogen-unlike" case, where the screening arguments described above (from the November 1916 paper) are reprised and shown to reproduce the Rydberg–Ritz form (with technical details relegated to Appendix 10). The association of the s, p, d, and b terms with angular momentum quantum numbers $n = 1, 2, 3$, and 4 is now possible, as the screening arguments make clear that the running terms in each series must correspond to a fixed angular momentum, and the starting values of the running index m are restricted by the requirement that the associated principal quantum number $n + n_r$ must be greater than or equal to the angular momentum number n.

Later, in Ch. 6 (p. 399), Sommerfeld, following Rubinowicz, derives a selection rule $\Delta n = \pm 1, 0$ for the possible change in the angular momentum quantum number from the requirement of energy and angular momentum conservation during spherical wave emission from the oscillating dipole produced by a radiating electron. This selection rule, further tightened to the more restrictive requirement $\Delta n = \pm 1$ (pp. 401–403, where Sommerfeld accepts the Bohr exclusion of $\Delta n = 0$ transitions, following the correspondence principle arguments of Bohr described in the previous chapter), then gave further support to the assignment of angular momenta to the series terms just given, as the four prominent series in the atomic spectra (again, primarily the elements of the first two groups of the periodic table are of concern here) correspond exactly to transitions $p \to s\,(\Delta n = -1)$, $d \to p\,(\Delta n = -1)$, $s \to p\,(\Delta n = +1)$, and $b \to d\,(\Delta n = -1)$ where this more restrictive selection rule is satisfied.

The treatment of the optical series spectra is considerably revamped and extended in the third edition of *Atombau und Spektrallinien* (Sommerfeld 1922a), where an entire chapter is devoted to a systematic exposition of the empirical data (now augmented by the new information provided by the Franck–Hertz method) and quantum-theoretical background. The discussion of electronic energy levels now becomes immediately

---

[11] By the third edition of *Atombau* (1922), the noisome factor of $\frac{1}{2}$ in the sharp series running terms is gone, reabsorbed in the Rydberg spectral defect s, and the running index written as whole integral for all series.

The annoying problem which immediately surfaced once Sommerfeld's screening theory of 1916 had provided a theoretical basis for the Rydberg–Ritz phenomenology, was that the relativistic corrections, surely present for the non-hydrogenic atoms as well, could not be pressed into service to account for the multiplet structure of the series spectra as well. In fact, the screening effects (resulting in a non-Coulombic effective potential for the valence electron(s)) had already been shown to yield the splitting of s from p, p from d, etc. terms, as is apparent in Fig. 7.1. Relativistic effects would certainly result in further (small!) shifts in these levels, but they had already been "dissected" by screening effects into different energies for different values of the angular momentum n, and the terms themselves would remain simple, just at slightly different energies.

In the case of the X-ray spectra, the inner electron states being probed were for all practical purposes exposed to the largely unscreened (and, for heavy elements, *large*) nuclear charge—screening effects were negligible, while the large nuclear charge amplified the energy splittings due to relativistic effects (by the enormous Z^4 factor, cf. Section 6.2) which could then be examined, uncontaminated, as it were, by inter-electron interactions. Exactly the reverse was true in the non-hydrogenic optical spectra, where the screening effects overwhelmed the relativistic ones, but exposing the presence of additional structure which could not be accounted for in any simple version of the Bohr–Sommerfeld methodology. In fact, by the early 1920s it was perfectly clear that a similarly inexplicable additional structure was also present in the X-ray spectra.

According to the Bohr–Sommerfeld theory, in other words, there should be a *single* energy level corresponding to any given value of the radial and azimuthal quantum numbers n and n_r, and hence only a single spectral line for any transition between two such levels. Instead, there was incontrovertible evidence (and had been for already close to a hundred years, from the time of Fraunhofer) that optical spectra contained series of lines appearing in duplicate or triplicate (and with even higher multiplicities). *This spectral "embarras du richesse" could not be explained in any natural way within the framework of the old quantum theory, as it involved, critically, a new physical ingredient (electron spin) which was unknown to the practitioners of this theory.*

In chapter 6, section 5 of the 3rd edition of *Atombau* (Sommerfeld 1922a, pp. 439–456) Sommerfeld reviews his strategy for arriving at an explanation for the complex structure of optical spectra within the framework of the Bohr–Sommerfeld approach. He begins, as Rydberg had three decades earlier, by reviewing the empirical information available given the most accurate spectroscopic data of the time. The situation described by Sommerfeld can be summarized briefly as follows:

1. The s-terms were known in all cases to be simple: there was a single energy level for each such term in the known spectra of all elements studied. Any multiplicity in spectral lines involving a transition to (e.g., principal series) or from (e.g., sharp series) such a level must therefore be attributed to the other participating term.

2. For alkali metals in the first group of the periodic table (Li, Na, K, etc.), all terms other than s (i.e., p, d, b (later f), etc.) were doublets, consisting of two distinct energy levels: denoted p_i, d_i, etc (with $i = 1, 2$, and $i = 1$ assigned to the higher

energy, i.e., less bound state). For heavier elements the doublet components could be quite well separated. For cesium, for example, the splitting in wavenumber of the lowest p doublet amounted to $\Delta v_p = 554$ cm^{-1} (compared to 0.34 cm^{-1} for lithium), for the lowest d doublet 98 cm^{-1}. This "permanence of multiplicity" for all terms (other than s) would seem to imply more complicated spectral multiplets: for the diffuse series corresponding to $d \rightarrow p$ transitions, for example, one would expect $2 \times 2 = 4$ possible lines. In fact, Rydberg had already detected the appearance of a "satellite" line close to the longer wavelength member of the pair of lines into which the diffuse series resolved (separated by the splitting Δv_p due to the final state pair), which partly accounted for the more "diffuse" appearance of these lines. In 1893 (Ann. d. Phys. 50 (1893) p 625) Rydberg had introduced the terminology "compressed doublet" (*zusammengesetztes Dublett*) for the peculiar structure of the multiplets in the alkali diffuse series. By the early 1920s, it was clear however that the diffuse series lines in the alkali metals could only be resolved into three, and not four, distinct components. Apparently, some new type of selection principle was at work here, distinct from the $\Delta n = \pm 1$ Bohr–Sommerfeld rule which allowed d to p (but not d to s) transitions.

3. For alkaline earth elements in the second group of the periodic table (Be, Mg, Ca, etc.) the terms were either singlets or triplets. To distinguish between these two types, Sommerfeld used the notation S, P, D, etc. for singlet terms, and p_i, d_i, b_i etc. (with the index i taking values 1,2, or 3) for the triplet terms. By convention, as with the doublets, the index values were assigned in decreasing order of energy (or increasing order of binding energy). Again, a careful study of the diffuse series triplet spectra showed only six of the nine possible transitions $d_i \rightarrow p_j, i,j = 1, 2, 3$.

4. In general, odd valence elements exhibited doublets, even valence elements triplets, but the identification of series became increasingly difficult as one progressed to the right in the periodic table, with the patterns of spectra of elements becoming practically "unclassifiable" (*unübersehbar*), in Sommerfeld's characterization, as one approached the rare gases.

5. The splittings of $p, d, \ldots$ multiplets (whether doublet or triplet) decreased rapidly as one increased the principal quantum number (or, equivalently, the radial quantum number) within a given series. The splittings of the lowest multiplets in each series also decreased with the azimuthal quantum number (thus, the lowest d splitting(s) were smaller than the lowest p splitting(s), etc.). On the other hand, the splittings increased as one went to heavier elements in each vertical column of the periodic table.

6. The intensity of the lines decreased as the index i of the initial (upper energy) state increased. Thus, the order of intensities was reversed in going from the principal (p to s) series to the sharp (s to p) series.

The absence of lines in the fully resolved diffuse series spectrum which would otherwise be expected on the basis of the angular momentum selection rule $\Delta n = \pm 1$ provided the clue which Sommerfeld exploited to uncover a previously hidden "inner" quantum number. The motivations for the introduction of this new quantum number had already been laid out two years earlier in a long paper entitled "General spectroscopic laws, in particular a magneto-optical decomposition relation" (Sommerfeld 1920). In this paper Sommerfeld surmises

> The distinguishing characteristic of the various *d*- and *p*-terms [i.e., levels] must rather lie in an inner quantum number, perhaps corresponding to a hidden rotation (p. 231).

This new ingredient is brought even more to the fore in the discussion of multiplet structure in the third edition of *Atombau*. In his typically colorful style, Sommerfeld writes

> The forbidden components of the complete doublets indicate the appropriate point where the fulcrum of the lever should be inserted. They indicate that the (previously introduced) azimuthal quantum number *n* is insufficient. It is equal to 2 for all p-levels, to 3 for all d-levels, and so on, and can therefore not provide a point of distinction between the [individual] p_i and d_j levels. If therefore we wish to exclude the forbidden lines via a selection rule, we must necessarily introduce a new quantum number—we will call this the *inner quantum number*,[13] and denote it by n_i (Sommerfeld 1922a, p. 446).

Sommerfeld then points out that, with the rule that the inner quantum number n_i is set equal to *n* for the highest (least bound) energy level, decreasing by one sequentially for the lower levels in each doublet or triplet of levels, the allowed lines correspond simply

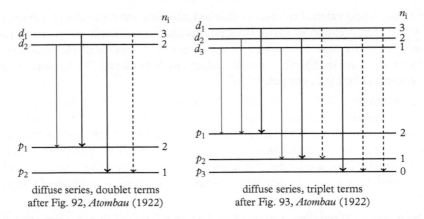

diffuse series, doublet terms
after Fig. 92, *Atombau* (1922)

diffuse series, triplet terms
after Fig. 93, *Atombau* (1922)

Figure 7.2 *Diffuse series transitions for doublet and triplet series. Sommerfeld's inner quantum number n_i is indicated to the right of the levels. Intensity is indicated roughly by thickness of line, forbidden transitions with $\Delta n_i > 1$ are dashed.*

---

[13] By this time, the original azimuthal quantum number *n* (*l* in modern notation) for the orbital angular momentum was now frequently referred to as an "outer" quantum number, associated exclusively with the outer valence electron undergoing a transition.

to transitions in which $\Delta n_i = \pm 1$ or 0, with the lines where n_i changes by one unit being more intense. The situation is summarized in Fig. 7.2, adapted from two of the illustrations in Sommerfeld (1922a). In contrast to the situation with angular momentum, transitions where the inner quantum number remained unchanged were not forbidden, just weaker than those with $\Delta n_i = -1$ (indicated by bold arrows in Fig. 7.2).

7.1.2   Heisenberg's core model and multiplets

A quantitatively successful model-theoretic explanation of the multiplet structure of series spectra—in particular, for the alkali doublets—was provided in late 1921 (and published early in 1922) by an article of Sommerfeld's new doctoral student in Munich, Werner Heisenberg, entitled "On the quantum theory of line structure and the anomalous Zeeman effect" (Heisenberg 1922). The stimulus for this model, as for so much other work in theoretical spectroscopy in the late teens and early 1920s, was the intricate structure of line splittings in the anomalous Zeeman effect, which we will discuss in great detail in Sections 7.2–7.3 (see Section 7.3.3 for a discussion of Heisenberg's core model in this context). A phenomenological description of these splittings employing Sommerfeld's new inner quantum number had already been provided by Alfred Landé (see Plate 24) in a series of articles (Landé 1921a, 1921b).[14] The important ingredient of Landé's work which was taken over by Heisenberg was the former's identification of the inner quantum number as the total (orbital) angular momentum of the atom, not just of the radiating electron. The basic physical picture which Heisenberg proposes can be summarized as follows:

1. The atom can be treated to a good approximation as consisting of a *core* (*Atomrumpf*), composed of the nucleus and some number of tightly bound inner electrons, and a set of outer "series-" or "valence-" electrons, which circulate outside the core. The optical series spectra arise from radiative transitions of the latter.

2. For an alkali atom (Li, Na, etc.), with a single valence electron, the orbital angular momentum of the core and valence electron are not separately conserved: rather there is a continual and rapid exchange of orbital angular momentum between the core and the valence electron. Since in the 1s ground state[15] (*Normalzustand*) the total angular momentum is one (in units of $\hbar$), the *average* orbital angular momentum of the core and the valence electron must add to one. *Heisenberg assumes that, in this ground state, the core and the valence orbital angular momentum are equal on average, i.e., to $\hbar/2$.* This assumption is raised almost immediately to a *quantization postulate*: the core, and consequently the valence electron angular momentum as well, are half-integral quantized. In the ground state, the core and valence electron angular momenta are always aligned, so as to reconstitute the

---

[14] We will later return to these important articles in connection with the Zeeman effect.

[15] Recall that in the spectroscopic notation introduced by Rydberg, the lowest term of the s-series for alkalis is always denoted 1s, so the term number does not coincide with the principal quantum number.

value $(\frac{1}{2} + \frac{1}{2})\hbar$ for the total orbital angular momentum. This situation therefore corresponds to a singlet term.

3. Heisenberg then makes an even more bold proposal: even in cases where the atom is excited to angular momentum states with azimuthal quantum number greater than one (i.e., $n = 2$ for p, $n = 3$ for d, etc.), the core angular momentum remains equal to $\hbar/2$, with the valence electron retaining $n - \frac{1}{2} \equiv n^*$ (times $\hbar$) for the magnitude of its angular momentum. However, in such cases (with $n > 1$), the valence and core angular momentum vectors can be either aligned or anti-aligned, giving a total angular momentum for the atom of either $n - \frac{1}{2} + \frac{1}{2} = n$ or $n - \frac{1}{2} - \frac{1}{2} = n - 1$, just the possible values which Sommerfeld had assigned to the inner quantum number n_i distinguishing the two members of the alkali doublet terms for p, d, f, etc. levels. The motivation for this alignment goes back to the idea of space quantization introduced by Sommerfeld in 1916: the inner magnetic field produced by the core magnetic moment (due to the core angular momentum) picks out a preferred direction with respect to which the valence angular momentum is space quantized, with the difference that the component of angular momentum in this direction is now half-integrally quantized. Similarly, the core angular momentum is space quantized relative to the angular momentum direction of the valence electron, with only two possibilities, alignment or anti-alignment.[16]

4. For divalent atoms in the second column of the periodic table (the alkaline earths, Be, Ca, etc., see Fig. 8.1), the core "borrows" on average a half unit $\hbar/2$ of angular momentum from each of the valence electrons. The two borrowed half-units then produce, if aligned, a net core angular momentum of one (times $\hbar$), and the triplet terms in the spectra of such elements arise from varying orientations of the core relative to valence angular momenta. If the two half-units are antiparallel, the core angular momentum is zero, there is no core angular momentum, and the resulting series consist of singlet terms, sorted as usual according to the orbital angular momentum of the outermost valence electron.

5. The complex multiplet structure of the series terms arises from the energy splitting generated, for different values of the inner quantum number, by the varying orientation of the core magnetic moment (generated by, and parallel to, the net angular momentum of the core) with respect to an "inner magnetic field" H_i

---

[16] Note that in both this case and the preceding case of the ground state $1s$ term, the new half-integral quantum numbers are taken seriously in Bohr–Sommerfeld terms, as altering the allowed quantized orbits for the valence electron. Preserving the energetic interpretation of the series levels (viewed as unsplit terms) in terms of screening, with the energy given at the leading order in terms of the principal quantum number $N = n_r + n$, required that the radial quantum number be correspondingly adjusted, with $n_r \to n_r + \frac{1}{2}$, to keep the principal quantum number fixed. Thus, the allowed eccentricities of the orbits of the valence electrons in the alkali metals are *not* the same as in hydrogen: in particular, circular orbits are excluded. Heisenberg comments on the peculiar feature that anti-alignment does not seem to occur in the $1s$ case, leading to a vanishing total angular momentum of the atom. The justification given for this is purely empirical, given the well established singlet character of the s-terms.

produced by the motion of the outermost valence electron. The orbit of a valence electron is thought of as supplying a steady current loop, generating a magnetic field at the location of the (effectively point-like, at least for smaller atoms) core. This field is calculated purely classically, by applying the Biot–Savart law to an (in general elliptical) current loop, and calculating the magnetic field at the focus of the ellipse.

As Sommerfeld comments in his description of Heisenberg's scheme in the appendix to Chapter 6 of Sommerfeld (1922a), one would hardly take all of this seriously were it not for the fact that it not only reproduces quite accurately, as we shall now see, the observed doublet splittings for the light alkalis, but also accounted *quantitatively* for the entire anomalous Zeeman structure of the alkali doublets, i.e., for weak, intermediate, and strong magnetic fields. We will return to the role of the core model in the Zeeman effect in the following section. For the moment, let us see, at least in the univalent case of the alkali metals, what Heisenberg's model has to say about the multiplet structure of the optical series terms. We begin by recalling some elementary results from classical electromagnetic theory.

The magnetic field generated by a particle of charge e executing a circular orbit of radius a with uniform speed v can be found in any elementary text, as it provides the most straightforward example of the use of the Biot–Savart Law. One finds that far from the loop formed by the orbit, the magnetic field is equivalent to that of a magnetic dipole, with vector magnetic moment $\vec{\mu}$ pointing in the direction normal to the plane of the orbit, and with magnitude equal to the product of the current and the area of the loop, divided by the speed of light (in Gaussian units):

$$\mu = \frac{1}{c}IA = \frac{e}{c} \cdot \frac{v}{2\pi a} \cdot \pi a^2 = \frac{eav}{2c} = \frac{e}{2mc} \cdot mva \equiv g_L L. \tag{7.21}$$

In the final equality, the magnetic moment is seen to be proportional to the orbital angular momentum of the particle $L = mva$: the proportionality constant $g_L = \frac{e}{2mc}$ between this rotational quantity and the magnetic moment is the *gyromagnetic ratio* for orbital motion of a charged particle. From this result, it follows that the orbital angular momentum of $\hbar/2$ assigned to the core in Heisenberg's model brings with it a magnetic moment of magnitude $\mu_R = g_L \hbar/2 = \frac{e\hbar}{4mc}$ (the subscript "R" referring here to "Rumpf", German for "core"). The energetic significance of the magnetic moment derives from the simple relation

$$E_{\text{mag}} = -\vec{\mu} \cdot \vec{H} \tag{7.22}$$

for the magnetic energy of a dipole immersed in a magnetic field $\vec{H}$. By assumption (item 3 above), the core magnetic moment ($\vec{\mu}_R$), parallel to the core angular momentum, is space quantized either parallel or anti-parallel to the inner magnetic field $\vec{H}_i$ due to the outer orbiting valence electron, so the splitting in energy of the two possible configurations is

$$\Delta E = 2\mu_R H_i = \frac{e\hbar}{2mc}H_i, \tag{7.23}$$

or, in terms of a frequency shift

$$\Delta v_0 = \frac{1}{h}\Delta E = \frac{eH_i}{4\pi mc}. \tag{7.24}$$

It remains to calculate the "inner magnetic field" H_i: namely, the magnetic field produced by the current loop formed by the orbiting valence electron at the location of the core, viewed in Heisenberg's model as a point-like object located at a focus of the elliptical path followed by the outer electron.[17] This requires an elementary calculation with the Biot–Savart law, giving

$$H_i = \frac{2\pi I}{ac(1-\epsilon^2)}, \tag{7.25}$$

where I is the current in the loop, a the semi-major axis ($= N^2 a_0$, with N the principal quantum number and $a_0 = \frac{\hbar}{\alpha mc}$ the Bohr radius for hydrogen), ϵ the eccentricity of the orbit, and c the speed of light. The current is given by the charge divided by the period T of the orbit, with the latter determined in terms of the angular momentum (now $n^*\hbar$) as $T = \frac{2\pi ma^2\sqrt{1-\epsilon^2}}{n^*\hbar}$. Using $\sqrt{1-\epsilon^2} = \frac{n^*}{N}$ (cf. Eq. (6.30)), one finally obtains

$$H_i = \frac{e\hbar}{mca_0^3 n^{*2}N^3}. \tag{7.26}$$

Inserting this in Eq. (7.23), one finds for the doublet splitting

$$\Delta E = \frac{1}{2}\alpha^4 mc^2 \frac{1}{n^{*2}N^3} = hc\Delta\left(\frac{1}{\lambda}\right). \tag{7.27}$$

Thus the p ($n^* = 3/2$, $N = 2$) and d ($n^* = 5/2$, $N = 3$) term splittings, expressed in wavenumber shifts $\Delta(\frac{1}{\lambda})$ could be immediately calculated, giving 0.325 cm^{-1} and 0.035 cm^{-1} respectively, close to the experimental observations.

As mentioned above, the immediate motivational context of the Heisenberg core model was the pressing need to bring the complex and perplexing regularities of the anomalous Zeeman effect into a model-theoretic framework compatible with the basic tenets of the Bohr–Sommerfeld theory. We will return to the extension of the core model to deal with externally applied magnetic fields in the next section, where we shall see that the further dynamical assumptions then needed to force the Zeeman phenomenology into the straitjacket of Bohr's quantized orbits in this model required a violation of

---

[17] Given that the radial quantum number of the valence electron was now half-integral, hence nonzero, circular paths were in fact excluded for the valence electron.

mechanical principles which were often "a bridge too far" for many of Heisenberg's theoretical coworkers in atomic theory, despite Sommerfeld's strong initial support. For this reason, the core model must ultimately be considered, despite its initial quantitative success, a failure *at the conceptual level* of the old quantum theory.

The quantitative success of Eq. (7.27), however, is not completely accidental. In modern quantum theory, the doublet splittings in the alkalis are known to arise from a spin-orbit coupling: the magnetic moment arising from the intrinsic spin of the electron interacts with the magnetic field generated in the electron's rest frame from the rest of the atom (the "core") rotating around the electron. The explanation is clearly very similar to Heisenberg's, with the role of the (as yet unknown) electron spin played by the assignment of net angular momentum $\hbar/2$ to the core. The rest of the calculation, assuming in the modern case that the inner electrons completely screen $Z - 1$ of the nuclear charge Z (a good assumption in the case of lithium, where the effective principal quantum numbers for p and d states are almost exactly 2 and 3 respectively), differs from the core model argument above in only three respects:

1. The insertion of the gyromagnetic ratio $g_e = 2$ for the electron, producing twice as much magnetic moment for given spin angular momentum as in the orbital case.
2. The inclusion of the Thomas factor of $\frac{1}{2}$, due to a relativistic kinematic effect.
3. The replacement of $n^{*2} = (l + \frac{1}{2})^2$ in the denominator of Eq. (7.27) by $l(l + 1)$ in the modern theory.

The first two alterations cancel out, of course, and $(l + \frac{1}{2})^2 = l(l + 1) + \frac{1}{4}$, so the effect of the last change is actually quite minor. If one uses $l(l + 1)$ instead of $(l + \frac{1}{2})^2$ for the p state splitting, for example, one would have found precisely 0.365 cm^{-1}, or Sommerfeld's famous $\Delta\nu_H$ from the relativistic fine structure, obtained by a completely different calculation. We recall the result of the fine structure calculation Eq. (6.53), with a fully-screened core ($Z = 1$, principal quantum number $n \to N$):

$$\Delta E_{Nl} = \frac{1}{2}\alpha^4 mc^2 \frac{1}{N^3 l} \tag{7.28}$$

with the splitting given by

$$\Delta E_{Nl} - \Delta E_{Nl+1} = \frac{1}{2}\alpha^4 mc^2 \frac{1}{N^3 l(l + 1)}, \tag{7.29}$$

exactly agreeing with Eq. (7.27), with the replacement $n^{*2} \to l(l + 1)$.

In the second (1921) edition of *Atombau und Spektrallinien*, Sommerfeld had commented on the peculiar closeness of the hydrogen fine structure to the doublet splitting of the lowest p term in lithium:

The size of the (lithium) doublet (splitting) agrees strangely enough quite closely with that of the hydrogen doublet... The approximate agreement between lithium and hydrogen in the size of their doublets would seem to indicate a common origin, as though the lithium doublet were also of relativistic origin. It is more appropriate however to associate lithium with the higher alkalis, sodium and potassium etc, with their much larger doublets, for which the influence of the atomic (electric) field overwhelms the relativistic effects, and to regard the agreement of the lithium and hydrogen doublets as accidental (Sommerfeld 1921, pp. 352–353).

In the third edition of *Atombau*, Sommerfeld is even more emphatic that the two types of splittings have *nothing* to do with each other. In chapter 8, section 4, he states:

> *The fine structure of the hydrogen spectrum does not correspond to the doublet or triplet structures of the non-hydrogenic spectra, but to their separate spectral series.* The doublet structure of the alkali lines or the triplet structure of the alkali earth lines has *no analogy* with hydrogen (Sommerfeld 1922a, p. 603).

In Volume 2 of this work, we shall see that the deciphering of the confusing similarity of relativistic and core-orbit (later spin-orbit) effects would play a critical role in the emergence of the exclusion principle and electron spin. But it would only become completely clear seven years later, with Dirac's relativistic treatment of the bound electron, that the close correspondence of these results is not accidental, but arises from the fact that these apparently unconnected calculations are in fact different representations of the same underlying physics.

7.2   The anomalous Zeeman effect

On September 2, 1896, Pieter Zeeman (see Plate 8), working in his laboratory in Leyden, observed a broadening of the sodium D lines arising from common salt inserted into a flame placed between the poles of a magnet. The idea of studying the effect of a magnetic field on spectral lines was not original to Zeeman: Faraday had tried unsuccessfully decades earlier to detect a magnetic influence on spectra, an effort of which Zeeman was aware through an article on Faraday written by Maxwell (Kox 1997). With the greatly improved gratings and much higher resolution available in the 1890s, the broadening of the spectral lines could clearly be observed, although the fact that the initially sharp lines were actually being *split* was not initially clear. The same broadening was found in both emission and absorption spectra. At the end of October, Zeeman presented his observations formally at a meeting of the Dutch Academy of Sciences, at which Hendrik Lorentz was present. A few days later, Lorentz presented Zeeman with a theoretical explanation of the broadening, based on Lorentz's theory of negatively charged "ion" particles[18] contained in atoms.

---

[18] The term "electron" came into common currency a little later: Lorentz was still referring to "mobile ions" in his article on "The influence of magnetic fields on the emission of light" (Lorentz 1897).

7.2.1 The Lorentz theory of the normal Zeeman effect

Lorentz's classical explanation of the broadening observed by Zeeman can be given here in a few lines, employing modern notation, and the terminology "electron", to lighten the load for a modern reader. Sharp spectral lines of a given (cyclic) frequency $\nu_0 = \omega_0/2\pi$ were assumed to arise when electrons of mass m and charge $-e$ execute harmonic oscillation of that frequency as a consequence of being bound by an isotropic three-dimensional potential $V(r) = \frac{1}{2}m\omega_0^2 r^2$ to the rest of the atom. In the absence of a magnetic field, Newton's second law for such an electron states

$$m\ddot{\vec{r}} = -\nabla V(r) = -m\omega_0^2 \vec{r}. \tag{7.30}$$

The general solution to this harmonic equation can be written (a, b complex, with $a = |a|e^{i\alpha}$, $b = |b|e^{-i\beta}$, c, α, β, γ real)

$$x + iy = ae^{i\omega_0 t} + be^{-i\omega_0 t} = |a|e^{i(\omega_0 t + \alpha)} + |b|e^{-i(\omega_0 t + \beta)}, \tag{7.31}$$

$$z = c\cos(\omega_0 t + \gamma). \tag{7.32}$$

Now suppose a magnetic field $\vec{H} = H\hat{z}$ is applied, introducing an additional Lorentz force on the right-hand side of Eq. (7.30):

$$m\ddot{\vec{r}} = -m\omega_0^2\,\vec{r} - \frac{e}{c}\,\dot{\vec{r}} \times \vec{H}. \tag{7.33}$$

The cross-product in the Lorentz force has no component in the z-direction, so the component Eq. (7.32) corresponding to linear oscillations in the z-direction is unaffected by the magnetic field, and gives rise to linearly polarized (in the z-direction) radiation at the original frequency ν_0, when viewed at right angles to the z axis.[19]

Taking real and imaginary parts of Eq. (7.31), we have that the transverse motion in the absence of the magnetic field takes the form

$$x = |a|\cos(\omega_0 t + \alpha) + |b|\cos(\omega_0 t + \beta), \tag{7.34}$$

$$y = |a|\sin(\omega_0 t + \alpha) - |b|\sin(\omega_0 t + \beta), \tag{7.35}$$

i.e., the superposition of a right-handed circular motion (of amplitude $|a|$) around the z-axis and a left-handed circular motion (of amplitude $|b|$). Once the magnetic field has been switched on, the frequency of these two circular motions will no longer be the same.

---

[19] The radiation emitted by a linear oscillator vanishes in the direction of the oscillation and is maximal orthogonal to this direction. Note that Maxwell's equations give a linear relation between the electric dipole moment $e\vec{r}$ of the oscillating charge and the radiation fields, which are therefore simply linear superpositions of the fields generated by different components of the coordinate vector $\vec{r}$.

To find the modified frequency ω of the right-hand motion, we set $b = 0$ in Eq. (7.31). The equation then reduces to

$$x + iy = ae^{i\omega t}. \tag{7.36}$$

Inserting this equation into the x-component plus i times the y-component of Eq. (7.33),

$$m(\ddot{x} + i\,\ddot{y}) = -m\omega_0^2(x + iy) + i\frac{e}{c}(\dot{x} + i\dot{y})H, \tag{7.37}$$

we find that

$$-m\omega^2 = -m\omega_0^2 - \frac{eH}{c}\omega. \tag{7.38}$$

It follows that

$$\omega^2 = \omega_0^2\left(1 + \frac{eH}{mc}\frac{\omega}{\omega_0^2}\right). \tag{7.39}$$

For a small magnetic field, with $\omega \approx \omega_0$, this amounts to

$$\omega = \omega_0 + \frac{eH}{2mc} \Rightarrow \nu = \nu_0 + \Delta\nu_{\text{norm}}, \quad \Delta\nu_{\text{norm}} \equiv \frac{eH}{4\pi mc}, \tag{7.40}$$

where we have introduced the "normal Zeeman shift" $\Delta\nu_{\text{norm}}$. Similarly, for the left-handed circular component, we find a frequency altered in the opposite direction, $\nu = \nu_0 - \Delta\nu_{\text{norm}}$. The radiation emitted from the right (resp. left) circular component of the motion appears as right (resp. left)-circularly polarized *when viewed along the z-axis*, but as linearly polarized when viewed at right angles to the field. The prediction of Lorentz's ion/electron theory for the effect of a magnetic field on a spectral line (indeed, on *any* spectral line) is therefore simply summarized as follows:

1. When viewed along the axis of the field, the original spectral line should split into two components, symmetrically around the location of the original line, with a frequency splitting proportional to the field. The two components of the split line are oppositely circularly polarized. By the end of November 1896, Zeeman was able to verify the circular polarization of the outer components viewed along the field (accomplished by drilling holes in the magnet poles) as predicted by Lorentz.

2. When viewed at right angles to the field, the line appears as a triplet, with the central line at the original (zero field) location, and the other two symmetrically displaced by the same amount as in the parallel case. All three components are linearly polarized, with the central line polarized along the direction of the field

(the "p", for "parallel polarized" case), the outer components transverse (or "s", for German "senkrecht") to this direction.

From Eq. (7.40), one sees that the measurement of the line broadening also allowed an estimate of the charge-to-mass ration e/m for the radiating charges, a value about one thousand times larger than that typical for electrolytic ions, but in line with the value measured by deflection of cathode rays. All of this was of course occurring more or less simultaneously with the identification of cathode rays as consisting of electrons—distinct particles of definite charge and mass.

7.2.2   Anomalous Zeeman effect: experimental results and pre-Bohr theoretical interpretations

Zeeman's further work on the effect that bears his name was seriously hampered by his move to Amsterdam in early 1897, where the practically ineradicable vibrations present in a laboratory located in a busy city proved a fatal obstacle to recording sharp photographic spectra, even with improved spectroscopic gratings. Leadership in the precision study of magnetically altered spectra passed to Thomas Preston, working at the University of Dublin, and simultaneously to Michelson in Chicago (using interferometric methods), who found that the sodium D_1 and D_2 lines initially studied by Zeeman actually split into a quartet and sextet of lines, respectively. In other cases, such as cadmium, *normal* triplets à la Lorentz were indeed seen, for example, in the sharp series line at 4680 Å, but in another line from the same multiplet (at 4800 Å), a quartet of lines appeared (later resolved into a sextet, with the two outer pairs narrowly split).

It became increasingly clear, as more and more elements were subjected to magnetic torture, that the normal Lorentz triplet was the exception rather the rule in the magnetic splittings of spectral lines. Instead, lines were found to split into doublets, triplets, quartets, etc.—indeed, in some cases, such as mercury, into as many as nine components! Each of these Zeeman patterns was also associated with a characteristic sequence of "p" (for radiation polarized parallel to the magnetic field, when viewed at right angles to the field) and "s" (polarized orthogonal to the field) lines. And even in the case where triplets were found, the separation of the lines was often different (by a peculiar factor of two) from the predicted Lorentzian result. In the early literature, these findings were typically referred to as the "complicated Zeeman types"; later, the terminology "anomalous Zeeman effect" emerged, which we shall henceforth follow.

By 1899 Preston had accumulated a considerable amount of empirical information concerning the Zeeman splittings found in a wide variety of spectral lines arising from a number of different elements. Two general patterns (later known as Preston's laws) emerged clearly from the study of this data, and were stated (without explicit empirical support) in a short article entitled "General law of the phenomena of magnetic perturbations of spectral lines" (Preston 1899). First, the form of the Zeeman splittings of the various lines in a given series multiplet (say, the sharp triplet at 5461, 4358, and 4047 Å in mercury, which split respectively into nine, six, and three lines) is replicated in all the higher members of the same series: indeed this persistence of the splitting pattern became an important aide to the identification of which series various lines should be assigned to in the heavier elements with more complex spectra. Not only the "character" of the splitting (the number and arrangement of split lines), but even the "magnitude", the actual separation in the frequency of split lines, was common to all the members of a Rydberg series (for a given magnitude of the applied magnetic field). Second, the similarity extended to the splitting patterns of homologous series in different elements of the same periodic table group (e.g., between alkali principal series doublets, alkaline earth sharp series triplets, etc.).

The hypothetical regularities proposed by Preston were soon subjected to extensive empirical tests in the work of Runge and Paschen (1900) on mercury and Runge (1902) on a variety of elements with both doublet and triplet series. Although the "Zeeman types" (i.e., the splitting pattern under a magnetic field) of almost all spectral lines did not follow the Lorentz prediction, it was confirmed that the separation of the various components of a magnetically split line, measured in frequency v (or wavenumber $\frac{1}{\lambda}$, but *not* in wavelength) units, was the same for all members of the same spectral series.[20] In his 1902 paper, Runge pointed out that the distinctive Zeeman patterns into which a specific spectral line was "dissected" by a magnetic field allowed one to confirm Rydberg's hypothetical association of the same spectral terms in different spectral series. In particular the reversal of the Zeeman quartet/sextet splittings in the alkali doublets in the sharp versus principal series confirmed Rydberg's hypothesis that the two series were related by an interchange of s and p spectral terms.[21] The three lines in the sharp series triplets were examined in a variety of even valence elements (Mg, Sr, Zn, Cd, Hg) and shown to confirm Preston's second law, by displaying an identical magnetic splitting pattern (Fig. 7.3(a)). The two lines of the sodium principal series doublet (D1, D2 lines) gave the characteristic (for doublets) splitting shown in Fig. 7.3(b). The graphical representations given by Runge, in which the split lines were displayed on the scale of

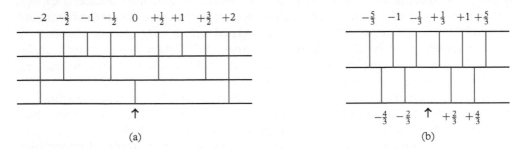

(a)   (b)

Figure 7.3 *(a) Zeeman pattern for sharp series triplet (Runge 1902, Fig. 1). (b) Zeeman pattern for principal series doublet (Na D lines) (Runge 1902, Fig. 2). Transverse (parallel) components in red (blue). Frequency shifts from original line (indicated by arrow) in units of* Δv_{norm}.

---

[20] Here, by "spectral series", we are referring to each "subseries" within a Rydberg series, e.g., the longest wavelength line in each group of three in a triplet sharp series. See Appendix B for a review of spectroscopic terminology.

[21] See Appendix B, Eq. (B.7).

wavenumbers $\frac{1}{\lambda}$, and arranged vertically above one another, immediately indicated that the displacements of the lines were integer multiples of a fixed quantity for each series multiplet. For example for the sharp triplets in Fig. 7.3(a), Runge states:

> If one regards the center of the line as the zero point, then the components of the first type have abscissas $-4, -3, -2, -1, 0, +1, +2, +3, +4$, provided one takes the separation of neighboring lines as unity. For the second line of the triplet the components have the abscissas $-4, -3, -1, +1, +3, +4$ and for the third line, $-4, 0, +4$. There is therefore a sort of law of multiple proportions. A single constant expresses the splitting of all three lines.

Even more remarkably, Runge found a simple relation between the smallest separations for the triplet and doublet types: the separation unit for the triplet was 3/2 of that for the doublet! But in the one case where a triplet of "Lorentzian" appearance appeared (the bottom triplet in Fig. 7.3(a)), the separation of the lines was twice that predicted in Eq. (7.40).

By April 1907, Runge (1907) was able to propose a further property obeyed by the magnetic splittings: the frequency displacements from the central line position (at zero magnetic field) were rational fractions of the normal Zeeman displacement $\Delta \nu_{\mathrm{norm}}$ in Eq. (7.40). Specifically, for each spectral series, one could identify an integer "Runge denominator" r such that the frequency displacements of each magnetically split line were integer multiples of $\Delta \nu_{\mathrm{norm}}/r$, with values $r = 2, 3, 4, 5, 6, 7, 11, 12$ "observed with certainty". Thus, Runge made the bold assertion (clearly contestable, as the rational numbers are dense in the reals) that Zeeman frequency splittings were always a rational fraction of the normal Lorentz splitting $\Delta \nu_{\mathrm{norm}}$. The correct explanation for this remarkable feature, as we shall see, would only be forthcoming with the emergence of electron spin and modern quantum mechanics in 1925–26.

The first serious theoretical attempt to come to terms with the bewildering variety of Zeeman splitting patterns found in multiplet spectra came shortly after Runge's 1907 paper, in September of that year, with the article of Woldemar Voigt (see Plate 19) entitled "Considerations on the more complicated forms of the Zeeman effect" (Voigt 1907). The presence of more than the canonical three lines predicted by the normal Lorentz effect in many (indeed, most) spectral multiplet lines suggested to early workers (e.g., A. A. Robb, whose work Voigt mentions in the introduction to his treatment) the cooperative action of more than one harmonically bound electron in producing the desired multiplicity of split lines. It must also be remembered that at this time, six years before the appearance of the Bohr model, the existence of sharp monochromatic spectral lines, without the simultaneous appearance of higher harmonics (of twice, three times, etc. the frequency), led to the perfectly natural conclusion (essentially unavoidable in a classical context) that the motion of the radiating charges must be simple harmonic in nature. In order to understand the motivations of Voigt's highly influential approach to the Zeeman effect, we must step back a decade or so, and review the electromagnetic view of the constitution of atoms and molecules championed by Lorentz in the last decade of the nineteenth century.

In the electromagnetic theory of matter propounded already in the early 1890s by Lorentz, and forming the basic agreed framework for most attempts to model radiation from atoms at this time, the "ions"[22] responsible for emitting a specific spectral line moved in a potential energy field[23] which (a) allowed for a static equilibrium configuration where the generalized coordinates q_i used to specify the locations of the radiating charges were constant (typically, chosen to be zero), and (b) excitations of the atom or molecule leading to emission of light corresponded to small displacements from equilibrium which resulted, by Taylor expansion around the equilibrium configuration, in a potential energy of harmonic form, $V(q_i) = \frac{1}{2} \sum_i m\omega_i^2 q_i^2$. Lorentz (1897) proposed a model of this type in an attempt to generalize his treatment of the normal Zeeman effect for a single ion/electron, described above, to the more general situation of an atom or molecule with several radiating "ions". The simple harmonic oscillatory motion associated with each generalized coordinate would then be responsible for the emission of a spectral line of frequency $\nu_i = \omega_i/2\pi$.[24] When an external magnetic field H is turned on, the equations of motion for the generalized coordinates q_i of the charge assembly acquire an extra magnetic force term which Lorentz showed would take the form (for the i'th coordinate) $\sum_j c_{ij} \dot{q}_j$, with the coefficients c_{ij} forming an anti-symmetric matrix linear in the magnetic field H, where the anti-symmetry follows from the requirement that the net work done by the magnetic field on the charges must be zero.[25] Lorentz's theory, which was advanced prior to the discovery of the anomalous Zeeman splittings, concerned itself solely with accounting for the normal triplet splitting.

Returning now to 1907, by which time empirical evidence existed for a baffling proliferation of Zeeman splitting patterns, almost all *not* of the normal triplet type, Voigt (1907) introduced a generalization of Lorentz's 1897 procedure which allowed for a more general form of coupling of the radiating charges (now explicitly called "electrons") to each other and to the imposed magnetic field. In contrast to Lorentz, the equations of motion are written directly in terms of the Cartesian coordinates (x_i, y_i, z_i) of the electrons, allowing direct conclusions to be drawn concerning the polarization of the emitted or absorbed light (with the magnetic field typically taken in the z direction). If we suppose that there exist q electrons in the atom subject, in the absence of a magnetic field, to the same harmonic potential $V(\vec{r}) = \frac{1}{2} m\omega_0^2 r^2 \equiv \frac{1}{2} k r^2$, the equation of motion for the set of q electrons is clearly (cf. Eq. (7.30)), separated into x, y, and z components

$$\begin{aligned}
m\ddot{x}_i + kx_i = 0, \quad i = 1, 2, \ldots, q, \\
m\ddot{y}_i + ky_i = 0, \quad i = 1, 2, \ldots, q, \\
m\ddot{z}_i + kz_i = 0, \quad i = 1, 2, \ldots, q.
\end{aligned} \tag{7.41}$$

[22] The term "electron" only came into wider usage after the turn of the century.

[23] The appearance of quasi elastic forces could be justified on the basis of "plum pudding" type atomic models, à la J. J. Thomson, where the positive charge in the atom was dispersed in a homogeneous sphere, for example. Of course, the stability of such a sphere required the presence of non-electromagnetic forces.

[24] Given the number of spectral lines, such models obviously implied the existence of a vast number of electrons in even the smallest atoms.

[25] As the work done on each charge by the magnetic field in time dt is $\frac{e}{c}(\dot{\vec{r}} \times \vec{H}) \cdot (\dot{\vec{r}} dt) = 0$.

If an external homogeneous magnetic field in the z-direction $\vec{H} = H\hat{z}$ is introduced, then the equations of motion are altered by the normal Lorentz force (cf. Eq. (7.33))

$$m\ddot{x}_i + kx_i + g\dot{y}_i = 0, \quad i = 1, 2, \ldots, q, \tag{7.42}$$

$$m\ddot{y}_i + ky_i - g\dot{x}_i = 0, \quad i = 1, 2, \ldots, q, \tag{7.43}$$

$$m\ddot{z}_i + kz_i = 0, \quad i = 1, 2, \ldots, q, \tag{7.44}$$

with $g = eH/c$. It is an important feature of the magnetic coupling that the external magnetic field *does no work* on the electrons (either individually, or in total), which therefore can maintain motion of a fixed amplitude indefinitely in the presence of the field. Voigt generalizes these equations[26] by introducing additional non-canonical terms, linear in the magnetic field, which couple the motion of *different electrons* to each other. The only requirement placed on these couplings (other than being linear in the magnetic field, like the conventional Lorentz force) is that the *net magnetic work on all q electrons* remains zero. Voigt also assumes that the new non-canonical terms do not couple the motion in the field direction (z) to the motion transverse to the field (x, y).

For example, for the transverse x and y components of the motion, responsible in the normal Zeeman effect for the shifted lines, Voigt now writes, instead of Eqs. (7.42)–(7.43),

$$m\ddot{x}_i + kx_i + \sum_j f_{ij}\dot{x}_j + \sum_j g_{ij}\dot{y}_j = 0, \tag{7.45}$$

$$m\ddot{y}_i + ky_i + \sum_j f_{ij}\dot{y}_j - \sum_j g_{ij}\dot{x}_j = 0. \tag{7.46}$$

Here, the diagonal elements of the g_{ij} matrix are just the normal Lorentz coupling,[27] $g_{jj} = g = eH/c$. The off-diagonal elements of the f_{ij} (we shall ignore damping terms here and set $f_{jj} = 0$) and g_{ij} matrices *have no explicit physical interpretation*, as Voigt frankly admits:

[T]he additional (coupling terms), which represent the effects of other electrons and come into being through couplings, are only introduced with the assistance of certain

---

[26] Voigt actually considers the forced harmonic analog of these equations, with right-hand side terms corresponding to imposed oscillating *electric* fields representing an incident light wave, the absorption of which by our harmonically oscillating system is then studied to deduce information about the absorption Zeeman spectrum. Voigt also includes terms linear in the electron velocities, absent the magnetic field, corresponding to damping of the harmonic motion, necessary to include the effect of line widths. We limit ourselves here to the motion of the free undamped electrons, resulting in emission, as the location of the shifted lines is the same in both cases. We will also make some innocent notational changes to render the argument more readily digestible by the modern reader. A much more complete account of Voigt's work on the anomalous Zeeman effect can be found in the web resource *Voigt Theory*, which in turn relies on some of the material presented in the web resource on *Classical Dispersion*.

[27] This association of the diagonal elements of g_{ij} with the canonical Lorentz force law couplings would later be relaxed by Voigt. See the web resource on *Voigt Theory*.

general principles (linear form of the terms, vanishing work of the magnetic field), and are completely devoid of intuitive meaning [*anschaulichen Deutung*] (Voigt 1907, p. 196).

The net magnetic work per unit time (=force times velocity) done on the assembly of q electrons by the additional "f_{ij}" forces in Eq. (7.45) is proportional to

$$\sum_{ij} \dot{x}_i f_{ij} \dot{x}_j \tag{7.47}$$

and vanishes for arbitrary $\dot{x}_i$ if and only if $f_{ij} = -f_{ji}$, i.e., the f_{ij} matrix is anti-symmetric. The net work done by the "g_{ij}" forces is proportional to

$$\sum_{ij} (\dot{x}_i g_{ij} \dot{y}_j - \dot{y}_i g_{ij} \dot{x}_j) = \sum_{ij} g_{ij} (\dot{x}_i \dot{y}_j - \dot{x}_j \dot{y}_i), \tag{7.48}$$

which vanishes provided the g_{ij} matrix is symmetric, $g_{ij} = g_{ji}$.

Eqs. (7.45)–(7.46) can be written more compactly by introducing, as previously in Eq. (7.31), the complex combination $\zeta = x + iy$ for each electron, whereupon we find the single equation (by taking Eq. (7.45) plus i times Eq. (7.46))

$$m\ddot{\zeta}_i + k\zeta_i + \sum_j h_{ij}^T \zeta_j = 0, \tag{7.49}$$

where the matrix $h_{ij}^T \equiv f_{ij} - ig_{ij}$ is anti-hermitian, $(h_{ij}^T)^* = -h_{ji}^T$ (the superscript "T" refers to the "transverse" components) . A completely analogous analysis for the z component of the motion of the electron assembly gives

$$m\ddot{z}_i + kz_i + \sum_j h_{ij}^L z_j = 0, \tag{7.50}$$

where the matrix h_{ij}^L is real and anti-symmetric (the superscript "L" for "longitudinal", i.e., the component in the direction of the magnetic field).

If we now seek monochromatic solutions of a definite angular frequency ω (which would then result in emission of light at that frequency, and hence a sharp spectral line), we may set, taking the longitudinal case first, $z_i(t) = z_{i0} e^{i\omega t}$ in Eq. (7.50), obtaining

$$(\lambda \delta_{ij} - ih_{ij}^L) z_{j0} = 0, \quad \lambda \equiv m \frac{\omega^2 - \omega_0^2}{\omega}, \quad \omega_0 \equiv \sqrt{\frac{k}{m}}. \tag{7.51}$$

The frequency ω_0 is the single frequency of the line, obtained by the unperturbed harmonic motion of all q electrons in the absence of the magnetic field (where $h_{ij}^L = 0$). The elements of the matrix h_{ij}^L are linear in the magnetic field H, which we now assume weak, so that $\Delta\omega \equiv \omega - \omega_0 \ll \omega_0$, and we may approximate

$$\lambda \approx 2m\Delta\omega. \tag{7.52}$$

The Zeeman frequency line shifts (for the longitudinal motion, giving rise to light polarized parallel to the magnetic field, the "p" modes) are thus, up to an unimportant factor of $2m$, simply the eigenvalues of the $q \times q$ complex hermitian matrix $\mathcal{H}^L_{ij} \equiv ih^L_{ij}$. In general, there will be q distinct eigenvalues, and as $\mathcal{H}^L$ is pure imaginary, they appear symmetrically around zero, as one can easily show that if v is an eigenvector with eigenvalue λ, v^* is an eigenvector with eigenvalue $-\lambda$. If q is odd, then the determinant of $\mathcal{H}^L$ vanishes, and there must be a central zero eigenvalue. The upshot is that one expects, in place of the single longitudinally polarized line in the normal Zeeman effect, a multiplet of q lines centered on the zero field line position, with one (at least) occupying this central location if q is odd. Indeed, as one sees in Fig. 7.3 (blue lines), this was exactly the empirical situation encountered in the anomalous Zeeman patterns.

For the transverse polarized ("s") lines, excited by the transverse motion described by the ζ_i coordinates, we have a very similar analysis. Here it is convenient to remove explicitly the "normal" Lorentz coupling term incorporated in the diagonal elements $g_{ii} = g = eH/c$ by defining a hermitian matrix $\mathcal{H}^T$ with zero diagonal matrix elements:

$$\mathcal{H}^T_{ij} \equiv ih^T_{ij} = g_{ij} + if_{ij}, \quad i \neq j, \quad \mathcal{H}^T_{ii} = 0. \tag{7.53}$$

Inserting the harmonic Ansatz $\zeta_i(t) = \zeta_{i0}e^{i\omega t}$ in Eq. (7.49) one then obtains, in analogy to Eq. (7.51),

$$(\lambda\delta_{ij} - \mathcal{H}^T_{ij})\zeta_{j0} = 0, \quad \lambda \equiv m\frac{\omega^2 - \omega_0^2}{\omega} - g, \quad \omega_0 \equiv \sqrt{\frac{k}{m}}. \tag{7.54}$$

Thus, if λ_i are the eigenvalues of $\mathcal{H}^T$, and again resorting to the weak field approximation with now $\lambda \approx 2m\Delta\omega - g$, the frequency shifts for the (right-circularly polarized when viewed along the field direction, transversely polarized when viewed at right angles, as we are considering $\zeta_i = x_i + iy_i$) transverse modes are given by

$$\Delta\omega_i = \frac{g}{2m} + \frac{1}{2m}\lambda_i = \frac{eH}{2mc} + \frac{1}{2m}\lambda_i. \tag{7.55}$$

The single line previously displaced upward in frequency $\Delta\omega = eH/2mc$ (or $eH/4\pi mc$ in cyclic frequency ν) of the normal Zeeman effect is now split into a multiplet of q lines, with locations determined by the eigenvalues of the coupling matrix $\mathcal{H}^T$. A similar set of q lines will appear with a common downward frequency displacement $\Delta\omega = -eH/2mc$ associated with the left-circularly polarized radiation generated by the $\zeta_i^* = x_i - iy_i$ coordinate. The pattern of lines is only symmetric around the (displaced) normal Zeeman position if the g_{ij} (off diagonal) couplings vanish, as then we return to the situation discussed above for the z-modes, with a pure imaginary and anti-symmetric $\mathcal{H}^T$ matrix.

The Voigt model accommodated simply an important feature of Zeeman phenomenology: provided the (non-canonical) coupling coefficients are taken to be the same for the electron sets responsible for spectral lines in the same spectral series, the anomalous Zeeman displacements will be identical (for weak fields) for all lines in a given series, as the expressions for $\Delta\omega$ above no longer depend on the zero field line position ω_0, but only on the eigenvalues of the $\mathcal{H}^L, \mathcal{H}^T$ matrices. Only the harmonic force constants k need to be varied appropriately to fix the zero field frequencies ω_0 for each series multiplet line.

However, an immediate objection which could be made to the theory concerned the *number of spectral lines* predicted once the magnetic field was turned on. For each assembly of q coupled electrons treated à la Voigt, there should be $3q$ lines, assuming the absence of the accidental degeneracies in the eigenvalues of the coupling matrices. And of course, in many cases the number of lines found were not multiples of three (e.g., the sodium D1 line, which split into four components). The matrix elements would be fixed by requiring the splittings to match the empirically observed values, and in general this would not lead to degeneracies (which could then be adduced as a reason for finding less than $3q$ distinct lines).

On the other hand, it might conceivably be the case that the *intensity* of an otherwise possible spectral line vanishes. Now, in the Lorentz electron theory, the intensity of *emission* lines was a matter of unknown initial conditions (i.e., the coordinates $\vec{r}_i$ and velocities $\dot{\vec{r}}_i$ of the electrons at time zero) which then determined the future oscillations, and concomitant radiation, of the electron set through the equations Eqs. (7.45)–(7.46). The intensity of absorption lines, by contrast, was calculable in this theory: one simply had to calculate the motion induced in initially quiescent electrons by an incident monochromatic light wave, with electric field vector $\vec{E} = (X, Y, Z)e^{i\omega t}$. The problem then became the standard one of calculating the response of an assembly of coupled forced harmonic oscillators, with equations of motion, for example for the x_i coordinates, now taking the form (in lieu of Eq. (7.45))

$$m\ddot{x}_i + kx_i + \sum_j f_{ij}\dot{x}_j + \sum_j g_{ij}\dot{y}_j = eXe^{i\omega t}. \tag{7.56}$$

The resulting polarization vector $e\sum_i(x_i, y_i, z_i)$ could now be related, as in the Drude dispersion theory,[28] to the index of refraction, the imaginary part of which would give the structure of the absorption spectrum of the theory (which, in the presence of a magnetic field, was referred to as the "inverse Zeeman effect").

The basic strategy of the Voigt theory, already fully in place in this first paper of 1907, would not, however, lead to a convincing "explanation" of the anomalous Zeeman patterns until further progress had been made on the empirical front—specifically, the

---

[28] See the web resource on *Classical Dispersion*.

discovery of the Paschen–Back effect in 1912. The additional insight provided by this study of the behavior of the Zeeman splittings for strong magnetic fields was crucial in providing Voigt with the missing information needed to complete the description of the anomalous patterns, especially in the case of the alkali doublets, where Voigt's completely classical coupling theory, ignorant not merely of electron spin and angular momentum quantization, but even of the basic ideas of the planetary Bohr model, quite amazingly[29] produced a formula which exactly coincides with the modern result, obtained with full knowledge of electron spin and wave mechanics!

7.2.3 The Paschen–Back transmutation of Zeeman lines

The discovery in 1912 of the transmutation of Zeeman types at "medium to strong fields" (exactly what this means will be explained shortly) by Paschen and his assistant Ernst Back (see Plate 24) was motivated by the appearance of apparent violations of Preston's laws in an increasing number of magnetically perturbed spectral lines. The reader will recall that early investigations of magnetic splittings had established the empirical rule that analogous series in different elements of the same column of the periodic table, e.g., the principal doublet series of the alkali metals, should display the same pattern of splittings (or "Zeeman type", in the language of the time). Instead, when the (very narrow) lithium doublet at 6708 Å (the analog for lithium of the sodium D lines) was subjected to a magnetic field, a normal Lorentz triplet was found, instead of the characteristic quartet/sextet pattern found in sodium (see Fig. 7.3(b)). The same held for the Balmer lines of hydrogen, at this time still not resolved (at zero field) into doublets. A quote from the paper of Paschen and Back illustrates the quandary these results represented:

> As with the other alkalis, one regards (the lithium lines) on the basis of the series laws as very narrow doublets, and expects correspondingly the magnetic type of the principal series lines at 6708 and 3233 to be an overlapping of the D_1 and D_2 types... Instead of this, the lines of lithium are all exactly normal triplets... It follows from this striking behavior of the lithium lines, that the normal triplet does not only appear in the case of a really singlet line, but also when instead one begins with a very narrow doublet. If this theoretically doubled line, e.g., lithium 6708, could be separated by sufficient resolution, then according to our point of view one would expect for very small magnetic fields a splitting of the two doublet components into the types D_1 and D_2. If the end result in a stronger magnetic field is a normal triplet, then we are in the presence of a new effect of magnetic fields on light radiation. The magnetic field must be capable of destroying series doublets and making simple lines out of them (Paschen and Back 1912, p. 902).

This paper of Paschen and Back and a follow-up paper the next year (Paschen and Back 1913) present several examples of this peculiar phenomenon—of the appearance of normal Lorentz triplets where one would expect the more complicated patterns of

---

[29] An explanation for this remarkable and largely, but not completely, accidental result can be found in the final section of the web resource on *Voigt Theory*.

the anomalous Zeeman effect. The important insight provided by their work was the realization that the strength of the magnetic field required to effect the transformation to normal behavior was entirely dependent on the size of the original (zero field) multiplet splitting. Specifically, the transition to normal triplets should be expected once the normal Lorentz magnetic frequency shift $(eh/4\pi mc)H$ was large compared to the initial (i.e., for zero magnetic field) splitting $\Delta\nu_0 = c\Delta(1/\lambda)$ of the multiplet. In other words, once the Zeeman splitting of the neighboring lines of a series multiplet reached the point where the lines began to overlap, the magnetically split lines clearly began to affect each other, and a rearrangement was set in motion which would eventually result in the (typically, considerably more than three) lines coalescing into a normal triplet.

The usual anomalous Zeeman types, which had originally been discovered in series multiplets of heavy atoms with fairly large splittings (at zero magnetic field), required much larger magnetic fields to reach this point. On the other hand, for the multiplets (such as in hydrogen, lithium, or the higher members of the alkali principal series) which were very narrow to start out with, the conventional anomalous Zeeman effect required very high resolving power to see, while even with magnetic fields of moderate size the transformation into normal triplets was well underway.

Paschen and Back explicitly set themselves the task of locating spectral lines where the process of transformation from the standard anomalous types to the large field triplets could be followed all the way through, from weak to strong fields. In fact, with the spectroscopic resolution available in 1912, a convincing demonstration of the weak field anomalous Zeeman types was not really possible for those cases where the magnetic fields necessary to reach the normal triplet configuration were available (up to about 32 000 gauss for Paschen and Back). Two years later, in the work of Norton Kent (1914) at Tübingen, the full evolution of the 6708 lithium line could be followed, confirming quantitatively the suggestions of Paschen and Back.

By this time, Voigt (1913a, 1913b) had extended the coupling theory described previously to the case of doublets, now imposing the requirement that the lines of the weak field Zeeman pattern (see Fig. 7.3(a)) coalesce into a normal triplet in the strong field limit. The obvious interaction of the split lines originating from the original doublet once they began to overlap clearly required a generalization of the original Voigt theory in which electrons with two different zero-field vibrational frequencies ω_{i0}, $i = 1, 2$ were coupled through the magnetic field. As pointed out earlier, in the Voigt theory one naturally obtains a splitting into a number of lines which is a multiple of three. At weak field, the ten observed lines (the D_1 quartet and D_2 sextet) of the doublet pattern required a coupling theory which would initially give at least twelve frequencies, of which two could perhaps be removed by arranging for their (absorption) lines to have zero intensity. The fact that (at zero field) the D_2 lines appear twice as intense as the D_1 lines suggested that twice as many electrons should be present with the natural frequency of the D_2 line as with that of the D_1 line. Moreover, the Zeeman pattern at weak field revealed six transversely polarized lines, which could be accommodated by choosing $q = 3$ transversely oscillating coupled electrons in Eq. (7.49) (giving the three right-circularly polarized lines in the field direction, with the three left-circularly polarized lines arising from the complex conjugate equations). The spring constants k_i of these three electrons were chosen to satisfy $k_1 \neq k_2 = k_3$, with the zero-field line frequencies given by $\omega_{i0} = \sqrt{k_i/m}$ (with $i = 1, 2$ for D_1, D_2).

The four observed parallel polarized lines were interpreted as arising from an original set of six (!) electrons capable of oscillation in the direction of the magnetic field (two with the natural frequency of the D_1 line, four with the natural frequency of the D_2 line), giving six lines, of which two could be arranged to have zero intensity by appropriate choice of the coupling constants. The latter were then further fixed (uniquely) by requiring agreement with the weak field Zeeman pattern shown in Fig. 7.3(b), as well as with the required Paschen–Back limit at large field giving a normal triplet. Voigt's equations were considerably simplified by Sommerfeld in early 1914 (Sommerfeld 1914).[30]

Figure 7.4 shows a comparison of the predictions of the Voigt–Sommerfeld theory with (some of) the actual data points of Kent. The curves represent the theoretical prediction, with red lines corresponding to transversely polarized, blue lines to parallel polarized light (viewed, as usual, at right angles to the magnetic field). The dashed lines represent spectral lines which were found to diminish in intensity at large magnetic fields (presumably vanishing altogether in the infinite field limit). The result, in the extreme large field limit, was a central p line (resulting from the coalescence of two Zeeman lines) surrounded symmetrically by a close doublet of transverse lines on either side, with the center of each doublet corresponding to exactly the normal Lorentz shift of $\Delta \nu_{\mathrm{norm}} = \pm eH/4\pi mc$.

For example, the lowest two components of the D_2 sextet (both transversely polarized) are found to correspond in the coupling theory to frequency shifts (defining the magnetic field dependence in terms of the abscissa coordinate $x = eH/4\pi mc$ in Fig. 7.4, and with the frequencies on the vertical axis measured in a scale where the zero field separation of D_1, D_2 is 3/2) given by the formulas[31]

$$\Delta \nu'_s = -x + \frac{1}{2}, \tag{7.57}$$

$$\Delta \nu''_s = -\frac{1}{4} - \frac{3}{2}x + \frac{x}{2}\sqrt{1 - \frac{1}{x} + \frac{9}{4x^2}} \rightarrow -x - \frac{1}{2} + O\left(\frac{1}{x}\right), \quad x \rightarrow \infty. \tag{7.58}$$

With the units chosen, the lower member of a normal Lorentz triplet would correspond precisely to a frequency shift of $-x = -eH/4\pi mc$, so we see that the two lines in question actually form a close doublet in the large field limit, with a constant frequency difference (with "the two perpendicular components superposed", in the actual observations of Kent for fields exceeding 36 000 gauss). Two parallel polarized lines, one each from D_1

---

[30] By this time, the purely phenomenological character of the Voigt approach, detached from any reasonable model-theoretic interpretation, had become unmistakable. It was certainly a bizarre feature that the Voigt doublet theory required six independent electron motions in the z direction, but only three in the transverse directions! We are also in the post-Bohr-atom era, where the idea of harmonically oscillating electrons in atoms could obviously not be taken seriously. For a detailed description of the Voigt–Sommerfeld coupling theory, see the web resource *Voigt Theory*.

[31] See the web resource *Voigt Theory* for a derivation of these formulas.

Bohr claims (citing Langevin[34]) that the total energy of the system is unaltered by this precession: he is therefore unable to derive a spectral line shift on the basis of the usual Bohr frequency condition, by equating the energy change in a transition to $h\nu$. Instead, he resorts to a correspondence principle argument: at least for the states at high quantum numbers, the frequency of emitted light will correspond to Fourier components of the motion (in the fixed coordinate system), which are obtained by beating the unperturbed Keplerian (cyclic) frequency ν with the Larmor frequency ν_L, thus obtaining radiation of frequency ν, $\nu \pm \nu_L$, exactly the normal triplet. For Bohr, the shifted components (corresponding to the transverse motion, hence giving rise to transverse polarized light viewed at right angles to the field) do not result from a shift in energy of the initial or final stationary states, but are due to a modification of the Bohr frequency condition:

> We must assume that a *magnetic* field exerts an influence on the mechanism of transition between the stationary states, and thereby on the relation between the frequency of the radiation and the amount of energy emitted (Bohr 1914b, p. 519).

In the final section of Bohr's paper, an attempt is made to explain the appearance of doublet lines. Bohr's suggestion for the origin of this structure, as due to the effect of varying configurations of the inner electrons on an outer one, is at first sight intriguing (given the later appearance of core models, as discussed above in Section 7.1). However, Bohr is insistent that the influence of the inner electrons on the outer is purely electrostatic in nature, whereas, as we have seen, the effect is in fact magnetic. In fact, the perturbations discussed by Bohr (leading to an effective non-Coulombian potential seen by the outer electron) are just the effects used later by Sommerfeld to account for the Rydberg–Ritz series structure (cf. Eq. (7.12)), which, as we discussed previously, cannot account for the additional doublet (or triplet) structure found in the optical series of multi-electron atoms.

Despite Sommerfeld's early interest in the place of the Zeeman effect in the new Bohr theory, as expressed in his letter to Bohr in September 1913, he does not explicitly refer to Bohr's treatment in the *Philosophical Magazine* article, either in his 1914 paper simplifying the classical Voigt doublet theory, or in a paper submitted in September 1916 to *Physikalische Zeitschrift*, entitled "On the theory of the Zeeman effect for hydrogen lines, with an appendix on the Stark effect" (Sommerfeld 1916d). In the latter work, a direct attack is made on the Zeeman effect, in the context of the Bohr–Sommerfeld procedure elaborated earlier in the year in his two-part paper in the *Annalen der Physik*. It follows, but is apparently independent of, an analysis by Peter Debye (see Plate 11), presented at a meeting of the Göttingen Academy in June, and submitted for publication in *Physikalische Zeitschrift* (Debye 1916) at roughly the same time as Sommerfeld's *Annalen* paper. Debye's treatment is purely nonrelativistic, and amounts to a straightforward application of the action-angle formalism à la Schwarzschild to the

---

[34] We have not been able to find a statement to this effect in the article by Langevin (1905) quoted by Bohr. In fact, the introduction of a magnetic field definitely changes the energy of an electron bound in an orbit in a central potential, as demonstrated explicitly by Sommerfeld (1916d).

problem of a particle in a central potential subjected to an applied magnetic field. Debye's results agree entirely with those obtained by Sommerfeld, who was apparently quite annoyed at being preempted in this way by his former assistant,[35] in the nonrelativistic case, so we will describe the procedure followed by the latter, which is both simpler and more intuitive.

In Sommerfeld's paper, the concept of space quantization assumes a central importance, and is combined with the Larmor theorem (which Sommerfeld rederives, but does not name explicitly) to produce a theory of normal multiplets (in which multiplets of an odd number of lines with the normal Lorentz spacing appear, but in some cases with more than the canonical three lines).

Sommerfeld's 1916 treatment of the Zeeman effect within the framework of a planetary atomic model includes both a nonrelativistic version, in which the electron kinematics is Newtonian, and a relativistic treatment including relativistic effects. There was clearly a hope that the additional fine structure present in the latter case would open the door to an understanding of the anomalous Zeeman splittings (if not in hydrogen, for which only normal splittings were as yet observed, for more complicated multi-electron atoms).

We begin with the nonrelativistic case, but give a simplified version of the argument (in which we explicitly appeal to Larmor's theorem), following the discussion in the first edition of *Atombau und Spektrallinien* (Sommerfeld 1919, pp. 431–439). We consider the motion of the single electron of a hydrogen atom in the presence of a magnetic field $\vec{H} = H\hat{z}$, referred first to a fixed spherical coordinate system (r, ϑ, χ). The coordinates of the electron in a spherical coordinate system precessing uniformly around the z-axis with angular velocity ω_L are then (r, ϑ, ψ) where $\chi = \psi + \omega_L t$. The (nonrelativistic) kinetic energy of the electron, expressed in the stationary frame

$$K = \frac{1}{2}m(\dot{r}^2 + r^2\dot{\vartheta}^2 + r^2\sin^2(\vartheta)\dot{\chi}^2), \tag{7.59}$$

thus acquires, given $\dot{\chi} = \dot{\psi} + \omega_L$, a first-order shift in the field

$$\Delta K = mr^2\sin^2(\vartheta)\omega_L\dot{\psi} = \omega_L p_\psi, \tag{7.60}$$

where the conjugate momentum to the azimuthal angle in the rotating frame p_ψ is given by

$$p_\psi = \frac{\partial K}{\partial\dot{\psi}} = mr^2\sin^2(\vartheta)\dot{\psi}. \tag{7.61}$$

---

[35] The resulting *contretemps* is described in Eckert (2013a, p. 52).

The imposition of the quantization conditions (see Eqs. (5.49)–(5.51)) then implies $p_\psi = n_1 \frac{h}{2\pi}$, or an energy shift[36]

$$\Delta E = n_1 \frac{h}{2\pi} \omega_L = \frac{eHh}{4\pi mc} n_1. \tag{7.62}$$

In a transition between states with azimuthal quantum numbers n_1 and m_1, one therefore expects a line frequency shift

$$\Delta\nu = \frac{eH}{4\pi mc}(m_1 - n_1) \equiv Z\Delta\nu_{\text{norm}}, \quad Z \equiv m_1 - n_1, \tag{7.63}$$

giving rise to uniformly spaced multiplets, with the spacing given by the normal Lorentz value $\Delta\nu_{\text{norm}} = eH/4\pi mc$.

At this point, Sommerfeld does not yet have at his disposal the selection rules (cf. Section 5.3) which he would later adopt (following the work of Rubinowicz (1918a, 1918b)), and which restrict the possible changes in the azimuthal quantum number $m_1 - n_1$ to zero or ± 1. Accordingly, one is led to lines with $Z = 0, \pm 1$ (a normal triplet) for the H_α Balmer line, $Z = 0, \pm 1, \pm 2$ (a quintet) for the H_β line, etc. Sommerfeld (1916d, p. 495) was admittedly not very happy about this (or, as we shall see, with the results in the relativistic case), but nevertheless insists that the essence of the argument is correct, insofar as establishing the origin of the Zeeman effect in an actual energy shift of the stationary states, and not—as for Bohr (1914b, p. 519), as we saw above— in a violation of the Bohr frequency condition due to some mysterious influence of a magnetic field on the transition from initial to final orbit. In his 1916 paper, after treating the nonrelativistic case, Sommerfeld goes on to examine the effect of a magnetic field on the energy levels of hydrogen, including the effects of relativity, which as we have seen, result in a splitting of the stationary state energies into N levels for principal quantum number N. Certainly, Sommerfeld hoped that the inclusion of relativistic effects could somehow be brought to bear to furnish an explanation of the peculiar anomalous Zeeman splittings which were now practically ubiquitous. The result was however disappointing, as the calculations stubbornly insisted in producing just normal Zeeman splittings superimposed on the fine-structure ones.

By the time of publication of the first edition of *Atombau und Spektrallinien* three years later, Sommerfeld had incorporated and extended the insights of Rubinowicz,[37] who showed on the basis of conservation of energy and angular momentum that certain restrictions, or selection rules, would necessarily apply to any electronic transition which corresponded classically to the emission of electromagnetic radiation in the form of a

---

[36] The radial coordinate, and hence the Coulomb potential energy, is unaffected by the precession. Note also that the quantity that is adiabatically invariant as the magnetic field is slowly switched on is p_ψ, the z-component of angular momentum in the precessing frame, in which the orbit remains fixed, not p_χ, the angular momentum in the fixed frame. This is the quantity which is subjected to integer quantization.

[37] For a detailed account of the work by Rubinowicz (1918a, 1918b), and of the evolution of Sommerfeld's approach to selection rules, see the web resource *Conservation Laws and Selection Rules*.

spherical wave. In particular, in the case of a system subject to space quantization, such as an atom in the presence of an external magnetic field, the change in azimuthal quantum number in any such transition was restricted in magnitude to zero or unity, i.e., in Eq. (7.63)

$$m_1 - n_1 = -1, 0, 1 \Rightarrow \Delta \nu = \pm \frac{eH}{4\pi mc}, 0. \tag{7.64}$$

The "superfluous components" with $Z = \pm 2, \pm 3, \ldots$ found by Sommerfeld in 1916 were thereby eliminated, with the result that the Bohr–Sommerfeld quantum theory was now seen to reproduce precisely the original "normal triplet" prediction of the Lorentz electron theory, but alas showed absolutely no indication of accommodating naturally the anomalous patterns which were the rule rather than the exception in atomic spectroscopy. Nevertheless, in his usual optimistic fashion, Sommerfeld deploys this confirmation of the *normal* Zeeman effect as added support for the validity of the recently derived selection rules:

> In this form [Eq. (7.64)] our quantum-theoretical result is in complete agreement with the result of the Lorentz theory: *we not only have both the lines of the Zeeman triplet displaced to lower and higher frequencies, but also the undisplaced line with $\Delta \nu = 0$.* In contrast to [Eq. (7.64)] our previous result [Eq. (7.63)] led to the expectation of additional components with twice, three times etc. the normal splitting. Insofar as such components are never observed in the normal Zeeman effect, one perceives a convincing confirmation of our selection principle in its application to the magnetic case (Sommerfeld 1919, pp. 436–437).

7.3.2 Empirical regularities and number mysticism (1919–1921)

In the previously cited paper in which he introduces the concept of inner quantum numbers, Sommerfeld (1920) proposes a sort of analog of the Ritz combination principle for Zeeman types, his "magneto-optical decomposition rule" (*magneto-optischer Zerlegungssatz*). The utility of this rule, in analogy to the original combination principle, would be that it would allow for the prediction of the structure of new Zeeman types from observed properties of previously observed multiplet splittings. Sommerfeld started from Runge's observation that the magnitude of the observed Zeeman splittings were in all cases, even for the most complicated Zeeman types, rational fractions q/r of the normal Zeeman splitting $\Delta \nu_{norm}$. Here, the least common denominator r needed to describe all the lines in the split Zeeman pattern of a line was called the "Runge denominator". The numerator q simply took integer values, positive or negative (though not necessarily consecutively). For lines displaying the normal Zeeman effect, one obviously had $r = 1$, leading to splittings which were integer multiples of $\Delta \nu_{norm}$. Sommerfeld assumed, in the spirit of the combination principle, that this rational character of the frequency shifts extended to the individual energy levels corresponding to each transition, i.e., to the individual terms, in spectroscopic language. This follows directly from the Ritz combination principle and the elementary observation that the difference between two rational numbers is itself a rational number. Thus a line arising from a transition from state 1 to state 2 would be shifted in frequency (in units of $\Delta \nu_{norm}$) by

$$\Delta \nu = \Delta \nu_1 - \Delta \nu_2 = \frac{q_1}{r_1} - \frac{q_2}{r_2} = \frac{q_1 r_2 - q_2 r_1}{r_1 r_2} \equiv \frac{q}{r}, \qquad (7.65)$$

where $h\Delta\nu_1, h\Delta\nu_2$ are the energy shifts of states 1 and 2 respectively. By this argument, the Runge denominator r of the *line* corresponding to the transition $1 \rightarrow 2$ satisfied the simple decomposition rule

$$r = r_1 r_2. \qquad (7.66)$$

In other words, it had to contain as factors the Runge denominators of the two terms from which it arose. In particular, singlet lines which were known to display the normal Zeeman effect would necessarily involve terms with $r_1 = r_2 = 1$. This included the s terms of all known spectral series. On the other hand, for spectral series involving a transition from an s term (e.g., the sharp, or second associated series, $s \rightarrow p$) or to an s term (e.g., the principal series, $p \rightarrow s$) the Runge number of the Zeeman type r would coincide with that of the other (non-s) term. This was known to be 3 for doublets involving a p state (e.g., principal or sharp lines of the alkali metals) and 2 for triplets involving a p state (principal or sharp lines of valence two alkaline earths). With the Runge numbers known for the p terms, Sommerfeld could now proceed to the determination of the numbers for d states, as the diffuse (second associated series) lines involved a transition from d to p states. In this way, it was possible to fill out the first three vertical columns, and the entire top row, of a table giving the Runge denominators for all singlet, doublet, or triplet series (see Table 7.1). The numbers for the b, x, and y (modern f, g, and h) series could not as yet be determined empirically, and Sommerfeld simply filled them in by the most obvious extrapolation of the first three terms (s, p, d) in each case. They are indicated in the Table by enclosure in parentheses. After completing the table, Sommerfeld pauses to wax lyrically on the mysterious numerical harmony it displays:

Table 7.1 *Runge denominators (after Sommerfeld 1920, Table 6)*

Multiplet	s	p	d	b (f)	x(g)	y(h)
Singlets	1	1	1	1	1	1
Triplets	1	2	3	(4)	(5)	(6)
Doublets	1	3	5	(7)	(9)	(11)

The number table completed in this way is perhaps the most complete example of that number-harmony, which the new theory of spectra has granted us. At least for the present, it represents a "number mystery". In fact, our table is essentially empirical in origin and just as theoretically incomprehensible as the question of line multiplicities themselves. The only certainty is that this harmony of integers of our Runge denominators must find its explanation in the operation of hidden quantum numbers and quantum relations (Sommerfeld 1920, p. 247).

The Zeeman line splittings observed empirically display very characteristic patterns of displacement and polarization: in each case, the parallel and transverse components follow a particular sequence of Runge numerator integers q (with the denominator r fixed for the entire pattern). If one allows each separate term to take on all possible values (i.e., $q_i = 0, \pm1, \pm2, \ldots$ in Eq. (7.65)), then the appearance of only certain values of q for the frequency differences implies the operation of selection rules. As Sommerfeld points out, these rules must go beyond the known ones for the azimuthal quantum number $\Delta n = \pm1$ and the new inner quantum number $\Delta n_i = 0, \pm1$ (cf. Fig. 7.2). He confesses to being unable to devise a suitable selection principle, capable of "imitating theoretically the empirical splitting patterns" (Sommerfeld 1920, p. 253). The gap pointed out by Sommerfeld, the lack, namely of a set of rules for determining a priori the actual number, location (i.e., Runge numerator), and polarization (parallel or transverse) of the Zeeman pattern for the lines of a given complex multiplet became the focus of an intense competition in the early months of 1921 between Ernst Back, Paschen's assistant in Tübingen, and Sommerfeld's former student Alfred Landé in Frankfurt.[38] Back advanced a new sum rule, restricting the number (though not the position, apart from some obvious symmetries with respect to the unperturbed line location) of parallel and transverse polarized lines which would appear for each split line in a given multiplet. Back (1921) found that the sum of the indices of the initial and final levels, labelled as in Fig. 7.2, plus the number of parallel components in the resulting Zeeman pattern, were constant within each multiplet. In the diffuse series triplet transitions in Fig. 7.2, for example, the $d_3 \to p_1$ transition, with an index sum of 4, was found to split into a pattern with 3 parallel components, giving a Back sum value of 7, while the $d_2 \to p_1$ transition, with an index sum of 3, was found to have 4 parallel components, again giving a total of 7 for the index sum + number of parallel lines. A similar rule was found to hold for the sum of one half the number of transverse components (taking only the right polarized components along the field direction, for example) with the index sum for the transition. It need hardly be pointed out that "number mysteries" of this type gave essentially no clues to the underlying physical phenomenon.

---

[38] Sommerfeld wanted the theorist Landé to hold off on publication until the experimentalist Back had published the results of his research, which he was using for his habilitation thesis. Sommerfeld was worried that otherwise Paschen and his Tübingen colleagues might no longer share their data as freely as they had up to this point. Paschen, however, did not mind Landé publishing before Back (Eckert 2013b, pp. 249–250; see also Forman 1970).

Simultaneously, and independently of the work of Back, Landé was concocting an armory of empirical rules to express the regularities found in the numerous empirically observed Zeeman patterns supplied by the spectroscopists (chief among which was Back himself). The Landé rules go beyond those of Back insofar as they place restrictions on the locations, and not just the number, of the parallel and transverse components in each Zeeman pattern. But, as with the Back sum rule, from a physical point of view, they appear completely artificial and hardly serve as the foundation for a model-theoretic explanation of the Zeeman effect. As an example, for the doublet pattern in Fig. 7.3(b), Landé gives the rule that every p component is accompanied on each side by a pair of equidistant s components, except in the case where two such overlap, in which case they cancel, and no line appears! Thus, the predicted right transverse partner of the D1 parallel line at $-2/3$ should lie at the center of the pattern (as the left partner is at $-4/3$), but is cancelled by a similar s partner of the parallel line at $+2/3$, so there is no unshifted (transverse) line in this case.

Landé, resisting Sommerfeld's coaxing (on behalf of Back) to delay publication until Back's latest results in Tübingen were also ready for submission, prepared and submitted on April 16, 1921 a paper to *Zeitschrift der Physik*. This paper, "On the Anomalous Zeeman Effect (Part 1)" (Landé 1921a), played a seminal role in further theoretical attempts to come to grips with the anomalous Zeeman problem in the old quantum theory. Unlike the previous "number-mystical" attempts of Sommerfeld, Back, and Landé himself, it makes a specific proposal for a new physical interpretation of Sommerfeld's inner quantum number, which is then used immediately to furnish an explanation of the anomalous Zeeman patterns, predicting both the locations and polarizations of the magnetically displaced lines.

Sommerfeld, in his 1920 *Annalen der Physik* paper, had been very cautious in advancing a specific physical interpretation for this new quantum number (suggesting at most, as we saw earlier, that some type of "hidden rotation" might be responsible). By contrast, in his April 1921 paper, Landé advances a quite specific "working hypothesis" for the meaning of the inner quantum number, which almost exactly reverses Sommerfeld's proposal, by transferring to the inner quantum number (now called k, rather than n_i) the association with the total angular momentum of the atom, while the "outer" azimuthal quantum number n remains the orbital angular momentum of the "one jumping outer electron" (Landé 1921a, p. 234) only. The latter quantum number of course still distinguishes the different series terms, $n = 1$ for s-terms, $n = 2$ for p-terms, etc. But now the possibility that "several crossed [i.e., non-coplanar] electron paths contribute to the total angular momentum of the atom" require a generalization of the Rubinowicz–Sommerfeld–Bohr selection rule, which for a single electron required $\Delta n = \pm 1$, but now had to be relaxed to allow, in the case of the total angular momentum k, the possibility $\Delta k = 0$, in addition to $\Delta k = \pm 1$, precisely as Sommerfeld had suggested in his 1920 paper. The structure of the complex multiplets, as we saw earlier, is therefore explained when both selection rules, $\Delta n = \pm 1$ for the outer, and $\Delta k = 0, \pm 1$ for the inner quantum number, are simultaneously applied.

The real importance of Landé's new interpretation of the inner quantum number appeared when a magnetic field is turned on, and the idea of space quantization is applied to the components of the *total angular momentum* in the direction (say, z) of the magnetic field. This component, previously called the "equatorial quantum number" m, now determines the energy shift of the state in the presence of the magnetic field. Moreover, we must have $|m| \leq k$, with the equality holding only if the total angular momentum is perfectly aligned (or anti-aligned) with the magnetic field. Landé now assumed that the emission of a quantum of radiation can only occur for transitions satisfying $\Delta m = 0, \pm 1$ for this "total equatorial" quantum number, taking over (with no clear physical justification) the Rubinowicz–Bohr–Sommerfeld selection rule previously derived for the orbital angular momentum of the radiating electron, but now applied to the component of the total atomic angular momentum (including the contribution of nonradiating electrons) in the direction of the field. The association of parallel polarized lines with transitions satisfying $\Delta m = 0$, and transverse polarized ones with $\Delta m = \pm 1$, was also assumed to carry over to the new situation.

Given the selection rule for m, allowed values for this quantum number would necessarily be spaced by intervals of unity. By symmetry under reflection of the z-direction, allowing for symmetric positive and negative values for m, this means that given a maximum possible value $\overline{m}$ (with $\overline{m} \leq k$), the allowed discrete values under spatial quantization of the z-component of the total angular momentum would be $-\overline{m}, -\overline{m} + 1, \ldots, \overline{m} - 1, \overline{m}$ ($2\overline{m} + 1$ values in all). Evidently, this evenly spaced *and symmetric* discrete series is only obtained by choosing $\overline{m}$ to be either integer, *or half-integer*. For the even valence elements in which singlets and triplets appeared, the maximum value $\overline{m}$ would simply be equal to the (integer) inner quantum number k, as naively expected. But for the odd valence elements with multiplets of even multiplicity, the Zeeman patterns indicated clearly a half-integral value $\overline{m} = k - \frac{1}{2}$, as though the total angular momentum in this case were forbidden from achieving perfect alignment with the magnetic field.

Landé's final assumption, and perhaps the most critical, is that the effective magnetic moment generated per unit of total angular momentum (the gyromagnetic ratio for the total angular momentum) be allowed to differ, by a dimensionless factor g, from the normal classical value $e/2m_e c$ (introducing the notation m_e for the electron mass to avoid potential confusion with m, the magnetic quantum number), corresponding to the orbital angular momentum of a single electron. Thus, the energy shift of the atom with angular momentum component $m\hbar$ along the direction of the field H would now be

$$\Delta E = gmh\frac{e}{2m_e c}H = gmh\Delta\nu_{\text{norm}}, \tag{7.67}$$

with m varying from $-\overline{m}$ to $+\overline{m}$ in steps of unity. The normal Zeeman effect corresponds to $g = 1$, m integer. For the anomalous patterns, as we shall now see, the *Landé g-factor* must be chosen in each case appropriately to yield the observed splittings, and the equatorial quantum number ranges as described above over either a symmetric integer or half-integer sequence. The mysterious Runge numbers evidently entered the theory precisely through the newly introduced g factors.

Remarkably, with these assumptions in place, Landé was able to account for the entire bewildering range of Zeeman types (for *weak fields*) encountered in the more than two decades of spectroscopic effort devoted to the Zeeman effect. And not just for the number of parallel and transverse components, but for their locations (in each case, after suitable choice of the g-factor for the line in question) and polarizations.

Let us illustrate this by examining the Landé treatment of the Zeeman pattern for the archetypal sodium D-lines. Here, the Runge fractions implied a simple formula for the g-factor,

$$g = \frac{2k}{2n-1},$$ (7.68)

where the factor $2n-1$ is simply Sommerfeld's extrapolation for the Runge denominator for doublets which we see from the final line of Table 7.1. Thus, for the lower energy initial p_2 state, with $k(=n_i)=1, n=2$ (see Fig. 7.2), one calculated a g-factor of $g = 2/3$, for the higher energy p_1 state with $k = n = 2$ (and therefore, with $\overline{m} = 3/2$, and magnetic quantum numbers m ranging from $-3/2$ to $+3/2$), a g-factor $g'' = 4/3$, and for the final s state with $k = n = 1$, a g-factor $g' = 2$. The frequency displacements of these terms once a magnetic field was introduced could therefore be indicated on a horizontal scale (as usual, in units of $\Delta\nu_{\text{norm}}$), as indicated in Fig. 7.5. The frequency shift of the actual observed line, as in Eq. (7.65), would be given by subtracting the frequency displacements of the initial and final state. One sees immediately that in the D_2 Zeeman pattern one predicts two parallel polarized lines shifted at $\pm\frac{1}{2}(g'-g'') = \pm 1/3$, arising from the vertical blue arrows in the top figure. From the slanted red lines, one finds two transverse components shifted by $\pm\frac{1}{2}(g'+g'') = \pm 5/3$ (central crossed lines), and two transverse components shifted by $\pm(\frac{3}{2}g'' - \frac{1}{2}g') = \pm 1$. For the D_1 line splittings, there

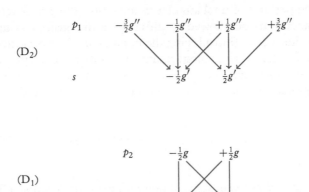

Figure 7.5 *Splittings and transitions in Landé theory of the sodium D line Zeeman pattern. Frequency displacements of terms given in units of $\Delta\nu_{norm}$. Red (resp. blue) arrows denote transversely (resp. parallel) polarized lines.*

are two parallel components with splittings of $\pm\frac{1}{2}(g'-g) = \pm 2/3$, and two transverse components at $\pm\frac{1}{2}(g'+g) = \pm 4/3$. The empirically observed patterns of Fig. 7.3(b) are thus precisely reproduced.

In a second paper submitted in the Fall of 1921, entitled "On the anomalous Zeeman effect (Part 2)," Landé (1921b) tried to reinterpret the anomalous g-factors needed to reproduce the various Zeeman patterns encountered in series spectra in terms of a violation of the Larmor theorem, examining the consequences of assuming either (a) that the Larmor precession frequency ω_L acquired an anomalous correction $\Delta\omega_L$, or (b) an anomalous additional rotation was present in the normally precessing frame, around the direction of the precessing axis. These speculations will not be further discussed here as they would soon be displaced by explicit physical models along the lines of the Heisenberg core model (see Sections 7.1.2 and 7.3.3). However, in sec. 3 of his paper, Landé makes a prescient association of his new rules with certain gyromagnetic phenomena which had recently yielded puzzling results. We recall that the formula Eq. (7.68) gives a Landé g-factor of 2 for s doublet states. Studies of the gyromagnetic ratio of ferromagnetic atoms, where the magnetism was of course still attributed to orbital motion of electrons, had suggested in fact a gyromagnetic ratio of 2, instead of the normal Lorentz value of 1. The two recent experiments in question were those of Einstein and de Haas (1915a, 1915b),[39] in which a suspended magnetizable rod developed a measurable angular momentum when its magnetization is changed by an applied magnetic field, and the experiment of Barnett based on the inverse effect, whereby a diamagnetic object acquires a magnetization when placed in motion.[40] As Landé points out, this doubling effect was a consequence of his formula Eq. (7.68). We now know, of course, that the anomalous value of two for the gyromagnetic ratio of the electron is actually a consequence of relativistic quantum mechanics: both electron spin and relativity, as in the Dirac theory, play a crucial role. As this value is also relevant in determining the anomalous Zeeman doublet splittings, it is not surprising that Landé's theory produces just such a factor.

At the same time (late Fall, 1921) that Landé was preparing the second part of his new systematization of Zeeman phenomenology, Sommerfeld returned to the Voigt theory of the anomalous doublet structure of the alkali D lines. As we saw earlier, in March 1914 he had been able to dramatically simplify the equations of this theory, which were by now (the early 1920s) known to reproduce the empirical data, all the way from weak magnetic fields (Zeeman splittings much smaller than the field-free separation of the doublet lines), through fields of intermediate strength to the extreme strong field Paschen–Back limit in which the normal Zeeman splittings reemerged. The original Voigt theory was based on the old Lorentz electron theory conception of electrons oscillating quasi-elastically within atoms. Now, in a paper entitled "Quantum theoretic Reinterpretation (*Umdeutung*) of

---

[39] For discussion of these experiments, see Galison (1987) and the editorial note, "Einstein on Ampère's molecular currents" (Einstein 1987–2018, Vol. 6, pp. 145–149).

[40] Ironically, the original experiment of Einstein and de Haas had yielded a value for g close to unity, the normal value: later, more accurate experiments by Beck in 1919 showed that the correct value was much closer to two.

the Voigt theory of the anomalous Zeeman effect of the D lines," Sommerfeld (1922b) turned to the problem of extracting the "quantum-theoretic" consequences of the classical Voigt results. In this paper, the *Umdeutung* Sommerfeld refers to involves the translation of the results obtained in an electron oscillation setting for *line frequencies* to the underlying quantized *Bohr energy levels* connected via the Bohr frequency condition to the line frequencies.[41]

The line frequencies v predicted by the Voigt theory of course determine only the difference $E_i - E_f = hv$ of the initial and final electron orbit in a given transition. For both D lines, the final state was an *s* state. The higher frequency D_2 line involved the $2p_1 \rightarrow 1s$ transition, the D_1 line the $2p_2 \rightarrow 1s$ transition. The *s* state, being a singlet (at zero field), was assumed to undergo a normal Zeeman splitting, hence to give rise to two levels with energy displaced by $mh\Delta v_{norm}$, $m = \pm 1$ from the zero field value. The quantum number m introduced here (Sommerfeld 1922b, p. 267) is called the "magnetic quantum number", an appellation which it retains to the present day.[42]

Once the energy levels of the final *s* states were identified, they could be combined with the Voigt line frequencies (using the Bohr frequency condition) to yield the energy levels of the initial $2p_1$ and $2p_2$ states. The resulting levels are shown in Table 7.2, in which the energy shifts of the various states (with the *p* state energies measured relative to the midpoint of the $2p_2$ and $2p_1$ states at zero field) are given in units of $h\Delta v_{norm}$, as a function of the dimensionless ratio v, given by

$$v = \frac{\Delta v_0}{\Delta v_{norm}}, \tag{7.69}$$

with Δv_0 the line frequency splittings of the D-doublet in the absence of a magnetic field.[43]

One notices immediately that Sommerfeld's magnetic quantum number m identifying the separate energy levels after magnetic splitting assumes *integer* values, just twice, in fact, the half-integral values which Landé had assigned them in the first part of his paper on anomalous doublet Zeeman patterns (of which Sommerfeld, of course, was perfectly aware). Sommerfeld at this time, in late 1921, was still constitutionally averse to admitting half-integral quanta into the theory, perhaps as a consequence of the embarrassing earlier misidentification of the Pickering helium lines as hydrogen lines with half-integral quanta. Unfortunately, this insistence on doubling Landé's values, came with a painful

---

[41] The term *Umdeutung*, of course, has acquired a much deeper historical resonance from the title of Heisenberg's groundbreaking 1925 paper, where the reinterpretation in question involves a formal translation of mechanical quantities into algebraic structures which also incorporate information about transitions and line frequencies. For a more detailed account of Sommerfeld's *Umdeutung* in the Zeeman effect, see the web resource on *Voigt Theory*.

[42] Sommerfeld's magnetic quantum number was twice that used by Landé, thus taking integer values, e.g., $2 \times (-\frac{1}{2}, +\frac{1}{2})$ for the singlet state.

[43] The square roots in Table 7.2 should be interpreted as positive in the weak field limit (i.e., for large $v \propto 1/\Delta v_{norm} \propto 1/H$; see Eqs. (7.40) and (7.69)) to ensure that in this limit the four p_1 levels in the third column converge to $\frac{1}{2}\Delta v_0$ and the two p_2 levels in the fourth column converge to $-\frac{1}{2}\Delta v_0$.

Table 7.2 *Magnetic energy shifts of 1s, 2p states, after Sommerfeld (1922b, p. 268), in units of $\Delta \nu_{norm}$. For a derivation of the formulas in this table see the web resource Voigt Theory.*

m	s	p_1	p_2
$+3$	—	$2 + \frac{v}{2}$	—
$+1$	$+1$	$\frac{1}{2}\left(1 + \sqrt{1 + \frac{2}{3}v + v^2}\right)$	$\frac{1}{2}\left(1 - \sqrt{1 + \frac{2}{3}v + v^2}\right)$
-1	-1	$\frac{1}{2}\left(-1 + \sqrt{1 - \frac{2}{3}v + v^2}\right)$	$\frac{1}{2}\left(-1 - \sqrt{1 - \frac{2}{3}v + v^2}\right)$
-3	—	$-2 + \frac{v}{2}$	—

price: Sommerfeld had to modify the selection rules so that transversely polarized lines were now associated with $\Delta m = \pm 2$ transitions, and not, as with Landé, with the usual $\Delta m = \pm 1$ condition. Sommerfeld's aversion to half-integral quantum numbers would however be overcome very shortly, with the (apparent) success of Heisenberg's physical core model, under development at just this period of late 1921, in explaining quantitatively, and for *arbitrary field values*, the doublet Zeeman splitting values.

As a simple example of how Sommerfeld's *Umdeutung* reproduces the earlier Voigt results, let us examine once again the lowest two components of the D_2 sextet, with frequency shifts given in the Voigt theory by Eqs. (7.57)–(7.58). According to Sommerfeld, the frequency shifts, measured from the midpoint of the zero field doublet in units of $\Delta \nu_{norm}$ for the two transitions in question are (see Table 7.2):

$$\Delta \nu_{p_1\,(m=-3) \to s(m=-1)} = \left(-2 + \frac{v}{2} - (-1)\right) \Delta \nu_{norm}$$
$$= \left(-1 + \frac{v}{2}\right) \Delta \nu_{norm}, \tag{7.70}$$

$$\Delta \nu_{p_1\,(m=-1) \to s(m=+1)} = \left(\frac{1}{2}\left(-1 + \sqrt{1 - \frac{2}{3}v + v^2}\right) - (+1)\right) \Delta \nu_{norm}$$
$$= \left(-\frac{3}{2} + \frac{1}{2}\sqrt{1 - \frac{2}{3}v + v^2}\right) \Delta \nu_{norm}. \tag{7.71}$$

In the frequency scale used previously (in Eqs. (7.57)–(7.58), and in Fig. 7.4), we used frequency units on the vertical axis in which $\Delta \nu_0 = 3/2$, and $x = \Delta \nu_{norm}$ to measure the magnetic field on the horizontal axis. Thus, Sommerfeld's $v = \Delta \nu_0 / \Delta_{norm} = 3/2x$. Moreover, we previously measured frequency shifts from the doublet "center of gravity", i.e., setting the zero field location of the D_2 line at $+1/2$ and the D_1 line at -1 (see Fig. 7.4), which is the location of the strong field limit of the normal parallel components. Sommerfeld's frequency shifts are therefore measured from the vertical ordinate $-1/4$ in Fig. 7.4. Setting $\Delta \nu_{norm} = x$ and $v = 3/2x$ in Eqs. (7.70)–(7.71), and adjusting for the vertical shift of $-1/4$, we obtain for the frequency shifts of the lines in question, in the previous notation,

$$\Delta v_s' = \left(-1 + \frac{3}{4x}\right)x - \frac{1}{4} = -x + \frac{1}{2}, \tag{7.72}$$

$$\Delta v_s'' = \left(-\frac{3}{2} + \frac{1}{2}\sqrt{1 - \frac{1}{x} + \frac{9}{4x^2}}\right)x - \frac{1}{4} = -\frac{1}{4} - \frac{3}{2}x + \frac{x}{2}\sqrt{1 - \frac{1}{x} + \frac{9}{4x^2}}, \tag{7.73}$$

exactly as in Eqs. (7.57)–(7.58).

7.3.3   Core models, unmechanical forces, and double-valuedness

We are now ready to return to the paper of Heisenberg (1922) we already encountered in Section 7.1, in which an explicit mechanical model was introduced to account for the multiplet structure in non-hydrogenic atoms. The reader will recall that, in this model, the line splittings in optical series multiplets were explained on the basis of a magnetic interaction between the magnetic moment produced by a net mean angular momentum of the core of the atom (i.e., atom minus valence electrons(s)), and the "inner" magnetic field H_i due to the orbital motion of the valence electron(s) themselves. We now know, of course, that the core angular momentum (taken to be half-integral for the odd valence atoms such as the alkali metals) was a necessary stand-in for the as yet unknown intrinsic spin of the valence electrons themselves, and that the (inert gas-like) core of the neutral atom, once valence electrons are removed, has in fact zero total angular momentum (of both spin and orbital varieties).

Nevertheless, the basic idea of the core model, that multiplet splittings involved a magnetic energy arising from the interaction of the orbital motion of the valence electron with some other source of angular momentum, had a sufficient element of the truth to produce (partially accidentally, as we have seen) a quantitatively almost correct result. Moreover, the success of the core model in reproducing the multiplet splittings, at least for the alkalis, in terms of an internal magnetic interaction, clearly suggested that the further shift in energy levels induced by applying an external magnetic field to the atom should be calculable along very similar lines to those employed in determining the energetic effects of the inner magnetic field H_i (cf. Eq. (7.23)).

In our treatment, we shall restrict our attention to the simpler case of alkali atoms with a single valence electron, and doublet spectra: in any event, the quantitative success of the Heisenberg model in reproducing precisely the detailed quantitative evolution of the doublet Zeeman splittings, all the way from weak to strong fields, was not matched for the more complicated case of even valence atoms, with more than one valence electron.

The situation envisaged by Heisenberg in the presence of an external magnetic field is indicated in Fig. 7.6. The angular momentum of the valence electron is denoted $\vec{L}_V$,

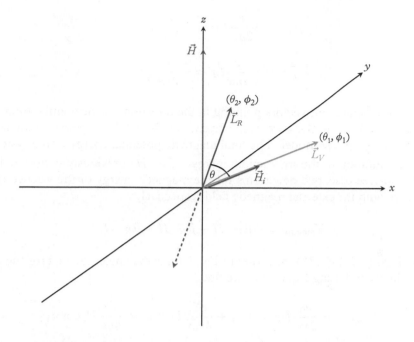

Figure 7.6 *Geometry of Heisenberg core model: $\vec{L}_V$ and $\vec{L}_R$ are the angular momenta of the valence electron and the core, respectively. $\vec{H}_i$ and $\vec{H}$ are the inner and the external magnetic field, respectively.*

parallel to the inner magnetic field $\vec{H}_i$ generated by this valence current, in a direction specified by polar angular coordinates (ϑ_1, φ_1). We recall that the magnitude of the valence angular momentum in the core model is now half-integrally quantized: $L_V = n^*\hbar$. The core angular momentum is denoted $\vec{L}_R$ and points in the direction specified by coordinates (ϑ_2, φ_2).[44] The Sommerfeld–Voigt parameter v is given by (see Eqs. (7.24) and (7.40)):

$$v \equiv \frac{\Delta v_0}{\Delta v_{\text{norm}}} = \left(\frac{eH_i}{4\pi mc}\right) \Big/ \left(\frac{eH}{4\pi mc}\right) = \frac{H_i}{H}, \qquad (7.74)$$

and has a simple physical interpretation as the ratio of the strength of the inner and outer (external) magnetic fields. The valence electron and the core give rise to magnetic moments (the minus signs from the negative electron charges)

---

[44] We have slightly altered Heisenberg's notation to smooth the way for modern readers.

$$\vec{\mu}_V = -\frac{e}{2mc}\vec{L}_V = -\frac{e}{2mc}n^*\hbar\hat{L}_V \qquad (7.75)$$

$$\vec{\mu}_R = -\frac{e}{2mc}\vec{L}_R = -\frac{e}{2mc}\frac{\hbar}{2}\hat{L}_R, \qquad (7.76)$$

where $\hat{L}_V$ and $\hat{L}_R$ are unit vectors pointing in the directions of the angular momenta of the valence electron and the core, respectively. Recall that $\vec{L}_R$ is permanently assigned the magnitude $\frac{1}{2}\hbar$ in the model. The total magnetic potential energy of the system now contains, in addition to the interaction energy $-\vec{\mu}_R \cdot \vec{H}_i$ previously considered in the external field-free case, two new terms for the magnetic energy of the valence electron and the core with the external magnetic field respectively:

$$V_{\text{mag,tot}} = -\vec{\mu}_V \cdot \vec{H} - \vec{\mu}_R \cdot \vec{H} - \vec{\mu}_R \cdot \vec{H}_i. \qquad (7.77)$$

Inserting Eqs. (7.75)–(7.76), referring to Fig. 7.6 for the angles needed for the various inner products, and using Eq. (7.74), we find

$$\begin{aligned}
V_{\text{mag,tot}} &= \frac{e\hbar}{2mc}Hn^*\cos\vartheta_1 + \frac{e\hbar}{4mc}H\cos\vartheta_2 + \frac{e\hbar}{4mc}H_i\cos\vartheta_{12} \\
&= h\Delta\nu_{\text{norm}}\left(n^*\cos\vartheta_1 + \frac{1}{2}\left(\cos\vartheta_2 + v\cos\vartheta_{12}\right)\right),
\end{aligned} \qquad (7.78)$$

where by a well known formula of spherical geometry

$$\cos\vartheta_{12} = \cos\vartheta_1\cos\vartheta_2 + \sin\vartheta_1\sin\vartheta_2\cos(\varphi_1 - \varphi_2). \qquad (7.79)$$

The extension of Heisenberg's core model to the situation in which an external magnetic field is present involves, in addition to the five basic tenets described earlier in Section 7.1.2, three additional assumptions, which we may summarize as follows:

1. The component of the angular momentum $\vec{L}_V$ of the valence electron along the direction of the external magnetic field is space quantized. In other words, in the new situation, the valence electron orbit orients itself discretely relative to the external field, in contrast to the field-free situation discussed previously, where the core angular momentum was space quantized relative to the inner magnetic field produced by the outer valence electron. Thus, the "inner magnetic quantum number" $m^* = n^*\cos\vartheta_1$ is forced to assume values in the discrete range $-n^*, -n^* +1, \ldots, +n^*$ (but see item 2 below).

2. Heisenberg adopts Landé's restriction that the maximum magnitude of the magnetic quantum number m^* cannot exceed the inner quantum number k $(=n_i)$, which both regard as the value of the *total* angular momentum of the atom. For Landé, this followed simply from the interpretation of the magnetic quantum number as the (z) component of the total angular momentum along the direction

of the field. But for Heisenberg, this meant an additional physical constraint (of unclear origin), since m^* was the z-component only of the valence electron's angular momentum. The latter was, for example, 3/2 in both p_1 and p_2 states, but in the case of the p_2 states, where the core and valence angular momentum were anti-aligned, giving total angular momentum 1, the $m^* = +3/2, -3/2$ states were excluded.

3. The orientation of the core angular momentum (of fixed magnitude $\hbar/2$) corresponds to either exact alignment or anti-alignment with the total resultant magnetic field it experiences, $\vec{H} + \vec{H}_i$. Alternatively, as we shall now see, one can simply demand that the net magnetic torque acting on the core magnetic moment vanish, for fixed orientations of $\vec{H}$ and $\vec{L}_V$.

The vanishing of the magnetic torque acting on the core (oriented at polar angles ϑ_2, φ_2, see Fig. 7.6) clearly requires

$$\frac{\partial V_{\text{mag,tot}}}{\partial \varphi_2} \propto \sin\left(\varphi_1 - \varphi_2\right) = 0, \tag{7.80}$$

so we must have $\varphi_2 = \varphi_1, \varphi_1 \pm \pi$ (so $\cos\left(\varphi_1 - \varphi_2\right) = \pm 1$), and the three vectors $\vec{L}_R, \vec{L}_V$, and $\vec{H}$ must lie in the same plane. We may now find the orientation of the core by requiring (see Eqs. (7.78)–(7.79)):

$$\frac{\partial V_{\text{mag,tot}}}{\partial \vartheta_2} \propto -\sin \vartheta_2 + v(-\cos \vartheta_1 \sin \vartheta_2 \pm \sin \vartheta_1 \cos \vartheta_2) = 0. \tag{7.81}$$

It follows that

$$\tan \vartheta_2 = \pm \frac{v \sin \vartheta_1}{1 + v \cos \vartheta_1}. \tag{7.82}$$

This geometrical condition corresponds precisely, as stated above, to the condition that the core angular momentum be either aligned (positive sign: solid red vector in Fig. 7.6) or anti-aligned (negative sign: dashed red vector in Fig. 7.6) with the vector resultant of the inner $\vec{H}_i$ and external $\vec{H}$ magnetic fields.[45]

We recall from our earlier discussion of the field-free case ($H = 0$), that the aligned configuration corresponds to the p_1 state (Sommerfeld's inner quantum number $n_i = 2$,

---

[45] The vector $\vec{H} + \vec{H}_i$ has components $(H_i \sin \vartheta_1 \cos \varphi_1, H_i \sin \vartheta_1 \sin \varphi_1, H + H_i \cos \vartheta_1)$ (see Fig. 7.6). The tangent of the angle this vector makes with the z-axis is

$$\frac{H_i \sin \vartheta_1}{H + H_i \cos \vartheta_1} = \frac{v \sin \vartheta_1}{1 + v \cos \vartheta_1},$$

where in the last step we used Eq. (7.74) to set $H_i = vH$. This result coincides with Eq. (7.82) for $\tan \vartheta_2$ with a plus sign. $\vec{L}_R$ is thus either aligned or anti-aligned with $\vec{H} + \vec{H}_i$.

or, $k = 2$ in Landé's notation), while the anti-aligned configuration holds for the p_2 state, with $n_i = k = 1$. The fact that only the aligned configuration actually corresponds to a *stable* mechanical equilibrium of the system (as can be seen by examining the second derivatives of $V_{\text{mag, tot}}$) is but the first of many peculiarities of the physical picture presented by the model.

We may now proceed to the evaluation of the total magnetic energy in the allowed configurations. The part involving the core is (cf. Eqs. (7.76)–(7.77)), with plus/minus signs for the aligned/anti-aligned configurations[46]

$$
\begin{aligned}
-\vec{\mu}_R \cdot (\vec{H} + \vec{H}_i) &= \pm \frac{e\hbar}{4mc} |\vec{H} + \vec{H}_i| \\
&= \pm \frac{e\hbar}{4mc} \sqrt{H^2 + 2HH_i \cos \vartheta_1 + H_i^2} \\
&= \pm \frac{eH\hbar}{4mc} \sqrt{1 + 2v \cos \vartheta_1 + v^2} \\
&= \pm \frac{1}{2} h \Delta \nu_{\text{norm}} \sqrt{1 + 2v \cos \vartheta_1 + v^2},
\end{aligned}
\tag{7.83}
$$

where we used Eq. (7.74) to set $H_i = vH$ and Eq. (7.40) for $\Delta \nu_{\text{norm}}$. Including the first term in Eq. (7.77) (or Eq. (7.78)) for the energy of interaction of the valence electron with the external field, we obtain for the total magnetic energy of the system (recalling that $n^* \cos \vartheta_1 = m^*$)

$$
V_{\text{mag,tot}} = h \Delta \nu_{\text{norm}} \left(m^* \pm \frac{1}{2} \sqrt{1 + 2 \frac{m^*}{n^*} v + v^2} \right),
\tag{7.84}
$$

with the upper (resp. lower) sign corresponding to inner quantum numbers 2 (resp. 1) for the p states with $n = 2$, $n^* = 3/2$.

With these results in hand, we may easily see that Sommerfeld's quantum-theoretic transcription of the Voigt doublet theory indicated in Table 7.2 is indeed reproduced. First, for the s (singlet) state, with $n = 1$, $n^* = 1/2$, we recall that in the field-free case the model only admits the "locked" orientation in which the valence and core angular momenta are aligned (and both equal to $\hbar/2$). The interaction of the core with the inner magnetic field produced by the valence electron is therefore always the same, so we may ignore the last term (containing v) in Eq. (7.78). The angles with the magnetic field thus satisfy $\vartheta_1 = \vartheta_2 = 0$ or π, leading to $V_{\text{mag, tot}} = \pm h \Delta \nu_{\text{norm}}$, or frequency shifts in units of $\Delta \nu_{\text{norm}}$ of ± 1, as in the second column of Table 7.2. Secondly, for the p_1 state, with $n^* = 3/2$ and inner quantum number $k = 2$, we must evaluate Eq. (7.84) with the four possible values $+3/2, +1/2, -1/2, -3/2$ for m^*, and with the positive sign in front of the

---

[46] The overall minus sign is cancelled by the negative charge of the electron.

square root, giving precisely the contents of the third column in Table 7.2.[47] Similarly, for the p_2 state with $n^* = 3/2$ and inner quantum number $k = 1$, we must take the negative sign in front of the square root, with only $m^* = \pm 1/2$ allowed (see item 2 above), thereby obtaining precisely the quantities listed in the final column of Table 7.2.

We shall shortly examine the multiple physical peculiarities and inconsistencies of this model, but it must be admitted that it incorporates in a very clever way a number of elements which work together to precisely reproduce the phenomenological Voigt doublet theory, by now known to correspond exactly to the empirical facts (cf. Fig. 7.4). As Sommerfeld puts it in his review of the Heisenberg model in the third edition of *Atombau und Spektrallinien,*

> Despite the striking agreement of the theoretical [i.e., core model] and empirical doublet splittings for lithium, one would hardly need to consider our model of doublet atoms convincing, were it not able to include the entire range of doublet Zeeman effects [i.e., all the way from weak fields to the Paschen–Back limit, including the complicated intermediate behavior]. Insofar, however, as it is possible to derive completely the previously only empirically or half-empirically deduced formula [Eq. (7.84)] from our model of the doublet atom, it would seem that our explanation of the field-free doublet has also been completely secured (Sommerfeld 1922a, p. 501).

By the Fall of 1922, Heisenberg's ingenious model had attracted a detailed theoretical response from Landé (1922), with whom Heisenberg had maintained close contact during his investigations. Landé summarized Heisenberg's assumptions as follows:

> The basic assumptions clearly appear to be: (a) half-integral sharing of quanta [*Quantenaufteilung*] between the atomic shell and the atomic core, (b) passivity [*Unwirksamkeit*] of the core in the spatial quantization due to the magnetic field, and (c) orientation of the axis of the core angular momentum along the direction of the resultant of the external and internal field, which cannot be reconciled as such with the general principles previously employed in the combined application of mechanics, electrodynamics and quantum theory, and which have been used in the recent systematic theory of atomic structure of Bohr (1922b) (Landé 1922, p. 353).

The failure of "general principles" (which for Landé is frequently expressed as a failure of Larmor's theorem) in Heisenberg's model can be easily seen as follows. A system with magnetic moment $\vec{\mu}$ in the presence of a magnetic field $\vec{H}$ experiences a torque $\vec{\tau} = \vec{\mu} \times \vec{H}$. The torque gives rise to a changing angular momentum $\vec{L}$, with $d\vec{L}/dt = \vec{\tau}$. For the core, assumption (c) implies that the magnetic moment of the core is always parallel to the net magnetic field it experiences, so we must have $d\vec{L}_R/dt = 0$. In particular, this implies that the direction of the net field $\vec{H} + \vec{H}_i$ is fixed. Except in the special case where all fields and angular momenta are parallel, this requires the direction of $\vec{H}_i$ also

---

[47] As in Table 7.2, the square roots in Eq. (7.84) should be interpreted as positive in the weak field limit (i.e., for large v; cf. note 43).

to be fixed (as its magnitude is fixed by the requirement that H_i reproduce the field-free splittings). But then this implies that the direction of the valence angular momentum $\vec{L}_V$ also be fixed, contradicting the equation of motion for the valence electron, which states (for nonparallel orientations)

$$\frac{d\vec{L}_V}{dt} = \vec{\mu}_V \times \vec{H} \neq 0. \tag{7.85}$$

Landé (1922, p. 362) nevertheless accepts the need for Heisenberg's "unmechanical orientation of the core" parallel to the sum of the external and internal fields, but tries in this paper (at least in the region of weak fields) to get rid of the need for the "passivity" assumption (b) by introducing anomalous precessional and rotational angular velocities. These allow the total angular momentum of the atom (core plus valence electron) to be restored to its central place as the object whose components along the direction of the field need to be spatially quantized, inevitably in the case of doublet spectra, given the empirical facts, in half-integral amounts (assumption(a)). But the "unmechanical" core orientation postulate remained stubbornly essential to account for the observed phenomena, both in the weak field Zeeman as well as the intermediate and strong (Paschen–Back) regimes.

Bohr strongly resisted the notion of half-integral orbital angular momenta for the valence electron that was such an integral part of the Heisenberg core model. His most specific argument against this feature was based on work he did with Dirk Coster in Copenhagen in late 1922 on a quantitative interpretation of X-ray spectra in the context of electron group models of atomic structure (Bohr and Coster 1923). The problem came in the application to X-ray spectra of the Sommerfeld fine-structure formula Eq. (6.53), appropriately generalized to include screening effects,

$$E = -\frac{e^4 m_0}{2\hbar^2 N^2}\left((Z - \gamma)^2 + (Z - \delta)^4 \alpha^2 \left(\frac{1}{Nn} - \frac{3}{4N^2}\right) + O(\alpha^4)\right), \tag{7.86}$$

where we have used N for the principal quantum number (instead of Bohr's n) and n, as previously in this section, for the azimuthal quantum number, instead of Bohr's k. The effects of screening of the nuclear charge by the inner electrons on the relativistic energy of an outer electron characterized by quantum numbers (N, n) are incorporated phenomenologically via the screening constants γ and δ, which were chosen independently for the leading (Balmer) term and the Sommerfeld fine-structure correction term.

The comparison with the empirical X-ray data (especially for the K levels, where the screening approximation should be most accurate) seemed to exclude values of the valence orbital angular momentum n as small as $\frac{1}{2}$, which occur in the Heisenberg core model for s states, and would result in too large a fine-structure correction in the second term in Eq. (7.86) (Bohr 1923, p. 542). In another paper published in *Annalen der Physik* in the Spring of the following year, Bohr (1923b, p. 639, note) points out that such small values of the orbital angular momentum would necessarily lead to collisions with the atomic nucleus in heavier elements. Unfortunately, the empirical agreement obtained in the Heisenberg core model with the observed multiplet splittings (cf. Eq. (7.27)) explicitly depends on accepting the half-integral values n^* in the model for the valence electron's orbital angular momentum.

On the other hand, Bohr had no objection to half-integral values of the *total* angular momentum of the atom, i.e., the vector sum of the core and valence electron angular momenta. Bohr (1923b, sec. 7) proposes an alternative to Heisenberg's model in an attempt to come to terms with the anomalous Zeeman effect. Bohr's argument started with the observation that the statistical weights of the doublet states in alkali atoms—in other words, the number of states of equal energy at some level of approximation (or, in modern terminology, the degeneracy factor)—could not be reconciled with a multiply periodic mechanical treatment of the atomic dynamics. For two weakly coupled multiply periodic systems (e.g., two distant hydrogen atoms) the combined system can be regarded as a new multiply periodic system with a total number of degrees of freedom equal to the sum of the degrees of freedom for the individual systems.[48] Furthermore, the degeneracy for a state of the combined system of given energy $E = E_1 + E_2$ where E_1 and E_2 are the energies of stationary states of the individual systems with degeneracy factors g_1 and g_2, respectively, is just the product $g_1 g_2$ of the degeneracy factors for the individual states.

However, the total degeneracy of the alkali p-states (with $n = 2$) was revealed in the Zeeman effect to be 6 (4 states from the p_1 term, 2 from the p_2 term), 10 for the d states, and in general, $4n - 2$ for the states of orbital angular momentum n. In the limit of very large quantum numbers, the valence electron can be regarded as very distant from, and therefore very weakly coupled to, the atomic core, which for the alkali metals was an inert gas closed shell configuration. Inert gas atoms were known to have no intrinsic magnetic moment (exhibiting diamagnetism rather than paramagnetism), and to be singlet states with statistical weight one. An electron of orbital angular momentum n in a hydrogen atom, on the other hand, had, according to Bohr, a degeneracy factor of $2n$, corresponding to orientations with component $\pm 1, \pm 2, \ldots, \pm n$ along the direction of the field, with the state of zero component excluded by stability arguments, as we saw in our discussion of the Stark effect. If a multiply periodic dynamics was in effect for the description of the system consisting of an inert gas core and a distant electron, one would therefore expect a total degeneracy of $1 \times 2n$, instead of the observed $4n - 2 = 2 \times (2n - 1)$.

This discrepancy was adduced by Bohr (1923b, p. 646) as evidence for a "non-mechanical constraint" [*unmechanischer Zwang*][49] acting between the core of the atom and the outer electron, incompatible with a multiply periodic description of the system. Taking the degeneracy $4n - 2$ seriously, Bohr argued, required that (due to the *Zwang*)

---

[48] By adiabatic variation of the coupling strength, the same result would hold even for strongly coupled systems.

[49] For discussion of this notion, see Serwer (1977).

the atomic core instead of only a single orientation in a constant external field is forced to assume two possible positions, while the outer electron... can only occupy $2n - 1$ possible orientations.

Under these circumstances, it was reasonable to suppose that the usual quantization principles for angular momentum (in terms of integer quanta) might have to be relaxed with regard to both the magnitude of the total angular momentum (j in Bohr's notation) of the atom (core plus electron), as well as for the space quantized components of this momentum along the direction of an external field. In particular, this meant for Bohr an acceptance of Landé's half-integral values for the components m (in units of $\hbar$) of the total angular momentum. In order to reproduce the observed degeneracies, Bohr proposed a value of $j = n + \frac{1}{2}$ for the upper members of the alkali doublets ($p_1, d_1, b(f)_1$ etc) and $j = n - \frac{1}{2}$ for the lower members ($p_2, d_2, \ldots$), with the additional supplementary rule that a perfect parallel (or anti-parallel) alignment of the total angular momentum of the atom with the external magnetic field was *forbidden*.

Thus, for the p_1 term with $j = \frac{5}{2}$, application of a magnetic field yielded four space quantized orientations with component $m = \pm\frac{3}{2}, \pm\frac{1}{2}$, but no states with $m = \pm\frac{5}{2}$. Similarly the p_2 state ($j = 3/2$) splits into two components with $m = \pm\frac{1}{2}$, with the $m = \pm\frac{3}{2}$ aligned configurations disallowed. Thus, the Landé values for the magnetic quantum number, and the observed Zeeman degeneracies are reproduced. Moreover, the singlet character of s-states with $n = 1$ is explained by the exclusion of the $j = 1 - \frac{1}{2} = \frac{1}{2}$ option, as the only half-integral values of m allowed in this case, $\pm\frac{1}{2}$, correspond to the forbidden fully aligned configuration. Instead, singlet states are assigned $j = \frac{3}{2}$, and under a magnetic field split, as with Landé, into two non-field-aligned terms corresponding to $m = \pm\frac{1}{2}$ (again, with the fully aligned $m = \pm\frac{3}{2}$ case excluded). All of this is presented by Bohr in a purely qualitative fashion: there are no explicit calculations, or any attempt to put these proposals on a firmer model-theoretic foundation by introducing and implementing quasi-mechanical postulates à la Heisenberg.[50]

At around the same time in the Spring of 1923 that Bohr was finishing his paper on line spectra, Landé (assisted by the availability of new data of Back's in Tübingen, where Landé was now working) proposed a generalized version of his formula for the anomalous g-factor (see Eq. (7.68)), valid for spectra displaying Zeeman multiplets of arbitrary multiplicity—quadruplets, quintets, etc., in addition to the long known doublets and triplets (Landé 1923a). The new Landé formula was

$$g = 1 + \frac{\mathcal{J}^2 - \frac{1}{4} + R^2 - K^2}{2(\mathcal{J}^2 - \frac{1}{4})} = \frac{3}{2} + \frac{1}{2}\frac{R^2 - K^2}{\mathcal{J}^2 - \frac{1}{4}}. \tag{7.87}$$

---

[50] The peculiar exclusion of aligned configurations can be seen in hindsight as a necessary compensation for the misidentification of $s, p, d, \ldots$ states as having angular momentum 1, 2, 3, \ldots, instead of, as we now know 0, 1, 2, \ldots

Here the quantum number $K = n - \frac{1}{2}$, with n the azimuthal quantum number of the valence electron, agreed with the half-integral value of angular momentum assigned in the Heisenberg model to the valence electron (with a half unit of angular momentum "stolen" by the core).[51] The quantum number R was equal to $\frac{1}{2}$ for singlets, 1 for doublets, $\frac{3}{2}$ for triplets, etc. In particular, for doublets and triplets, it was a half unit larger than the angular momentum assigned to the core (*Rumpf*, hence the letter "*R*") by Heisenberg. The quantum number $\mathcal{J}$ appearing in the denominator was, for odd valence systems (yielding doublets, quadruplets, etc.) just the original Sommerfeld inner quantum number, associated with the total angular momentum of the atom. However, the empirical data showed that the allowed values for $\mathcal{J}$ were not the "natural" values one would expect from combining angular momenta K and R, ranging in integer steps from $|K - R|$ to $|K + R|$, but the values half-way in between these. For example, taking the (p_1, p_2) terms (alkali principal series doublet), with $K = 2 - \frac{1}{2} = \frac{3}{2}$ and $R = 1$, one would expect values $\frac{5}{2}$, $\frac{3}{2}$, and $\frac{1}{2}$ for the vector sum of core and valence angular momentum, whereas the values of $\mathcal{J}$ needed in Eq. (7.87) were 2 and 1 (the original Sommerfeld inner quantum numbers for p_2 and p_1). This was hardly surprising of course given the discrepancy of one half between Landé's core quantum number R and the Heisenberg model value. Nevertheless, the simple and intriguing formula Eq. (7.87) acted as a clear signpost for further model development, as we shall see below.

The final stage of the attempts to come to terms with the anomalous Zeeman effect within the overall conceptual framework of the old quantum theory, while still missing the critical element of electron spin, begins with Pauli's extended stay visiting Bohr's institute in Copenhagen from September 1922 to the Fall of 1923. In addition to numerous discussions with Bohr,[52] Pauli expended considerable intellectual effort on the Zeeman effect. The course of this work can be traced in letters to Landé, as well as in a two-part paper of Landé (1923a, 1923b) and in papers by Pauli (1923, 1924).

By this point, one had more or less given up on the ambitious Heisenberg strategy to reproduce model-theoretically the entire quantitative description of the Voigt–Sommerfeld theory, valid for all values of the magnetic field, from the weak field regime where the essential role is played by the Landé g-factor, through intermediate fields where the term levels are described by more complex formulas (cf. Eq. (7.84)), to the Paschen–Back strong field regime where the normal Zeeman effect reasserts itself. Instead, one adopted the more modest goal of understanding the two limiting cases (weak and strong field) in terms of model assumptions consistent with each other.

The easiest place to start was clearly at the strong field end, where the appearance of the normal Zeeman effect suggested that the usual mechanical conceptions should be on the firmest footing. In particular, the complicating effects of the magnetic coupling of the core to the inner magnetic field H_i produced by the valence electron could be ignored

---

[51] Landé uses the quantum number k for the azimuthal quantum number, hence the choice of K for the valence orbital angular momentum.

[52] Pauli also assisted Bohr in the preparation of various publications, especially the paper on line spectra and atomic structure discussed above (Bohr 1923b), in which the assistance of Pauli and Kramers is explicitly acknowledged.

in this case, and both the core and the valence electron angular momentum would be space quantized with reference to the dominant external field $\vec{H} = H\hat{z}$. In the limit $v = H_i/H \to 0$, the Voigt–Sommerfeld–Heisenberg magnetic energies become, from Table 7.2,

$$V_{\text{mag,tot}} = h\Delta\nu_{\text{norm}}\left(m \pm \frac{1}{2}\right) = h\Delta\nu_{\text{norm}}(m + \mu), \qquad (7.88)$$

where, now following Pauli's (and earlier, Landé's) notation, the magnetic quantum number m (previously m^*) takes half-integral values $\frac{3}{2}, \frac{1}{2}, -\frac{1}{2}, -\frac{3}{2}$ (just one-half the Sommerfeld assignments in Table 7.2), and the quantum number μ can take on the values $(+\frac{1}{2}, -\frac{1}{2})$, so that the total magnetic energy once again involves integer multiples of the normal splitting,[53] as expected in the Paschen–Back limit. In his first paper on the Zeeman problem, Pauli (1923) incorporates the Bohr degeneracy requirements described above by further decomposing the magnetic quantum number m into an integral part m_1, allowed to take on the $2n-1$ values $-(n-1), \ldots, -1, 0, 1, \ldots, (n-1)$, and the two-valued μ (for doublet spectra),

$$m = m_1 + \mu. \qquad (7.89)$$

The total degeneracy was therefore $2(2n-1)$, obtained by precisely the decomposition Bohr had pointed out. Pauli was also able to formulate a remarkable sum rule, *valid for all values of the magnetic field*, including the complicated intermediate region. The rule states that the sum of term energies for a fixed value of m (and of course n) is a *linear function* of the magnetic field, for all magnetic field values. How this comes about in the doublet case is easily seen from the Voigt–Sommerfeld–Heisenberg formula Eq. (7.84), where the sum of the two terms in question eliminates the nonlinear square root term, leaving a result which is a multiple of the normal splitting $\Delta\nu_{\text{norm}}$, which in turn is of course linearly proportional to the magnetic field H.

Pauli refuses in this first paper to advance any specific model-theoretic interpretation of the various quantum numbers he introduces. He is more forthcoming in a letter to Landé (Pauli 1979, Doc. 35), written on 23 May, 1923, a few weeks after the paper was completed. Here he explicitly interprets the quantum number μ as taking values $\mu = +\frac{1}{2}$ (resp. $\mu = -\frac{1}{2}$) accordingly as the core magnetic moment aligns or anti-aligns with the (strong) external field. Pauli, with Bohr, agreed that the total angular momentum j of the atom was half-integral, and that the quantum number m should be interpreted as the component of this quantity along the field (z) direction. The quantum number m_1 was to be interpreted as the component of the orbital angular momentum of the outer electron in the direction of the external field, which, in contrast to the case

---

[53] Namely, $+2, +1, 0$, and -2 for the p_1 states, 0 and -1 for the p_2 state. In the modern theory these values arise from the expression $M_L + 2M_S$, where $M_L = 0, \pm 1$ ($M_S = \pm 1/2$) are the magnetic quantum numbers associated with the orbital (resp. spin) angular momentum of the valence electron.

for the Heisenberg model, would be integral, and, following Bohr's degeneracy factor arguments given above, should be allowed to assume, as posited in the paper, $2n - 1$ values, namely $0, \pm 1, \ldots, \pm(n - 1)$. Thus, the old rule banning $m_1 = 0$ is thrown out the window, and the maximum values $\pm n$ allowed previously in the space quantization of an electron's orbital angular momentum are now disallowed! Since the z-component of the total angular momentum m is just the sum $m_1 + \mu$, Eq. (7.88) may be written (Pauli 1923, p. 160, Eq. (4)),[54]

$$V_{\text{mag,tot}} = h\Delta\nu_{\text{norm}}(m_1 + 2\mu) = h\Delta\nu_{\text{norm}}m_1 + 2h\Delta\nu_{\text{norm}}\mu. \tag{7.90}$$

The observed selection rules in the large field Zeeman effect could be summarized quite simply as the requirements $\Delta m_1 = 0, \pm 1$, precisely the original Rubinowicz–Sommerfeld rule, plus the requirement of core "passivity" in radiative transitions, $\Delta\mu = 0$.

The equation Eq. (7.90) had an obvious interpretation, which Pauli points out in his letter to Landé:

> The energy values for strong fields, given this assignment of the quantum numbers, can be interpreted through the assumption that the magnetic moment of the inner system (i.e., core) is twice as large as would be expected on the basis of its angular momentum according to the classical electron theory of magnetism, while the outer electron generates a magnetic moment which is in agreement with the Biot–Savart Law (Pauli 1979, Doc. 35).

The idea of double magnetism of the core was the key to a renewed attempt to build an explicit physical model for the behavior of the core and valence components of an atom in the presence of an external magnetic field. The aims would be more modest than in the Heisenberg model, which had succeeded, at least for the doublet spectra, in producing a complete, and empirically successful, theory of the line splittings for all values of the magnetic field. Pauli instead contented himself with an analysis of the weak field regime only, but retaining as input the single new hypothesis that the gyromagnetic ratio (magnetic moment per unit angular momentum) for the core would be twice the classical Biot–Savart value. Whereas Heisenberg had assumed conventional gyromagnetic ratios both for the core and the valence electron, but introduced mechanical hypotheses which violated the classical dynamics of magnetic moments subjected to imposed magnetic fields, Pauli's model—later dubbed the "Pauli–Landé vector model"—would retain the expected dynamics of magnets immersed in magnetic fields.

The result given by this model for the g-factor was stated (without proof) in Pauli's May 23 letter to Landé, and the model itself was explored in considerable detail in subsequent papers of the latter. Here we will explain the basic idea, retaining for simplicity the notation used previously in our discussion of the Heisenberg model. Thus, we use $\vec{L}_V$ (instead of $\vec{K}$) for the angular momentum of the valence electron, $\vec{L}_R$ (instead

---

[54] With m_1 taking the values $+1, 0, -1$, and $\mu = \pm\frac{1}{2}$, one thus finds the previously obtained $v = 0$ limit values $+ 2,1,0,0,-1,-2$.

of $\vec{R}$) for the core angular momentum, and introduce in addition $\vec{\mathcal{J}} = \vec{L}_V + \vec{L}_R$ for the total angular momentum of the atom. In place of Eqs. (7.75)–(7.76), we now have for the magnetic moments of the valence electron and core, incorporating Pauli's double magnetism of the latter,

$$\vec{\mu}_V = -\frac{e}{2mc}\vec{L}_V, \tag{7.91}$$

$$\vec{\mu}_R = -\frac{e}{mc}\vec{L}_R. \tag{7.92}$$

On the other hand, in contrast to the Heisenberg model, the magneto-mechanical equations are assumed to hold exactly as in classical theory. The torque exerted on the entire atom by the external magnetic field $\vec{H}$ is equated to the rate of change of the total angular momentum $\vec{\mathcal{J}}$,

$$\vec{\tau}_{\text{atom}} = (\vec{\mu}_V + \vec{\mu}_R) \times \vec{H} = \frac{d\vec{\mathcal{J}}}{dt}, \tag{7.93}$$

while the torque exerted on the core by the external and internal magnetic field gives the rate of change of the angular momentum of the core,

$$\vec{\tau}_{\text{core}} = \vec{\mu}_R \times (\vec{H} + \vec{H}_i) = \frac{d\vec{L}_R}{dt}. \tag{7.94}$$

As we are interested in the g-factor we may assume throughout that the applied magnetic field H is weak, i.e., much smaller than the inner magnetic field H_i produced by the valence electron at the core. Thus, the rate of change of $\vec{\mathcal{J}}$ in Eq. (7.93) is much smaller than the rate of change of $\vec{L}_R$ in Eq. (7.94). The core angular momentum therefore precesses (neglecting the external field to a first approximation) according to

$$\frac{d\vec{L}_R}{dt} = -\frac{e}{mc}\vec{L}_R \times \vec{H}_i. \tag{7.95}$$

The cross-product on the right-hand side is proportional to

$$\vec{L}_R \times \vec{L}_V = \vec{L}_R \times (\vec{L}_V + \vec{L}_R) = \vec{L}_R \times \vec{\mathcal{J}}, \tag{7.96}$$

where the vector $\vec{\mathcal{J}}$ is approximately constant due to the weakness of the external field. The proportionality constant involves the large inner magnetic field H_i so the core angular momentum $\vec{L}_R$ precesses rapidly around $\vec{\mathcal{J}}$ (see the red dashed loop in Fig. 7.7), while the total angular momentum vector $\vec{\mathcal{J}}$ itself precesses slowly (according to Eq. (7.93)) around the external field (i.e., the z-axis). The rapid precession of $\vec{L}_R$ implies that, on the timescale of the precession of $\vec{\mathcal{J}}$, we may replace the core angular momentum $\vec{L}_R$ by its time average $\langle \vec{L}_R \rangle$, which is just the projection of $\vec{L}_R$ along the direction of $\vec{\mathcal{J}}$:

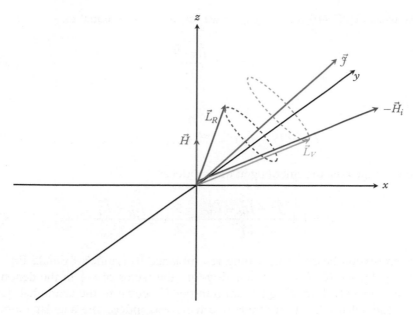

Figure 7.7 *Geometry of Pauli–Landé vector model.*

$$\langle \vec{L}_R \rangle = \frac{\vec{L}_R \cdot \vec{\mathcal{J}}}{\vec{\mathcal{J}}^2}\vec{\mathcal{J}}. \tag{7.97}$$

Using Eqs. (7.91)–(7.92), we can write the portion of the total magnetic energy due to the external field (cf. Eq. (7.77)) as

$$V_{\mathrm{mag,ext}} = -\vec{\mu}_V \cdot \vec{H} - \vec{\mu}_R \cdot \vec{H} = \frac{e}{2mc}(\vec{L}_V + 2\vec{L}_R) \cdot \vec{H} = \frac{e}{2mc}(\vec{\mathcal{J}} + \vec{L}_R) \cdot \vec{H}. \tag{7.98}$$

Replacing $\vec{L}_R$ by its time average and using Eq. (7.97), we find

$$\langle V_{\mathrm{mag,ext}} \rangle = \frac{e}{2mc}(\vec{\mathcal{J}} + \langle \vec{L}_R \rangle) \cdot \vec{H} = \frac{e}{2mc}\left(1 + \frac{\vec{L}_R \cdot \vec{\mathcal{J}}}{\vec{\mathcal{J}}^2}\right)\vec{\mathcal{J}} \cdot \vec{H}. \tag{7.99}$$

Space quantization of the total angular momentum $\vec{\mathcal{J}}$ along the direction of the external field implies $\vec{\mathcal{J}} \cdot \vec{H} = M\hbar H$, with M the magnetic quantum number labelling the split energy levels in a weak magnetic field. We can thus write the average total magnetic energy in the form

$$\langle V_{\mathrm{mag,ext}} \rangle = g\frac{e\hbar H}{2mc}M = g(h\Delta \nu_{\mathrm{norm}})M \tag{7.100}$$

(where we used Eq. (7.40) for $\Delta\nu_{\text{norm}}$), if we set the g-factor equal to:

$$g = 1 + \frac{\vec{L}_R \cdot \vec{\jmath}}{\vec{\jmath}^2}.$$

(7.101)

Using that

$$\vec{L}_V^2 = (\vec{\jmath} - \vec{L}_R)^2 = \vec{\jmath}^2 + \vec{L}_R^2 - 2\vec{L}_R \cdot \vec{\jmath},$$

(7.102)

we can rewrite the g-factor emerging in this model as

$$g = 1 + \frac{\vec{\jmath}^2 + \vec{L}_R^2 - \vec{L}_V^2}{2\vec{\jmath}^2} = \frac{3}{2} + \frac{1}{2}\frac{\vec{L}_R^2 - \vec{L}_V^2}{\vec{\jmath}^2}.$$

(7.103)

This last expression bears an intriguing resemblance to Landé's formula Eq. (7.87), if we identify $\vec{L}_R^2 \rightarrow R^2, \vec{L}_V^2 \rightarrow K^2$, and ignore the factor of $-\frac{1}{4}$ in the denominator of the Landé formula. Landé's quantum number K (equal to the azimuthal quantum number of the valence electron minus one half) was indeed the angular momentum assigned to the outer electron in the Heisenberg model, but as we saw earlier, his quantum number R disagreed with the core angular momentum required in that model. The most disagreeable feature of the Pauli result was, however, the noisome $-\frac{1}{4}$ needed in the denominator of the formula. The huge amount of empirical data on Runge denominators accumulated up to this point showed quite definitively that what was needed in the denominator of any putative formula for the g-factor necessarily involved, not the square of an (integrally or half-integrally) quantized angular momentum, but that quantity displaced by the mysterious factor of $-\frac{1}{4}$. With the Pauli–Landé vector model we have reached in some sense the point of closest approach of the old quantum theory to the "correct" treatment of the anomalous Zeeman effect, which would follow it within three years, with the introduction of the concept of electron spin and the new quantum kinematics of matrix and wave mechanics. The formula furnished by the new theory is in fact structurally identical to Eq. (7.103), with the core angular momentum $\vec{L}_R$ replaced by the total spin vector of the outer shell electrons $\vec{S}$, the valence electron angular momentum $\vec{L}_V$ by the total orbital angular momentum vector $\vec{L}$, and, as in the old theory, the total angular momentum of the shell electrons $\vec{\jmath}$ as the sum of these. Pauli's "double magnetism of the core" is replaced by the gyromagnetic ratio $g_s = 2$ (to high accuracy) of intrinsic spin, which would not be explained until Dirac's relativistic electron theory appeared on the scene, but plays an essentially identical role as in the argument just given. The final "corrective" needed to obtain the modern formula is the realization, quite impossible within the terms of the old quantum theory, that the squared angular momenta take values which differ from the square of the maximum component in any given direction: thus, $\vec{\jmath}^2 \rightarrow \jmath(\jmath + 1)\hbar^2$, with $\jmath$ the (integral or half-integral) total angular momentum quantum number, which turns out to be just Landé's $\jmath$, minus one half (whence $\jmath(\jmath+1) \rightarrow \vec{\jmath}^2 - \frac{1}{4}$, giving the peculiar factor of $-\frac{1}{4}$ which so puzzled Pauli).

Similarly, for the spin and orbital squared momenta, one sets $\vec{S}^2 = S(S+1)\hbar^2, \vec{L}^2 = L(L+1)\hbar^2$. The post-1925 formula for the Landé g-factor thus takes the form

$$g = 1 + \frac{\vec{\mathcal{J}}^2 + \vec{S}^2 - \vec{L}^2}{2\vec{\mathcal{J}}^2} = 1 + \frac{\mathcal{J}(\mathcal{J}+1) + S(S+1) - L(L+1)}{2\mathcal{J}(\mathcal{J}+1)}. \tag{7.104}$$

In fact, one can find exactly the vector-model derivation of the g-factor, with figures essentially identical to Fig. 7.7, in early post-1925 quantum texts such as Herzberg's 1937 monograph (incorporating of course the post-1925 understanding that squared angular momentum took the form $\mathcal{J}(\mathcal{J}+1)$ etc.) (Herzberg 1945, pp. 109–111). But the crucial missing physical ingredient—the intrinsic spin of the electron—necessarily condemned all attempts of the old quantum theory to come to grips with the Zeeman effect to ultimate failure.

The frustration caused by the inability to come to terms with the Zeeman effect in the old quantum theory is illustrated by the following oft-repeated anecdote. As Pauli recalled in an article based on a speech given at a dinner at the Institute for Advanced Study in Princeton to celebrate his 1946 Nobel prize for the discovery of the exclusion principle:

> A colleague who met me strolling rather aimlessly in the beautiful streets of Copenhagen [in the early 1920s] said to me in a friendly manner: "You look very unhappy," whereupon I answered fiercely, "How can one look happy when he is thinking of the anomalous Zeeman effect" (Pauli 1946, p. 214; quoted, for instance, in Mehra and Rechenberg 1982a, p. 494).

7.4　The problem of helium

The question of how to generalize the quantization procedure which had led to such stunning success in the case of the hydrogen atom to atoms with more than one electron was addressed head on by Bohr (1913c), in the second part of his famous trilogy. In this paper it was assumed that in the normal (ground) state of the atom the electrons occupy a single coaxial ring, are arranged at equal angular intervals, and rotate in the same sense around the nucleus. Moreover, each electron is posited to carry exactly one unit of angular momentum $\hbar = h/2\pi$, as in the hydrogen atom. For the case of neutral helium, the simplest atom with more than one electron, one therefore arrives at the simple picture indicated in Fig. 7.8(a). The electrons rotate on a circle (the boundary of the shaded disc in the figure) of radius a with speed v, occupying opposite locations on the circle and with opposite velocities.

The total energy of such a configuration is easily calculated. Each electron experiences an electrostatic attraction $2e^2/a^2$ towards the nucleus and an electrostatic repulsion $e^2/4a^2$ in the opposite direction from the other electron, giving a total centripetal force of $7e^2/4a^2$. Thus, from Newton's second law, we have

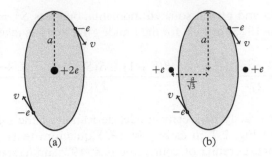

Figure 7.8 *Bohr's 1913 models of (a) Helium , (b) Hydrogen molecule (cf. Fig. 4.2).*

$$\frac{mv^2}{a} = \frac{7e^2}{4a^2},$$

(7.105)

while the quantization condition is

$$mva = \hbar.$$

(7.106)

Solving for the radius and speed, we find

$$a = \frac{4}{7}\frac{\hbar^2}{me^2} = \frac{4}{7}a_0,$$

(7.107)

$$v = \frac{7}{4}\frac{e^2}{\hbar}.$$

(7.108)

Here $a_0 = \hbar^2/me^2$ is the Bohr radius for the hydrogen atom, with $e^2/2a_0$ the ionization energy, 13.6 electron volts, of the latter.

Using Eqs. (7.107)–(7.108) and the expression for a_0, we can write the total electronic energy of the helium atom as

$$\begin{aligned} E &= mv^2 - \frac{4e^2}{a} + \frac{e^2}{2a} \\ &= m\left(\frac{7e^2}{4\hbar}\right)^2 - \frac{7e^2}{2}\left(\frac{7}{4a_0}\right) \\ &= \frac{e^2}{2}\cdot\frac{49}{8a_0} - \frac{e^2}{2a_0}\cdot\frac{49}{4} = -\frac{49}{8}\frac{e^2}{2a_0}. \end{aligned}$$

(7.109)

The numerical value of this energy is

$$E = -6.125 \times 13.6\text{eV} = -83.3\text{eV}.$$

(7.110)

The binding energy of a single electron in the ground state of ionized helium, He^+, is given immediately by the Bohr theory as $4 \times 13.6 = 54.4$ eV, so that the ionization energy of neutral helium, i.e., the energy needed to remove just one electron, must be the difference $83.3 - 54.4 = 28.9$ eV, if Bohr's simple model is to be believed.

The ionization energy in this model is so high that the corresponding term limit corresponds to inaccessible (at the time) lines in the far ultraviolet. Instead, ionization energies for helium were first estimated in experiments by Franck and Hertz, in which one attempted to measure ionization potentials by passing accelerated electrons through a gas of the atom or molecule under study and increasing the energy of the electrons until evidence appeared of positive ions (resulting from ionization of the atom or molecule) collected on an electrode. The electrode current being seen by Franck and Hertz in their early experiments was in fact a photocurrent generated by light emitted in the return of electrons raised to an excited level (but not completely ejected) to the ground state, and their typical value of 20.5 eV was initially considerably lower than the correct modern value of 24.6 eV, and even further away from the Bohr value.[55] However, by the third edition of *Atombau und Spektrallinien*, Sommerfeld could state definitively that "experience has not confirmed this ionisation energy (i.e., the 28.9 eV value), but instead... the value $\mathcal{J} = 25.3$ which has been found in various ways and multiply confirmed—the Bohr Helium model can therefore not be correct" (Sommerfeld 1922a, p. 727).

Given the early difficulty of pinning down the ionization energies, the first serious objections to the Bohr picture for helium actually came from another direction, namely optical dispersion calculations by Debye (1915). Debye's principal objective was a calculation of the dispersion (i.e., variation of refractive index with wavelength) of hydrogen gas using the model for the hydrogen molecule H_2 proposed by Bohr in the third part of his 1913 trilogy.[56] The model is illustrated in Fig. 7.8(b), and is obviously related to Bohr's model for helium: the electrons move at opposite points of a single circular orbit of radius a, with the two stationary hydrogen nuclei located symmetrically on the axis of the plane of the orbit, at a distance $a/\sqrt{3}$ from the center of the electronic orbit, which ensures balancing of electrostatic forces, as a simple application of Coulomb's Law shows. The actual size of the orbit a (and the angular frequency ω of the electron motion) is determined by a combination of the mechanical equations and the requirement, just as in the original Bohr model of the hydrogen ground state, that each electron carries a single unit of angular momentum $\hbar$.

Calculations of optical dispersion prior to the advent of the Bohr model had relied on the Drude–Lorentz theory, in which harmonically bound electrons were driven into forced oscillation by the electric field of an incident electromagnetic wave, and the induced (oscillating) electric dipole moment used to compute a polarization, and thence,

---

[55] The first "semi-accurate" determination (using this method) of the ionization energy of helium came a few years later, in work by Franck and Knipping (1919), who found a value of 25.5 eV.

[56] As we saw in Chapter 4, Bohr first proposed this model in the Rutherford Memorandum. See Eqs. (4.10)–(4.12) for the analysis of this model.

a formula for the refractive index as a function of the frequency (or wavelength) of the incident wave.[57]

For the moment, we note that a phenomenological formula for the dependence of refractive index on wavelength, which closely approximates the empirical data for a wide variety of gases, had already been given in the nineteenth century by Cauchy. Cauchy's formula gives the deviation from unity of the refractive index n as the first two terms of an expansion in inverse powers of the squared wavelength (always expressed in centimeters):

$$n - 1 = A\left(1 + \frac{B}{\lambda^2}\right). \tag{7.111}$$

The empirical constants A and B were already available to an accuracy of a few percent or less for gases such as hydrogen (H_2) or helium. The general structure of the Cauchy formula is easily understood as follows. The response of an undamped charged rotating (or oscillating) particle, with natural (angular) frequency ω_0, to an applied periodic force of frequency ω typically contains a resonance pole of the form $K/(\omega_0^2 - \omega^2)$. In the case of light impinging on an atomic system, the natural frequency ω_0 of the electron's rotational motion is much greater than the frequency ω ($= 2\pi c/\lambda$) of the incident (optical or near optical) light. We may therefore expand

$$\frac{K}{\omega_0^2 - \omega^2} = \frac{K}{\omega_0^2}\left(1 + \frac{\omega^2}{\omega_0^2} + \ldots\right) \approx \frac{K}{\omega_0^2}\left(1 + \frac{4\pi^2 c^2}{\omega_0^2}\frac{1}{\lambda^2}\right). \tag{7.112}$$

The polarization, and hence the deviation from unity of the refractive index, contains this factor, which thus explains the Cauchy form in Eq. (7.111). Note that both constants A and B contain the factor $1/\omega_0^2$, which cancels out in the ratio B/A. Typically, the ratio is fixed by the purely classical mechanics operative in the unperturbed system, while the natural frequency ω_0 is fixed by the further imposition of a quantization condition (e.g., the requirement that the electron angular momentum be $\hbar$).

The calculation of the Cauchy constants A and B on the basis of proposed Bohrian models (such as those in Fig. 7.8) clearly provided a possible detailed check on the validity of these models. Debye (1915) and subsequently Sommerfeld (1918a) were able to generalize the Drude–Lorentz theory to the new planetary picture of atoms or molecules. The calculations are messy in practice but simple in principle: one examines the response of the unperturbed motion (the two electrons circulating on a circular path) to a weak oscillating electric field. The displacements from the unperturbed motion (to first order in the applied field) are then used to yield an induced electric dipole moment of

---

[57] We will return to dispersion in Vol. 2, in our discussion of the origins of matrix mechanics, at which point we will present the technical details in full detail (Duncan and Janssen 2007). See also the web resource *Classical Dispersion Theory*.

Table 7.3 *Dispersion coefficients for H_2, He, and H. The results of contemporary measurements (exp) are compared with model-theoretic (theor) calculations by Debye (1915) and Sommerfeld (1918a).*

System	A	B	B/A
H_2 molecule (exp)	1.36×10^{-4}	7.6×10^{-11}	5.6×10^{-7}
Helium (exp)	3.47×10^{-5}	2.4×10^{-11}	6.9×10^{-7}
H atom (exp)	–	–	–
H_2 molecule (theor)	1.32×10^{-4}	7.35×10^{-11}	5.57×10^{-7}
Helium (theor)	5.15×10^{-5}	1.53×10^{-10}	2.97×10^{-6}
H atom (theor)	6.9×10^{-5}	3.57×10^{-11}	5.17×10^{-7}

the system, whence the polarization, and index of refraction, can be obtained as indicated previously.[58]

The results of the dispersion calculations of Debye (helium and hydrogen molecule) and Sommerfeld (hydrogen atom) are shown in Table 7.3. For the hydrogen molecule case, employing Bohr's model as depicted in Fig. 7.8(b), the agreement is quite good, for the individual Cauchy coefficients as well as the ratio B/A (sensitive to the spatial configuration but not the validity of the quantization condition fixing the unperturbed frequency). But for helium, the Bohr model completely fails to yield the observed dispersion coefficients. There was the further problem of the observed diamagnetism of helium, while the existence of a permanent magnetic moment in the model of Fig. 7.8(a), due to the rotation of both electrons in the same sense around the nucleus, suggested that helium gas should exhibit paramagnetism.

Following the work of Debye, Epstein had pointed out (early Spring of 1916) to Sommerfeld[59] that the dispersion coefficients for helium suggested a model of the helium ground state in which the two electrons followed very different orbits, one close to the nucleus (and, being bound tightly, inactive in dispersion) while the outer electron (responsible for dispersion, the optical spectra, etc.) followed a considerably larger orbit. Such a picture for the excited states of helium responsible for its spectrum (of which much more below) had already been used by Sommerfeld (cf. Section 7.1) to explain the Rydberg–Ritz term formulas in terms of the screening effect of the nuclear potential by an inner ring of electron(s), assumed coplanar with the orbit of the outer electron whose stationary state energies gave the term values for optical spectra, both for helium and for the alkali metals. Assuming the same situation for the ground state of helium (relevant for dispersion), with one electron in a tight orbit, the other much further out,

---

[58] For a detailed introduction to classical dispersion theory, see the web resource on *Classical Dispersion Theory*.

[59] As acknowledged by Sommerfeld (1918a, p. 547) in the final section of his paper, "Appendix on the putative constitution of helium."

gave of course a system very close to a hydrogen atom, with a single electron orbiting a core with a net charge of $+ e$. If only the outer electron participated in optical dispersion one should find similar dispersion for helium and atomic hydrogen. The results for the dispersion calculation for the hydrogen atom (not directly measurable, as hydrogen gas contains molecules not atoms), as given by Sommerfeld (1918a) are shown in the final row of Table 7.3. It is clear that, for the more robust ratio B/A, testing the geometry of the model, the agreement is much closer between the experimental helium results and the theoretical hydrogen atom calculation than with the original Bohr model of Fig. 7.8(a).

In the nine years from 1916 to the appearance of modern quantum mechanics in 1925, a huge amount of effort was devoted to detailed calculations of the atomic structure of helium in the old quantum theory, ultimately without success. As for the hydrogen atom, any successful model had to account at least qualitatively for the spectral lines of helium, which had been found very early to display a very peculiar dual character. The story of the helium spectrum, intimately intertwined as it is with the discovery of the element itself, is a fascinating one, so we shall pause here to fill in the relevant historical details.[60]

The approximately 20 electron volts needed to excite an atom of helium into the lowest excited state is so large that the corresponding absorption line would be situated in the far ultraviolet, at wavelengths of about 60 nm. There was no question, then, of Fraunhofer detecting absorption lines of unexcited neutral helium in the optical part of the solar spectrum. However, by the mid to late 1860s, there was considerable interest in studying the nature of solar sunspots, as well as of the solar protuberances (or flares) which were particularly prominent in solar eclipses, once the main light of the solar disc had been blocked by the interposed moon. Here, one expected to find bright emission lines, as one was viewing directly the light emitted by incandescent gas, without an intervening cooler absorbing region. From Kirchhoff's Laws (cf. Appendix B) one certainly expected that the dark Fraunhofer lines would find their counterparts, as bright emission lines at the same wavelength.

On October 20, 1868, Norman Lockyer observed,[61] by focusing his spectroscope on the rim of the sun's disc, three emission lines in the red, blue, and yellow part of the optical spectrum. The first two coincided with the well known hydrogen Balmer lines, the third yellow line fell close to the Fraunhofer D line (the sodium doublet) at 5890/5895 Å, but was definitely a separate line, at about 5876 Å (a conclusion which Lockyer considered definitive by November 15 (Nath 2013, p. 182)). Moreover, the line did not coincide with any spectral line known from flame studies of terrestrial elements. To distinguish it from the sodium D lines (D_1 and D_2), it was dubbed the D_3 line.

Around 1870, Lockyer began referring to the new element presumed responsible for the D_3 line as "helium" (from Greek "helios", for sun), a name which spread informally, and as we now know, permanently. It would take another quarter of a century for the culprit to be uncovered in a terrestrial source, specifically, in the gas emitted from the

---

[60] An entertaining, and highly detailed, account of the discovery of helium can be found in the book of Biman Nath (2013).

[61] This discovery did not occur *in vacuo*, as it were: for the fascinating background, see Nath (2013).

mineral cleveite (mainly, uranium oxide) when heated, originally thought to be nitrogen, but after careful collection and purification by William Ramsay (University College, London), found in 1895 to have a completely different and quite unusual spectrum, including a bright yellow line close to, but distinct from, the sodium D lines. On informing William Crookes of his findings, Ramsay received a few days later a telegram from Crookes explaining that the yellow line was none other than the D_3 helium line discovered decades earlier by Lockyer.

Almost immediately after Ramsay's discovery, Runge and Paschen in Hannover began a systematic spectral analysis of the "cleveite gas" (i.e., terrestrial helium). Their results, summarized in a paper submitted to the *Astrophysical Journal* in late 1895 (Runge and Paschen 1896), indicated that the spectrum of helium could be separated into *two distinct sets of principal, sharp, and diffuse series terms*,[62] with *no transitions observed between the terms of one set with the other*. Qualitatively, each set resembled the three main series of an alkali metal such as lithium, but for one set, the lines were all singlets, in the other doublets (just as in the alkali case). For the diffuse and sharp series (1st and 2nd associated/subordinate series), just as for alkali metal spectra, the doublet splittings were constant, while they decreased rapidly for the higher members of the principal series. Moreover, the lines of these two series approach the same term limit (corresponding to the energy of the same final p stationary state, as would later become clear in the old quantum theory).

This phenomenon of *two* sets of sharp/diffuse series had never been previously seen in spectrum of a pure element, so the hypothesis could hardly be avoided that one was dealing with a mixture of two elements, one of which had the yellow line, which was found to be a doublet in the terrestrial observations (though it was difficult to resolve in solar observations), and therefore had to be Lockyer's original helium. The other three (singlet) series were associated with a new element dubbed parhelium by George Stoney (also responsible for the term "electron"). The evidence for the existence of a distinct element was considered so strong that Rydberg (1896) even attempted to fit it into the periodic table, as an element of atomic weight three. But within a few years it became apparent that other elements with an even number of valence electrons (e.g., oxygen) also displayed spectra containing sets of noncommunicating series terms (such as singlet and triplet series). Moreover, attempts to separate, either chemically or by diffusion, the parhelium (later, *parahelium*) and helium (later, *orthohelium*) components were ultimately unsuccessful. By the end of the first decade of the twentieth century, physicists thus grudgingly began to accept that they were dealing with a single atom with a more complex type of spectrum.

The term diagram for helium is shown in Fig. 7.9, with the three series discovered by Runge and Paschen (1896) shown in red (principal), green (sharp), and blue (diffuse). The famous Lockyer yellow D_3 line leading to the discovery of helium in the solar

---

[62] For a history of spectroscopic terminology, see Appendix B.

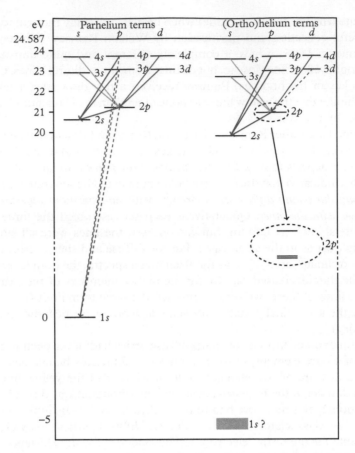

Figure 7.9 *Term diagram for Parhelium and (Ortho)helium. Principal, sharp, and diffuse series shown in red, green, and blue respectively.*

spectrum is due to the dashed blue diffuse series line in orthohelium.[63] The $2p$ term is shown magnified by a factor of $10\,000$ to reveal a triplet structure (see dashed ellipse). However, one pair of levels in the triplet is very close, so the triplet structure in orthohelium was for quite a while misidentified as a doublet structure. This was especially plausible as the overall arrangement of the levels (for either par(a)helium or orthohelium) is very similar to that in an alkali metal such as lithium (see Fig. 7.1). The main principal series for parahelium (dashed red lines) involves transitions of more than 20 eV and lies in the far ultraviolet. Runge and Paschen instead studied the second principal series of parahelium ($2p, 3p, 4p, \ldots \rightarrow 2s$ transitions, solid red lines).

---

[63] This is the later designation of the set of transitions containing the Lockyer line, which we shall henceforth adopt to avoid confusion. The second set was renamed parahelium, which we shall also henceforth employ.

The clear similarity of the helium spectra terms—both for the parahelium or orthohelium components—to the alkali metal levels suggested that the former could also be understood, at least qualitatively, with the help of the nuclear screening ideas introduced by Sommerfeld (cf. Section 7.1). Following Epstein's suggestion, Sommerfeld had already begun the task of treating the helium spectrum along similar lines to the alkalis in his November 1916 paper (Sommerfeld 1916c) (Section 3, "The spectra of the simpler elements, in particular of lithium and helium …"), by adopting a model in which one electron orbited close to the nucleus, and the second (valence) electron further out. The energy levels of the outer electron would be calculated by standard Bohr–Sommerfeld quantization, but in an electrostatic potential consisting of the sum of the attractive nuclear term $-2e^2/r^2$ and the repulsive potential felt by the outer electron due to the inner one, where the rapid rotation of the inner electron is used to justify treating its charge as a uniformly smeared ring of charge, of radius $a = a_0/2$, with a_0 the Bohr radius for hydrogen, and the factor of two due to the fact that the inner electron sees the full unshielded charge of $+2e$ of the helium nucleus.

In his 1916 paper, Sommerfeld had considered only configurations in which the inner and outer electron rotate in the same plane (coplanar orbits). The additional complexity (or duplexity, if one will) of the helium case suggested the need for more general configurations, in which the inner electron ring could also be tilted at an angle ϑ to the orbit of the valence electron. In 1918 Sommerfeld addressed the problem of multiple electron rings, not necessarily coplanar (Sommerfeld 1918b). For two rings with a single electron each, the geometry is displayed in Fig. 7.10, where the outer electron circulates in a circular ring of radius r (in the x-y plane), while the inner electron occupies a circular orbit of radius a, tilted at an angle ϑ to the x-y plane. For $r \gg a$ one finds, by

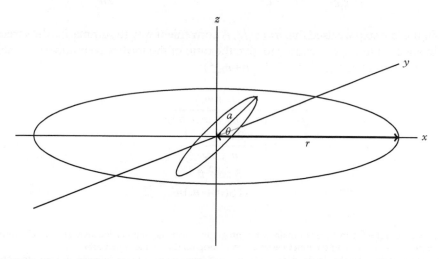

Figure 7.10 *Sommerfeld–Landé model of helium.*

a simple integration, for the average electrostatic potential generated by the core ring at the radius r of the valence electron[64]

$$V_c(r) = e^2 \left(\frac{1}{r} + \frac{a^2}{8r^3}(3\cos^2\vartheta - 1) + O\left(\frac{a^4}{r^5}\right) \right), \tag{7.113}$$

where the second (and higher) terms in the expansion can be interpreted as corrections to the naive screening of the nuclear charge by a single unit of charge effected by the inner electron. The total electrostatic potential seen by the outer electron, assuming the inner electron to have the center of its orbit coincident with the nucleus, thus takes the form discussed previously in Eq. (7.1), with the coefficients $b_1, b_2, \ldots = 0$ and

$$c_1 = \frac{e^2 a^2}{8}(3\cos^2\vartheta - 1). \tag{7.114}$$

Sommerfeld's result from 1916, $c_1 = e^2 a^2/4$, thus just corresponds to $\vartheta = 0$. If we include the dependence on the tilt angle, we can find the correction term $a(n, \vartheta)$ in the term wavenumbers[65]

$$\epsilon_{He} = -\frac{E_{He}}{hc} = \frac{N}{(n + n_r + a(n, \vartheta))^2} \tag{7.115}$$

by using Eq. (7.10) that Sommerfeld found for the relation between $a(n)$ and c_1 and Eq. (7.114) for c_1:

$$a(n, \vartheta) = \frac{c_1}{e^2 a_0^2 n^3} = \frac{1}{8n^3}\left(\frac{a}{a_0}\right)^2 (3\cos^2\vartheta - 1) = \frac{3\cos^2\vartheta - 1}{32n^3}, \tag{7.116}$$

where in the last step we used that $a = a_0/2$. A convenient way to summarize the screening effect in spectral terms was simply to give the ratio of the term wavenumbers for helium (Eq. (7.115)) and hydrogen ($\epsilon_H = N/(n+n_r)^2$):

$$\begin{aligned}
\frac{\epsilon_{He}}{\epsilon_H} &= \frac{(n + n_r)^2}{(n + n_r + a(n, \vartheta))^2} \\
&\approx 1 - \frac{2a(n, \vartheta)}{n + n_r} \\
&= 1 - \frac{3\cos^2\vartheta - 1}{16(n + n_r)n^3} + \ldots
\end{aligned} \tag{7.117}$$

---

[64] The average referred to here is obtained by integrating over the angular location of both electrons, the outer one in the x-y plane, and the inner electron circulating on the smaller tilted orbit.

[65] We remind the reader that in the notation being used here, n refers to the angular, n_r the radial, and $n + n_r$ to the principal quantum number.

In a paper in early 1920 summarizing the results of his habilitation thesis—and correcting some inadequacies of a previous paper on the same subject (Landé 1919a)—Landé (1920) extended the Sommerfeld calculations, both including the effect of the relative orbit tilt of core and valence electron, as well as calculating (perturbatively) the polarizing effect of the outer electron on the inner one, which tends to displace the center of the core electron orbit away from the nuclear position (and from the outer electron).

The procedures used by Landé are taken from celestial mechanics, in which the Lagrangian for a system of two electrons circulating around a helium nucleus in the same plane (and the same sense) is split into (i) an unperturbed piece containing the kinetic energy of both electrons, the full electrostatic potential energy $-2e^2/r$ of the inner electron, and the screened electrostatic potential energy $-e^2/R$ of the outer electron, and (ii) a perturbation term obtained by expanding the inter-electron repulsive potential energy in powers of the ratio r/R of radial distances, assumed to be small. From all possible motions of such a system, one can select consistently a simply periodic class of motions in which the radial and angular parameters of the inner and outer electron r, φ, R, φ can all be expanded in Fourier series of a single frequency.

The quantization condition was applied first to the unperturbed orbits, and the energy change from unperturbed to the actual orbits computed in terms of an adiabatic process by which the inner electron is brought adiabatically from a location close to the nucleus (e.g., by making the nuclear-inner electron interaction initially large and then decreasing it to the physical value), and the outer electron adiabatically brought in from infinity. This procedure was a special case of a general method for computing adiabatically the energy of quantized perturbed electron orbits which Landé (1919b) had published just prior to the appearance of his second paper on the helium spectrum (Landé 1920).

The result of Landé's calculations was the appearance of two new terms in the helium-to-hydrogen ratio formula Eq. (7.117), which now becomes

$$\frac{\epsilon_{He}}{\epsilon_H} = 1 + \frac{n}{n+n_r}\left(\cos\vartheta\left(\frac{3}{4}n^{-10/3} - \frac{1}{6}n^{-11/3}\right) - \frac{3\cos^2\vartheta - 1}{16n^4} + \cdots\right). \quad (7.118)$$

Landé obtained this result for coplanar orbits ($\vartheta = 0$) and generalized it by the insertion of geometrical factors involving $\cos\vartheta$. This could be done uncontroversially for the final term, which coincides with the result in Eq. (7.117). For the terms linear in $\cos\vartheta$, which are to be ascribed to the displacement of the inner electron's orbit due to the outer electron and which were not taken into account in the calculation leading to Eq. (7.117)), it could only be done by analogy to Landé's earlier work on helium.

The result Eq. (7.118) contains, in addition to the discrete quantum numbers n_r and n (which determine the size and eccentricity of the valence electron orbit), a so far continuous angle variable ϑ, specifying the relative orientation of the core and valence orbits. To obtain a discrete spectrum, this quantity must be subjected to the Sommerfeld procedure of spatial quantization. In an earlier paper, entitled "A quantum rule for the spatial orientation of electron orbits," Landé (1919c) had proposed that the spatial quantization be carried out in such a way that the vector sum of the core

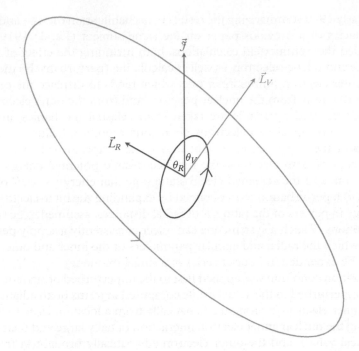

Figure 7.11 *Landé vector model of helium.*

and valence orbital angular momenta ($\vec{L}_R$ and $\vec{L}_V$ respectively), i.e., the *conserved* total angular momentum $\vec{\mathcal{J}}$ of the atom, be subjected to integer quantization, so that $|\vec{\mathcal{J}}| \equiv k\hbar$, with $k = 1, 2, 3, \ldots$ The geometry envisaged by Landé is shown in Fig. 7.11.

The magnitude of the inner electron angular momentum was one unit of $\hbar$ (i.e., $|\vec{L}_R| = \hbar$), that of the valence electron n (i.e., $|\vec{L}_V| = n\hbar$), so in terms of components along and perpendicular to the direction of $\vec{\mathcal{J}}$ (shown as the vertical z-axis in Fig. 7.11), this means

$$k = \cos \vartheta_R + n \cos \vartheta_V, \tag{7.119}$$

$$\sin \vartheta_R = n \sin \vartheta_V, \tag{7.120}$$

with the relative tilt of core and valence orbits given by $\vartheta = \vartheta_R + \vartheta_V$.

For coplanar orbits with $\vartheta_V = \vartheta_R = \vartheta = 0$, one has $k = n + 1$, and the formula Eq. (7.118) was found to give at least a qualitative agreement with the terms of the orthohelium spectrum, with the agreement best for the p ($n = 2$) and d ($n = 3$) levels. Various non-zero choices for ϑ were found to roughly match the parahelium levels, e.g., $n = k = 2$, giving $\cos \vartheta_V = \frac{7}{8}$, $\cos \vartheta_R = \frac{1}{4}$, $\vartheta = 104.5°$, which gave values for ϵ_{He}/ϵ_H very close to the parahelium p levels. However, the s levels of parahelium, corresponding to $n = k = 1$ (in which case $\vartheta_R = \vartheta_V = 60°$ and $\vartheta = 120°$), were very badly described in the model, with the theoretical prediction for ϵ_{He}/ϵ_H less than unity and increasing with the level number (i.e., the radial quantum number n_r), while the experimental values were greater than unity and decreased. Moreover, the a priori perfectly allowable coplanar orbits with counter-rotating core and valence electrons (i.e., $\vartheta = 180°$) gave levels with much smaller binding energy which were not observed and which Landé simply excluded by fiat on "stability grounds".

To summarize, the bottom line conclusion of the Landé calculations in 1920 was that parahelium levels corresponded to the core and valence electrons circulating in orbits at various tilt angles, all fairly close to ninety degrees, while the orthohelium levels corresponded to coplanar orbits, with both electrons circulating in the same sense. As these more detailed spectral considerations came to the fore, the old problem of the nature of the "normal" (i.e., ground state) of helium, and the associated question of the correct value of the ionization energy (i.e., the energy required to remove one electron from the atom in its ground state) reemerged, with the more accurate determinations of the latter energy giving 25.4 eV (Franck and Knipping 1919) or 25.7 eV (Horton and Davies 1919).[66] Empirical extrapolations of the Rayleigh–Ritz spectral defect $a(n, 0)$ (giving rise to the second term in Eq. (7.118) with $\vartheta = 0$ for orthohelium) gave a value (Landé 1919a) of about 0.32,[67] so that the energy required to remove the outer electron in the lowest 1s ($n_r = 0$, $n = 1$) state should be about $1/(1-0.32)^2 = 2.16$ Rydbergs, while removing the remaining inner (now solitary) electron costs 4 Rydbergs. The estimated binding energy of neutral orthohelium in its lowest state would then be about 6.16 Rydbergs, or about 84 eV (corresponding to a single electron ionization energy of almost 30 eV), close to Bohr's original estimate using the single ring model of Fig. 7.8(a). This would locate the lowest orthohelium state at roughly the location indicated by a shaded block (and question mark) in Fig. 7.9.

Unfortunately, the new results, as we saw above, indicated an ionization energy (for removing the outer electron) of only about 1.88 Rydbergs (25.6 eV). A glance at Fig. 7.9 certainly suggests that, given that the parahelium levels were consistently higher than the ortho ones, this might also hold true for the lowest (1s) para state. However, one must keep in mind that the actual location of this lowest parahelium state was not predicted theoretically at this point: the utter failure of Landé's model to predict the 1s spectral terms for parahelium led him to a prudent avoidance of any attempt to estimate this energy. A convincing argument for identifying the quantum-theoretic status of the helium ground state (as either para- or ortho-) finally appeared with an important paper of James Franck (see Plate 26) and Fritz Reiche (1920), which we now describe briefly. After reviewing the recent results on ionization energy described above, the low value of which suggests that unexcited helium was to be found in a para state above the putative ortho ground state (cf. Fig. 7.9), Franck and Reiche go on to discuss an interesting observation of infrared resonant scattering by Paschen which clearly supported the suggestion that

---

[66] The question of the accurate determination of the ionization energy for helium would finally be settled spectroscopically in mid-1922, with Lyman's (1922) measurement of the far ultraviolet line at 584.4 Å for the $2p \rightarrow 1s$ first principal line of parahelium (shortest dashed line in Fig. 7.9). Given the previously known term value for the $2p$ parahelium state, this fixed the ionization energy at 24.59 eV.

[67] From the empirical orthohelium terms, one found $a(1, 0) = 0.298, 0.299, 0.303,$ and 0.311 for $n_r = 4, 3, 2, 1$ respectively, so one could assume a value somewhere in the neighborhood of 0.32–0.33 for $n_r = 0$.

the ground state of helium must be a para state. Paschen found that when infrared light was passed through helium gas excited by an electrical discharge, the first absorption lines to appear (as the discharge strength is increased) corresponded to the first principal series lines ($2p \rightarrow 2s$) for parahelium (at 20 582 Å) and orthohelium (at 10 830 Å) (cf. Fig. 7.9). The orthohelium line showed strong resonant absorption—i.e., the incident light at 10 830 Å was being absorbed and reradiated at the same wavelength—while the parahelium line showed a weaker absorption with only a fraction of the incident light reradiated at 20 582 Å. This can only be explained by the presence of a lower $1s$ state on the parahelium side to which the majority of atoms excited to the parahelium $2p$ state can de-excite before being able to return to the $2s$ state. On the orthohelium side, the strong resonant absorption indicated that the $2s$ state had to be metastable, with no lower $1s$ orthohelium state, which if it existed would have the same effect as for parahelium, by depopulating the $2p$ state, thereby weakening the resonant $2s \rightarrow 2p \rightarrow 2s$ transitions. Franck and Reiche's conclusion is unambiguous:

> It therefore follows that the single quantum (i.e., principal quantum number $= 1$) coplanar state absolutely is not present . . . in the normal (i.e., ground) state only the single quantum state of helium with crossed electron orbits (parahelium) exists (Franck and Reiche 1920, pp. 158–159).

In presenting their analysis of the principal series absorption lines of helium, Franck and Reiche rely heavily on the empirically established separation of the para- and ortho- spectral terms: namely, transitions from a higher state of ortho (resp. para) helium to a lower state of para (resp. ortho) helium were never observed. Franck and Reiche claim that this prohibition actually follows from a selection rule of Sommerfeld, based on a combination of the Rubinowicz energy and angular momentum conservation arguments and the requirement that the angular momentum of the final orbit of an electron after a radiative transition be space quantized relative to the orientation of the initial orbit angular momentum (Sommerfeld 1919, p. 417). The upshot of these arguments was that the direction of the total angular momentum (though not of course the magnitude) of an atom would have to be unchanged by a radiative transition:

> The invariable plane (i.e., the plane to which the total conserved angular momentum is orthogonal) of the atomic planetary system (in the simplest Keplerian case the plane of the orbit) remains constant not only in stationary states, but also during transitions from one such state to another (Sommerfeld 1919, p. 396).

The nonexistence of para- to ortho- transitions (and vice versa) is adduced by Franck and Reiche (1920, p. 157) as confirmation of this Sommerfeld rule. Of course, the rule refers to the total angular momentum direction, so it is not at first sight obvious why transitions are forbidden in which $\vec{L}_V$ and $\vec{L}_R$, the angular momenta of valence and core electrons, simultaneously move into alignment from an initially non-collinear (para) configuration, while preserving the direction of $\vec{\mathcal{J}} = \vec{L}_R + \vec{L}_V$ (cf. Fig. 7.11). The language of the Sommerfeld rule obscures, however, the actual way in which selection rules were employed in the old quantum theory. The entire understanding of the alkali series spectra, for example, rested on the application of these rules *to the valence electron* (or *Leuchtelektron*) *only*, with the core of the atom inert during the radiative transition. Otherwise, one would simply not have an understanding of why only transitions with $\Delta n = \pm 1$, with n the azimuthal (orbital angular momentum) quantum number of the valence electron, were allowed. Clearly, a transition from a para- to an ortho-configuration of helium in which the direction of the total angular momentum remained fixed requires both the inner and outer electron to change orbits, a circumstance which was implicitly excluded by Franck and Reiche.

By late 1920, Bohr had concluded that the ground state of helium corresponded to a configuration in which the two electrons were *not* coplanar (as in his original 1913 model), but that "crossed orbits" were needed. It was by then clear (as Franck and Reiche had argued) that the ground state belonged to the parahelium spectrum, for which the most detailed spectral studies, such as Landé's, suggested such crossed configurations for excited parahelium states, although Landé was extremely careful to indicate the failure of his perturbation theory for the lowest *s*-states, where the interaction of the valence and core electrons would be strongest. For Bohr, correspondence-principle arguments required that transitions within each of the helium spectral types (para- or ortho-) preserved the qualitative orbital geometry, so that if the excited parahelium states were crossed, and the ground state was a parahelium state, the latter would also have to be a crossed-orbit state. There remained the question of whether the two electrons occupied similar orbits in the ground state, or whether (as in Sommerfeld's earlier models of the ground state) there was still an "inner" and "outer" electron, with orbits of different dimensions.

Independently of each other, and more or less at the same time (late 1920-early 1921), Bohr and his assistant Hans Kramers (see Plate 25) in Copenhagen and Edwin C. Kemble (1921) at Harvard adopted a symmetric crossed-orbit model for the ground state of the helium atom. The chemical inertness of unexcited helium, which seemed incompatible with the presence of a more loosely bound "valence" electron in the ground state, certainly provided a strong motivation for this symmetric picture. The geometry of the proposed model is indicated in Fig. 7.12.

As in the Landé vector model (see Fig. 7.11), the total angular momentum $\vec{\mathcal{J}}$, the sum of the orbital momenta $\vec{L}_1 + \vec{L}_2$ of the two electrons, is quantized in integer multiples of $\hbar$, with the individual electrons in 1s states ($n = 1$) in the unperturbed limit in which the repulsion between the electrons is neglected. In this lowest-level approximation, the electron angular momenta, of magnitude $\hbar$, are at an angle of 60° to the z-axis, with $\vec{\mathcal{J}} = \hbar \hat{z}$. Of all possible motions of the two electrons one selects then the class of symmetric motions in which the two electrons execute orbits separated by 180° in azimuthal angle around the z-axis, with equal values of the z coordinates, as indicated in Fig. 7.12. If the first electron has Cartesian coordinates (x_1, y_1, z_1), the coordinates of the second electron are then

$$x_2 = -x_1, \quad y_2 = -y_1, \quad z_2 = z_1. \tag{7.121}$$

of the two electrons once the electrostatic repulsion is fully switched on, subject to the quantization conditions $L_{1z} = L_{2z} = \hbar/2$, and Eq. (7.127).

If we treat the transverse variables $\vec{r}_\perp = (x, y), \vec{p}_\perp = (p_x, p_y)$ separately from the z-variables (z, p_z), the Hamilton equations of motion for one of the electrons are

$$\vec{p}_\perp = m\dot{\vec{r}}_\perp,$$

$$p_z = m\dot{z},$$

$$\dot{\vec{p}}_\perp = -\frac{Ze^2}{r^3}\vec{r}_\perp + \xi\frac{e^2}{4\rho^3}\vec{r}_\perp,$$

$$\dot{p}_z = -\frac{Ze^2}{r^3}z, \tag{7.128}$$

where Z is the nuclear charge ($= 2$ for helium),

$$\rho = |\vec{r}_\perp| = \sqrt{x^2 + y^2}, \quad r = \sqrt{x^2 + y^2 + z^2}, \tag{7.129}$$

and ξ is an adiabatic parameter multiplying the inter-electron repulsion term: $\xi = 0$ for the unperturbed motion in which the electrons circulate independently in Bohr orbits and $\xi = 1$ for the desired exact motion. The motion generated by the dynamics of Eq. (7.128) is multiply periodic for the symmetric configurations considered here, with two independent frequencies (for $\xi \neq 0$)—the librational frequency in the meridional plane, and the rotational (precessional) frequency around the z-axis. Unfortunately the motion does not separate by choice of a suitable coordinate system, which would allow an analytic solution along the lines of Hamilton–Jacobi theory. Thus one is forced to employ either approximate analytic approaches (the only option available in 1920), or precise numerical methods, which yield an accurate result in a minute or two on a modern laptop computer (see Fig. 7.13).

The analysis of the crossed-orbit motion elicited heroic calculational efforts from John H. Van Vleck, a doctoral student of Kemble at Harvard,[70] and Kramers in Copenhagen, culminating in publications by Van Vleck (1922b) submitted in March 1922, and by Kramers (1923), submitted in December of that year. Van Vleck's paper summarizes the extensive calculations he performed in his doctoral dissertation, in which the multiply periodic motion of the two electrons in an axially symmetric configuration (with the symmetry proposed by Kemble) was approximated by a systematic expansion in a parameter $w = \sin^2\vartheta_i$, where ϑ_i is the inclination angle of the plane of either electron orbit to the invariant plane orthogonal to the total angular momentum vector (thus, $\vartheta_i = 60°$ for the unperturbed motion, giving $w = \frac{3}{4}$, the factor in the square root in Eq. (7.123)). The desired solution was obtained by expanding the deviations of the perturbed from the unperturbed motion $\rho(t) - \rho_0(t)$, $z(t) - z_0(t)$ as power series in w, with the series expansion carried as far as practically feasible (which turned out to be

---

[70] Van Vleck's early career is discussed in the dissertation of Fred Fellows (1985) and in Midwinter and Janssen (2013).

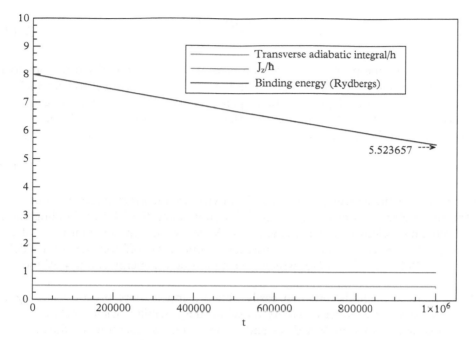

Figure 7.13 *Numerical simulation of crossed-orbits model for helium.*

the seventh order!). Imposing the quantum conditions then gave the value $w = 0.7216$, which on substitution into the expression for the total energy (also given as an expansion in w) gave the result for the binding energy of both electrons (in Rydbergs)[71]

$$W_{\text{VanVleck}} = 74.9 \ eV = 5.53 \ \text{Ry.} \tag{7.130}$$

Kramers' calculation is quite different from Van Vleck's. He uses a clever decomposition of the Hamiltonian in which the electrostatic potential energy,

$$V = -\frac{Ze^2}{r} + \frac{e^2}{2\rho}, \tag{7.131}$$

is decomposed into an unperturbed part,

$$V_0 = -\frac{Z^*e^2}{r} + \frac{C}{\rho^2}, \tag{7.132}$$

plus a perturbation

[71] We have converted Van Vleck's result of 74.9 eV to Rydbergs using the then current value of a Rydberg, as used by Kramers, of 13.54 eV.

We have undertaken the task of systematically examining *all possible orbital types of the excited helium atom*, to separate out the quantum-theoretically allowed solutions and to calculate their energies, in order to determine whether or not orbits are present which correctly give the empirical term values. The result of our calculation is negative: one cannot arrive at an explanation of the helium spectrum by consistent application of the known quantum rules (Born and Heisenberg 1923, p. 229).

In particular, Born and Heisenberg's analysis leads to the prediction of *three* distinct spectral series, corresponding to the orbital angular momentum k $(= n)$ of the outer electron combining with the unit angular momentum of the core electron to give a total angular momentum j equal to either $k + 1$ (coplanar orbits in the same sense), k (for large k, almost perpendicular orbital planes), or $k - 1$ (coplanar orbits with the two electrons orbiting in opposite senses). The Rydberg–Ritz spectral defects δ, defined by equating the binding energy W of the valence electron to

$$W = \frac{Z - 1}{(n + \delta)^2} \qquad (7.136)$$

(with n the principal quantum number), was computed theoretically for the $j = k + 1$ and $j = k$ cases (which were the closest to the empirical ortho- and parahelium spectra, respectively), for paths with radial quantum number zero (i.e., circular orbits with the inter-electron force switched off). The results, which were not that different from Landé's, deviated unacceptably from the experimental results, especially for the higher angular momenta (d, f, etc. terms), where the computed spectral defects were an order of magnitude larger than the measured ones (in fact, Born and Heisenberg's results were further from the experimental data than Landé's). Moreover, there was no explanation for the absence of the $j = k - 1$ terms (which had also been excluded by Landé), or for the strict absence of combinations of para- with ortho- terms which remained the most striking feature of the helium spectrum.[74]

On the threshold of the new quantum theory, Born's lectures in Göttingen in the Winter semester of 1923/1924, published in book form (*Vorlesungen über Atommechanik, Erster Band*) in early 1925, ends with this pessimistic assessment:

We arrive at the conclusion: *the consistent application of the principles of quantum theory, as expounded in the second chapter,* namely the calculation of the motion according to the laws of classical mechanics and the selection of stationary states by fixing the action variables to integer multiples of the Planck constant, *leads to agreement with experience only in the case where we are concerned with the motion of a single electron: the calculation fails already in the treatment of the motion of the two electrons of the helium atom* (Born 1925, p. 341; emphasis in the original).

---

[74] This noncombination also posed a huge problem in Bohr's correspondence-principle approach to atomic structure (which we will discuss in Volume 2), as both the ortho- and para- orbits were now shown by Born and Heisenberg to be continuously related. For further discussion see the section, "The catastrophe of helium," in Darrigol (1992, pp. 175–179).

Appendices

Appendices

A

Classical Mechanics

The account of the old quantum theory presented in Chapters 5 through 7 relies heavily on applications of some of the most sophisticated formulations of analytical mechanics accessible to physicists in the early twentieth century: Hamilton–Jacobi theory, action-angle formalism, canonical perturbation theory, and so on. These techniques are no longer part of the working apparatus of modern physicists. If mentioned at all, they are given short shrift even in advanced graduate courses on classical mechanics (or, to give the more fashionable modern term, dynamical systems). There is a natural temptation to assume that theoretical physicists of the early twentieth century received a thorough training in these methods and were fluent in their application, a fluency which has eroded as the subject matter of physics became almost entirely imbued with the techniques of the new quantum theory: linear algebra, differential equations, Hilbert space, and the like.

In fact, an examination of the standard texts on introductory classical mechanics for physicists at the turn of the twentieth century and the two decades following shows only a superficial attention to the analytical approach to mechanics pioneered by Lagrange and brought to perfection by Hamilton and Jacobi. Kirchhoff's *Vorlesungen über Mechanik* (1897) ignores the Lagrangian and Hamiltonian approaches to mechanics altogether (although Hamilton's principle of least action is briefly mentioned). Mechanical problems are treated entirely with Newtonian methods, by analysis of vectorial forces (with equations appearing everywhere in triplicate, as modern vector notation is not used). August Föppl's monumental six-volume treatise *Vorlesungen über Technische Mechanik* (1897–1910) is almost entirely "force oriented" (thankfully, with vector notation employed more or less systematically), with Lagrange's equations finally making a brief appearance in Vol. 4, Ch. 4 (before the author hastens on to hydrodynamics), and again briefly in Vol. 6, Ch. 2, where the double pendulum is analyzed using Lagrangian methods. The notions of canonical coordinates, phase space, and Hamilton's equations of motion are nowhere to be found (although, as in Kirchhoff, Hamilton's principle of least action does get a brief mention). Similar statements hold for Georg Hamel's *Elementare Mechanik* (1912).

One exception to this general tendency can be found in the second volume of Boltzmann's *Vorlesungen über die Prinzipe der Mechanik* (1904), where the classical dynamical equations are discussed thoroughly in both Lagrangian and Hamiltonian form, as well as an introduction to the Hamilton–Jacobi technique (not including, however, the treatment of conditionally periodic systems in terms of action-angle variables that would take a central role in the old quantum theory). This volume also contains a statement of the adiabatic theorem (Ch. 4, sec. 48), which was to play an important role in the conceptual development of the old quantum theory.[1] Nevertheless, the orientation of Boltzmann's treatment of mechanics, with statistical-mechanical considerations always lurking in the immediate background, meant that these lectures were probably mostly studied by a few physicists such as Paul Ehrenfest with a deep interest in the conceptual foundations of heat theory. And Ehrenfest later admitted to his student Jan Burgers that he was "horribly ignorant, and inept" in the "theory of integration for canonical systems" (Klein 1970a, p. 291; cf. Section 1.3.3 and Section 1.3.2, note 14).

The situation is quite different when one examines the texts in wide use among theoretical astronomers concerned with problems in celestial mechanics. Charlier's *Die Mechanik des Himmels* (1902–1907) contains a thorough introduction to the Lagrangian, Hamiltonian, and Hamilton–Jacobi formalisms in the first appendix (entitled "Hilfssätze aus der Mathematik und der Mechanik"). Here already one sees laid out (Ch. 2, sec. 3) the complete formalism needed for the treatment of conditionally periodic Hamiltonian systems which would be taken over wholesale in the development of the old quantum theory in the nine years from 1916 to the appearance of matrix mechanics in late 1925. It is not surprising that the introduction of these techniques to the audience of physicists working on the quantum theory was primarily due to the agency of an astronomer, Karl Schwarzschild (see Sections 1.3.2 and 5.1.4; Plate 21).

In this appendix we give a highly abbreviated account of the elements of classical mechanics available to physicists and astronomers in the first decade of the twentieth century. Only those methods and techniques that played an important role in the development of quantum theory will be discussed. In a perhaps somewhat arbitrary fashion, we have divided the discussion into two parts, outlining first those aspects of classical mechanics which formed part of the lingua franca of most theoretical physicists up to about the year 1916, and then moving on to the more sophisticated techniques employed by astronomers in celestial mechanics, which became widely used in the old quantum theory after about 1916. The use of these methods would culminate in 1924 with the Kramers dispersion theory, which is widely regarded as the immediate precursor of the *Umdeutung* paper with which Heisenberg (1925) laid the basis for matrix mechanics (Duncan and Janssen 2007).

Because of their importance for the development of quantum theory, these advanced topics in classical mechanics entered into physics handbooks and textbooks. In the former category, we mention two appendices in Vol. 5 of the *Handbuch der Physik* published in 1927: Ch. 3 on Hamilton–Jacobi theory (Nordheim and Fues 1927) and Ch. 4 on canonical perturbation theory (Fues 1927). In both appendices, the importance of these

---

[1] Also, the late work of J. Willard Gibbs, *Elementary Principles in Statistical Mechanics* (1902) is firmly grounded in the use of the Hamiltonian formalism, with the statistical dynamics of ensembles formulated entirely in terms of the multidimensional phase space coordinatized by the collection of coordinates and momenta of all the molecules in the system.

topics for quantum mechanics is emphasized. Nordheim and Fues conclude Ch. 4 with a reference to Schrödinger's use of the optical-mechanical analogy;[2] Fues notes at the beginning of Ch. 5 that before the advent of the Bohr model of the atom physicists were not interested in the techniques covered in the appendix (see Section A.2.4 for the relevant passage). These articles by Fues and Nordheim are among the main references for Chs. 8 and 9, on canonical transformations and Hamilton–Jacobi theory, respectively, of the first edition of Goldstein's (1950) well-known textbook on classical mechanics. Goldstein also lists *The Mechanics of the Atom* (Born 1927), *Atomic Structure and Spectral Lines* (Sommerfeld 1923), and *Quantum Principles and Line Spectra* (Van Vleck 1926), the NRC Bulletin on the old quantum theory by one of Goldstein's senior colleagues at Harvard. In the preface, Goldstein explains that he included these more advanced topics precisely because of their importance for quantum theory:

> [C]lassical mechanics, in one or another of its advanced formulations, serves as the springboard of the various branches of modern physics. Thus, the technique of action-angle variables is needed for the older quantum mechanics, the Hamilton–Jacobi theory and the principle of least action provide the transition to wave mechanics, while Poisson brackets and canonical transformations are invaluable in formulating the newer quantum mechanics (Goldstein 1950, p. ix).

Almost a century after action-angle variables and the Hamilton–Jacobi equation helped give birth to quantum mechanics, graduate textbooks in classical mechanics dutifully continue to cover these topics.

A.1   The physicist's mechanical toolbox (ca 1915)

A.1.1   Newtonian mechanics

The standard introduction to mechanics in almost all introductory physics texts at the beginning of the twenty-first century follows Newton closely, just as the corresponding texts used by apprentice physicists a hundred years previously when the quantum theory was in its infancy. There are, of course, concessions to modern mathematical notation (especially in calculus, where Newton's notation has been completely superseded by that of Leibniz). The central concept employed is that of the response of massive point particles to applied vectorial forces. The critical link is provided by Newton's second law

$$\vec{F} = m\vec{a}. \tag{A.1}$$

Here $\vec{F}$ is the net force (i.e., the vectorial sum of all applied forces) acting on a point particle of mass m, resulting in the acceleration $\vec{a}$ of that particle. A simple example, which

---

[2] For discussion of the role of the optical-mechanical analogy in the development of wave mechanics, see Joas and Lehner (2009). We shall return to this subject in depth in Volume 2.

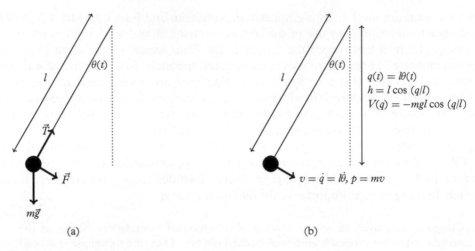

Figure A.1 *The swinging pendulum, analyzed according to (a) Newtonian, or (b) Lagrangian/Hamiltonian principles.*

will serve us as an archetypal illustrative case for many of the more advanced methods of analytical mechanics to be described below, the freely swinging pendulum, is illustrated in Figure A.1(a). A small metal ball of mass m is suspended on a linearly extended string (or rigid massless rod) of length l as indicated, and subject to two external forces, the tension $\vec{T}$ in the string, pulling towards the point of attachment of the string, and the force of gravity $m\vec{g}$ pulling the mass vertically down. If we resolve the gravitational force along the direction perpendicular to the string (along which the particle is constrained to move as the string is inextensible), we find for the magnitude of the force in this direction (there is no contribution from the tension, which is perpendicular to this direction)

$$F = mg\sin(\theta) = -ml\frac{d^2\theta(t)}{dt^2}. \tag{A.2}$$

The second equality follows from Newton's second law, given that the distance travelled by the pendulum bob along its circular path is $l\theta$. The minus sign appears as the force evidently acts in the direction to decrease the angle θ (Fig A.1(a)). Canceling the mass factor, we find the "equation of motion"

$$\ddot{\theta}(t) \equiv \frac{d^2\theta(t)}{dt^2} = -\frac{g}{l}\sin(\theta(t)). \tag{A.3}$$

We have introduced a notation that will reappear on many occasions in this book: time derivatives are indicated with dots above the function to be differentiated (in this case, two dots for a second derivative). If the maximum extension of the pendulum bob is a small angle, we may use the usual approximation $\sin(\theta) \approx \theta$ and obtain a simpler, *linear* equation

$$\ddot{\theta}(t) = -\frac{g}{l}\theta(t), \tag{A.4}$$

known as the *simple harmonic equation*. The general solution, discussed in exhaustive detail in all freshman texts, is

$$\theta(t) = A\sin\left(\sqrt{\frac{g}{l}}t + \alpha\right) = A\sin(\omega t + \alpha). \tag{A.5}$$

The amplitude A and phase α here are arbitrary: they must be fixed by the initial conditions of the problem (for example, knowing that the maximum angle A is attained at time zero, in which case $\alpha = \pi/2$).

A.1.2 Lagrangian mechanics

Lagrange devised an essentially automatic procedure for turning any and all problems of Newtonian mechanics into equivalent differential equations. The formalism he introduced is so general that it handles with equal ease discrete (particle) or continuous (field) dynamics, and is still a central part of modern field theory. In the method of Lagrange the state of the system is specified by giving the value of a set of N "generalized coordinates" $q_i(t)$, $i = 1, 2, \ldots N$ that together uniquely determine the location of all the particles (and/or rigid bodies) in the system. The choice of these coordinates is quite arbitrary (apart from the fact that they together determine the configuration of the mechanical system uniquely), and is generally made for convenience. For a single mass particle moving on a line, a single coordinate suffices, which we could choose to be simply $q = x$, the coordinate of the particle measured from an arbitrarily chosen origin on the line. For the pendulum of Figure A.1, the coordinate could be $q = l\theta$, giving the (signed) distance traversed by the swinging mass from the lowest point to the location in question. For each generalized coordinate $q_i(t)$, there is an associated "generalized velocity" $\dot{q}_i(t) = dq_i/dt$. The kinetic energy of all the particles in the system can evidently be computed in terms of these velocities, while the potential energy is typically a function of the coordinates themselves.

Lagrange introduced a function $\mathcal{L}(q_i, \dot{q}_i)$ of the coordinates $q_i(t)$ and the velocities $\dot{q}_i(t)$ of the particles in the system, equal in simple cases to the kinetic energy minus the potential energy. For each such coordinate, there is a Lagrangian equation of motion, given simply by

$$\frac{d}{dt}\frac{\partial \mathcal{L}}{\partial \dot{q}_i} = \frac{\partial \mathcal{L}}{\partial q_i}. \tag{A.6}$$

These differential equations are clearly second order in time as they involve the time derivative of a function of first time derivatives, and a unique solution presupposes the specification of both the $q_i(t)$ and their first derivatives $\dot{q}_i(t)$ at some initial time $t = t_s$.

As an example of the application of the Lagrangian approach, let's return to our previous example of the pendulum. The relevant aspects for a Lagrangian analysis of the problem are displayed in Figure A.1(b). For our "generalized coordinate", we will take the distance $q(t) = l\theta(t)$ along the particle path, measured clockwise from the bottom point. Our first task is to determine the kinetic energy K as a function of the coordinate and its first time derivative. This is easy:

$$K = \frac{1}{2}mv^2 = \frac{1}{2}m\dot{q}^2. \tag{A.7}$$

The potential energy V is just $-mgh$ where $h = l\cos(\theta) = l\cos(q/l)$ is the distance *downwards* (hence the minus sign) from the attachment point:

$$V = -mgl\cos(q/l). \tag{A.8}$$

The Lagrangian function is therefore

$$\mathcal{L} = K - V = \frac{1}{2}m\dot{q}^2 + mgl\cos(q/l). \tag{A.9}$$

The Lagrange equation Eq. (A.6) now requires us to calculate the derivatives

$$\frac{\partial \mathcal{L}}{\partial \dot{q}} = m\dot{q}, \tag{A.10}$$

$$\frac{\partial \mathcal{L}}{\partial q} = -mg\sin(q/l), \tag{A.11}$$

and the Lagrangian equation of motion gives us, in agreement with Eq. (A.2),

$$m\ddot{q} = ml\ddot{\theta}(t) = -mg\sin(\theta(t)). \tag{A.12}$$

This nonlinear "anharmonic" equation is replaced in the small angle approximation by the much more tractable linear simple harmonic equation (again, just Eq. (A.4)),

$$\ddot{\theta}(t) = -\frac{g}{l}\theta(t). \tag{A.13}$$

The power of the Lagrangian approach to mechanics really becomes apparent in more complicated systems with tricky geometrical constraints (for example, a ball rolling down a hoop, various rigid bodies attached by hinges, etc.), where some of the forces are only specified indirectly. Also, the formalism handles non-Cartesian coordinate systems with great ease. In fact, the Lagrangian equations of motion retain their form if we change to a new set of coordinates

$$Q_i(q_1(t), q_2(t), \ldots, q_n(t)) \equiv Q_i(q_j), \tag{A.14}$$

as long as the new coordinates Q_i are differentiable and invertible functions of the old ones q_j. We assume, in other words, that we may turn Eq. (A.14) "inside out" (repeated indices summed over!):

$$q_i(t) = q_i(Q_j), \quad \dot{q}_i(t) = \frac{\partial q_i}{\partial Q_j}\dot{Q}_j \quad \Rightarrow \quad \frac{\partial \dot{q}_i}{\partial \dot{Q}_j} = \frac{\partial q_i}{\partial Q_j}. \tag{A.15}$$

If we now form a new Lagrangian function $\hat{\mathcal{L}}(Q_j, \dot{Q}_j) \equiv \mathcal{L}(q_i(Q_j), \dot{q}_i(Q_j, \dot{Q}_j))$ by simply writing the old variables in terms of the new ones, one sees that

$$\frac{\partial \hat{\mathcal{L}}}{\partial Q_j} = \frac{\partial \mathcal{L}}{\partial q_i}\frac{\partial q_i}{\partial Q_j} + \frac{\partial \mathcal{L}}{\partial \dot{q}_i}\frac{\partial \dot{q}_i}{\partial Q_j}, \tag{A.16}$$

$$\frac{\partial \hat{\mathcal{L}}}{\partial \dot{Q}_j} = \frac{\partial \mathcal{L}}{\partial \dot{q}_i}\frac{\partial \dot{q}_i}{\partial \dot{Q}_j}. \tag{A.17}$$

From Eq. (A.15) we have the identity,

$$\frac{d}{dt}\frac{\partial \dot{q}_i}{\partial \dot{Q}_j} = \frac{d}{dt}\frac{\partial q_i}{\partial Q_j} = \frac{\partial \dot{q}_i}{\partial Q_j}, \tag{A.18}$$

(where in the last step we used that we can change the order of d/dt and $\partial/\partial Q_j$) and the validity of the Lagrangian equations in the new coordinates now follows directly. Differentiating Eq. (A.17), we find that

$$\frac{d}{dt}\frac{\partial \hat{\mathcal{L}}}{\partial \dot{Q}_j} = \left(\frac{d}{dt}\frac{\partial \mathcal{L}}{\partial \dot{q}_i}\right)\frac{\partial \dot{q}_i}{\partial \dot{Q}_j} + \frac{\partial \mathcal{L}}{\partial \dot{q}_i}\frac{d}{dt}\frac{\partial \dot{q}_i}{\partial \dot{Q}_j} \tag{A.19}$$

$$= \frac{\partial \mathcal{L}}{\partial q_i}\frac{\partial q_i}{\partial Q_j} + \frac{\partial \mathcal{L}}{\partial \dot{q}_i}\frac{\partial \dot{q}_i}{\partial Q_j} \tag{A.20}$$

$$= \frac{\partial \hat{\mathcal{L}}}{\partial Q_j} \tag{A.21}$$

where we have used the old Lagrangian equations (A.6) plus Eqs. (A.15) and (A.18) in going from Eq. (A.19) to Eq. (A.20), and Eq. (A.16) in the final line.

The rather involved demonstration just given for the coordinate independence of the Lagrangian formalism obscures the very simple physical basis for this invariance, which is most easily understood in the context of the formulation of the theory in terms of an extremal principle (i.e. "the system evolves dynamically in such a way as to minimize/maximize some integral over the motion"). For each possible dynamical evolution of our system between an initial time t_s and a final time t_f,[3] specified by the functions $q_i(t)$ between these limits, we may associate a number, evidently a function of the initial and final times, and coordinates $q_i(t_s)$ and $q_i(t_f)$,

$$S_L \equiv \int_{t_s}^{t_f} \mathcal{L}(q_i, \dot{q}_i) dt. \tag{A.22}$$

Hamilton's principle asserts that the path actually taken by a system satisfying the Lagrangian equations of motion produces an extremal value for this number. Small variations of the path, wherein $q_i(t) \rightarrow q_i(t) + \delta q_i(t)$ (while preserving the initial and final locations $q_i(t_s)$ and $q_i(t_f)$), therefore induce a variation δS_L which vanishes to first order in the variations δq_i. For infinitesimal variations, we have

$$\delta S_L = \int_{t_s}^{t_f} \left(\frac{\partial \mathcal{L}}{\partial q_i} \delta q_i(t) + \frac{\partial \mathcal{L}}{\partial \dot{q}_i} \delta \dot{q}_i(t) \right) dt \tag{A.23}$$

$$= \int_{t_s}^{t_f} \left(\frac{\partial \mathcal{L}}{\partial q_i} \delta q_i(t) + \frac{\partial \mathcal{L}}{\partial \dot{q}_i} \frac{d}{dt} \delta q_i(t) \right) dt \tag{A.24}$$

$$= \int_{t_s}^{t_f} \left(\frac{\partial \mathcal{L}}{\partial q_i} - \frac{d}{dt} \frac{\partial \mathcal{L}}{\partial \dot{q}_i} \right) \delta q_i(t) dt, \tag{A.25}$$

where in going from the second to the third line we have performed an integration by parts and neglected the boundary terms

$$\frac{\partial \mathcal{L}}{\partial \dot{q}_i} \delta q_i(t) \Big|_{t_s}^{t_f}, \tag{A.26}$$

which vanish at the initial and final times, as we consider only paths which begin and end at fixed chosen configurations of the system. Hence, the first-order variation δS_L vanishes for arbitrary choices of the path variation $\delta q_i(t)$ if (and only if) the expression in parentheses in Eq. (A.25) vanishes, i.e., if (and only if) the Lagrange equations Eq. (A.6) hold.

The reason for the generality of the Lagrange equations, independent of choice of coordinates, is now clear. The extremum of S_L is simply not affected if we rewrite the Lagrangian function $\mathcal{L}(q_i, \dot{q}_i)$ in terms of a new set of variables Q_i as in Eq. (A.14), and equate our new Lagrangian $\hat{\mathcal{L}}(Q_i, \dot{Q}_i)$ to the old one (but reexpressed in the new

---

[3] We use the subscript 's' for 'start' instead of 'i' for initial to avoid confusing with the index 'i' we will continue to use for the components q_i and $\dot{q}_i$ of position and velocity. One can likewise read 'finish' instead of 'final' for the subscript 'f'.

variables). The actual numerical value of the Lagrangian at any given time is not changed by this procedure and neither, a fortiori, is the value of the integral in Eq. (A.22). Accordingly, the extremum found with the new Lagrangian $\hat{\mathcal{L}}$ represents exactly the same physical path (expressed in the new variables Q_i) as the one found with the old.

However important a role it played in the development of classical physics in the late eighteenth and through the nineteenth century, the Lagrangian approach to mechanics was largely supplanted in the development of quantum mechanics by an alternative procedure, due to Hamilton, in which particle coordinates and momenta are treated on an equal footing.

A.1.3 Hamiltonian mechanics

For each coordinate $q_i(t)$ in a Lagrangian system, introduce a conjugate momentum

$$p_i \equiv \frac{\partial \mathcal{L}}{\partial \dot{q}_i}. \tag{A.27}$$

Except for some important exceptions (where so-called "gauge symmetries" are present), which will not concern us here, the definition of the canonical momentum allows us to solve for the velocities in terms of the momenta, as we shall see shortly with explicit examples. If one now defines a new function, this time of the coordinates and the *momenta* (instead of the velocities, as previously in the Lagrangian approach)

$$H(p_i, q_i) \equiv \sum_i p_i \dot{q}_i - \mathcal{L}(q_i, \dot{q}_i), \tag{A.28}$$

where in H, all occurrences of the velocities $\dot{q}_i$ on the right-hand side are replaced by their expressions in terms of associated canonical momenta p_i as defined in Eq. (A.27), then Hamilton showed that the original Lagrange equations split into a pair of first-order equations (now called Hamilton's equations):

$$\frac{\partial H}{\partial p_i} = \dot{q}_i, \tag{A.29}$$

$$\frac{\partial H}{\partial q_i} = -\dot{p}_i. \tag{A.30}$$

To illustrate how this works we use the same example that we used to illustrate the Lagrangian approach to mechanical problems, the freely swinging pendulum. There is only a single coordinate (which we chose to be the distance travelled by the pendulum bob from the lowest point of its path) so the index i may be dropped. From Eq. (A.9) we find first the derivative giving the conjugate momentum

$$p = \frac{\partial \mathcal{L}}{\partial \dot{q}} = \frac{\partial}{\partial \dot{q}} \left(\frac{1}{2} m \dot{q}^2 + mgl \cos(q/l) \right) = m\dot{q} \quad \Rightarrow \quad \dot{q} = \frac{p}{m}. \tag{A.31}$$

We may now calculate the Hamiltonian function H, as defined in Eq. (A.28), remembering to use $\dot{q} = p/m$ to eliminate the velocity in favor of the momentum:

$$H = p\dot{q} - \mathcal{L} = p\left(\frac{p}{m}\right) - \frac{1}{2}m\left(\frac{p^2}{m^2}\right) - mgl\cos(q/l) = \frac{p^2}{2m} - mgl\cos(q/l). \quad \text{(A.32)}$$

Note that the H function is just the total energy (*sum* of the kinetic and potential energy of the pendulum bob).

The Hamilton equations of motion now become

$$\frac{\partial H}{\partial p} = \dot{q} = \frac{p}{m}, \quad \text{(A.33)}$$

$$\frac{\partial H}{\partial q} = -\dot{p} = mg\sin(q/l). \quad \text{(A.34)}$$

The first of these equations agrees with Eq. (A.31). Using that $p = m\dot{q}$ and that $q = l\theta$, we can rewrite the second equation as

$$m\ddot{q} = ml\ddot{\theta} = -mg\sin(\theta), \quad \text{(A.35)}$$

in agreement with Eq. (A.12).

That H (equal to the total energy) is conserved in time is not an accident: note that as a consequence of the Hamiltonian equations,

$$\frac{dH(p,q)}{dt} = \frac{\partial H}{\partial p}\frac{dp}{dt} + \frac{\partial H}{\partial q}\frac{dq}{dt} = \dot{q}\dot{p} - \dot{p}\dot{q} = 0, \quad \text{(A.36)}$$

so H is automatically a constant of the motion (provided H only depends on time through the coordinates and the momenta, and not explicitly).

In the Hamiltonian framework, the state of a mechanical system is specified at any time by giving the value of the $2n$ coordinates and momenta $(q_1, q_2, \ldots, q_n, p_1, p_2, \ldots, p_n)$. Eqs. (A.29) and (A.30) are first order in time and so allow us, at least in principle, to determine the value of these variables at any later time. Thus the dynamics of the whole system can be thought of as a unique trajectory in the $2n$-dimensional *phase space* coordinatized by the variables (q_i, p_i), passing through the specified initial configuration point $(q_i(t_s), p_i(t_s))$.

The Hamilton differential equations of motion, like the Lagrangian ones, are completely equivalent to an integral extremal principle (also due to Hamilton): the path actually followed by a mechanical system subject to Hamilton's equations is an extremum of the *action functional*, defined by

$$S_H = \int (p_i\dot{q}_i - H(p_i, q_i))dt. \quad \text{(A.37)}$$

We imagine the integral extending from an initial point in phase space $(p_i(t_s), q_i(t_s))$ at time t_s to a final point $(p_i(t_f), q_i(t_f))$ at a later time t_f. Now consider all (time-differentiable) paths in phase space connecting the system at initial coordinates $(q_i(t_s))$ to points at final coordinates $(q_i(t_f))$ (but with no constraints on the $p_i(t_s)$ or $p_i(t_f)$) and search for the path that makes the value of S_H extremal (typically, a minimum). Nearby paths $p_i(t) + \delta p_i(t), q_i(t) + \delta q_i(t)$ will therefore have an altered action $S_H + \delta S_H$, where for the desired extremal path, δS_H should vanish to first order in the infinitesimal variations $(\delta p_i, \delta q_i)$. From Eq. (A.37), the first-order variation of the action is

$$\delta S_H = \delta \int_{t_s}^{t_f} (p_i \dot{q}_i - H(p_i, q_i)) dt \tag{A.38}$$

$$= \int_{t_s}^{t_f} \left(\delta p_i \dot{q}_i + p_i \delta \dot{q}_i - \frac{\partial H}{\partial p_i} \delta p_i - \frac{\partial H}{\partial q_i} \delta q_i \right) dt \tag{A.39}$$

$$= \int_{t_s}^{t_f} \left(\delta p_i \dot{q}_i - \dot{p}_i \delta q_i - \frac{\partial H}{\partial p_i} \delta p_i - \frac{\partial H}{\partial q_i} \delta q_i \right) dt \tag{A.40}$$

$$= \int_{t_s}^{t_f} \left(\delta p_i \left(\dot{q}_i - \frac{\partial H}{\partial p_i} \right) - \delta q_i \left(\dot{p}_i + \frac{\partial H}{\partial q_i} \right) \right) dt. \tag{A.41}$$

In performing the integration by parts to go from Eq. (A.39) to Eq. (A.40),

$$\int_{t_s}^{t_f} p_i \frac{d}{dt} \delta q_i dt = [p_i \delta q_i]_{t_s}^{t_f} - \int_{t_s}^{t_f} \frac{dp_i}{dt} \delta q_i dt = - \int_{t_s}^{t_f} \dot{p}_i \delta q_i dt, \tag{A.42}$$

we have used the constraint that only paths between the identified initial and final coordinate points $q_i(t_s)$ and $q_i(t_f)$ are considered, so $\delta q_i(t)$ vanishes at the end points of the integral (note that no such constraint is required for the $\delta p_i(t)$). Returning to Eq. (A.41), we see that the first-order variation δS_H indeed only vanishes, for arbitrary variations $(\delta p_i(t), \delta q_i(t))$ of the path, provided Hamilton's equations (A.29)–(A.30) hold throughout the motion. The extremal property of S_H is sometimes referred to as the "modified Hamilton's principle".

We may now consider, as previously for the Lagrangian formalism, the possibility of changing from one set of coordinates (q_i, p_i) used to specify the state of the system (as a point in $2n$-dimensional phase space) to another, perhaps more convenient, one.[4] Unlike the situation for Lagrange's equations, which, as we saw, continue to hold under essentially arbitrary (subject to mild smoothness constraints) point transformations Eq. (A.14), the preservation in form of Hamilton's equations under a general change[5] of phase space coordinates $(q_i, p_i) \rightarrow (Q_i, P_i)$, with

---

[4] A reparametrization of a dynamical system purely in terms of the generalized coordinates of the system is called a "point transformation".

[5] We here allow as well for an explicit time-dependence in the change of canonical variables.

$$Q_i = Q_i(q_1, q_2, \ldots, q_n, p_1, p_2, \ldots, p_n, t), \tag{A.43}$$

$$P_i = P_i(q_1, q_2, \ldots, q_n, p_1, p_2, \ldots, p_n, t), \tag{A.44}$$

is *not* guaranteed. We shall now examine the conditions under which a new Hamiltonian $\tilde{H}(Q_i, P_i)$ can be found such that the Hamilton equations of motion for $\tilde{H}$,

$$\frac{\partial \tilde{H}}{\partial P_i} = \dot{Q}_i, \tag{A.45}$$

$$\frac{\partial \tilde{H}}{\partial Q_i} = -\dot{P}_i, \tag{A.46}$$

are exactly equivalent to the original ones in Eqs. (A.29)–(A.30) in terms of H. If these conditions hold, we say that the new variables (Q_i, P_i) are related by a *canonical transformation* to the old ones (q_i, p_i).

For the special case of Hamiltonians without explicit time dependence, there exists a simple analytic criterion for ensuring that, as in the Lagrangian case, simply reexpressing the old Hamiltonian $H(q_i, p_i)$ in terms of the new variables (Q_i, P_i) yields the desired new Hamiltonian $\tilde{H}(Q_i, P_i)$. In order to demonstrate this, it is convenient to introduce a single set of $2n$ coordinates for the phase space of the system. Thus, we set

$$
\begin{aligned}
z_i &= (q_1, \ldots, q_n, p_1, \ldots, p_n), \quad i = 1, \ldots, 2n, \\
Z_i &= (Q_1, \ldots, Q_n, P_1, \ldots, P_n), \quad i = 1, \ldots, 2n.
\end{aligned}
\tag{A.47}
$$

Next, we define a constant $2n \times 2n$ matrix as follows

$$\mathcal{J} = \begin{pmatrix} 0 & 1 \\ -1 & 0 \end{pmatrix} \tag{A.48}$$

where the symbols "1" refer to $n \times n$ unit matrices. One immediately sees that the Hamilton equations (A.29)–(A.30) can be combined into the single equation[6]

$$\dot{z}_i = \mathcal{J}_{ij} \frac{\partial H}{\partial z_j}. \tag{A.49}$$

---

[6] In the one-dimensional case, Eq. (A.49) can be written as:

$$\begin{pmatrix} \dot{q} \\ \dot{p} \end{pmatrix} = \begin{pmatrix} 0 & 1 \\ -1 & 0 \end{pmatrix} \begin{pmatrix} \partial H / \partial q \\ \partial H / \partial p \end{pmatrix}.$$

Now under a general differentiable change of variables $z_i \to Z_i(z_1, \ldots, z_{2n})$, we have

$$\dot{Z}_i = \frac{\partial Z_i}{\partial z_j}\dot{z}_j \equiv \mathcal{T}_{ij}\dot{z}_j. \tag{A.50}$$

The original Hamilton equations Eq. (A.49) imply

$$\dot{z}_j = \mathcal{J}_{jk}\frac{\partial H}{\partial z_k} = \mathcal{J}_{jk}\frac{\partial \tilde{H}}{\partial Z_l}\frac{\partial Z_l}{\partial z_k}. \tag{A.51}$$

Inserting this result in Eq. (A.50), we find

$$\dot{Z}_i = \mathcal{T}_{ij}\mathcal{J}_{jk}\mathcal{T}_{lk}\frac{\partial \tilde{H}}{\partial Z_l} = (\mathcal{T}\mathcal{J}\mathcal{T}^T)_{il}\frac{\partial \tilde{H}}{\partial Z_l}, \tag{A.52}$$

so we see that the form of Hamilton's equations is preserved in the new variables (with the new Hamiltonian $\tilde{H}$ simply equal to the old one reexpressed in the new variables) provided

$$\mathcal{T}\mathcal{J}\mathcal{T}^T = \mathcal{J}. \tag{A.53}$$

This is a nonlinear analytic constraint on the Jacobian $\mathcal{T}_{ij} \equiv \partial Z_i/\partial z_j$ of the transformation for the old to the new variables, which unfortunately does not provide an easily implementable and explicit procedure for actually *generating* suitable canonical transformations. However, it does allow one to verify trivially that the determinant of the Jacobian of canonical transformations must be unity, as Eq. (A.53) immediately implies, taking the determinant of both sides

$$\det(\mathcal{T}\mathcal{J}\mathcal{T}^T) = \det(\mathcal{T})^2\det(\mathcal{J}) = \det(\mathcal{J}) \Rightarrow |\det(\mathcal{T})| = 1. \tag{A.54}$$

Consequently, in performing a valid canonical transformation, the volume element in phase space is preserved

$$\int dq_1\,dq_2\ldots dq_n dp_1\,dp_2\ldots dp_n = \int dQ_1\,dQ_2\ldots dQ_n dP_1\,dP_2\ldots dP_n. \tag{A.55}$$

Another immediate consequence of the simplified form Eq. (A.49) of the Hamilton equations is *Liouville's theorem*. Imagine that we consider an ensemble of classical systems all evolving according to the same Hamiltonian H, as a dense array of points in phase space, with density $\rho(z_i)$. We imagine that at time zero, the density of ensemble points is constant in phase space. Each point then follows the appropriate phase space orbit determined by the Hamilton equations Eq. (A.49), leading to a current flow $\vec{j} = \rho\vec{v}$, where $v_i = \dot{z}_i$:

$$j_i = \rho \mathcal{J}_{ij} \frac{\partial H}{\partial z_j}. \tag{A.56}$$

The divergence of the current vector at time zero (as ρ is constant) is

$$\vec{\nabla} \cdot \vec{j} = \frac{\partial j_i}{\partial z_i} = \rho \mathcal{J}_{ij} \frac{\partial^2 H}{\partial z_i \partial z_j} = 0, \tag{A.57}$$

due to the anti-symmetry of $\mathcal{J}$ and the symmetry of the second partial derivatives of H. By the continuity equation,

$$\frac{\partial \rho}{\partial t} + \vec{\nabla} \cdot \vec{j} = 0, \tag{A.58}$$

this implies $\partial \rho / \partial t = 0$, so stepping the system forward in time, we see that if the ensemble points are distributed uniformly in phase space at time zero, they continue to be so distributed at any later time. Alternatively, the volume occupied by any bounded region of points is preserved as these points are evolved separately along their orbits as determined by the Hamiltonian equations of the system. This is Liouville's theorem: together with the preservation of phase space volumes implied by Eq. (A.55), it ensures the primary role of phase space as the appropriate theater of operations in the statistical mechanics of Gibbs and Boltzmann.

Fortunately, there exists an essentially automatic procedure for generating proper canonical transformations which preserve the form of Hamilton's equations. The easiest approach to the relevant equations begins from the pair of equations expressing the simultaneous validity of the modified Hamilton principle for both the old and new variables:

$$\delta \int_{t_s}^{t_f} (p_i \dot{q}_i - H(q_i, p_i)) dt = \delta \int_{t_s}^{t_f} (P_i \dot{Q}_i - \tilde{H}(Q_i, P_i))\, dt = 0. \tag{A.59}$$

It follows that

$$\delta \int_{t_s}^{t_f} (p_i \dot{q}_i - H(q_i, p_i) - (P_i \dot{Q}_i - \tilde{H}(Q_i, P_i)))\, dt = 0. \tag{A.60}$$

In Eq. (A.60), despite the presence of $4n$ variables (q_i, p_i, Q_i, P_i), we must regard the integrand as really only a function of $2n$ independent variables, given the implicit $2n$ constraints given by Eqs. (A.43)–(A.44). The variation in Eq. (A.60) can only vanish for arbitrary path variations if the integrand on the right-hand side is a complete derivative dF/dt, where F can be chosen to be a function of any set of $2n$ independent variables chosen from the $4n$ set (q_i, p_i, Q_i, P_i) (more on this freedom of choice below). The integral on the right then becomes simply $F(t_f) - F(t_s)$, the variation of which vanishes as we consider only paths beginning and ending at the same point in phase space. We must therefore have

$$p_i \dot{q}_i - H - P_i \dot{Q}_i + \tilde{H} = \frac{dF}{dt}. \tag{A.61}$$

It has become standard procedure to choose the $2n$ independent coordinates on which the function F can depend from entire sets of either coordinates or momenta (as well as a possible explicit time dependence), so there are essentially four possibilities:[7]

1. $F = F_1(q_i, Q_i, t)$,
2. $F = F_2(q_i, P_i, t)$,
3. $F = F_3(p_i, Q_i, t)$,
4. $F = F_4(p_i, P_i, t)$.

If we take the first possibility, $F = F_1(q_i, Q_i, t)$, we have

$$\frac{dF_1}{dt} = \frac{\partial F_1}{\partial q_i} \dot{q}_i + \frac{\partial F_1}{\partial Q_i} \dot{Q}_i + \frac{\partial F_1}{\partial t}. \tag{A.62}$$

Substituting this expression for dF_1/dt on the right-hand side of Eq. (A.61), and multiplying by dt, we find

$$p_i dq_i - H dt - P_i dQ_i + \tilde{H} dt = \frac{\partial F_1}{\partial q_i} dq_i + \frac{\partial F_1}{\partial Q_i} dQ_i + \frac{\partial F_1}{\partial t} dt. \tag{A.63}$$

Comparing coefficients of dq_i, dQ_i and dt on both sides, we obtain

$$\frac{\partial F_1(q_i, Q_i, t)}{\partial q_i} = p_i, \tag{A.64}$$

$$\frac{\partial F_1(q_i, Q_i, t)}{\partial Q_i} = -P_i, \tag{A.65}$$

$$\frac{\partial F_1(q_i, Q_i, t)}{\partial t} = \tilde{H} - H. \tag{A.66}$$

Knowledge of F_1 implies that (at least in principle) we may invert Eq. (A.64) to find the new coordinates Q_i as functions of the old coordinates and momenta, (q_i, p_i), as indicated in Eq. (A.43). Inserting this result in Eq. (A.65) then gives the new momenta P_i in terms of the old canonical variables, as in Eq. (A.44). The function F_1 therefore encodes the complete information needed to implement a canonical transformation from the old to the new variables: it is called the "generating function" for the transformation. The appropriate new Hamiltonian is then given from Eq. (A.66) as

---

[7] For many more details of the theory of canonical transformations, see Goldstein, Poole, and Safko (2002, Ch. 9).

$$\tilde{H} = H + \frac{\partial F_1}{\partial t}, \tag{A.67}$$

where on the right-hand side (q_i, p_i) are reexpressed in terms of the new canonical variables (Q_i, P_i).

Before continuing to a discussion of further aspects of the canonical formalism, an explicit example which turns out to be of great significance in the development of quantum theory will be helpful to illustrate the rather abstract preceding discussion. This is the one-dimensional oscillator, with Hamiltonian

$$H = \frac{p^2}{2m} + \frac{1}{2}m\omega^2 q^2. \tag{A.68}$$

As our generating function (of type "F_1"), we choose

$$F_1(q, Q) = \frac{1}{2}m\omega q^2 \cot(2\pi Q). \tag{A.69}$$

Note that in this case the generating function does not explicitly depend on time, so by Eq. (A.67) the new Hamiltonian $\tilde{H}$ will just be the old one H reexpressed in terms of the new variables (Q, P). These can be found using Eqs. (A.64)–(A.65),

$$\frac{\partial F_1}{\partial q} = p = m\omega q \cot(2\pi Q), \tag{A.70}$$

$$-\frac{\partial F_1}{\partial Q} = P = \frac{\pi m\omega q^2}{\sin^2(2\pi Q)} \tag{A.71}$$

(where in the last line we used that the derivative of $\cot(x)$ is $-1/\sin^2(x)$). From Eq. (A.70) we can solve for the new coordinate Q in terms of the old phase space variables:[8]

$$Q = \frac{1}{2\pi}\cot^{-1}\left(\frac{p}{m\omega q}\right) \quad \Rightarrow \quad \sin^2(2\pi Q) = \frac{m^2\omega^2 q^2}{m^2\omega^2 q^2 + p^2} \tag{A.72}$$

The new momentum variable P now follows from Eq. (A.71),

$$P = \frac{\pi}{m\omega}\left(m^2\omega^2 q^2 + p^2\right) = \frac{2\pi}{\omega}\left(\frac{p^2}{2m} + \frac{1}{2}m\omega^2 q^2\right) = \frac{2\pi}{\omega}H. \tag{A.73}$$

---

[8] Here $\cot^{-1}$ refers to the functional inverse cotangent, or arc cot, *not* the algebraic inverse. To understand Eq. (A.72), imagine a right triangle with hypotenuse c and remaining sides a and b. Let α be the angle for which $\cot(\alpha) = b/a$. Hence, $\cot^{-1}(b/a) = \alpha$ and $\sin(\cot^{-1}(b/a)) = \sin(\alpha)$. Hence, $\sin^2(\alpha) = a^2/c^2 = a^2/(a^2 + b^2)$. In this case, $a = m\omega q$ and $b = p$.

The new Hamiltonian is thus given by:

$$\tilde{H} = H = \frac{\omega}{2\pi}P = \nu P, \qquad (A.74)$$

where $\nu = \omega/2\pi$ is the cyclic frequency of the motion. The new Hamilton equations of motion are

$$\frac{\partial \tilde{H}}{\partial P} = \dot{Q} = \frac{\omega}{2\pi}, \qquad (A.75)$$

$$\frac{\partial \tilde{H}}{\partial Q} = -\dot{P} = 0. \qquad (A.76)$$

Note that the new coordinate Q does not appear in the new Hamiltonian (it is a "cyclic" variable), which has the immediate consequence in Eq. (A.76) that the associated canonical variable P, which in fact is simply the energy itself up to a constant rescaling, is a constant of the motion. The equation of motion for Q can be solved trivially: by appropriate choice of the zero of time, we can just set

$$Q(t) = \frac{\omega}{2\pi}t. \qquad (A.77)$$

Setting the conserved energy to the constant value E and substituting $(2\pi/\omega)E$ for P and Eq. (A.77) for Q in Eq. (A.71), we arrive at

$$\frac{2\pi E}{\omega} = \frac{\pi m \omega q^2}{\sin^2(\omega t)}. \qquad (A.78)$$

Solving for the old coordinate $q(t)$, we find

$$q(t) = \sqrt{\frac{2E}{m\omega^2}} \sin(\omega t). \qquad (A.79)$$

Using Eq. (A.72), we find that the old momentum $p(t)$ is given by:

$$p(t) = m\omega q \cot(2\pi Q) = \sqrt{2Em} \cos(\omega t). \qquad (A.80)$$

Eqs. (A.79)–(A.80) are the familiar equations of simple harmonic motion.[9]

We mentioned above the possibility of choosing other sets of variables in terms of which to write the generating function for the canonical transformation. These turn out

---

[9] The new variables (P, Q) given in this simple illustration are our first examples of the famous "action-angle" variables which will play a prominent role in the old quantum theory. We will present a more general treatment of the action-angle formalism in Section A.2.3.

to be simple Legendre transforms of the basic function $F_1(q, Q)$ discussed above. For example, we can define

$$F_2(q, P) = F_1(q, Q(q, P)) + PQ(q, P), \tag{A.81}$$

where we can determine Q as a function of q and P by inverting $P = -\partial F_1(q, Q)/\partial Q$ (cf. Eq. (A.65)). For the derivative of F_2 with respect to q, we find:

$$\frac{\partial F_2(q, P)}{\partial q} = \frac{\partial F_1(q, Q)}{\partial q} + \frac{\partial F_1(q, Q)}{\partial Q}\frac{\partial Q(q, P)}{\partial q} + P\frac{\partial Q(q, P)}{\partial q} = p, \tag{A.82}$$

where in the last step we used that $\partial F_1(q, Q)/\partial q = p$ and, again, that $\partial F_1(q, Q)/\partial Q = -P$ (cf. Eqs. (A.64)–(A.65)). We similarly find that

$$\frac{\partial F_2(q, P)}{\partial P} = \frac{\partial F_1(q, Q)}{\partial Q}\frac{\partial Q(q, P)}{\partial P} + Q(q, P) + P\frac{\partial Q(q, P)}{\partial P} = Q. \tag{A.83}$$

Once again, the new variables can be obtained in terms of the old ones by solving first Eq. (A.82) to obtain P in terms of q and p, and then Eq. (A.83) is used directly to obtain Q in terms of q and p. The new Hamiltonian as before is given by $\tilde{H} = H$ (i.e., one simply reexpresses the old Hamiltonian in terms of the new variables) or, in case F_2 explicitly depends on time, by $\tilde{H} = H + (\partial F_2/\partial t)$. This version of the canonical transformation procedure lies at the heart of the Hamilton–Jacobi theory to be discussed in Section A.2.1.

A.1.4   The adiabatic principle

One of the most important conceptual strands running through the development of the old quantum theory, especially in connection with the Copenhagen school led by Niels Bohr, was the adiabatic principle in classical mechanics. The relevance of this principle to the identification of suitable quantization conditions was first realized clearly by Paul Ehrenfest (see Section 5.2). The principle goes back to a theorem already clearly stated in the second volume of Boltzmann's lectures on classical mechanics (Boltzmann 1904, Sec. 48).

Here Boltzmann identifies changes to the energy of a mechanical system under periodic motion which result from variations in the parameters of the system on a timescale much longer than the typical periods of the motion with the adiabatic work performed by or on the system, whereas other changes on a much smaller timescale, which directly change the kinetic energy of the particles (for example, by contact with particles in a heat bath), are identified as the heat change δQ.

For ideal gas particles contained in a cubical box, for example, the dimensions of the box $a(t) \times a(t) \times a(t)$ may be considered to be increasing very slowly, so that the fractional change in $a(t)$ over the typical time required for a particle to bounce back and forth is very much smaller than one. This is precisely the situation usually termed a "reversible quasistatic expansion" in thermodynamics. The theorem of Boltzmann, also known as the Boltzmann–Clausius–Szily theorem (Klein 1970a, pp. 260–262), states simply that under such circumstances, if the motion is periodic between times t_0 and t_1, and T is the kinetic energy of the system, then the heat exchanged with the environment is

$$\delta Q = \frac{2}{t_1 - t_0} \delta \left(\int_{t_0}^{t_1} T dt \right) = \frac{2}{\tau} \delta \left(\tau \overline{T} \right), \tag{A.84}$$

where $\tau = t_1 - t_0$ is the period and $\overline{T}$ the average kinetic energy over a period. This result shows that the inverse average kinetic energy is indeed the integrability factor for the heat change, as

$$\frac{\delta Q}{\overline{T}} = \frac{2}{\tau \overline{T}} \delta \left(\tau \overline{T} \right) = \delta \left\{ 2 \ln \left(\tau \overline{T} \right) \right\} = \delta S, \tag{A.85}$$

with S now identified (up to a factor of Boltzmann's constant) with the entropy.

The adiabatic principle championed by Ehrenfest begins with the special case of this result with $\delta Q = 0$. We examine the response of the system to very slow changes in the parameters of the system (relative to a typical period of the motion) and prohibit all other sources of energy change of the system. Introducing the frequency ν of the motion as the inverse of the period τ, we find that, under such adiabatic transformations, the ratio $\overline{T}/\nu$ remains constant. This ratio is thus a so-called *adiabatic invariant*.

Boltzmann's proof of this result is somewhat involved. Instead, we shall shortly give a different demonstration modeled on the work of Ehrenfest's student Burgers in 1917, which takes full advantage of the canonical formalism discussed in the preceding section. In the meantime, the essential content of the theorem can easily be displayed with two examples where it can be illustrated with elementary techniques.

In the first example, we consider a particle bouncing elastically in one dimension between two walls (see Figure A.2). The left wall, labeled **L** in the figure, is stationary and located at $x = 0$, while the right wall, labeled **R**, is located at $x = a(t) = a_0 + \epsilon t$. At time zero, we imagine our particle moving with speed $v(t) = v_0$, and choose the speed

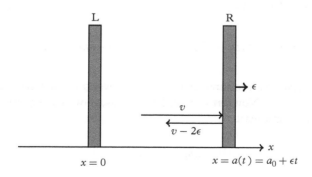

Figure A.2 *Particle in a one-dimensional box that is undergoing an adiabatic expansion.*

$$H = \frac{p(t)^2}{2m} + \frac{1}{2}ma(t)^2 q(t)^2. \tag{A.95}$$

We perform a canonical transformation of type F_1, as before (see Eq. (A.69)), but now with a generating function that explicitly depends on time (due to $a(t)$). Instead of using the notation (P, Q) for the new canonical variables, we use the notation $(\mathcal{J}, w)$, so that $F_1(q, Q, t)$ becomes $F_1(q, w, t)$. We perform the same canonical transformation at each instant of time as before, but with the previously fixed frequency ω replaced by the corresponding instantaneous value $a(t)$. Again suppressing the time dependence for the moment, we thus arrive at (cf. Eqs. (A.69)–(A.71)):

$$F_1 = \frac{1}{2}maq^2 \cot(2\pi w), \tag{A.96}$$

$$p = \frac{\partial F_1}{\partial q} = maq \cot(2\pi w), \tag{A.97}$$

$$\mathcal{J} = -\frac{\partial F_1}{\partial w} = \frac{\pi maq^2}{\sin^2(2\pi w)}. \tag{A.98}$$

Solving as before for the old coordinates in terms of the new, we find

$$q = \sqrt{\frac{\mathcal{J}}{\pi ma}} \sin(2\pi w), \tag{A.99}$$

$$p = \sqrt{\frac{m\mathcal{J}a}{\pi}} \cos(2\pi w), \tag{A.100}$$

$$H = \frac{a}{2\pi}\mathcal{J}. \tag{A.101}$$

These are just Eqs. (A.79), (A.80), and (A.74) with $(w, \mathcal{J})$ for (Q, P), a for ω, $(a/2\pi)\mathcal{J}$ for E and $2\pi w$ for ωt (cf. Eq. (A.77)).

From Eq. (A.101) we see that the new canonical momentum variable $\mathcal{J}$ has dimensions of *action* (i.e., energy × time), whence the term "action variable", while the new (dimensionless) coordinate w appears in trigonometric functions of periodicity $w = 1$, whence the term "angle variable" (much more on this in Section A.2.3). The new Hamiltonian,

$$\tilde{H}(\mathcal{J}, w, t) = H(\mathcal{J}, w, t) + \frac{\partial F_1}{\partial t}, \tag{A.102}$$

contains an additional term due to the explicit time dependence of the generating function via the slowly changing angular frequency $a(t)$:

$$\tilde{H}(\mathcal{J}, w, t) = \frac{a}{2\pi}\mathcal{J} + \frac{1}{2}mq^2\dot{a}\cot(2\pi w). \tag{A.103}$$

Using Eq. (A.99) for q, we can rewrite the new Hamiltonian as

$$\tilde{H}(\mathcal{J}, w, t) = \frac{a}{2\pi}\mathcal{J} + \frac{1}{2}m\left(\frac{\mathcal{J}}{\pi ma}\sin^2(2\pi w)\right)\dot{a}\cot(2\pi w)$$

$$= \frac{\mathcal{J}}{4\pi}\left(2a + \sin(4\pi w)\frac{\dot{a}}{a}\right). \tag{A.104}$$

The Hamilton equations giving the time dependence of the new momentum and coordinate variables are

$$\dot{\mathcal{J}} = -\frac{\partial\tilde{H}}{\partial w} = -\mathcal{J}\frac{\dot{a}}{a}\cos(4\pi w), \tag{A.105}$$

$$\dot{w} = \frac{\partial\tilde{H}}{\partial\mathcal{J}} = \frac{a}{2\pi} + \frac{1}{4\pi}\frac{\dot{a}}{a}\sin(4\pi w). \tag{A.106}$$

Suppose that at time t_0 the frequency $a(t_0) \equiv \omega_0$ corresponds to a period $\tau = 2\pi/\omega_0$. From Eq. (A.105) we find by integrating from t_0 to $t_1 = t_0 + \tau$,

$$\mathcal{J}(t_1) - \mathcal{J}(t_0) = -\int_{t_0}^{t_1}\dot{a}\frac{\mathcal{J}}{a}\cos(4\pi w)dt. \tag{A.107}$$

We now regard $\dot{a}$ as a first-order small quantity, taking $\dot{a}\tau/a \ll 1$, so that up to terms of higher order in $\dot{a}$, we may remove $\dot{a}$ from the integral (replacing it by its value at the initial time t_0, for example) and obtain

$$\frac{\mathcal{J}(t_1) - \mathcal{J}(t_0)}{\dot{a}(t_0)} \simeq -\int_{t_0}^{t_1}\frac{\mathcal{J}}{a}\cos(4\pi w)dt. \tag{A.108}$$

We note that in Eq. (A.108), if $a(t)$ were constant, rather than varying by terms of first order in $\dot{a}$ over a period, the factor $\mathcal{J}(t)/a(t)$ would be constant, and the cosine function would execute a whole number of periods and therefore integrate to zero. The critical property of action-angle variables established by Burgers in the general conditionally-periodic case is that *the integrand only contains sine or cosine functions of the angle variables which oscillate a nonzero integer number of times* (the constant term in a general Fourier expansion being removed by the derivative with respect to w in Eq. (A.105)). The slow variation of $a(t)$ over the period τ therefore means that the integral on the right gives a result of order $f(\mathcal{J}_0, \omega_0)\dot{a}(t_0)$, where $f(\mathcal{J}_0, \omega_0)$ is assumed to be a function which remains bounded over the motion, independently of the slow variation of $a(t)$. After $N = T/\tau$ such periods, we accumulate a contribution to $(\mathcal{J}(t_0 + T) - \mathcal{J}(t_0))/\dot{a}$ of order $N\dot{a} \simeq T\dot{a}$ so

$$\frac{\mathcal{J}(t_0 + T) - \mathcal{J}(t_0)}{\dot{a}} \simeq T\dot{a}. \tag{A.109}$$

The new coordinate is given by $Q = \partial S/\partial E$ (cf. Eq. (A.83)), so

$$
\begin{aligned}
Q &= \sqrt{2m} \int \frac{1}{2\sqrt{E - \frac{1}{2}m\omega^2 q^2}} \, dq \\
&= \sqrt{\frac{m}{2E}} \int \frac{1}{\sqrt{1 - \frac{1}{2}\frac{m\omega^2 q^2}{E}}} \, dq \qquad\qquad \text{(A.117)} \\
&= \frac{1}{\omega} \arcsin\left(\sqrt{\frac{m\omega^2}{2E}} q\right)
\end{aligned}
$$

On the other hand, the Hamilton equation of motion for the new Hamiltonian $\tilde{H} = H = P$ is trivially solvable:

$$
\dot{Q} = \frac{\partial \tilde{H}}{\partial P} = 1 \Rightarrow Q(t) = t + \text{constant}. \qquad\qquad \text{(A.118)}
$$

We can choose to start our clock so that this constant is zero. Inserting $Q(t) = t$ in Eq. (A.117) and rearranging, we find

$$
q(t) = \sqrt{\frac{2E}{m\omega^2}} \sin(\omega t), \qquad\qquad \text{(A.119)}
$$

exactly the solution Eq. (A.79) previously obtained by more mundane methods.

Our first generalization will be to systems with more than one degree of freedom. We still assume that the Hamiltonian has no explicit time-dependence and is therefore again a constant of the motion. As before, we denote the initial set of canonical variables as $(q_1, q_2, \ldots, q_n, p_1, p_2, \ldots, p_n)$. We search for a generating functional $S(q_i, P_i)$ with the property that the new Hamiltonian is cyclic in all the new coordinates Q_i, in other words, that it is a function only of the P_i, which must therefore be constants of the motion, as

$$
\dot{P}_i = -\frac{\partial \tilde{H}}{\partial Q_i} = 0. \qquad\qquad \text{(A.120)}
$$

The Hamilton–Jacobi equation in this case is (cf. Eq. (A.114))

$$
H\left(q_i, \frac{\partial S}{\partial q_i}\right) = E, \qquad\qquad \text{(A.121)}
$$

which is now a first-order partial differential equation for S in n independent variables. Of the corresponding n integration constants, one will be a trivial additive integration constant in S. Since the canonical transformation we are interested in only involves derivatives of S, this constant is of no physical significance. The $n - 1$ remaining constants will be denoted by $\alpha_2, \ldots, \alpha_n$, for reasons that will soon become clear. A general solution of Eq. (A.121) can therefore be written

$$S = S(q_1, q_2, \ldots, q_n, E, \alpha_2, \alpha_3, \ldots, \alpha_n). \tag{A.122}$$

By convention, for reasons of symmetry, the conserved energy constant E is usually denoted by α_1, and one writes

$$S = S(q_1, q_2, \ldots, q_n, \alpha_1, \alpha_2, \ldots, \alpha_n). \tag{A.123}$$

The constants α_i, or indeed any n independent functions of them,

$$P_i = \gamma_i(\alpha_1, \ldots, \alpha_n), \tag{A.124}$$

can be regarded as the new (a fortiori time-independent) momenta P_i, on which a generating function of type F_2 must depend. In general, the energy $E = \alpha_1$ is obtained as a function of the P_i by inverting the system of equations Eq. (A.124)

$$E = E(P_1, P_2, \ldots P_n). \tag{A.125}$$

The Hamilton equations for the new coordinates $Q_i(t)$ read (cf. Eq. (A.118) and recall that $H = \tilde{H} = E$)

$$\dot{Q}_i = \frac{\partial E}{\partial P_i} \equiv v_i \Rightarrow Q_i(t) = v_i t + \beta_i. \tag{A.126}$$

The "velocities" (or "frequencies", depending on the dimensions of the Q_i) v_i here are, of course, time-independent as are all the new momenta P_i. The constants β_i are arbitrary phase constants, or integration constants for the n first-order equations of motion of the Q_i.

We thus see that the essence of the Hamilton–Jacobi procedure is to reduce the Hamilton equations to trivially solvable differential equations, in which the new coordinates are at most linear functions of time and the new momenta are all constants of the motion. The complexity of the original dynamical problem (which can be considerable!) is entirely packed into obtaining a sufficiently explicit solution S of the Hamilton–Jacobi equation Eq. (A.121). By "sufficiently explicit", we mean that the solution of the original problem—i.e., the explicit time-dependence of the original coordinates $q_i(t)$ and momenta $p_i(t)$—can be obtained by

1. Determining the constant values of the new momenta P_i in terms of the initial values $(q_i(0), p_i(0))$ by solving the n equations

$$p_i(0) = \frac{\partial S(q_i, P_i)}{\partial q_i}\Big|_{q_i=q_i(0)} \tag{A.127}$$

for the P_i.

2. Inverting the n equations (with the P_i now set to their determined constant values)

$$Q_i(t) = \frac{\partial S(q_i, P_i)}{\partial P_i} \tag{A.128}$$

to give the $q_i(t)$ in terms of the $Q_i(t)$ (now known as simple linear functions of time, as in Eq. (A.126)).

In practice, the Hamilton–Jacobi method does not turn out to be a particularly useful tool for dealing with intrinsically complicated dynamical systems. The utility of the approach is, perhaps not coincidentally given its origins in celestial mechanics, essentially limited to exactly the set of problems of interest to astronomers calculating planetary orbits and to physicists in the 1910s exploring quantum ideas in atomic systems (which, after all, were pictured as "planetary systems", with light electrons circulating around a heavy nucleus, subject to an inverse-square attractive force).

Mathematically, the Hamilton–Jacobi equation only supplies a practically useful technique in the subset of cases corresponding to so-called *separable systems*. These are systems in which the form of the Hamiltonian allows a choice of initial variables q_i, p_i such that a solution $S(q_i, P_i)$ of the Hamilton–Jacobi equation (A.121) can be found as a sum of functions of each coordinate q_i separately:

$$S(q_i, P_i) = S_1(q_1, P_i) + S_2(q_2, P_i) + \dots + S_n(q_n, P_i). \tag{A.129}$$

Note that each of the functions S_i, while only depending on the single corresponding coordinate variable q_i, can in general depend on all the (constant) new momentum variables P_i.

Let us illustrate the situation with a simple example which is important for our discussion of both the Bohr–Sommerfeld quantization conditions and the adiabatic principle (see Sections 5.1 and 5.2). A two-dimensional isotropic (equal frequencies in both x and y directions) harmonic oscillator is defined by the Hamiltonian

$$H = \frac{1}{2m}(p_x^2 + p_y^2) + \frac{1}{2}m\omega^2(x^2 + y^2) \tag{A.130}$$

$$= \frac{1}{2m}\left(p_r^2 + \frac{p_\theta^2}{r^2} + m^2\omega^2 r^2\right), \tag{A.131}$$

where in the second line we have introduced polar coordinates, $(x = r\cos\theta, y = r\sin\theta)$, with p_r the radial component of momentum and p_θ the angular momentum of the particle (so that p_θ/r is the tangential component of the momentum). In the Cartesian coordinates (x, y), the separability of the system is immediately obvious, as the Hamiltonian can clearly be written as a sum of terms, each involving only one of the coordinates separately (which is, as we shall see, a sufficient but not a necessary condition for separability of the Hamilton–Jacobi equation). If we write

$$S(x, y, P_i) = S_x(x, P_i) + S_y(y, P_i) \tag{A.132}$$

and use the form Eq. (A.130), the Hamilton–Jacobi equation becomes

$$E = \left\{ \frac{1}{2m} \left(\frac{\partial S}{\partial x} \right)^2 + \frac{1}{2} m\omega^2 x^2 \right\} + \left\{ \frac{1}{2m} \left(\frac{\partial S}{\partial y} \right)^2 + \frac{1}{2} m\omega^2 y^2 \right\} \tag{A.133}$$

$$= \left\{ \frac{1}{2m} \left(\frac{\partial S_x}{\partial x} \right)^2 + \frac{1}{2} m\omega^2 x^2 \right\} + \left\{ \frac{1}{2m} \left(\frac{\partial S_y}{\partial y} \right)^2 + \frac{1}{2} m\omega^2 y^2 \right\}. \tag{A.134}$$

The quantity between the left set of curly brackets on the second line clearly can only depend on x, but is equal to terms which are independent of x, so must in fact be equal to a constant α_1 (independent of both x and y).[12] The same argument shows that the right pair of curly braces contain a constant quantity α_2. Thus

$$\alpha_1 = \frac{1}{2m} \left(\frac{\partial S_x}{\partial x} \right)^2 + \frac{1}{2} m\omega^2 x^2, \tag{A.135}$$

$$\alpha_2 = \frac{1}{2m} \left(\frac{\partial S_y}{\partial y} \right)^2 + \frac{1}{2} m\omega^2 y^2, \tag{A.136}$$

$$E = \alpha_1 + \alpha_2. \tag{A.137}$$

The constants α_1 and α_2 are (for obvious reasons) called the *separation constants* of the Hamilton–Jacobi equation in this case. Their physical interpretation is clear: α_1 and α_2 represent the contributions to the total energy E from the x and y components of the motion, respectively, which in this model happen to be separately conserved.

At this point, we have reduced the problem to a pair of one-dimensional problems, so the solution for Eq. (A.135) and Eq. (A.136) proceeds exactly along the lines indicated previously for the one-dimensional oscillator. The solutions for $x(t)$ and $y(t)$ then take the usual trigonometric form (with arbitrary phase shifts in the sine functions), generating the familiar Lissajous ellipses in the x-y plane.

The Hamilton–Jacobi equation for the isotropic 2D oscillator can also be separated in polar coordinates (r, θ), where the Hamiltonian is given by the form Eq. (A.131), even though the expression for the energy does not take the form of a sum of terms individually containing only one of the coordinates. With the usual replacements

$$p_r \rightarrow \frac{\partial S}{\partial r}, \qquad p_\theta \rightarrow \frac{\partial S}{\partial \theta}, \tag{A.138}$$

[12] In this example, we temporarily abandon the policy of reserving "α_1" for the energy E itself.

the Hamilton–Jacobi equation takes the form

$$E = \frac{1}{2m}\left(\left(\frac{\partial S}{\partial r}\right)^2 + \frac{1}{r^2}\left(\frac{\partial S}{\partial \theta}\right)^2 + m^2\omega^2 r^2\right). \tag{A.139}$$

Through multiplication by $2mr^2$ and some rearrangement of terms, this equation can be separated into terms involving only r and θ separately:

$$2mEr^2 - r^2\left(\frac{\partial S}{\partial r}\right)^2 - m^2\omega^2 r^4 = \left(\frac{\partial S}{\partial \theta}\right)^2. \tag{A.140}$$

The separation Ansatz

$$S(r,\theta) = S_r(r) + S_\theta(\theta) \tag{A.141}$$

then leads to

$$2mEr^2 - r^2\left(\frac{dS_r(r)}{dr}\right)^2 - m^2\omega^2 r^4 = \left(\frac{dS_\theta(\theta)}{d\theta}\right)^2. \tag{A.142}$$

By the same logic as above (cf. Eqs. (A.134)–(A.136)), we have that each side of this equation must be constant. In particular, we may choose

$$\alpha_2 \equiv \frac{dS_\theta(\theta)}{d\theta}, \tag{A.143}$$

while for the other separation constant we may choose, as discussed previously, the energy itself: $\alpha_1 \equiv E$. The Hamilton–Jacobi equation has now been reduced to a one-dimensional ordinary differential equation in the radial variable r (obtained through substitution of $\partial S/\partial r = dS_r/dr$ and $\partial S/\partial \theta = dS_\theta/d\theta = \alpha_2$ in Eq. (A.139)):

$$\alpha_1 = E = \frac{1}{2m}\left(\frac{dS_r(r)}{dr}\right)^2 + \frac{\alpha_2^2}{2mr^2} + \frac{1}{2}m\omega^2 r^2. \tag{A.144}$$

It follows that

$$p_r = \frac{dS_r(r)}{dr} = \sqrt{2m\alpha_1 - \frac{\alpha_2^2}{r^2} - m^2\omega^2 r^2}. \tag{A.145}$$

The solution $S_r(r)$ of Eq. (A.144) is thus given by:

$$S_r(r) = \int \sqrt{2m\alpha_1 - \frac{\alpha_2^2}{r^2} - m^2\omega^2 r^2} \, dr. \tag{A.146}$$

One further generalization of the Hamilton–Jacobi procedure should be mentioned briefly at this point. In cases where the Hamiltonian $H(q_i, p_i, t)$ contains an explicit time dependence (due, say, to externally imposed time-varying forces), it is common to choose a type F_2 canonical transformation $\hat{S}(q, P, t)$ (in this case, called *Hamilton's principal function*) with the requirement that the new Hamiltonian *vanishes identically*:

$$\tilde{H} = H\left(q_i, \frac{\partial \hat{S}}{\partial q_i}, t\right) + \frac{\partial \hat{S}}{\partial t} = 0. \tag{A.147}$$

This is a first-order partial differential equation in $n + 1$ variables (the coordinates q_1, $q_2, \ldots, q_n$ plus the time t) with, apart from the trivial additive constant in $\hat{S}$, n nontrivial integration constants $\alpha_1, \alpha_2, \ldots, \alpha_n$. Hamilton's equation in the new variables imply that *both* the Q_i and the P_i are constants of the motion, as

$$\dot{Q}_i = \frac{\partial \tilde{H}}{\partial P_i} = 0, \qquad \dot{P}_i = -\frac{\partial \tilde{H}}{\partial Q_i} = 0. \tag{A.148}$$

The new momenta can be chosen as any function of the α_i, or simply as the α_i themselves.[13]

A.2.2   Poisson brackets

In 1809 Poisson introduced what has come to be known as the *Poisson bracket*. This analytic construction was to play a profound role both in the advancement of the technology of celestial mechanics and in the conceptual formulation of the new quantum theory. Poisson was motivated by the desire to obtain an automatic procedure for generating constants of the motion in complicated dynamical systems (such as the three- or higher-number body problem in celestial mechanics). If we consider any two functions $f(q_i, p_i)$ and $g(q_i, p_i)$, on phase space, then the Poisson bracket of these two functions is defined as

---

[13] For further details, we refer the reader to Goldstein, Poole, and Safko (2002, Sec. 10.3).

$$\{f,g\} \equiv \frac{\partial f}{\partial q_i} \frac{\partial g}{\partial p_i} - \frac{\partial g}{\partial q_i} \frac{\partial f}{\partial p_i}. \tag{A.149}$$

If the functions f and g are already known to be constants of the motion, then Poisson's theorem asserts that their Poisson bracket automatically supplies a constant of the motion (a result that we shall prove below). Combining q_i's and p_i's into $z_i = (q_1, \ldots, q_n, p_1, \ldots, p_n)$ as we did before (see Eq. (A.47)) and using the matrix $\mathcal{J}$ defined in Eq. (A.48), we can rewrite the definition of Poisson brackets as[14]

$$\{f,g\} = \frac{\partial f}{\partial z_i} \mathcal{J}_{ij} \frac{\partial g}{\partial z_j}. \tag{A.150}$$

Equation (A.53), the condition for the invariance of the Hamilton equations under a canonical change of coordinates, is also the condition for the invariance of the Poisson bracket under such a change of variables. Let $\mathcal{T}_{ij} = \partial Z_i / \partial z_j$ be the Jacobian for the transformation from the old z_i variables to the new Z_i ones. Defining $\hat{f}(Z_i) \equiv f(z_i(Z_i))$ and $\hat{g}(Z_i) \equiv g(z_i(Z_i))$, we can rewrite Eq. (A.150) as

$$\{f(z_i), g(z_i)\} \equiv \frac{\partial f(z_i)}{\partial z_i} \mathcal{J}_{ij} \frac{\partial g(z_i)}{\partial z_j} = \frac{\hat{f}(Z_i)}{\partial Z_k} \mathcal{T}_{ki} \mathcal{J}_{ij} \mathcal{T}_{lj} \frac{\partial \hat{g}(Z_i)}{\partial Z_l}. \tag{A.151}$$

Comparing this last expression to the definition of Poisson brackets in the new coordinates

$$\{\hat{f}(Z_i), \hat{g}(Z_i)\} \equiv \frac{\partial \hat{f}(Z_i)}{\partial Z_k} \mathcal{J}_{kl} \frac{\partial \hat{g}(Z_i)}{\partial Z_l}, \tag{A.152}$$

we see that $\{f(z_i), g(z_i)\} = \{\hat{f}(Z_i), \hat{g}(Z_i)\}$ if and only if $\mathcal{T}\mathcal{J}\mathcal{T}^T = \mathcal{J}$, which is just Eq. (A.53).

---

[14] In the one-dimensional case, Eq. (A.150) can be written as (cf. note 6):

$$\{f,g\} \equiv \left(\frac{\partial f}{\partial q}, \frac{\partial f}{\partial p} \right) \begin{pmatrix} 0 & 1 \\ -1 & 0 \end{pmatrix} \begin{pmatrix} \partial g/\partial q \\ \partial g/\partial p \end{pmatrix}.$$

The Hamiltonian equation of motions can also be written in a very simple and illuminating form using Poisson brackets,[15]

$$\dot{z}_i = \mathcal{J}_{ij} \frac{\partial H}{\partial z_j} = \{z_i, H\}. \tag{A.153}$$

More generally, the time derivative of any function on phase space $f(z_i)$ becomes

$$\frac{d}{dt} f(z_i) = \frac{\partial f}{\partial z_i} \dot{z}_i = \frac{\partial f}{\partial z_i} \mathcal{J}_{ij} \frac{\partial H}{\partial z_j} = \{f, H\}, \tag{A.154}$$

so the constants of the motion are simply those functions on phase space with vanishing Poisson brackets with the Hamiltonian.

We are now in a position to prove Poisson's theorem. It emerges most simply from the identity

$$\{\{f, g\}, h\} + \{\{h, f\}, g\} + \{\{g, h\}, f\} = 0, \tag{A.155}$$

now associated with the name of Jacobi, who made extensive use of Poisson brackets in his 1842 Königsberg lectures, later published in Jacobi (1884). The proof of the Jacobi identity requires nothing but the definition of Poisson brackets in Eq. (A.150), the symmetry of the second partial derivatives of the functions f, g, and h, and some relabeling of indices. We will leave this as an exercise for the reader. To prove Poisson's theorem, we choose the function h to be the Hamiltonian H and take f and g to be known already as constants of the motion. According to Poisson's theorem, their Poisson bracket $\{f, g\}$ must then be a constant of the motion as well. This follows directly from the Jacobi identity (A.155). Since f and g are constants of the motion, their Poisson brackets with $h = H$ vanish (see Eq. (A.154)). If $\{f, h\} = \{g, h\} = 0$, Eq. (A.155) reduces to $\{\{f, g\}, h\} = 0$. So $\{f, g\}$ has a vanishing Poisson bracket with the Hamiltonian as well. Like f and g, $\{f, g\}$ is thus a constant of the motion, which is just what Poisson's theorem says.

The Poisson brackets of the fundamental canonical coordinates themselves have a particularly simple form. Going back to separate notations for generalized coordinates q_i and their conjugate momenta p_i, one has

$$\{q_i, q_j\} = \{p_i, p_j\} = 0, \qquad \{q_i, p_j\} = \delta_{ij}. \tag{A.156}$$

[15] In the one-dimensional case, Eq. (A.153) can be written as (cf. note 14):

$$\{z_i, H\} \equiv \left(\frac{\partial z_i}{\partial q}, \frac{\partial z_i}{\partial p} \right) \begin{pmatrix} 0 & 1 \\ -1 & 0 \end{pmatrix} \begin{pmatrix} \partial H / \partial q \\ \partial H / \partial p \end{pmatrix}.$$

For $z_i = q$, $\left(\frac{\partial z_i}{\partial q}, \frac{\partial z_i}{\partial p} \right) = (1, 0)$ and $\{z_i, H\} = \{q, H\} = \partial H / \partial p$.

For $z_i = p$, $\left(\frac{\partial z_i}{\partial q}, \frac{\partial z_i}{\partial p} \right) = (0, 1)$ and $\{z_i, H\} = \{p, H\} = -\partial H / \partial q$.

These equations as well as Eq. (A.154) were to play a fundamental role in the conceptual development of matrix mechanics and Dirac's q-number formalism in the 1920s.

A.2.3 Action-angle variables

We indicated previously that the Hamilton–Jacobi method is really only technically feasible for separable systems, where the differential equation allows solution by a Hamilton characteristic function that can be written as a sum of terms each dependent on only a single coordinate variable: in other words, in cases where the original partial differential equation is reducible to a system of n decoupled ordinary differential equations. The power of the method is even more apparent for the subset of separable mechanical systems that exhibit periodic (or multiply periodic) motion. These systems correspond to problems in which a set of canonical coordinates $(q_1, \ldots, q_n, p_1, \ldots, p_n)$ can be found in which (a) the Hamilton–Jacobi equation separates, and (b) the motion in the two-dimensional plane of each canonical pair (q_i, p_i) *separately* is of either (*i*) *librational* or (*ii*) *rotational* type.

We must first explain the classification of motion as either libration or rotation. The pendulum defined by the Hamiltonian in Eq. (A.32) fortunately provides us with a simple system displaying both types of motion. For smaller values of the total energy, the pendulum swings between maximum angular displacements $-\pi < \theta_{\min} < 0 < \theta_{\max} < +\pi$ (with $\theta_{\min} = -\theta_{\max}$). The coordinate q (here chosen as the signed arc length displacement $l\theta$ from the bottom point, where l is the length of the string) is then said to exhibit *libration*: a complete cycle of the motion corresponds to the coordinate sequence $l\theta_{\min} \to 0 \to l\theta_{\max} \to 0 \to l\theta_{\min}$, with the phase space point of the system executing traversals of the closed orbits indicated in red in Figure A.3.

Note that for smaller energies (the smaller orbits) the orbit paths approximate ellipses, as in the small angle approximation the equation of the orbits is simply

$$\frac{1}{2m}p^2 + \frac{mg}{2l}q^2 = E + mgl, \tag{A.157}$$

where E is the conserved total energy (with the potential energy measured from the attachment point of the pendulum). As the energy increases, the anharmonicity of the oscillator (q^4 and higher terms in the expansion of the cosine) leads to a distortion of the orbits from elliptical form, but they remain closed paths in the (p, q) plane. Once the energy E exceeds $+ mgl$, the pendulum has sufficient energy to rotate up to and through the vertical position, and then continues to execute full rotations around the point of attachment, either in a clockwise (corresponding to the upper green orbits in Fig. A.3) or anti-clockwise (lower green orbits) direction. In this case, the momentum variable becomes a periodic function of the coordinate q, which is, up to an overall constant, an angle variable, $q = l\theta$, so that the state of the system at two values of θ differing by a multiple of 2π is the same. The system therefore traverses the upper green curves in Fig. A.3 from left to right, reentering the figure on the left at $\theta_{\min} = -\pi$ once $\theta_{\max} = +\pi$ is reached on the right. For obvious reasons, this type of phase space orbit is referred to as *rotation*.

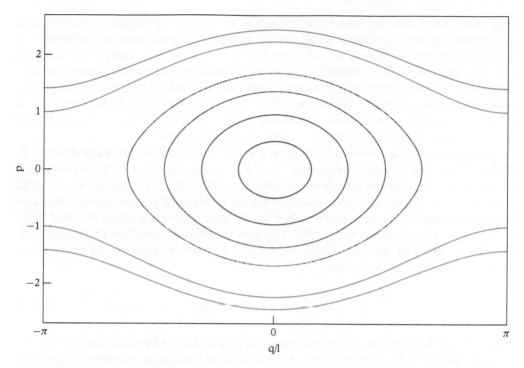

Figure A.3 *Orbits of the swinging pendulum (m = g = l = 1), showing (a) libration (red), (b) rotation (green). Based on Fig. 1 in Ehrenfest (1913c, p. 595).*

The applicability of the action-angle formalism requires that a set of separated coordinates be found in which all canonical pairs (q_i, p_i) correspond to either librational or rotational motion.

Let us therefore assume that a solution $S(q_i, \alpha_i)$ to the Hamilton–Jacobi equation Eq. (A.121) has been found in terms of n nontrivial separation constants, with

$$S(q_i, \alpha_i) = S_1(q_1, \alpha_1, \ldots, \alpha_n) + \ldots + S_n(q_n, \alpha_1, \ldots, \alpha_n), \qquad (A.158)$$

where each momentum variable p_i is now given as a function only of the single corresponding coordinate q_i (and, in general, *all* separation constants $\alpha_1, \ldots, \alpha_n$):

$$p_i = p_i(q_i, \alpha_1, \ldots, \alpha_n) = \frac{\partial S_i(q_i, \alpha_1, \ldots, \alpha_n)}{\partial q_i}. \qquad (A.159)$$

Eq. (A.159) for $i = 1, \ldots, n$ are the *orbit equations* of the system in the original canonical variables.

As noted previously, one is free to use directly the separation constants α_i as the new momentum variables P_i. A different choice, in which the P_i are chosen as very particular functions of the α_i, turns out to be of enormous importance both in celestial mechanics and in the old quantum theory. These special P_i's are called *action variables*. In celestial mechanics, they allow for a highly efficient way to evaluate by systematic perturbative approximation the corrections to Keplerian motion in a multiplanetary solar system. In the old quantum theory, the adiabatic invariance of action variables (see Section A.1.4) singles them out as natural candidates for quantization in a wide class of atomic systems. The action-angle formalism also played a central role in the development of the Kramers dispersion theory (Kramers 1924a, 1924b, Kramers and Heisenberg 1925), which, in turn, led to the famous *Umdeutung* paper with which Heisenberg (1925) laid the foundation for matrix mechanics (Duncan and Janssen 2007).

Conventionally the canonically conjugate action and angle variables are denoted $\mathcal{J}_i$ and w_i. They are chosen by requiring (a) the $\mathcal{J}_i$ to be constants of the motion, and therefore functions of the separation constants α_i; and (b) the motion to be rotational with respect to *all* the new variables, with the w_i normalized so that each w_i increases by exactly unity when the corresponding (q_i, p_i) pair executes a single traversal of its orbit (either vibrational or rotational). The new momentum variables that accomplish this remarkable feat can very simply be defined in terms of the original phase space coordinates (q_i, p_i) of the system:

$$\mathcal{J}_i \equiv \oint p_i(q_i, \alpha_1, \ldots, \alpha_n) dq_i, \tag{A.160}$$

where the cyclic notation on the integral indicates that it is to be taken exactly once (in the direction in which $p_i dq_i$ is always positive) around the closed (libration) or periodic (rotation) orbit for that coordinate alone. It is apparent from the definition in Eq. (A.160) that the $\mathcal{J}_i$ are indeed just functions of the separation constants α_i and hence constants of the motion (which, by definition, are necessarily positive).

Alternatively, these equations may be inverted to give the α_i's in terms of the $\mathcal{J}_i$'s. If we choose the first separation constant α_1, as usual, to be the constant energy E (i.e., H for time-independent Hamiltonians), we have

$$E = \alpha_1 = H(\mathcal{J}_1, \ldots, \mathcal{J}_n). \tag{A.161}$$

We may also rewrite the Hamilton–Jacobi functions $S_i(q_i, \alpha_1, \ldots, \alpha_n)$ as functions of the new action variables instead, i.e., as $S_i(q_i, \mathcal{J}_1, \ldots, \mathcal{J}_n)$.

We still need to check that the action variables $\mathcal{J}_i$ defined as in Eq. (A.160) are conjugate to new coordinate variables w_i which have the unit period property mentioned above. The new coordinate variables are given in the usual way for a type-F_2 canonical transformation (see Eqs. (A.81)–(A.83)):

$$w_i = \frac{\partial S(q_1, \ldots, q_n, \mathcal{J}_1, \ldots, \mathcal{J}_n)}{\partial \mathcal{J}_i} = \sum_j \frac{\partial S_j(q_j, \mathcal{J}_1, \ldots, \mathcal{J}_n)}{\partial \mathcal{J}_i}. \tag{A.162}$$

In principle, for given fixed values of the action variables $\mathcal{J}_i$, these equations can be inverted to give the original coordinates q_i in terms of the new w_i, which, as we shall now see, have a very simple time dependence.

The Hamilton equation of motion for the new coordinates,

$$\dot{w}_i = \frac{\partial H(\mathcal{J}_1,\ldots,\mathcal{J}_n)}{\partial \mathcal{J}_i} \equiv \nu_i(\mathcal{J}_1,\ldots,\mathcal{J}_n), \tag{A.163}$$

implies, as the $\mathcal{J}_i$ are constants of the motion, simple linear time evolution for the w_i:

$$w_i(t) = \nu_i t + \beta_i. \tag{A.164}$$

If we imagine a particular w_i expressed as a function of the original coordinates q_j $(j = 1,\ldots, n)$ and consider the change in w_i under an infinitesimal change $q_j \to q_j + dq_j$, we have (using Eq. (A.82))

$$\delta w_i = \sum_j \frac{\partial w_i}{\partial q_j} dq_j = \sum_j \frac{\partial^2 S_j}{\partial \mathcal{J}_i \partial q_j} dq_j$$

$$= \sum_j \frac{\partial p_j}{\partial \mathcal{J}_i} dq_j = \frac{\partial}{\partial \mathcal{J}_i}\left(\sum_j p_j dq_j\right). \tag{A.165}$$

The total accumulated change in w_i as the system is passed through n_j complete cycles of the (q_j, p_j) orbit (either libration or rotation) is therefore, by the definition of the action variables Eq. (A.160),

$$\Delta w_i = \frac{\partial}{\partial \mathcal{J}_i}\sum_j n_j \mathcal{J}_j = n_i. \tag{A.166}$$

In other words, the w_i variable increases precisely by unity for each complete cycle of the q_i, p_i orbit, or equivalently, the original phase space coordinates q_i, p_i must, when expressed in terms of the w_i (for any fixed values of the new conserved momenta $\mathcal{J}_i$), be periodic with unit period.

The terminology "angle variables" for the w_i is now clear: the system motion can be regarded as taking place in each $w_i, \mathcal{J}_i$ plane along horizontal lines of constant $\mathcal{J}_i$ between two vertical lines corresponding to values of w_i differing by unity (or by 2π, if we take $2\pi w_i$ for the new coordinates), analogous to the phase space orbits of rotational type discussed previously.

The separate periodicity of the original coordinates with respect to the angle variables implies, by standard Fourier analysis, the representation

$$q_i(\mathcal{J}_1,\ldots,\mathcal{J}_n; t) = \sum_{\vec{\tau}} A_{i\vec{\tau}}(\mathcal{J}_1,\ldots,\mathcal{J}_n)\exp\left(2\pi i(\tau_1 w_1 + \ldots + \tau_n w_n)\right)$$

$$= \sum_{\vec{\tau}} A_{i\vec{\tau}}(\mathcal{J}_1,\ldots,\mathcal{J}_n)\exp\left(2\pi i(\vec{\tau}\cdot\vec{w})\right) \tag{A.167}$$

where $\vec{\tau} = (\tau_1, \tau_2, ...\tau_n)$ is a n-vector of integer (positive, negative, or zero) components. A function representable by a multiple Fourier expansion of the form Eq. (A.167) is said to be *multiply periodic*. The original canonical momenta p_i share the periodicity of the q_i, so we also have

$$p_i(\mathscr{I}_1, \mathscr{I}_2, \ldots, \mathscr{I}_n; t) = \sum_{\vec{\tau}} B_{i\vec{\tau}}(\mathscr{I}_1, \ldots, \mathscr{I}_n) \exp\left(2\pi i (\vec{\tau} \cdot \vec{w})\right) \qquad (A.168)$$

If the frequencies v_i in Eq. (A.163) are *commensurate* (i.e., if there exist an integer N and a fundamental frequency v_0 such that $N v_i$ is an integer multiple of v_0 for all i) then the motion of the entire system is periodic. If the condition for the v_i's to be commensurate is satisfied, the exponentials in Eq. (A.167) and Eq. (A.168) all return to their original values after a time $T = N/v_0$. The entire system traces out a closed orbit in the original $2n$-dimensional phase space. If the frequencies are *not* commensurate in this sense, then the trajectory of the system in phase space typically fills out densely a bounded region in phase space, coming arbitrarily close, given sufficient time, to any point of this region.[16]

At the end of Section A.1.3, we already encountered an example of action-angle variables in disguise (cf. Eqs. (A.68)–(A.80)). This was in our analysis of the one-dimensional harmonic oscillator, with Hamiltonian (see Eq. (A.68))

$$H = \frac{p^2}{2m} + \frac{1}{2}m\omega^2 q^2. \qquad (A.169)$$

The orbit of the system in the (q, p)-plane for any value E of its energy given by this Hamiltonian is thus an ellipse,

$$\frac{p^2}{a^2} + \frac{q^2}{b^2} = 1, \qquad (A.170)$$

with semi-major/minor axes $a = \sqrt{2mE}$ and $b = \sqrt{2E/m\omega^2}$. The action variable measures the area contained within this ellipse

$$\mathscr{I} = \oint p\,dq = \pi ab = \frac{2\pi E}{\omega}. \qquad (A.171)$$

This is precisely the new momentum variable $P = 2\pi E/\omega$ obtained previously in Eq. (A.73) as an example of a type-F_1 canonical transformation. The conjugate coordinate variable $Q(t) = (\omega/2\pi)t$ in Eq. (A.77) clearly increments by exactly unity for each cycle of period $2\pi/\omega$ and is thus indeed an angle variable.

---

[16] There may exist a commensurability among a proper subset of the frequencies, in which case the motion densely fills a region of dimensionality strictly smaller than the full phase space.

A.2.4 Canonical perturbation theory

In the introduction to his appendix on perturbation theory (*Störungsrechnung*) in Vol. 5 of the *Handbuch der Physik* (1927), Erwin Fues, after summarizing the development of approximation techniques for dealing with the many-body problem in planetary motion, points out that:

> Physics has previously had little occasion to be interested in this particular calculational approach, until the development of the Bohr atomic model suddenly produced a close connection to astronomy. Bohr was really the first to become aware of the tools which, readied for use by astronomers, were available for use in atomic research. The influence of an external electric field [i.e., the Stark effect] and of the relativistic inertial law on the Kepler orbits of the hydrogen atom could be calculated using the method of secular perturbations (Fues 1927, p. 132).[17]

Although Schwarzschild and not Bohr deserves the credit for the injection of sophisticated techniques of analytical mechanics into atomic research,[18] the overall point made here reinforces our previous assertion: prior to the middle of the second decade of the twentieth century, physicists concerned with quantum theory (even those with as extensive a classical training as Planck and Sommerfeld) were limited to the use of fairly elementary methods in classical mechanics. This situation underwent a radical alteration in the period from 1916 to 1925, when physicists rapidly learned to adapt the methods of celestial mechanics to problems in atomic and molecular physics.

In this section we shall give a short introduction to canonical perturbation theory, using as our example a problem that played a critical role in the transition from the old to the new quantum theory: the scattering of light from atomic systems. This is the essential problem of dispersion theory, the calculation of the frequency dependence of the index of refraction of a transparent medium, typically a gas. The problem amounts to calculating the response of a charged electron executing a Bohr orbit to a weak applied oscillating electric field, with a period considerably longer than the orbital period of the electron.

The connection to the analogous astronomical problem is easy to see: one needs to calculate the gradual shift in the Kepler orbit of an inner planet due to the slowly varying ("secular") external force due to the gravitational pull of a distant outer one (e.g., Saturn) with a much longer period (i.e., planetary year). For example, the Earth makes about 29.5 revolutions around the Sun for each revolution of Saturn; an electron in the ground state of the Bohr hydrogen atom makes about 13 revolutions in one period of a light wave (and therefore, one full oscillation of the electric field of the wave at the position of the atom) in the mid-visible range passing over the atom.

---

[17] At the end of his introduction, Fues (Fues 1927, p. 133) emphasizes that his treatment of perturbation theory builds on the exposition of Hamilton–Jacobi theory and canonical transformations in the preceding appendix of the *Handbuch*, which he co-authored (Nordheim and Fues 1927).

[18] See Section 5.1.4 for the letter with which Schwarzschild alerted Sommerfeld to these methods and Section 6.3 for the application of these methods by Schwarzschild (1916) and Epstein (1916b) to the Stark effect.

The general problem faced in both these examples can be phrased as follows. We have a system subject to a Hamiltonian which can be written as the sum of a time-independent part H_0 (for the inner planet/electron plus Sun/nucleus), and a time-dependent part $H_{ext}(t)$ due to the external perturbation (outer planet/light wave). It is also assumed that the amplitude of the external perturbation is small, allowing us to compute its effects in a formal series in which first-order terms in the perturbation dominate.

The beautiful feature of the action-angle formalism is that the perturbation theory can be carried out essentially independently of the form of the unperturbed Hamiltonian. This feature was emphasized by John Van Vleck (1924b, p. 350) in a paper containing some of the most advanced applications of canonical perturbation theory in the old quantum theory (Duncan and Janssen 2007, p. 630). We shall exploit this generality to illustrate the method using our by now familiar toy model of a one-dimensional harmonic oscillator as the unperturbed system.[19] We realize that we are using the proverbial sledgehammer on a fly, as the problem under consideration is just that of a periodically forced undamped harmonic oscillator, described in every introductory physics text! Nevertheless, the final result Eq. (A.190), when expressed in terms of a derivative with respect to action variables, was precisely the ingredient needed for the insertion of correspondence-principle arguments due to Kramers (1924b), Born (1924), and Van Vleck (1924a) in dispersion theory in the run-up to Heisenberg's (1925) *Umdeutung* paper and matrix mechanics. These matters will be discussed in Volume Two.

We attach an electric charge $-e$ to the particle of mass m in our one-dimensional harmonic oscillator. The particle is subjected to a weak oscillating electric field $E(t) = F \cos(\omega t)$. The resulting total Hamiltonian is

$$H = H_0 + H_{ext}(t) = \frac{p^2}{2m} + \frac{1}{2}m\omega_0^2 x^2 + eFx \cos(\omega t). \qquad (A.172)$$

We have replaced the general coordinate q by x here, and assume that the charged particle is subject to the influence of an x-polarized incident electromagnetic wave. Also, the natural frequency of the oscillator is denoted by ω_0, to distinguish it from the frequency ω of the incident light.

Our task is to determine the perturbation $\Delta x(t)$ in the orbit $x(t)$ of the particle due to the impressed perturbation $H_{ext}(t) = eFx \cos(\omega t)$. First, note that if we switch off temporarily the perturbation (so $F = 0$, $H = H_0$), we can write $x(t)$ in terms of the action-angle variables $\mathcal{J}$ and $w = \omega_0 t/2\pi = \nu_0 t$ (cf. Eq. (A.167)):

$$x(t) = \sum_{\tau = \pm 1} A_\tau(\mathcal{J})e^{2\pi i \tau w}, \qquad (A.173)$$

---

[19] Our presentation follows Duncan and Janssen (2007, pp. 631–634). The generalization of the resulting formulas to the case of Kepler orbits in the Bohr model will be unproblematic (ibid., pp. 648–651).

where A_τ has to satisfy the conjugacy relation $A_\tau = A^*_{-\tau}$ to ensure that $x(t)$ in Eq. (A.173) is real $(x(t) = x^*(t))$.[20] The action-angle variables $\mathcal{J} = H_0/\nu_0$ and $w = \nu_0 t$ satisfy Hamilton's equations (cf. Eqs. (A.75)–(A.76), with $(\mathcal{J}, w)$ instead of (P, Q)):

$$0 = -\dot{\mathcal{J}} = \frac{\partial H_0}{\partial w}, \quad \frac{\partial H_0}{\partial \mathcal{J}} = \dot{w} = \nu_0. \tag{A.174}$$

It is a special feature of the simple harmonic (as opposed to anharmonic) oscillator that the frequency ν_0 is independent of the amplitude of motion (and hence of the action variable). The generating function Eq. (A.69) for the contact transformation from (x, p) to $(w, \mathcal{J})$ is time-independent, so the old and new Hamiltonians coincide in value (i.e., one simply reexpresses the original Hamiltonian in the new variables).

Even with the perturbation turned on *we shall continue to use the same contact transformation*,[21] computing the perturbations $(\Delta w, \Delta \mathcal{J})$ induced by the applied field in the action-angle variables $(w, \mathcal{J})$ as an expansion in the electric field amplitude F. These are *not* action-angle variables for the full Hamiltonian $H_0 + H_{\text{ext}}$, only for the unperturbed Hamiltonian H_0.

Eventually, we are interested in the displacement Δx in the particle coordinate (to first order in F) induced by the applied field. To first order, Δx is given by

$$\Delta x = \frac{\partial x}{\partial \mathcal{J}} \Delta \mathcal{J} + \frac{\partial x}{\partial w} \Delta w. \tag{A.175}$$

Using Eq. (A.173) to evaluate $\partial x/\partial \mathcal{J}$ and $\partial x/\partial w$, we can rewrite this as:

$$\Delta x = \sum_{\tau = \pm 1} \left(\frac{\partial A_\tau}{\partial \mathcal{J}} \Delta \mathcal{J} + 2\pi i\tau A_\tau \Delta w \right) e^{2\pi i\tau w}. \tag{A.176}$$

Assuming the external field to be switched on at time zero, the first-order shifts Δw and $\Delta \mathcal{J}$ are given by:

$$\Delta \mathcal{J} = \int_0^t \Delta \dot{\mathcal{J}} \, dt, \quad \Delta w = \int_0^t \Delta \dot{w} \, dt. \tag{A.177}$$

where the integrands $\Delta \dot{\mathcal{J}}$ and $\Delta \dot{w}$ are determined by Hamilton's equations.

---

[20] Substituting $A_\tau = |A_\tau| e^{i\varphi_\tau}$ in Eq. (A.173), we can rewrite $x(t)$ as:

$$x(t) = |A_1| e^{2\pi i w + i\phi_1} + |A_{-1}| e^{-2\pi i w + i\phi_{-1}}.$$

If $A_\tau = A^*_{-\tau}$, it follows that $|A_{-1}| = |A_1|$ and $\varphi_{-1} = -\varphi_1$. Inserting this into the expression for $x(t)$ above, we find:

$$x(t) = |A_1| e^{2\pi i w + i\phi_1} + |A_1| e^{-(2\pi i w + i\phi_1)} = 2|A_1| \cos(2\pi w + \phi_1)$$

The condition $A_\tau = A^*_{-\tau}$ thus indeed ensures that $x(t)$ is real.

[21] An alternative approach, used, for instance, by Born (1924), employs the generating function taking us from the action-angle variables of H_0 to those of H, expanded in powers of the perturbation. The method described here is the one used by Van Vleck (1924b).

The perturbation in Eq. (A.172) will induce a time-dependence in the action variable, as Hamilton's equation for the action variable in the presence of the perturbing field now reads

$$\dot{\mathcal{J}} = -\frac{\partial H_0}{\partial w} - eF\frac{\partial x}{\partial w}\cos(2\pi\nu t) = -eF\frac{\partial x}{\partial w}\cos(2\pi\nu t). \tag{A.178}$$

Note that we still have $\partial H_0/\partial w = 0$, so $\Delta\dot{\mathcal{J}} = \dot{\mathcal{J}}$. At this point it is convenient to go over to complex exponentials and replace $\cos(2\pi\nu t)$ by $\frac{1}{2}\left(e^{2\pi i\nu t} + e^{-2\pi i\nu t}\right)$. Eq. (A.178) then turns into:

$$\Delta\dot{\mathcal{J}} = \dot{\mathcal{J}} = -\frac{eF}{2}\frac{\partial x}{\partial w}\left(e^{2\pi i\nu t} + e^{-2\pi i\nu t}\right). \tag{A.179}$$

Inserting Eq. (A.173) for $x(t)$ into Eq. (A.179), we find

$$\Delta\dot{\mathcal{J}} = -\pi ieF\sum_{\tau=\pm 1}\tau A_\tau\left(e^{2\pi i(\tau w+\nu t)} + e^{2\pi i(\tau w-\nu t)}\right). \tag{A.180}$$

To obtain the displacement of the orbit, which is a linear effect in the applied field, of amplitude F, we only need $\Delta\dot{\mathcal{J}}$ and Δw to first order in F. This means that the angle variables w in the exponents in Eq. (A.181) can be taken to zeroth order, i.e., $w = \nu_0 t$.

Integrating Eq. (A.180) for $\Delta\dot{\mathcal{J}}$, we find:

$$\Delta\mathcal{J} = \int_0^t \Delta\dot{\mathcal{J}}dt \tag{A.181}$$

$$= -\pi ieF\sum_{\tau=\pm 1}\tau A_\tau\left\{\left.\frac{e^{2\pi i(\tau\nu_0 t+\nu t)}}{2\pi i(\tau\nu_0 + \nu)}\right|_0^t + \left.\frac{e^{2\pi i(\tau\nu_0 t-\nu t)}}{2\pi i(\tau\nu_0 - \nu)}\right|_0^t\right\}$$

$$= \frac{eF}{2}\sum_{\tau=\pm 1}\tau A_\tau\left\{\frac{1 - e^{2\pi i(\tau\nu_0 t+\nu t)}}{\tau\nu_0 + \nu} + \frac{1 - e^{2\pi i(\tau\nu_0 t-\nu t)}}{\tau\nu_0 - \nu}\right\}.$$

Next, we need to compute the first-order shift Δw in the angle variable w. Hamilton's equation for the angle variable w in the presence of the perturbation is:[22]

---

[22] It is a special feature of the simple harmonic oscillator that the characteristic frequency ν_0 is independent of the amplitude and thus of the action variable $\mathcal{J}$. In general, ν_0 will be a function of $\mathcal{J}$. The first term on the right-hand side of Eq. (A.182) would then become $\partial H_0/\partial\mathcal{J} = \nu_0(\mathcal{J}) = \nu_0 + (\partial\nu_0/\partial\mathcal{J})\Delta\mathcal{J}$.

$$\dot{w} = \frac{\partial H_0}{\partial \mathcal{J}} + eF \frac{\partial x}{\partial \mathcal{J}} \cos(2\pi \nu t)$$

$$= \nu_0 + \frac{eF}{2} \sum_{\tau = \pm 1} \frac{\partial A_\tau}{\partial \mathcal{J}} \left(e^{2\pi i (\tau w + \nu t)} + e^{2\pi i (\tau w - \nu t)} \right), \tag{A.182}$$

where in the second line we substituted Eq. (A.173) for $x(t)$. Once again, w may be replaced by $\nu_0 t$ in the exponentials in Eq. (A.182). Integrating the second term in Eq. (A.182), which gives the shift $\Delta \dot{w}$ due to H_{int}, we find:

$$\Delta w = \int_0^t \Delta \dot{w} dt \tag{A.183}$$

$$= \frac{ieF}{4\pi} \sum_{\tau = \pm 1} \frac{\partial A_\tau}{\partial \mathcal{J}} \left\{ \frac{1 - e^{2\pi i (\tau \nu_0 t + \nu t)}}{\tau \nu_0 + \nu} + \frac{1 - e^{2\pi i (\tau \nu_0 t - \nu t)}}{\tau \nu_0 - \nu} \right\}.$$

Substituting Eq. (A.181) for $\Delta \mathcal{J}$ and Eq. (A.183) for Δw into Eq. (A.176) (with $w = \nu_0 t$), we find

$$\Delta x = \sum_{\tau' = \pm 1} \left(\frac{\partial A_{\tau'}}{\partial \mathcal{J}} \Delta \mathcal{J} + 2\pi i \tau' A_{\tau'} \Delta w \right) e^{2\pi i \tau' \nu_0 t} \tag{A.184}$$

$$= \frac{eF}{2} \sum_{\tau' = \pm 1} \sum_{\tau = \pm 1} \left\{ \frac{\partial A_{\tau'}}{\partial \mathcal{J}} \tau A_\tau - \tau' A_{\tau'} \frac{\partial A_\tau}{\partial \mathcal{J}} \right\} \frac{1 - e^{2\pi i (\tau \nu_0 t - \nu t)}}{\tau \nu_0 - \nu} e^{2\pi i \tau' \nu_0 t}$$

$$+ (\nu \to -\nu)$$

where "$(\nu \to -\nu)$" is shorthand for: "the same term with ν replaced by $-\nu$ everywhere."

If $\tau = \tau'$, the expression in curly brackets in Eq. (A.184) vanishes. Only terms with $\tau' = -\tau$ survive and the double sum over τ and τ' reduces to one sum of τ. Moreover, since $A_\tau = A^*_{-\tau}$, we have that $|A_\tau|^2 = A_\tau A^*_\tau = A_\tau A_{-\tau}$. Hence, for $\tau' = -\tau$, the expression in curly brackets in Eq. (A.184) reduces to:

$$\frac{\partial A_{\tau'}}{\partial \mathcal{J}} \tau A_\tau - \tau' A_{\tau'} \frac{\partial A_\tau}{\partial \mathcal{J}} = \tau \left(\frac{\partial A_{-\tau}}{\partial \mathcal{J}} A_\tau + A_{-\tau} \frac{\partial A_\tau}{\partial \mathcal{J}} \right) = \tau \frac{\partial |A_\tau|^2}{\partial \mathcal{J}}. \tag{A.185}$$

The terms with $\tau' = -\tau$ in Eq. (A.184) have two distinct time-dependent factors. Terms with the time dependence $e^{\pm 2\pi i \nu_0 t}$ of the unperturbed oscillator correspond to transients excited by the sudden turning on of the perturbation at time zero: in a realistic system, these terms would eventually disappear due to dissipative effects. The remaining nontransient terms have the time dependence of the perturbing field $e^{\pm 2\pi i \nu t}$, and are typically the ones of interest. In the case of light scattering, these are the "coherent" terms.

Substituting Eq. (A.185) for the expression in curly brackets in Eq. (A.184) and making the additional substitution

$$\frac{1 - e^{2\pi i(\tau \nu_0 t - \nu t)}}{\tau \nu_0 - \nu} e^{2\pi i \tau' \nu_0 t} \quad \longrightarrow \quad \frac{-e^{\mp 2\pi i \nu t}}{\tau \nu_0 \mp \nu}, \tag{A.186}$$

to select the coherent part of Δx, we arrive at:

$$\Delta x_{\text{coh}} = \frac{eF}{2} \sum_{\tau = \pm 1} \tau \frac{\partial |A_\tau|^2}{\partial \mathcal{J}} \left\{ \frac{e^{-2\pi i \nu t}}{\nu - \tau \nu_0} - \frac{e^{2\pi i \nu t}}{\nu + \tau \nu_0} \right\}. \tag{A.187}$$

The imaginary part of this expression is a sum over the product of odd and even functions of the index τ,

$$-\frac{eF}{2} \sum_{\tau = \pm 1} \tau \frac{\partial |A_\tau|^2}{\partial \mathcal{J}} \left(\frac{1}{\nu - \tau \nu_0} + \frac{1}{\nu + \tau \nu_0} \right) \sin 2\pi \nu t, \tag{A.188}$$

and therefore vanishes, leaving only the real part:

$$\Delta x_{\text{coh}} = \frac{eF}{2} \sum_{\tau} \tau \frac{\partial |A_\tau|^2}{\partial \mathcal{J}} \left(\frac{1}{\nu - \tau \nu_0} - \frac{1}{\nu + \tau \nu_0} \right) \cos (2\pi \nu t)$$

$$= \frac{eF}{2} \sum_{\tau} \tau \frac{\partial |A_\tau|^2}{\partial \mathcal{J}} \left(\frac{2\tau \nu_0}{\nu^2 - \tau^2 \nu_0^2} \right) \cos (2\pi \nu t). \tag{A.189}$$

Since $|A_\tau|^2 = |A_{-\tau}|^2$, and since τ only takes on the values ± 1 in the case of the simple harmonic oscillator, $\tau^2 = 1$ and the two terms in the summation over τ are identical. Although in this special case the derivative with respect to $\mathcal{J}$ only acts on $|A_\tau|^2$, we are free to include the expression $2\nu_0/(\nu^2 - \nu_0^2)$ within the scope of the derivative (recall that ν_0 does not depend on $\mathcal{J}$ in this case). Eq. (A.189) then becomes[23]

$$\Delta x_{\text{coh}} = 2eF \frac{\partial}{\partial \mathcal{J}} \left(\frac{\nu_0}{\nu^2 - \nu_0^2} |A_{\tau=1}|^2 \right) \cos (2\pi \nu t). \tag{A.190}$$

This result, in which the induced displacement of a bound charge due to an applied electric field is expressed as a derivative with respect to an action variable, was in an

---

[23] Although in the present simple case of a simple harmonic oscillator, the natural frequency ν_0 is independent of amplitude, so that the derivative with respect to $\mathcal{J}$ does not affect the $\nu_0/(\nu^2 - \nu_0^2)$ factor, this is not the case for more general periodic or multiply periodic systems, where the formula for Δx_{coh} must be written as in Eq. (A.190). In this connection, see footnote 22. As we shall see in Volume Two, this formula is the indispensable starting point for the derivation of the Kramers dispersion formula (Kramers 1924b, p. 310; cf. Duncan and Janssen 2007, p. 634).

ideal form for transcription to the quantum theory, via a sharpened version of the correspondence principle introduced by Born, Kramers, and van Vleck, whereby the continuous classical derivative is replaced by a discrete quantum difference (as the action variable(s) are replaced by discretely quantized quantities, $\mathcal{J}_i = n_i h$, in the old quantum theory).

Of course, in our toy example, from Eqs. (A.78)–(A.79), we have (as $\mathcal{J} = P = 2\pi E/\omega_0$)

$$A_{\tau=1} = \frac{1}{2}\sqrt{\frac{\mathcal{J}}{\pi\, m\omega_0}}, \tag{A.191}$$

so in this case the coordinate displacement becomes

$$\Delta x_{\mathrm{coh}} = 2eF\frac{1}{4\pi\, m\omega_0}\frac{v_0}{v^2 - v_0^2}\cos(2\pi v t) = \frac{eF}{m}\frac{1}{\omega^2 - \omega_0^2}\cos(2\pi v t), \tag{A.192}$$

a result which can be found in many introductory treatments of harmonic motion and resonance phenomena.[24]

---

[24] See, for instance, Eq. (23.5) in Vol. 1 of the Feynman lectures (Feynman, Leighton and Sands 1964).

B

Spectroscopy

Although the phenomenon of the rainbow, and even the appearance of separate colors when white light passed through certain transparent media, were clearly known to the ancients, it was the discovery by Isaac Newton in 1666 of the variable bending encountered by light of different colors passing through such media that marks the beginning of the road leading to modern spectroscopy.[1] Newton projected sunlight passing through a hole in a screen onto a glass prism and observed that the emergent rays formed an orderly sequence of colors (which he named a "spectrum"), with the violet rays bent (or *refracted*) the most, the red rays at the other end of the spectrum bent the least. The terminology "more (or less) refrangible" was consequently adopted to indicate this feature. Until the advent and full acceptance of the wave theory, the degree of "refrangibility" served as a characterizing feature of "homogeneous"—i.e., monochromatic—rays, instead of wavelength.

Unfortunately, Newton did not realize the importance of a collimated source of light: the light from the hole in the screen impinged at various angles on the prism, with the result that the emergent spectrum was considerably blurred out, with a mixture of light rays of different colors arriving at the same point on the observation screen. When William Hyde Wollaston, almost a century and a half later, noticed that sunlight passing through a *narrow* slit and then refracted through a prism displayed a spectrum with additional features—specifically, a number of black lines parallel to the orientation of the slit—the path to a quantitatively useful spectroscopy was opened.

B.1   Early quantitative spectroscopy

The founder of spectroscopy as a quantitative discipline is undoubtedly Joseph Fraunhofer, a Munich optician who for the last fifteen years of his life (he died in 1826) produced optical equipment at the very limits of the technology available at the time. His earlier work involved refraction of the light from various sources through a prism,

---

[1] A particularly helpful source for the history of spectroscopy up to Rydberg is Baly (1912). This appendix relies heavily on Baly's book.

like Newton and Wollaston, but with the improvement that the light incident on the prism from a slit was further collimated by use of a convex lens. With such an arrangement the dark lines in the solar spectrum became extremely prominent and sharp. They have ever since been referred to as "Fraunhofer lines". Fraunhofer found several hundred such lines, but picked the eight most prominent for designation with the initial letters of the alphabet, A through H. Fraunhofer did not propose any explanation for the existence of these dark lines, now called absorption lines, but realized their practical importance as reliable and precise standard lines. In particular, the construction of achromatic lenses for telescopes, one of the chief activities of the Optical Institute of which Fraunhofer was technical director, required careful measurement of the refractive index of various types of lens glass as a function of wavelength of light, clearly only possible if "homogeneous rays" were available.

In the course of his investigations, Fraunhofer made several discoveries of seminal importance. We list here only the most important:

1. The light from incandescent laboratory sources (e.g., candles) did not display the dark lines characteristic of the solar spectrum.

2. The strong yellow color imparted to flames into which common salt (sodium chloride) was introduced produced a bright line coincident in location with the fourth (D) dark line in the solar spectrum. Moreover, in both cases, the line in question actually consisted of a pair of two very close lines.

3. The light from various bright stars (studied by attaching his prism spectrometer to the Institute's telescope) contained different lines (as well as some of the same) from the solar spectrum. This observation marks the beginning of modern stellar spectroscopy.

It should be recalled that we are now in the era of firm acceptance of the wave theory of light, following the work of Young and Fresnel, and Fraunhofer was fully aware of the phenomenon of interference of light from two sources when, around 1821, he invented the diffraction grating which was to become for more than a century the indispensable apparatus for the determination of absolute wavelengths of light rays. The essential point was to use the light diffracted through a series of many evenly spaced (at a separation d) parallel slits, which, for incident monochromatic light of wavelength λ, would produce diffracted lines at angles ϑ_n given by $n\lambda = d \sin \vartheta_n, n = 1, 2, 3$, etc.[2] With his first diffraction grating, Fraunhofer was able to determine the wavelength of the D line in the solar spectrum to be 5890 Å, with an error of a few Å (1 Ångstrom $= 10^{-10}$ m; 1 nm $= 10$ Å).

By the early 1830s, a variety of types of spectra had been distinguished by spectral analysis, both by the venerable prism spectrometer and by the more recent grating method introduced by Fraunhofer. Incandescent solids, such as quicklime (calcium

---

[2] The reasoning here is exactly analogous to that which produces the Bragg formula discussed in Chapter 6 in connection with X-ray diffraction.

oxide) heated to a high temperature, emitted a white light which yielded a continuous and featureless spectrum, lacking either bright or dark lines. Flames typically produced a basically continuous spectrum, occasionally with some superimposed bright lines (for example, due to sodium impurities). Metallic salts introduced into flames produced bright line spectra characteristic of the particular salt. The extremely bright yellow D line of sodium[3] often overpowered other lines when only trace amounts of sodium were present, which caused considerable confusion, and delayed the acceptance of the notion that spectra could be used as reliable markers of chemical composition.[4] And then there was the ubiquitous sunlight, with its continuous spectrum interrupted by the mysterious dark Fraunhofer lines.

The 1830s also saw the emergence of a technology that was to become extremely important in the study of atomic spectra, the use of electric sparks and arcs to generate light characteristic of either the electrodes themselves or the gas in the gap between the electrodes (or both). The first use of this approach seems to have been Wheatstone's demonstration in 1835 of the appearance of spectra of volatilized metal atoms (from the electrodes) in an electrical discharge at atmospheric pressure (where the voltage required to initiate the discharge was very high). The use of electric arcs soon became common. Here, a continuous discharge (i.e., electric current) between two electrodes to which a high voltage difference was applied could be achieved by bringing the electrodes into momentary contact and then gradually separating them. A related approach, passing electric sparks generated by an induction coil through the material to be studied, gave spectra with new lines (as well as some of the same lines seen in arc spectra), presumably evoked by the even higher voltages present in the spark apparatus.

Electric arcs with a metal cathode (negative terminal) and carbon anode (positive terminal) contaminated with sodium could be used, for example, to provide an intense source of the yellow sodium D line. In 1849 Foucault observed that if sunlight is passed through an arc discharge of this kind, and viewed side by side with direct sunlight, the Fraunhofer D line is relatively *darkened* by the passage through the sodium-containing vapor of the arc. The spectrum of the arc discharge viewed directly, on the other hand, displayed a bright line spectrum, dominated by the extremely strong D line. He concluded that the medium in the arc (i.e., the sodium) was simultaneously capable of emitting and absorbing light of the specific wavelength of the D line. This observation seems to have mostly spread by word of mouth, and was apparently unknown to Kirchhoff when he brought order to the subject in his seminal paper of 1859 on emission and absorption, to which we now turn.

---

[3] A sodium D line can often be found in the light emerging from a blast furnace, the quintessential continuous black-body spectrum!

[4] The confusion abated once William Swan, in 1857, was able to identify the culprit, by introducing ever more dilute amounts of sodium salts until the yellow line could finally be eliminated. Nevertheless, by the early 1830s, William Talbot was already certain of the association of distinct bright lines with specific chemical agents, and suggested the adoption of optical analysis as an analytical tool of chemistry. With Kirchhoff and Robert Bunsen, he deserves to be considered a founding father of spectrochemistry.

B.2   Kirchhoff's Laws

Gustav Kirchhoff made fundamental contributions in both electricity and heat theory.[5] In the paper of concern to us here, "On the connection between the emission and absorption of light and heat," Kirchhoff (1859b) used thermodynamic arguments (specifically, the second law of thermodynamics) to show that for bodies maintained at a fixed temperature T, and for light of a given wavelength λ, the ratio of the emissive power (basically, energy emitted per unit area per unit time from the surface of the body at that wavelength) to the absorption coefficient (the fraction of incident light energy of that wavelength absorbed) is a universal function of λ and T.[6] In plain English, materials that strongly absorb light of a certain wavelength must also strongly emit light of that wavelength, and similarly for materials which absorb/emit only weakly, or not at all. It follows that materials that are opaque, or at least absorb a finite fraction of incident light at all wavelengths, must, when heated sufficiently (so that the universal function becomes sufficiently sizable) emit a continuous spectrum. Kirchhoff went further:

> It further follows, that when a glowing gas gives a discontinuous [i.e., line] spectrum, and one passes through this gas rays of sufficient intensity, lacking bright or dark lines [i.e., a continuous spectrum], dark lines must appear at exactly those places in the spectrum where the bright lines of the glowing gas are located. This provides the theoretical foundation for the procedure which I suggested in my earlier communication was adapted to the chemical analysis of the sun's atmosphere (p. 9).

In this earlier communication, "On the Fraunhofer lines," Kirchhoff (1859a) had effectively repeated Foucault's observation, by passing sunlight through a sodium-impregnated flame and observing a darkening of the Fraunhofer D line. The insights provided by Kirchhoff (1859b, p. 786) in the later communication have become sanctified as "Kirchhoff's Three Laws of Spectroscopy." When heated,

1. solid (but nontransparent) bodies emit continuous radiation.

2. gases (as long as they are not too dense) emit light with a discrete spectrum.

3. light with a continuous spectrum when passed through a (dilute) gas displays a spectrum with dark absorption lines located at the wavelengths where the same gas emits light.

In the last case, the continuous light needs to be of a sufficient intensity for the absorption effect to become apparent (in comparison to the nearby wavelengths which are not absorbed), given that there will inevitably also be some emitted radiation at the given line

---

[5] This section relies on Brand (1995).

[6] In a subsequent paper, Kirchhoff (1860) identifies this function with the now famous black-body function described in Chapter 2.

position. Thus, an intense continuous beam passed through a glowing gas will exhibit dark absorption lines if viewed head on, but if the gas is then viewed from the side (out of the way of the continuous beam), one sees a bright line spectrum, with the emission lines at the same locations as the previous dark absorption lines.

Kirchhoff's seminal theoretical work was soon followed by an extensive experimental collaboration with the chemist Robert Bunsen, in which flame spectra of numerous elements were studied and compared to solar lines. In rapid succession, cesium (1860), rubidium (1860), thallium (1861), and indium (1863) were identified and isolated with the help of flame spectroscopy. In all cases, their names derive from references to the color of prominent spectral lines in their flame spectra (Brand 1995, p. 63).

B.3   Technological advances and the emergence of analytic spectroscopy

Precision spectroscopy (both of laboratory emission and solar absorption lines) begins with the work of Anders Ångström in Sweden. In the early 1850s, he showed that the light generated by an electric spark passed through a gas between metal electrodes produced bright lines, now called emission lines, characteristic both of the metal atoms composing the electrodes and of the gas between them. He was already aware at this point, several years before Kirchhoff's work, that a body emitting light of definite wavelengths when heated would preponderantly absorb the same wavelengths at normal temperatures, a conclusion arrived at by analogy to mechanical resonance. In 1853 he observed the optical emission lines of hydrogen (subsequently called the Balmer lines) in an electric spark discharge. In the next decade and a half, he undertook an enormous extension of Fraunhofer's work on the solar spectrum, publishing in 1868 a map of the entire visible range (between Fraunhofer's A and H lines), with wavelengths measured in the now familiar Å unit ($= 10^{-10}$ m), to an accuracy of six significant digits.

Ångström was able to distinguish lines due to absorption in the solar atmosphere from the so-called "telluric" lines due to absorption in the Earth's atmosphere. Moreover, in almost all cases, the solar absorption lines could be associated with a known element, by comparison with laboratory spectral studies. In particular, the four hydrogen lines H_α, H_β, H_γ, and H_δ at 6562, 4861, 4340, and 4101 Å respectively were clearly identified, the first being Fraunhofer's C line.

The late 1850s saw the appearance, and rapid adoption for spectroscopic purposes, of yet another technology for producing intense bright line sources for study—the Geissler tubes, as improved by Plücker, who introduced a small capillary tube connecting two regions of the glass container with a gas at low pressure, each region supplied with an electrode. The electric discharge (i.e., a current passing between the electrodes through the gas, and exciting the gas atoms to a luminous state) was achievable at lower voltages than needed in an electric arc or spark apparatus as a consequence of the low pressure of the gas. The intense current density achieved in the capillary section resulted in a very intense source of light from that part of the tube.

Another technical improvement around this time was the introduction of photo-graphic methods to record the spectrum into the ultraviolet: silver emulsion was sensitive to wavelengths less than 3000 Å where the human eye failed. By 1888, the new concave grating introduced by the American physicist Henry Rowland allowed him to publish a comprehensive photographic map of the solar spectrum, with very accurate *relative* wavelength measurements, using the sodium D1 line as a reference.

B.4   The numerology of spectra: Balmer and Rydberg

The first successful effort to provide a useful mathematical description of spectra appeared in a paper by the Swiss high-school teacher Johann Balmer (1885) (see Plate 17), who showed that a very simple formula could be devised to reproduce the four prominent visible lines (H_α through H_δ) in the emission spectrum of hydrogen. He found that for

$$\lambda_m = h \frac{m^2}{m^2 - 4}, m = 3, 4, 5, 6 \tag{B.1}$$

with the constant h (called by Balmer the *Grundzahl* or base number) chosen as 3645.6 Å the four calculated wavelengths agreed with the best values obtained by Ångstrom to 0.1 Å or less. The fifth line, $m = 7$, at the edge of the visible spectrum, could not be easily located (although there was a weak nearby line in the solar spectrum, about 1.5 Å from the calculated value). Balmer's confidence that the agreement for the first four lines was not purely accidental seems to have been considerably amplified by the additional data of the astronomical spectroscopists William Huggins and Hermann Carl Vogel on ultraviolet hydrogen lines found in the spectrum of white stars. For the first four ultraviolet lines found by these workers, Balmer could show agreement with Eq. (B.1) for $m = 8, 9, 10, 11$ with discrepancies of the order of 1 Å. All in all, the likelihood of a simple formula like (B.1) reproducing so many distinct lines by pure accident seemed very low indeed.

It had been realized quite early that many of the lines seen in the solar spectrum, or in laboratory measurements, were actually resolvable into a pair, or in some cases, a triplet of nearby lines. For example, by 1868 Ångstrom had resolved the D line produced by sodium into a pair of lines 6 Å apart (the D_1 line at 5895.13 Å, D_2 at 5889.12 Å). It further emerged that doublets (pairs of close lines) were ubiquitous in the spectrum of the elements in the first column of the periodic table (the alkali metals, Li, Na, K, etc.) while triplets were found in the spectrum of the alkaline earths in column II (Be, Mg, Ca, etc.), or more generally in elements of even valence, such as zinc or cadmium.

George Downing Liveing and James Dewar (1879) at Cambridge performed a detailed study of the spectra of sodium and potassium, sorting the lines in the optical (i.e., visible) region of the spectrum into two distinct alternating series (once the much brighter D line was set aside). The lines in one of these series were called "sharp", the lines in the other "diffuse", reflecting their appearance in the spectrum. In both series, the lines appeared at decreasing separations as one went toward smaller wavelengths, giving the appearance of an approach to a limiting value.

An advance of absolutely critical importance for further progress in understanding regularities of spectra came with the work of Walter Noel Hartley (1883), who found that if, instead of using the wavelengths themselves, the lines were characterized in terms of their *wave numbers* (i.e., the inverse of the wavelength, typically measured in units of numbers per centimeter, cm^{-1}), groups (or series) of lines could be identified in which the separation between the first and second, or second and third, member of the triplet remained essentially constant. This empirical rule would soon be applied to great advantage by the Swedish physicist Johannes Rydberg (see Plate 16), and used to identify lines which belonged in the same multiplet, which were sometimes separated quite considerably, with, confusingly, other unrelated lines intervening.

Rydberg's (1890) seminal paper, "Study of the constitution of the emission spectra of the chemical elements," was the starting point for a huge amount of organizational work that followed, trying to bring order to the enormous variety of spectra observed in atoms other than the quintessentially simple hydrogen. Rydberg begins (Rydberg 1890, pp. 6–7) with a nod to the work of Hartley in discovering the constancy of wavenumber separation in line series, and to the work of Liveing and Dewar in identifying homologous line series (i.e., series of lines which could be grouped by use of some common property). He then emphasizes the importance of the work of Balmer in finding a simple algebraic representation of the hydrogen spectrum, including the continuation of the Balmer series into the ultraviolet. After a survey of the most well known spectra in chapter 2, Rydberg proceeds in chapter 3 to a general discussion of line series. Two short quotes will serve to indicate his choice of terminology, which would become a permanent part of the language of atomic spectroscopy:

> [Liveing and Dewar] have found various series of analogous characteristics in the elements Li, Na, K, Mg, Ca and Zn. The successive lines of these series approach one another [towards the violet], at the same time decreasing in intensity. A very important observation is that the lines are alternately *sharp* and *diffuse* [the French original has: "*étroite* ('sharp')" and "*nébuleuse* ('diffuse')"], which allows them to distinguish the two types of series. In the ultraviolet part of the spectrum of Li, Na, K they have also found another type of series of which they have not been able to identify the first member. These first lines, which are also the most intense lines in the entire spectrum of these elements, are located in the visible part of the spectrum at quite a large distance from the others, which explains the difficulty in finding the connection between successive lines without already knowing the form of the series function. As we have already said, it was Balmer who was the first to find this function in calculating the spectrum of hydrogen. As fas as I know, this is the only spectrum for which one possesses an equation which represents in a satisfactory way the observed wavelengths (Rydberg 1890, p. 27).

With regard to the third series, composed of the brightest visible line (the D line in the case of sodium) with the remaining lines in the ultraviolet, Rydberg comments at the beginning of Ch. 4:

We have found three different types of series which I will provisionally term the *sharp series, diffuse series*, and *principal series*. I have already mentioned the two first types discovered by Liveing and Dewar. We also owe the first recognition of the third type to these experts, who were however unable to identify the most intense lines which constitute their first member. It is on account of these strong lines, the red line of Li, the lines D_1 and D_2 of Na, etc., that I have named these series the "principal series" (Rydberg 1890, p. 54–55).

The terminology introduced by Rydberg for the various line series would eventually lead to the electron state designations s,p,d which are still encountered by modern students of quantum mechanics in their first exposure to atomic theory.

In Ch. 3, Rydberg begins his search for a function, or set of functions, which would generalize the Balmer formula to the cases where one had clearly identified homologous series of lines, in particular, in the alkali metals in group I of the periodic table, where the spectra were the most well known. He begins by listing the twelve most prominent lines (each a narrow doublet) of the sodium spectrum, from a line in the infrared at 8199 Å, to an indigo line at 4423 Å. The third line, at 5895 Å, is the very intense yellow-orange D line (actually, a doublet, but for the time being the fine structure is being ignored as Rydberg is focusing on the location of the doublets viewed as single entities). These lines are assigned ordinal numbers $m = 1$ through 12, and then the wavelength is plotted as a function of m, as indicated in Fig. B.1. In this range, only the first line of the principal series is visible,

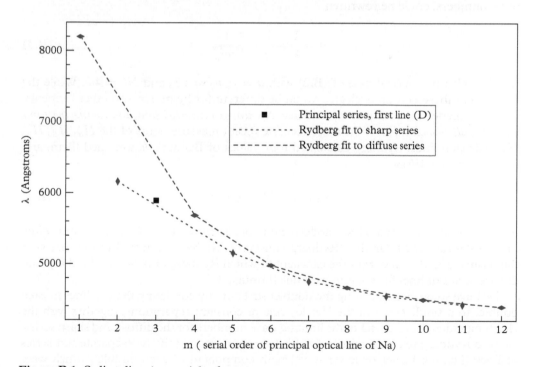

Figure B.1 *Sodium lines, λ vs serial order m.*

at $m = 3$. Setting this line aside, Rydberg could see that the wavelengths of the diffuse series could be connected by a curve that lay systematically above a corresponding curve for the sharp series.

Recognizing the importance of wavenumber $n \equiv 1/\lambda$ from the Hartley rule of constant wavenumber differences within the multiplets, Rydberg then studies the sequences of n values within the sharp and diffuse series *separately*, reordering the lines sequentially within each series, giving six diffuse lines with wave numbers $n(m)$, labelled $m = 1$ to $m = 6$, and five sharp lines, labelled $m = 1$ to $m = 5$. By studying the sequential separations $\Delta n = n(m + 1) - n(m)$ within each series, Rydberg found that plots of $\Delta n(m)$ revealed curves which were of similar shape, i.e., congruent when translated in the abscissa variable m. Accordingly, one could write $\Delta n = F(m + \mu)$, or, by summation (discrete integration) $n = n_0 - f(m + \mu)$. The function f (identical for the different series of a given element, differing only in the horizontal shift μ) was immediately found to be an inverse quadratic to high accuracy, so the final result was

$$n \equiv \frac{1}{\lambda} = n_0 - \frac{N_0}{(m + \mu)^2}. \tag{B.2}$$

The constant n_0 corresponds to the limiting accumulation point of the series in the violet direction. Rydberg recognized that Balmer's formula Eq. (B.1), reexpressed in terms of wave numbers, could be rewritten

$$n = \frac{1}{\lambda} = \frac{1}{h} - \frac{4}{h}\frac{1}{m^2}, \tag{B.3}$$

which is clearly a special case of (B.2) with $\mu = 0$, $n_0 = 1/h$, and $N_0 = 4/h$. While the first two of these constants clearly would be different for hydrogen and other elements, Rydberg made the bold assumption that N_0 *was a universal constant, taking the same value for all elements*. From the most accurate recent measurements of the $H_\alpha, H_\beta, H_\gamma$, H_δ hydrogen lines, Rydberg calculated four values of Balmer's h, averaged them, and then computed $4/h$ as

$$N_0 = 109721.6 \text{ cm}^{-1}. \tag{B.4}$$

This constant, renamed R in modern notation, is now, with complete justification, universally referred to as the "Rydberg constant". We have indicated in Fig. B.1 with the dashed and dotted curves the excellent fit which Rydberg obtained to the sharp and diffuse series of lines for sodium with this formula.

We have so far been ignoring the doublet structure by conflating the two lines in each doublet to a single one, but the Hartley rule of constant separations, together with the clear coincidence in Fig. B.1 of the limiting wave numbers for the diffuse and sharp series, allowed Rydberg to quickly generalize his formula to each of the two separate sub-series of lines (i.e., the higher or lower wavelength component of each doublet, which were known to correspond to stronger or weaker lines respectively), for both the sharp and the diffuse series. The μ values would of course be different for the sharp and diffuse series, and Rydberg (1890, p. 58) adopted the notation $\mu \to \sigma$ for the sharp, $\mu \to \delta$ for the diffuse series. One thus obtained the formulas

$$n = n_1 - \frac{N_0}{(m + \sigma)^2}, \quad \text{strong sharp doublet line;}$$

$$n = n_2 - \frac{N_0}{(m + \sigma)^2}, \quad \text{weak sharp doublet line;}$$

$$n = n_1 - \frac{N_0}{(m + \delta)^2}, \quad \text{strong diffuse doublet line;}$$

$$n = n_2 - \frac{N_0}{(m + \delta)^2}, \quad \text{weak diffuse doublet line.} \qquad \text{(B.5)}$$

Similar sets of equations (three for each series, instead of two) were found to apply for the divalent elements, such as the alkaline earths Mg, Ca, Sr in the second column of the periodic table, where the series lines resolved into triplets.[7] The starting value of the running ordinal m in these formulas was chosen so that σ, δ were positive numbers less than unity. For example, for lithium,[8] the sharp and diffuse series began at $m = 2$, with $\sigma = 0.5951$ and $\delta = 0.9974$. It is apparent that in the case of the diffuse series, the denominator factor involves numbers 2.9974, 3.9974, etc., very close to integer values: an explanation (of sorts) for this would in fact be forthcoming from the Bohr–Sommerfeld theory (cf. Section 7.1).

The principal series, represented by the strongest (hence "characteristic") lines of the spectrum (and the *only* lines seen reversed, i.e., as absorption lines, if white light is passed through the cold dilute vapor of an alkali metal), was known also to consist of doublets, but with rapidly decreasing separation of the lines as one went to shorter wavelengths, so that both sub-series converged to the same term limit. Moreover, in the principal series, in contrast to the sharp and diffuse cases, the more intense line was the one with shorter wavelength. There was also the mysterious (apparent) coincidence that the wave-number splitting of the first line of the principal series was almost identical to the constant splittings ν of the doublets in the sharp and diffuse series ($\nu \equiv n_2 - n_1$ in (B.5)). Incorporating these features, Rydberg suggested the following Ansatz for the principal series

---

[7] As we shall later see, in the case of the diffuse series, the alkali doublets actually consisted of three lines, and the alkaline earth triplets of six lines. The extra "satellite" lines, just on the violet side of the strongest doublet or triplet lines, were first noticed by Rydberg in the case of thallium.

[8] It should be pointed out that the doublet lines of lithium had not yet been resolved, owing to the much smaller splitting (e.g., 0.15 Å for the leading line of the principal series, compared to 6 Å for sodium and 34 Å for potassium). Given the large increase in doublet splittings between sodium and potassium however, there was no doubt at the time that the lithium lines would also reveal a doublet structure with improved optical resolution.

$$n = n_0 - \frac{N_0}{(m + \mu_1)^2}, \quad \text{strong principal doublet line;}$$

$$n = n_0 - \frac{N_0}{(m + \mu_2)^2}, \quad \text{weak principal doublet line;}$$

$$v = n_2 - n_1 = N_0 \left(\frac{1}{(1 + \mu_1)^2} - \frac{1}{(1 + \mu_2)^2} \right), \tag{B.6}$$

which gave decreasing doublet splittings of the order $(2N_0/m^3)(\mu_2 - \mu_1)$ for large m.

Rydberg then noticed a remarkable relation between the *terms* (*membres* in the original French) appearing in the expressions for the sharp and principal series, which for lithium could be written (ignoring the as yet unresolved doublet structure), with $\mu \approx 0.96$ and $\sigma \approx 0.59$,

$$\frac{n}{N_0} \left(= \frac{1}{R\lambda} \right) = \frac{1}{(1 + \mu)^2} - \frac{1}{(m + \sigma)^2}, \quad \text{sharp series } (m = 2, 3, 4, \ldots);$$

$$\frac{n}{N_0} \left(= \frac{1}{R\lambda} \right) = \frac{1}{(1 + \sigma)^2} - \frac{1}{(m + \mu)^2}, \quad \text{principal series } (m = 1, 2, 3, \ldots). \tag{B.7}$$

In other words, the fixed term (giving the limiting wavenumber of the series) of each series was just the running term of the other, with $m = 1$. Moreover, the unphysical "first" line of the sharp series, with $m = 1$, giving a negative wavenumber, corresponds precisely, on reversing the sign, to the first line of the principal series, which is typically the brightest line in the spectra of the alkali elements.

On viewing Eq. (B.7), it is impossible of course for the modern reader to avoid making the immediate connection with the notion of transitions between discrete energy levels which would appear a quarter century later in the Bohr model, and to recognize in this remarkable formula the existence of two sets of states ("s" and "p", say), with principal series lines arising from "p" to "s" transitions, and sharp series lines from "s" to "p" transitions. But the individual "terms" in Eq. (B.7) were for Rydberg exactly and only that: algebraic terms in an Ansatz which Rydberg hoped would generalize the Balmer formula and bring numerical order to the vast profusion of spectral data which the spectroscopists were now producing in massive quantities. Nowhere in the 155 pages of his article does Rydberg advance, even provisionally, a physical interpretation of his remarkable formulae, much as we can hardly fail now to see in them the clear embryo of the Bohr model *avant la lettre*.

A little after the publication of Rydberg's seminal paper, a detailed study of alkali spectra appeared, authored by the prominent German spectroscopists Heinrich Kayser and Carl Runge (1890). These authors presented new data, using the recently developed Rowland grating and photographic techniques, with a precision claimed to be an order of magnitude better than preceding measurements. The primary relevance of this paper for the later development of *quantum theory* was terminological: they introduced the terminology "Hauptserie" (principal series), "Nebenserie I" (associated/subordinate series, type 1: the diffuse series), and "Nebenserie II" (associated/subordinate series, type 2: the sharp series) which would become standard in the German literature, in particular in Sommerfeld's "bible" of atomic spectroscopy, *Atombau und Spectrallinien.*

Kayser and Runge also attempted to develop simple formulas which would reproduce the wavenumbers of series lines, but unfortunately, chose the Ansatz

$$\frac{1}{\lambda} = A - \frac{B}{m^2} - \frac{C}{m^4},$$
(B.8)

with the constants A, B, C all varying in incomprehensible ways from one series to another. In particular, they completely missed the remarkable feature, discovered by Rydberg, that the single *universal* Rydberg constant N_0 (instead of the varying B in Eq. (B.8)) was shared in formulas of type Eqs. (B.5)–(B.7) by all the series of a wide range of elements. This failure, especially telling given, as Bohr was to discover, the role of N_0 in expressing the fundamental fact of atomic structure—electrons bound by Coulomb forces to a point-like nucleus—was instead obscured by the fact that, given three (A, B, C) fitting constants at their disposal, rather than Rydberg's two (n_0 and the shift constant μ, σ or δ), they were able to obtain a somewhat more accurate fit to the measured data in many cases. Their condescending description of Rydberg's seminal work reads rather painfully at this point: "On the whole it seems to us that Rydberg is too much inclined to promote features that appear in particular elements to general laws" (Kayser and Runge 1890, p. 319).

B.5   The Zeeman effect

In 1896, the Dutch physicist Pieter Zeeman (see Plate 8) made a discovery of critical importance in the development of both spectroscopy and the quantum theory. He found that the bright D line of sodium, obtained by introducing common salt into a flame, underwent a distinct broadening when the flame was placed in the magnetic field of a strong electromagnetic. This effect was not unexpected: the current theory of Lorentz, which attributed spectral lines to the emission of electromagnetic radiation from charged electrons (at the time, called "mobile ions") in an atom undergoing harmonic vibrations (i.e., subject to an isotropic three-dimensional harmonic oscillator potential) around a point of equilibrium, actually predicted the appearance of a triplet of equally spaced lines in the presence of a magnetic field, with the central line at the location of the original field-free line.

This purely classical prediction (we are four years still from Planck, and seventeen from the Bohr atom!), which nevertheless implies the emergence of a *discrete* set of lines, is due to the fact that the radiation fields of a charged accelerated particle depend in the first approximation linearly on the second time derivative of the dipole moment (and thus, the coordinate vector) of the particle, which can be decomposed geometrically into rectilinear motion in the direction of the field $\vec{H}$, and a superposition of clockwise and anti-clockwise motions in the plane orthogonal to the field (see Section 7.2). Each of these component motions then produces radiation of a definite frequency, with the rectilinear motion unaltered by the field (as $\vec{r} \times \vec{H} = 0$ if $\vec{r}$ is parallel to $\vec{H}$) and emitting radiation at the original (zero-field) frequency, and the two circular motions producing radiation shifted equally and oppositely from the original central value. This classical theory moreover gave a precise prediction for the shift in wavenumbers of the two shifted lines, namely $\pm (eH/4\pi mc^2)$.

This simple and intuitive account of the influence of magnetic fields on spectral lines was unfortunately not borne out by further experiments. With the improved resolution of a Rowland grating, Preston two years later was able to show that the expected (the terminology "normal" for the classical prediction soon became standard) splittings of each line into three were the exception rather than the rule: for a triplet in the sharp series of cadmium, for example, he found a "diffuse triplet" for the first line, a *quartet* for the second line, and a simple triplet for the third line. With even further improved resolution, these lines were found to actually split into nine, six, and three components respectively! Even for the case of a triple splitting, the Lorentz theory could not be said to have survived intact, as the splittings were twice the expected value. Before long, the sodium D line doublet was subjected to the same scrutiny, with the D_1 line found to split magnetically into a quartet of lines, and the D_2 line into a sextet.

The only clear features that could be adduced in these early days of the Zeeman effect were that (i) the splitting of the frequencies (or wavenumbers) of a given line was always proportional to the magnetic field strength, (ii) the pattern of splittings was maintained for all the lines in a given series for elements of the same periodic group (e.g., for the alkali principal series, etc.). The latter feature ("Preston's law") turned out to serve a very practical purpose, in simplifying the identification of lines of the same series in heavier elements with very complex spectra. But the almost ubiquitous appearance of the "anomalous" (rather than "normal", i.e., Lorentzian triplet) Zeeman effect would turn out to present insuperable difficulties for the old quantum theory of Bohr and Sommerfeld, and continue to exercise the efforts (and patience) of the most prominent quantum theorists, right up to the appearance of modern quantum theory in 1925. The complex and twisted story of the struggles to understand the Zeeman effect in the old quantum theory are the subject of Sections 7.2 and 7.3 of the text.

B.6   A troublesome red herring

About the same time that Zeeman was discovering the broadening of spectral lines in a magnetic field, Pickering (1896) identified a series of lines in the spectrum of the star ζ Puppis which followed exactly the Balmer formula, but with *half-integer* values for the running quantum number. The apparent validity of a formula of the Balmer type led to the initial assumption that these were hydrogen lines as well. We have already seen in Chapter 4 that following the appearance of the Bohr model in 1913 these lines were soon correctly reinterpreted as due to singly ionized helium. Up to this point, however, they were regarded as providing additional support for the Rydberg division of alkali (or more generally, including hydrogen, elements of group I) spectra into principal, sharp, and diffuse series. The lines found by Pickering took the form (using Rydberg's notation)

$$\frac{n}{N_0} = \frac{1}{2^2} - \frac{1}{(m + \frac{1}{2})^2}, \quad m = 2, 3, 4, \ldots \tag{B.9}$$

which clearly corresponds to a series of lines which alternate with the usual Balmer lines, satisfying

$$\frac{n}{N_0} = \frac{1}{2^2} - \frac{1}{m^2}, \quad m = 3, 4, 5, \ldots \tag{B.10}$$

Moreover, Eq. (B.9) looks exactly like Rydberg's expression in Eq. (B.7) for the sharp series, with $\mu = 1$ and $\sigma = 0.5$ exactly (recall that the corresponding values for Li, the next higher element in group I, were $\mu = 0.96$ and $\sigma = 0.59$). The Balmer series, intertwined with the Pickering lines, would therefore correspond to the diffuse lines of the hydrogen spectrum. Finally, using the reciprocal connection established by Rydberg between sharp and principal series, the latter would necessarily be given by lines satisfying

$$\frac{n}{N_0} = \frac{1}{(\frac{3}{2})^2} - \frac{1}{m^2}, \quad m = 2, 3, 4, \ldots \tag{B.11}$$

The first, and brightest, line of this principal series should therefore be found at $m = 2$, corresponding to a wavelength of 4688 Å. Such a line was indeed found in several stellar (and even, nebular) spectra. Of course, this "hydrogen principal series" (see Section 4.6 for further details) really corresponds to He^+ transitions from levels 4, 5, 6 etc. to level 3, while the so-called "hydrogen sharp series" are the 5, 7, 9 ... to level 4 transitions of ionized helium. The importance of this episode, apart from serving as yet another example of the role played by misleading numerical accidents in the development of the quantum theory, is that it led to a persistent but confusing notation involving half-integral quantum numbers for sharp series, even after the proper interpretation of the Pickering lines had been reached. In order to see how this occurred, we must turn briefly to the work of Walter Ritz.

B.7   Ritz and the combination principle

Walter Ritz (see Plate 18) received his undergraduate training in mathematical physics in the Zurich polytechnic in 1897–1901, attending class at the same time as Einstein. He then moved to Göttingen, defending his doctoral work, carried out under the supervision of Voigt and Riecke, "On the theory of spectral series," in December of 1902. The thesis was published as a long paper (46 pages) in the following year (Ritz 1903). Ritz's primary goal was to improve the accuracy of the Rydberg series formulas, which continued to show systematic discrepancies from the increasingly accurate data available (discrepancies which were reduced, as we saw above, by Kayser and Runge's device of simply adding more fitting constants, but at the cost of losing any connection between the series). Unlike Rydberg, Ritz proposed to do this by finding an explicit physical model explaining the form of Rydberg formulas of the type in Eq. (B.7).

Even though Planck's work on black-body radiation had already appeared at this point, Ritz took a purely classical approach.[9] The appearance of a discrete spectrum of oscillation frequencies was not of course unknown in classical theory: one has only to think of the proper vibrations (*Eigenschwingungen* in German) of a string or membrane, characterized by a single integer, or pair of integers, respectively. For example, the Lagrangian (= kinetic minus potential energy) for a square membrane confined to the interior of the square $-a < x < +a$, $-a < y < +a$, with $w(x, y, t)$ representing the transverse displacement of the membrane, can be written, for small transverse displacements,

$$\mathcal{L} = \int_{-a}^{+a} dx \int_{-a}^{+a} dy \left(\frac{\rho}{2} (\frac{\partial w}{\partial t})^2 - \frac{\tau}{2} |\vec{\nabla} w|^2 \right), \tag{B.12}$$

where ρ is the mass per unit area of the membrane, and τ characterizes its elasticity. Variation of the action arising from Eq. (B.12) then gives the well known wave equation

$$\rho \frac{\partial^2 w(x, y, t)}{\partial t^2} = \tau \Delta w(x, y, t), \quad \Delta \equiv \frac{\partial^2}{\partial x^2} + \frac{\partial^2}{\partial y^2}. \tag{B.13}$$

With the membrane pinned ($w = 0$, Dirichlet boundary conditions) at the edges, one finds the familiar doubly discrete set of solutions

$$w_{m,n}(x, y, t) = A \sin (2\pi \nu t) \sin \left(\frac{m\pi x}{a} \right) \sin \left(\frac{n\pi y}{a} \right), m, n = \pm 1, \pm 2, \ldots, \tag{B.14}$$

which, when inserted in Eq. (B.13), give the discrete frequency spectrum

$$\nu^2 = \frac{\tau}{4\rho a^2} (m^2 + n^2). \tag{B.15}$$

An alternative choice of boundary conditions (Neumann condition, normal derivative of w vanishing at the edge) would give the same results with the simple replacement of half-integral rather than integral values for m and n. Of course, the result Eq. (B.15) cannot be considered useful for atomic spectra: the frequencies grow without limit, rather than tending to a finite accumulation point, and worse yet, the Rydberg formulas for ν involve a *difference*, rather than sum of two terms, with each term varying inversely as the square of the (shifted) ordinal number.

---

[9] Planck is not referenced by Ritz at any point, either in the 1903 *Annalen* paper or in the 1908 *Astrophysical Journal* paper introducing the combination principle (Ritz 1908b, see below).

Ritz was not deterred by this clear discrepancy in the vibrational spectra of familiar mechanical objects and the optical spectra of atoms. He concluded, from the presence of a doubly discrete spectrum in the atom case, that one was clearly dealing with a two-dimensional system (membrane) of some kind. All that was needed was an appropriate modification of the dynamics, specifically, of the potential energy term in Eq. (B.12). He succeeded in concocting a potential energy expression, dependent on a peculiar asymmetric curvature term $\Delta' w \equiv \left(\frac{\partial^2}{\partial x^2} - \frac{\partial^2}{\partial y^2} \right) w$, which resulted in the astonishing equation of motion for the membrane (v a constant), in place of Eq. (B.13),

$$\rho \frac{\partial^{10} w}{\partial t^2 \partial x^4 \partial y^4} = 32 v^2 \Delta' \Delta' w. \tag{B.16}$$

Taking the usual wave function Ansatz Eq. (B.14), and introducing membrane wavenumbers k_x, k_y (so $w = A \sin(2\pi v t) \sin(k_x x) \sin(k_y y)$), one finds on substituting the Ansatz into Eq. (B.16)

$$\rho (2\pi v)^2 k_x^4 k_y^4 = 32 v^2 (k_x^2 - k_y^2)^2,$$

$$\Rightarrow v = \frac{1}{2\pi} \sqrt{\frac{32}{\rho}} v \left(\frac{1}{k_x^2} - \frac{1}{k_y^2} \right). \tag{B.17}$$

If the usual Dirichlet boundary conditions ($w = 0$ at the boundaries) are imposed, we must take $k_x = m(\pi/a)$, $k_y = n(\pi/a)$ and we obtain precisely a Balmer-type spectrum, (here v is the cyclic frequency, differing from the wave-number by a factor of the speed of light)

$$v = Nc \left(\frac{1}{m^2} - \frac{1}{n^2} \right), \quad N \equiv \sqrt{\frac{32}{\rho}} \frac{v a^2}{2 \pi^3 c}. \tag{B.18}$$

With Neumann boundary conditions, the result is the same, with half-integral values appearing (i.e., $m \to m + \frac{1}{2}, n \to n + \frac{1}{2}$).

Ritz makes a remark, just prior to his derivation of the equation of motion Eq. (B.16), which appears extraordinarily prescient, in light of the underlying philosophy which would emerge a decade later in the old quantum theory:

> In order to apply this form of Hamilton's principle (in derivation of the mechanical equations of motion), *we must be allowed to neglect emitted radiation, as far as the calculation of the location of spectral lines is concerned* (Ritz 1903, p. 278, emphasis in original).

Precisely an assumption along these lines, asserting the possibility of calculation of the mechanical motion *ignoring the classically required simultaneous radiation*, lay at the heart of Bohr's model and its subsequent elaborations in the old quantum theory.

As Ritz points out clearly (in the paragraph following this quote), such an assumption is unavoidable if one wishes to explain the existence of *narrow* spectral lines.

By generalizing the allowed boundary conditions, Ritz was able to obtain an altered condition for the k_x, k_y wavenumbers characterizing the membrane's oscillation. In the high frequency limit, the new boundary conditions implied an asymptotic expansion (for either k_x or k_y)

$$ka = m\pi + c_0 + \frac{c_1 a}{m\pi + c_0} + \frac{c_2 a^2}{(m\pi + c_0)^2} + \dots \tag{B.19}$$

If only the c_0 term is included, we recover Eq. (B.18) with the replacement $m \to m + c_0/\pi \equiv m + \mu$ (and similarly for the other running index n), and we have recovered the Rydberg formulas of type Eq. (B.7). Ritz's study of the discrepancies between the basic Rydberg formulas and the spectral data indicated that much closer agreement was possible if the terms in the expansion (B.19) with odd inverse negative powers (with coefficients c_1, c_3, ... etc.) were absent, an exclusion which he was able to justify on the basis of further constraining the boundary conditions. The first improvement beyond Rydberg therefore resulted, after setting $c_1 = c_3 = c_4 \dots = 0$ in the expansion, in the term expression (soon to become known as the "Rydberg–Ritz" Ansatz)

$$\frac{N}{(m + a + b/(m+a)^2)^2}, \tag{B.20}$$

with the constants a, b characteristic of the particular series under examination (and N replacing N_0 for the Rydberg constant).

The extremely contrived elastic membrane model which Ritz used to motivate his modification of the Rydberg term structure was abandoned after a few years in favor of an equally contrived atomic model (Ritz 1908a) involving magnetic dipoles of quantized magnetic moment,[10] with which Ritz hoped to arrive at an understanding of the complex multiplet structure and even the anomalous Zeeman effect. But the basic structure of the new Ansatz (B.20) was preserved, and in a famous paper in the Astrophysical Journal in 1908 (Ritz 1908b), Ritz introduced the extremely influential *combination principle*, which elevated the individual terms in Rydberg's formulas to primary physical significance—though, of course, not yet recognizing them as the energies of quantized electronic states in the atom.

---

[10] Ritz imagines electrons deployed along the axis of symmetry of a magnetic dipole μ, viewed as two magnetic poles separated by a distance l, executing Larmor rotations of frequency $\nu = eH/mc$, where the magnetic field

$$H = \mu \left(\frac{1}{r^2} - \frac{1}{(r+l)^2} \right)$$

for an electron on the axis at a distance r from the nearest pole. Assuming both distances r and l to be integer multiples of a fixed elementary distance then evidently produces frequency expressions of Balmer type. Generalizations of this basic idea are then invoked to provide an "explanation" of the Rydberg–Ritz series formulae for complex spectra, and even for the anomalous Zeeman splittings!

According to the combination principle, spectral frequencies could now be sought *by taking the difference of any two terms in previously established spectral series*. And indeed, Ritz was able to account for a number of new lines which did not fall into the previous classification of principal, sharp, diffuse (or principal, first/second subordinate), and so on. Some of these lines were weak lines which correspond (in modern language) to dipole forbidden lines and involve higher multipole transitions. Others correspond to taking higher ordinal numbers in the fixed term of the series (thus, in post-Bohr language, to transitions down to a fixed higher excited state). The importance of this paper for our purposes is that it established the spectroscopic notation for spectral terms used throughout the period of the old quantum theory, and in particular in the "bible" of the old theory, Sommerfeld's *Atombau und Spektrallinien*.

Recognizing that the new term involving b in the denominator of (B.20) is inverse quadratic in the ordinal number n, and hence basically proportional to the term value itself, Ritz adopted the Ansatz, asymptotically equivalent to (B.20),

$$(m, \alpha, \beta) = \frac{N_0}{(m + \alpha + \beta(m, \alpha, \beta))^2}. \tag{B.21}$$

This is an implicit (actually a cubic) equation for the term value (m, α, β), which can however easily be solved iteratively for small β. With the Rydberg constant N_0 fixed from its hydrogen values, there are now two constants α, β to be fixed for each term. For the "s" terms appearing in the principal and sharp series, Rydberg uses the notation $(m + \frac{1}{2}, s, \sigma)$, so the Rydberg shift ($\sigma$ in (B.5)) has been split into a "standard" $\frac{1}{2}$ component and a residual shift s. This move is clearly motivated by (i) the fact that in the (fictional) hydrogen sharp series, $s = 0$, and (ii) even for lithium, the Rydberg shift (0.59) is quite close to one-half. For the "p" terms appearing in the principal series, the notation is (m, p_i, π_i), $m = 2, 3, 4, \ldots; i = 1, 2$, with the index i identifying the doublet line (for the alkalis). The "d" terms appearing in the diffuse (first subordinate) series are written $(m, d, \delta), (m, d', \delta')$, as by now a doublet structure had been detected in the d-terms as well.

In Section 7.1 we show that the old quantum theory was in fact able to account plausibly for precisely the term structure indicated in Eq. (B.21), *at least at the level of simple lines* (i.e., ignoring the fact that the lines were in most cases resolvable into doublets, triplets, etc.). The existence, however, of a *complex* multiplet structure, and especially, the further dissociation of the component lines in this complex structure into a baffling array of patterns under the influence of an external magnetic field, would turn out to be fundamentally inexplicable within the conceptual boundaries of the old quantum theory.

Bibliography

Aaserud, Finn. 2013. "Niels Bohr's Private Sphere Revealed Through Unpublished Family Correspondence." In Aaserud and Heilbron (2013, pp. 1–101).

Aaserud, Finn, and John L. Heilbron. 2013. *Love, Literature, and the Quantum Atom. Niels Bohr's 1913 Trilogy Revisited.* Oxford: Oxford University Press.

Aaserud, Finn, and Helge Kragh, eds. 2015. *One Hundred Years of the Bohr Atom. Proceedings from a Conference.* Copenhagen: Det Kongelige Danske Videnskabernes Selskab.

Abraham, Max. 1902. "Dynamik des Elektrons." *Königliche Gesellschaft der Wissenschaften zu Göttingen. Mathematisch-physikalische Klasse, Nachrichten*: 20–41.

——. 1903. "Prinzipien der Dynamik des Elektrons." *Annalen der Physik* 10: 105–179.

Andrade, Edward Neville da Costa. 1964. *Rutherford and the Nature of the Atom.* Gloucester, MA: Peter Smith.

Back, Ernst. 1921. "Ein weiteres Zahlenmysterium in der Theorie des Zeemaneffektes." *Die Naturwissenschaften* 9: 199–204.

Badino, Massimiliano. 2009. "The Odd Couple: Boltzmann, Planck and the Application of Statistics to Physics (1900–1913)." *Annalen der Physik* 18: 81–101.

——. 2011. "Mechanistic Slumber vs. Statistical Insomnia: The Early phase of Boltzmann's H-theorem (1868–1877)." *European Physical Journal H* 36: 353–378.

——. 2015. *The Bumpy Road. Max Planck from Radiation Theory to the Quantum (1896–1906).* New York: Springer.

Badino, Massimiliano, and Jaume Navarro, eds. 2013. *Research and Pedagogy: A History of Quantum Physics through Its Textbooks.* Berlin: Edition Open Access.

Balmer, Johann Jakob. 1885. "Notiz über die Spectrallinien des Wasserstoffs." *Annalen der Physik* 261: 80–87.

Baly, Edward Charles Cyril. 1912. *Spectroscopy.* 2nd ed. London, New York: Longmans, Green & Co.

Barkan, Diana Kormos. 1993. "The Witches' Sabbath: The First International Solvay Congress in Physics." In Beller, Cohen, and Renn (1993, pp. 59–82).

——. 1999. *Walther Nernst and the Transition to Modern Physical Science.* Cambridge: Cambridge University Press.

Beller, Mara, Robert S. Cohen, and Jürgen Renn, eds. 1993. *Einstein in Context.* Special issue of *Science in Context.* Cambridge: Cambridge University Press.

Bernstein, Jeremy. 1991. "Besso." In *Quantum Profiles,* pp. 143–165. Princeton: Princeton University Press.

Bethe, Hans. 1929. "Berechnung der Elektronenaffinität des Wasserstoffs." *Zeitschrift für Physik* 57: 815–821.

Bethe, Hans, and Roman Jackiw. 1986. *Intermediate Quantum Mechanics.* Boulder: Westview Press.

Biedenharn, Lawrence C. 1983. "The 'Sommerfeld puzzle' Revisited and Resolved." *Foundations of Physics* 13: 13–34.

Birks, J. B., ed. 1962. *Rutherford at Manchester.* London: Heywood. Reprint: New York: Benjamin, 1963.

Bjerrum, Niels. 1911. "Über die spezifische Wärme der Gase." *Zeitschrift für Elektrochemie und angewandte physikalische Chemie* 17: 731–735.

———. 1912. "Über die spezifische Wärme der Gase. II." *Zeitschrift für Elektrochemie und angewandte physikalische Chemie* 18: 101–104.

Boguslawski, Sergei. 1914. "Pyroelektrizität auf Grund der Quantentheorie." *Physikalische Zeitschrift* 15: 569–572.

Bohr, Niels. 1909. Master's thesis, University of Copenhagen. Published in translation in Bohr (1972–2008, Vol. 1, pp. 131–161).

———. 1911. *Studier over metallernes elektrontheori.* PhD diss., University of Copenhagen. Reprinted in facsimile in Bohr (1972–2008, Vol. 1, pp. 167–290). Page references to English translation ("Studies on the Electron Theory of Metals") in ibid., pp. 291–395.

———. 1912. *Rutherford Memorandum.* In Bohr (1972–2008, Vol. 2, pp. 135–158). Transcription: pp. 136–143; facsimile: pp. 145–158.

———. 1913a. "On the Theory of the Decrease of Velocity of Moving Electrified Particles on Passing through Matter." *Philosophical Magazine* 25: 10–31. Reprinted in facsimile in Bohr (1972–2008, Vol. 2, pp. 18–39).

———. 1913b. "On the Constitution of Atoms and Molecules (Part I)." *Philosophical Magazine* 26: 1–25. Reprinted in facsimile in Bohr (1972–2008, Vol. 2, pp. 161–185) and in Aaserud and Heilbron (2013, pp. 203–227); also reprinted in Birks (1962, pp. 228–256) and Ter Haar (1967, pp. 132–159).

———. 1913c. "On the Constitution of Atoms and Molecules (Part II)." *Philosophical Magazine* 26: 476–502. Reprinted in facsimile in Bohr (1972–2008, Vol. 2, pp. 188–214) and in Aaserud and Heilbron (2013, pp. 228–254).

———. 1913d. "On the Constitution of Atoms and Molecules (Part III)." *Philosophical Magazine* 26: 857–875. Reprinted in facsimile in Bohr (1972–2008, Vol. 2, pp. 215–233) and in Aaserud and Heilbron (2013, pp. 255–273).

———. 1913e. *Draft of a Letter to "Nature" Concerning Nicholson's theory of spectra.* In Bohr (1972–2008, Vol. 2, pp. 269–271).

———. 1913f. "The Spectra of Helium and Hydrogen." *Nature* 92: 231–232. Reprinted in facsimile in Bohr (1972–2008, Vol. 2, pp. 274–275).

———. 1914a. "Atomic Models and X-ray Spectra." *Nature* 92: 553–554. Reprinted in facsimile in Bohr (1972–2008, Vol. 2, p. 304).

———. 1914b. "On the Effect of Electric and Magnetic Fields on Spectral Lines." *Philosophical Magazine* 27: 506–524. Reprinted in facsimile in Bohr (1972–2008, Vol. 2, pp. 350–368).

———. 1914–1915. *Answer to Nicholson.* In Bohr (1972–2008, Vol. 2, pp. 311–316).

———. 1915. "On the Quantum Theory of Radiation and the Structure of the Atom." *Philosophical Magazine* 30: 394–415. Reprinted in facsimile in Bohr (1972–2008, Vol. 2, pp. 392–415).

———. 1916. "On the Application of the Quantum Theory to Periodic Systems." Intended for publication in *Philosophical Magazine* in April 1916 but withdrawn. Published in Bohr (1972–2008, Vol. 2, pp. 433–461).

———. 1918. "On the Quantum Theory of Line Spectra. Parts I and II." *Det Kongelige Danske Videnskabernes Selskab. Skrifter. Naturvidenskabelig og Matematisk Afdeling* 8 (*Raekke*, IV.1): 1–100. Reprinted in facsimile in Bohr (1972–2008, Vol. 3, pp. 67–166). Introduction and Part I reprinted in Van der Waerden (1967, pp. 95–136).

———. 1920. "Über die Serienspektra der Elemente." *Physikalische Zeitschrift* 2: 423–469. Based on a lecture at the *Deutsche Physikalische Gesellschaft,* Berlin, April 27, 1920. Reprinted in translation in Bohr (1972–2008, Vol. 3, pp. 241–282).

——. 1921. "Constitution of Atoms." Unpublished report for the Solvay Congress. Page references to Bohr (1972–2008, Vol. 4, pp. 100–174).

——. 1922a. "On the Spectra of Elements of Higher Atomic Number." *Det Kongelige Danske Videnskabernes Selskab. Skrifter. Naturvidenskabelig og Matematisk Afdeling* 8 (*Raekke,* IV.1): 101–118. Part III of Bohr (1918). Based on a manuscript from 1918 with an appendix added in 1922. Reprinted in facsimile Bohr (1972–2008, Vol. 3, pp. 167–184).

——. 1922b. *Drei Aufsätze über Spektren und Atombau.* Braunschweig: Vieweg. English translation: Bohr (1922c).

——. 1922c. *The Theory of Spectra and Atomic Constitution. Three Essays.* Cambridge University Press. Translation of Bohr (1922b).

——. 1923a. "Über die Anwendung der Quantentheorie auf den Atombau. I. Die Grundpostulate der Quantentheorie." *Zeitschrift für Phyik* 13: 117–165. Page reference to English translation, Bohr (1924).

——. 1923b. "Linienspektren und Atombau." *Annalen der Physik* 71: 228–288. Page references to the English translation in Bohr (1972–2008, Vol. 4, pp. 611–656).

——. 1924. "On the Application of the Quantum Theory to Atomic Structure. Part I. The Fundamental Postulates." *Proceedings of the Cambridge Philosophical Society* Supplement:1–42. Translation of Bohr (1923a). Reprinted in facsimile in Bohr (1972–2008, Vol. 3, pp. 458–499).

——. 1961. "Reminiscences of the Founder of Nuclear Science and of Some Developments Based on His Work." *Proceedings of the Physical Society* 78: 1083–1115. Based on Bohr's 1958 Rutherford Memorial Lecture. Reprinted in facsimile in Bohr (1972–2008, Vol. 10, pp. 383–415; unpublished references: pp. 416–420) and in Birks (1962, pp. 114–167).

——. 1963. *On the Constitution of Atoms and Molecules. Papers of 1913 Reprinted from the Philosophical Magazine with an Introduction by L. Rosenfeld.* Copenhagen/New York: Munksgaard/ W. A. Benjamin.

——. 1972–2008. *Collected Works.* Edited by Léon Rosenfeld, Finn Aaserud, Erik Rüdiger, et al. Amsterdam: North-Holland.

Bohr, Niels, and Dirk Coster. 1923. "Röntgenspektren und periodisches System der Elemente." *Zeitschrift für Physik* 12: 342–374. Page references to the English translation in Bohr (1972–2008, Vol. 4, pp. 520–548).

Bokulich, Alisa. 2008. *Reexamining the Quantum-Classical Relation: Beyond Reductionism and Pluralism.* Cambridge: Cambridge University Press.

Boltzmann, Ludwig. 1871. "Analytischer Beweis des 2. Hauptsatzes der mechanischen Wärmetheorie aus den Sätzen über das Gleichgewicht der lebendigen Kraft." *Kaiserliche Akademie der Wissenschaften* (Vienna). *Mathematisch-Naturwissenschaftliche Classe. Zweite Abteilung. Sitzungsberichte* 63: 712–732. Reprinted in Boltzmann (1909, Vol. 1, pp. 288–308).

——. 1872. "Weitere Studien über das Wärmegleichgewicht unter Gasmolekülen." *Kaiserliche Akademie der Wissenschaften* (Vienna). *Mathematisch-Naturwissenschaftliche Classe. Zweite Abteilung. Sitzungsberichte* 66: 275–370. Reprinted in Boltzmann (1909, Vol. 1, pp. 316–402). English translation in Brush (1965–1966, Vol. 2, pp. 88–175).

——. 1877. "Über die Beziehung zwischen dem zweiten Hauptsatze der mechanischen Wärmetheorie und der Wahrscheinlichkeitsrechnung respektive den Sätzen über das Wärmegleichgewicht." *Kaiserliche Akademie der Wissenschaften* (Vienna). *Mathematisch-Naturwissenschaftliche Classe. Zweite Abteilung. Sitzungsberichte* 76: 373–435. Reprinted in Boltzmann (1909, Vol. 2, pp. 164–223).

——. 1883. "Über das Arbeitsquantum, welches bei chemischen Verbindungen gewonnen werden kann." *Kaiserliche Akademie der Wissenschaften* (Vienna). *Mathematisch-Naturwissenschaftliche Classe. Zweite Abteilung. Sitzungsberichte* 88: 861–869. Page reference to reprint in Boltzmann (1909, Vol. 3, pp. 66–100).

Boltzmann, Ludwig. 1884. "Über eine von Hernn Bartoli entdeckte Beziehung der Wärmestrahlung zum zweiten Hauptsatze." *Annalen der Physik* 22: 31–39. Reprinted in Boltzmann (1909, Vol. 3, pp. 110–117).

——. 1896. "Entgegnung auf die wärmetheoretische Betrachtungen des Hrn. E. Zermelo." *Annalen der Physik* 57: 773–784. Page references to reprint in Boltzmann (1909, Vol. 3, pp. 567–578).

——. 1896–1898. *Vorlesungen über Gastheorie*. 2 Vols. Leipzig: Barth. Translation (by Stephen G. Brush): *Lectures on Gas Theory* (Berkeley and Los Angeles: University of California Press, 1964).

——. 1897a. "Zu Hrn. Zermelo's Abhandlung 'Über die mechanische Erklärung irreversibler Vorgänge'." *Annalen der Physik* 60: 392–398. Reprinted in Boltzmann (1909, Vol. 3, pp. 579–586).

——. 1897b. "Über irreversible Strahlungsvorgänge, I." *Königlich Preussische Akademie der Wissenschaften* (Berlin). *Sitzungsberichte*: 660–662. Reprinted in Boltzmann (1909, Vol. 3, pp. 615–617).

——. 1897c. "Über irreversible Strahlungsvorgänge, II." *Königlich Preussische Akademie der Wissenschaften* (Berlin). *Sitzungsberichte*: 1016–1018. Reprinted in Boltzmann (1909, Vol. 3, pp. 618–621).

——. 1898. "Über vermeintlich irreversible Strahlungsvorgänge." *Königlich Preussische Akademie der Wissenschaften* (Berlin). *Sitzungsberichte*: 182–187. Reprinted in Boltzmann (1909, Vol. 3, pp. 622–628).

——. 1904. *Vorlesungen über die Prinzipe der Mechanik, II. Teil*. Leipzig: Barth.

——. 1909. *Wissenschaftliche Abhandlungen*. 3 Vols. Edited by Friedrich Hasenöhrl. Leipzig: Barth.

Born, Max. 1924. "Über Quantenmechanik." *Zeitschrift für Physik* 26: 379–395. English translation in Van der Waerden (1967, pp. 181–198).

——. 1925. *Vorlesungen über Atommechanik*. Berlin Heidelberg: Springer.

——. 1927. *The Mechanics of the Atom*. London: Bell. Translation of Born (1925).

——. 1949. "Einstein's statistical theories." In *Albert Einstein: Philosopher-Scientist*, edited by Paul A. Schilpp, pp. 163–177. Evanston, IL: Library of Living Philosophers.

Born, Max, and Werner Heisenberg. 1923. "Die Elektronenbahnen im angeregten Heliumatom." *Zeitschrift für Phyik* 16: 229–243.

Born, Max, Werner Heisenberg, and Pascual Jordan. 1926. "Zur Quantenmechanik II." *Zeitschrift für Physik* 35: 557–615. English translation in van der Waerden (1967, 321–385).

Born, Max, and Pascual Jordan. 1930. *Elementare Quantenmechanik*. Berlin: Springer.

Brand, John C. D. 1995. *Lines of Light: The Sources of Dispersive Spectroscopy, 1800–1930*. Luxembourg: Gordon and Breach.

Brillouin, Léon. 1926. "La mécanique ondulatoire de Schrödinger: une méthode générale de resolution par approximations successives." *Comptes Rendus de l'Academie des Science* 183: 24–26.

Broglie, Louis de. 1924. *Recherche sur la théorie des quanta*. PhD diss., University of Paris (Sorbonne).

Brown, Harvey R., Wayne Myrvold, and Jos Uffink. 2009. "Boltzmann's H-theorem, its Discontents, and the Birth of Statistical Mechanics." *Studies in History and Philosophy of Modern Physics* 40: 174–191.

Brush, Stephen G., ed. 1965–1966. *Kinetic theory. Selected Readings in Physics*. 2 Vols. Oxford: Pergamon.

——. 1967. "Foundations of Statistical Mechanics, 1845–1915." *Archive for History of Exact Sciences* 4: 145–183.

———. 1976. *The Kind of Motion We Call Heat: A History of the Kinetic Theory of Gases in the 19th Century.* Vol. 1, *Physics and the Atomists.* Vol. 2, *Statistical Physics and Irreversible Processes.* Amsterdam: North-Holland.

Bucherer, Alfred Heinrich. 1908. "Messungen an Becquerelstrahlen. Die experimentelle Bestätigung der Lorentz-Einsteinschen Theorie." *Physikalische Zeitschrift* 9: 755–762.

Buchwald, Jed Z. 1985. *From Maxwell to Microphysics. Aspects of Electromagnetic Theory in the Last Quarter of the Nineteenth Century.* Chicago: University of Chicago Press.

Buchwald, Jed Z., and Andrew Warwick, eds. 2001. *Histories of the Electron. The Birth of Microphysics.* Cambridge, MA: The MIT Press.

Burgers, Johannes M. 1917a. "Adiabatische invarianten bij mechanische systemen." Pts. 1–3. *Koninklijke Akademie van Wetenschappen te Amsterdam. Wis- en Natuurkundige Afdeeling. Verslagen van de Gewone Vergaderingen* 25: 849–857 (I), 918–927 (II), 1055–1061 (III). English translation: "Adiabatic invariants of mechanical systems." *Koninklijke Akademie van Wetenschappen te Amsterdam. Section of Sciences. Proceedings* 20: 149–157 (I), 158–162 (II), 163–169 (III).

———. 1917b. "Adiabatic Invariants of Mechanical Systems." *Philosophical Magazine* 33: 514–520.

———. 1917c. "Die adiabatischen Invarianten bedingt periodischer Systeme." *Annalen der Physik* 52: 195–202. Abridged version of Burgers (1917a, Part I).

Büttner, Jochen, Jürgen Renn, and Matthias Schemmel. 2001. "Exploring the Limits of Classical Physics: Planck, Einstein, and the Structure of a Scientific Revolution." *Studies in History and Philosophy of Modern Physics* 34: 35–59.

Cairns-Smith, Alexander Graham. 1985. *Seven Clues to the Origin of Life: A Scientific Detective Story.* Cambridge: Cambridge University Press.

Charlier, Carl Ludwig. 1902–1907. *Die Mechanik des Himmels.* 2 Vols. Leipzig: Veit. Vol. 1 (1902), Vol. 2 (1907).

Cheng, Ta-Pei. 2013. *Einstein's Physics: Atoms, Quanta, and Relativity Derived, Explained, and Appraised.* Oxford: Oxford University Press.

Clausius, Rudolf. 1850. "Über die bewegende Kraft der Wärme." *Annalen der Physik* 79: 368–397, 500–524. English translation: "On the Moving Force of Heat and the Laws of Heat which may be Deduced Therefrom." *Philosophical Magazine* 2 (1851): 1–21, 102–119.

Conn, G. K. T., and H. D. Turner. 1965. *The Evolution of the Nuclear Atom.* London: Iliffe Books.

Cuthbertson, Clive, and Maude Cuthbertson. 1909. "On the Refraction and Dispersion of Air, Oxygen, Nitrogen, and Hydrogen, and Their Relation." *Proceedings of the Royal Society of London A* 83: 151–171.

Darrigol, Olivier. 1988. "Statistics and Combinatorics in Early Quantum Theory, I." *Historical Studies in the Physical Sciences* 19: 17–80.

———. 1991. "Statistics and Combinatorics in Early Quantum Theory, II: Early Symptoms of Indistinguishability and Holism." *Historical Studies in the Physical Sciences* 21: 237–298.

———. 1992. *From c-Numbers to q-Numbers: The Classical Analogy in the History of Quantum Theory.* Berkeley: University of California Press.

———. 2000a. *Electrodynamics from Ampère to Einstein.* Oxford: Oxford University Press.

———. 2000b. "Continuities and Discontinuities in Planck's *Akt der Verzweiflung*." *Annalen der Physik* 9: 951–960.

———. 2001. "The Historians' Disagreement Over the Meaning of Planck's Quantum." *Centaurus* 43: 219–239.

———. 2002. "Quantum Theory and Atomic Structure, 1900–1927." In *The Cambridge History of Science,* Vol. 5, *The Modern Physical and Mathematical Sciences,* edited by Mary Jo Nye, pp. 331–349. Cambridge: Cambridge University Press.

Darrigol, Olivier. 2009. "A Simplified Genesis of Quantum Mechanics." *Studies In History and Philosophy of Modern Physics* 40: 151–166.

——. 2014. "The Quantum Enigma." In Janssen and Lehner (2014, pp. 117–142).

——. 2018. *Atoms, Mechanics, and Probability. Ludwig Boltzmann's Statistico-Mechanical Writings. An Exegesis.* Oxford: Oxford University Press.

Darwin, Charles Galton. 1912. "A Theory of the Absorption and Scattering of the α Rays." *Philosophical Magazine* 23: 901–920.

Debye, Peter. 1910. "Der Wahrscheinlichkeitsbegriff in der Theorie der Strahlung." *Annalen der Physik* 33: 1427–1434.

——. 1912. "Zur Theorie der spezifischen Wärmen." *Annalen der Physik* 39: 789–839.

——. 1914. *Vorträge über die kinetische Theorie der Materie und der Elektrizität.* Leipzig: Teubner.

——. 1915. "Die Konstitution des Wasserstoffmoleküls." *Königlich Bayerische Akademie der Wissenschaften zu München. Mathematisch-physikalische Klasse. Sitzungsberichte*: 1–26.

——. 1916. "Quantenhypothese und Zeeman-Effekt." *Physikalische Zeitschrift* 17: 507–512.

Delaunay, Charles-Eugène. 1860–1867. "Théorie du mouvement de la lune." *Mémoires de l'Académie des Sciences* (Paris). 28 (1860); 29 (1867).

Dirac, Paul Adrien Maurice. 1927. "The Physical Interpretation of the Quantum Dynamics." *Proceedings of the Royal Society of London. Series A* 113: 621–641.

——. 1928a. "The Quantum Theory of the Electron." *Proceedings of the Royal Society of London. Series A* 117: 610–624.

——. 1928b. "The Quantum Theory of the Electron. II." *Proceedings of the Royal Society of London. Series A* 118: 351–361.

Dorling, Jon. 1971. "Einstein's Introduction of Photons: Argument by Analogy or Deduction from the Phenomena." *British Journal for the Philosophy of Science* 22: 1–8.

Dresden, Max. 1987. *H. A. Kramers: Between Tradition and Revolution.* New York: Springer.

Drude, Paul. 1900. "Zur Elektronentheorie der Metalle." *Annalen der Physik* 1: 566–613.

Dulong, Pierre Louis, and Alexis Thérèse Petit. 1819. "Recherches sur quelques points importants de la théorie de la chaleur." *Annales de Chimie et de Physique* 10: 395–413.

Duncan, Anthony, and Michel Janssen. 2007. "On the Verge of *Umdeutung* in Minnesota: Van Vleck and the Correspondence Principle." 2 Pts. *Archive for History of Exact Sciences* 61: 553–624, 625–671.

——. 2008. "Pascual Jordan's Resolution of the Conundrum of the Wave–Particle Duality of Light." *Studies in History and Philosophy of Modern Physics* 39: 634–666.

——. 2009. "From Canonical Transformations to Transformation Theory, 1926–1927: The Road to Jordan's *Neue Begründung.*" *Studies In History and Philosophy of Modern Physics* 40: 352–362.

——. 2013. "(Never) Mind your p's and q's: Von Neumann versus Jordan on the Foundations of Quantum Theory." *The European Physical Journal H* 38: 175–259.

——. 2014. "The Trouble with Orbits: The Stark effect in the Old and the New Quantum Theory." *Studies In History and Philosophy of Modern Physics* 48: 68–83.

——. 2015. "The Stark Effect in the Bohr–Sommerfeld Theory and in Schrödinger's Wave Mechanics." In Aaserud and Kragh (2015, pp. 217–271).

Duncan, Anthony, and Enric Pérez. 2016. "The Puzzle of Half-integral Quanta in the Application of the Adiabatic Hypothesis to Rotational Motion." *Studies In History and Philosophy of Modern Physics* 54: 1–8.

Earman, John, and Michel Janssen. 1993. "Einstein's Explanation of the Motion of Mercury's Perihelion." In *The Attraction of Gravitation,* edited by John Earman, Michel Janssen, and John D. Norton, pp. 129–172. Boston: Birkhäuser.

Eckert, Michael. 1993. *Die Atomphysiker. Eine Geschichte der theoretischen Physik am Beispiel der Sommerfeldschule.* Braunschweig: Vieweg.

——. 2013a. *Historische Annäherung.* In Sommerfeld (2013, pp. 1–60).

——. 2013b. *Arnold Sommerfeld. Atomphysiker und Kulturbote 1868–1951. Eine Biographie.* Göttingen: Wallstein. Page references to English translation (by Tom Artin): *Arnold Sommerfeld. Science, Life and Turbulent Times.* New York: Springer, 2013.

——. 2013c. "Sommerfeld's *Atombau und Spektrallinien.*" In Badino and Navarro (2013, pp. 117–135).

——. 2014. "How Sommerfeld Extended Bohr's Model of the Atom." *European Physical Journal H* 39: 141–156.

——. 2015a. "From Aether Impulse to QED: Sommerfeld and the Bremsstrahlen Theory." *Studies In History and Philosophy of Modern Physics* 51: 9–22.

——. 2015b. "From X-rays to the *h*-Hypothesis: Sommerfeld and the Early Quantum Theory 1909–1913." *The European Physical Journal Special Topics* 224: 2057–2073.

Ehrenfest, Paul. 1905. "Über die physikalischen Voraussetzungen der Planck'schen Theorie der irreversiblen Strahlungsvorgänge." *Kaiserliche Akademie der Wissenschaften* (Vienna). *Mathematisch-Naturwissenschaftliche Classe. Zweite Abteilung. Sitzungsberichte* 114: 1301–1314. Reprinted in facsimile in Ehrenfest (1959, pp. 88–101).

——. 1906. "Zur Planckschen Strahlungstheorie." *Physikalische Zeitschrift* 7: 528–532. Reprinted in facsimile in Ehrenfest (1959, pp. 120–124).

——. 1911. "Welche Züge der Lichtquantenhypothese spielen in der Theorie der Wärmestrahlung eine wesentliche Rolle?" *Annalen der Physik* 36: 91–118. Reprinted in facsimile in Ehrenfest (1959, pp. 185–212).

——. 1913a. *Zur Krise der Lichtaether-hypothese.* Leyden: Eduard Ijdo. [Berlin: Julius Springer, 1913]. Reprinted in facsimile in Ehrenfest (1959, pp. 306–327).

——. 1913b. "Bemerkung betreffs der spezifischen Wärme zweiatomiger Gase." *Deutsche Physikalische Gesellschaft. Verhandlungen* 15: 451–457. Reprinted in facsimile in Ehrenfest (1959, pp. 333–339).

——. 1913c. "A Mechanical Theorem of Boltzmann and its Relation to the Theory of Energy Quanta." *Koninklijke Akademie van Wetenschappen te Amsterdam, Section of Sciences, Proceedings* 16: 591–597. Reprinted in facsimile in Ehrenfest (1959, 340–346).

——. 1916a. "On Adiabatic Changes of a System in Connection With the Quantum Theory." *Koninklijke Akademie van Wetenschappen te Amsterdam, Section of Sciences, Proceedings* 19: 576–597. Reprinted in facsimile in Ehrenfest (1959, pp. 378–399).

——. 1916b. "Adiabatische Invarianten und Quantentheorie." *Annalen der Physik* 51: 327–352. German translation of Ehrenfest (1916a).

——. 1925. "Energieschwankungen im Strahlungsfeld oder Kristallgitter bei Superposition quantisierter Eigenschwingungen." *Zeitschrift für Physik* 34: 362–373. Reprinted in facsimile in Ehrenfest (1959, pp. 513–524).

——. 1959. *Collected Scientific Papers.* Edited by Martin J. Klein. Amsterdam: North-Holland.

Ehrenfest, Paul, and Tatiana Ehrenfest-Afanassjewa. 1911. "Begriffliche Grundlagen der statistischen Auffassung in der Mechanik." In *Encyklopädie der Mathematischen Wissenschaften mit Einschluss ihrer Anwendungen.* Vol. 4, Pt. 4, edited by Felix Klein and Conrad Müller, pp. 3–90. Leipzig: Teubner. English translation: *The Conceptual Foundations of the Statistical Approach in Mechanics* (New York: Dover, 1990).

Ehrenfest, Paul, and Heike Kamerlingh Onnes. 1914. "Simplified Deduction of the Formula from the Theory of Combinations which Planck Uses as the Basis of his Radiation Theory." *Proceedings of the Koninklijke Nederlandse Akademie van Wetenschappen. Series B. Physical sciences* 17: 870–873. Reprinted in facsimile in Ehrenfest (1959, pp. 353–356).

Ehrenfest, Paul, and Heike Kamerlingh Onnes. 1915. "Vereinfachte Ableitung der kombinatorischen Formel, welche der Planckschen Strahlungstheorie zugrunde liegt." *Annalen der Physik* 46: 1021–1024. German translation of Ehrenfest and Kamerlingh Onnes (1914).

Einstein, Albert. 1902. "Kinetische Theorie des Wärmegleichgewichtes und des zweiten Hauptsatzes der Thermodynamik." *Annalen der Physik* 9: 417–433. Reprinted in Einstein (1987–2018, Vol. 2, Doc. 3).

——. 1903. "Eine Theorie der Grundlagen der Thermodynamik." *Annalen der Physik* 11: 170–187. Reprinted in Einstein (1987–2018, Vol. 2, Doc. 4).

——. 1904. "Zur allgemeinen molekularen Theorie der Wärme." *Annalen der Physik* 14: 354–362. Reprinted in Einstein (1987–2018, Vol. 2, Doc. 5).

——. 1905a. "Über eine die Erzeugung und die Verwandlung des Lichtes betreffenden heuristischen Gesichtspunkts." *Annalen der Physik* 17: 132–148. Reprinted in Einstein (1987–2018, Vol. 2, Doc. 14). Page references to English translation in Stachel (2005, pp. 177–197).

——. 1905b. *Eine neue Bestimmung der Moleküldimensionen.* PhD diss., University of Zurich. Reprinted in Einstein (1987–2018, Vol. 2, Doc. 15). English translation in Stachel (2005, pp. 45–65).

——. 1905c. "Über die von der molekularkinetischen Theorie der Wärme geforderte Bewegung von in ruhenden Flüssigkeiten suspendierten Teilchen." *Annalen der Physik* 17: 549–560. Reprinted in Einstein (1987–2018, Vol. 2, Doc. 16). English translation in Stachel (2005, pp. 85–98).

——. 1905d. "Zur Elektrodynamik bewegter Körper." *Annalen der Physik* 17: 891–921. Reprinted in Einstein (1987–2018, Vol. 2, Doc. 23). English translation in Stachel (2005, pp. 123–159).

——. 1905e. "Ist die Trägheit eines Körpers von seinem Energieinhalt abhängig?" *Annalen der Physik* 18: 639–641. Reprinted in Einstein (1987–2018, Vol. 2, Doc. 24). English translation in Stachel (2005, pp. 161–164).

——. 1906a. "Eine neue Bestimmung der Moleküldimensionen." *Annalen der Physik* 19: 289–305. Slightly revised version of Einstein (1905b).

——. 1906b. "Zur Theorie der Lichterzeugung und Lichtabsorption." *Annalen der Physik* 20: 199–206. Reprinted in Einstein (1987–2018, Vol. 2, Doc. 34).

——. 1906c. "Review of Max Planck (1906a)." *Beiblätter zu den Annalen der Physik* 30 (15): 764–766. Reprinted in Einstein (1987–2018, Vol. 2, Doc. 37).

——. 1907. "Die Plancksche Theorie der Strahlung und die Theorie der spezifischen Wärme." *Annalen der Physik* 22: 180–190. Reprinted in Einstein (1987–2018, Vol. 2, Doc. 38).

——. 1909a. "Zum gegenwärtigen Stand des Strahlungsproblems." *Physikalische Zeitschrift* 10: 185–193. Reprinted in Einstein (1987–2018, Vol. 2, Doc. 56).

——. 1909b. "Über die Entwicklung unserer Anschauungen über das Wesen und die Konstitution der Strahlung." *Physikalische Zeitschrift* 10: 817–825. Reprinted in Einstein (1987–2018, Vol. 2, Doc. 60).

——. 1910. "Theorie der Opaleszenz von homogenen Flüssigkeiten und Flüssigkeitsgemischen in der Nähe des kritischen Zustandes." *Annalen der Physik* 33: 1275–1298. Reprinted in Einstein (1987–2018, Vol. 3, Doc. 9).

——. 1911a. "Bemerkungen zu den P. Hertzschen Arbeiten: 'Über die mechanischen Grundlagen der Thermodynamik'." *Annalen der Physik* 34: 175–176. Reprinted in Einstein (1987–2018, Vol. 3, Doc. 10).

——. 1911b. "Elementare Betrachtungen über die thermische Molekularbewegung in festen Körpern." *Annalen der Physik* 35: 679–694. Reprinted in Einstein (1987–2018, Vol. 3, Doc. 21).

——. 1912. "L'état actuel du problème des chaleurs spécifiques." In Langevin and de Broglie (1912, pp. 407–435). Translation of Einstein (1914a).

——. 1913. "Zum gegenwärtigen Stande des Gravitationsproblems." *Physikalische Zeitschrift* 14: 1249–1262. Reprinted in Einstein (1987–2018, Vol. 4, Doc. 17).

——. 1914a. "Zum gegenwärtigen Stande des Problems der spezifische Wärme." In Eucken (1914, pp. 330–352). French translation published as Einstein (1912). Reprinted in Einstein (1987–2018, Vol. 3, Doc. 26).

——. 1914b. "Beiträge zur Quantentheorie." *Deutsche Physikalische Gesellschaft. Verhandlungen* 16: 820–828. Reprinted in Einstein (1987–2018, Vol. 6, Doc. 5).

——. 1916a. "Strahlungs-Emission und -Absorption nach der Quantentheorie." *Deutsche Physikalische Gesellschaft. Verhandlungen* 18: 318–323. Reprinted in Einstein (1987–2018, Vol. 6, Doc. 34).

——. 1916b. "Zur Quantentheorie der Strahlung." *Physikalische Gesellschaft Zürich. Mitteilungen* 18: 47–62. Reprinted as Einstein (1917a). Reprinted in Einstein (1987–2018, Vol. 6, Doc. 38).

——. 1917a. "Zur Quantentheorie der Strahlung." *Physikalische Zeitschrift* 18: 121–128. Reprint of Einstein (1916b). Page references to English translation in van der Waerden (1967, 63–77).

——. 1917b. "Zum Quantensatz von Sommerfeld und Epstein." *Deutsche Physikalische Gesellschaft. Verhandlungen* 19 (82–92). Reprinted in Einstein (1987–2018, Vol. 6, Doc. 45).

——. 1925. "Quantentheorie des einatomigen idealen Gases. Zweite Abhandlung." *Preußische Akademie der Wissenschaften* (Berlin). *Physikalisch-mathematische Klasse. Sitzungsberichte*: 3–14. Reprinted in Einstein (1987–2018, Vol. 14, Doc. 385).

——. 1949. "Autobiographical Notes." In *Albert Einstein: Philosopher-Scientist*, edited by Paul A Schilpp, pp. 1–96. Evanston, IL: Library of Living Philosophers.

——. 1987–2018. *The Collected Papers of Albert Einstein*. 15 Vols. Edited by John Stachel, Martin J. Klein, Robert Schulmann, Diana Barkan Buchwald, et al. Princeton: Princeton University Press.

Einstein, Albert, and Wander Johannes de Haas. 1915a. "Experimenteller Nachweis der Ampèreschen Molekularströme." *Deutsche Physikalische Gesellschaft. Verhandlungen* 17: 152–170. Reprinted in Einstein (1987–2018, Vol. 6, Doc. 13).

——. 1915b. "Experimental Proof of the Existence of Ampere's Molecular Currents." *Koninklijke Akademie van Wetenschappen te Amsterdam. Section of Sciences. Proceedings* 18: 696–711. Reprinted in Einstein (1987–2018, Vol. 6, Doc. 14).

Einstein, Albert, and Ludwig Hopf. 1910. "Statistische Untersuchung der Bewegung eines Resonators in einem Strahlungsfeld." *Annalen der Physik* 33: 1105–1115. Reprinted in Einstein (1987–2018, Vol. 3, Doc. 8).

Einstein, Albert, Hendrik Antoon Lorentz, Hermann Minkowski, and Hermann Weyl. 1952. *The Principle of Relativity. A Collection of Original Papers on the Special and General Theory of Relativity. Notes by A. Sommerfeld*. New York: Dover. Translation of Lorentz et al. (1922).

Einstein, Albert, Max Planck, Heinrich Rubens, Johannes Stark, and Hans Ziegler. 1909. "Diskussion." *Physikalische Zeitschrift* 10: 825–826. Discussion following Einstein (1909b). Reprinted in Einstein (1987–2018, Vol. 2, Doc. 61).

Einstein, Albert, and Otto Stern. 1913. "Einige Argumente für die Annahme einer molekularen Agitation beim absoluten Nullpunkt." *Annalen der Physik* 40: 551–560. Reprinted in Einstein (1987–2018, Vol. 4, Doc. 11).

Elzinga, Aant. 2006. *Einstein's Nobel Prize: A Glimpse behind Closed Doors: The Archival Evidence*. Sagamore Beach, MA: Science History Publication.

Epstein, Paul. 1916a. "Zur Theorie des Starkeffektes." *Physikalische Zeitschrift* 17: 148–150.

——. 1916b. "Zur Theorie des Starkeffektes." *Annalen der Physik* 50: 489–521.

Epstein, Paul. 1916c. "Versuch einer Anwendung der Quantenlehre auf die Theorie des lichtelektrischen Effekts und der β-Strahlung radioaktiver Substanzen." *Annalen der Physik* 50: 815–840.

———. 1916d. "Zur Quantentheorie." *Annalen der Physik* 51: 168–188.

———. 1918. "Über die Struktur des Phasenraumes bedingt periodischer Systeme." *Königlich Preussische Akademie der Wissenschaften* (Berlin). *Sitzungsberichte*: 435–446.

———. 1926. "The Stark Effect From the Point of View of Schrödinger's Quantum Theory." *Physical Review* 28: 695–710.

Eucken, Arnold, ed. 1914. *Die Theorie der Strahlung und der Quanten. Verhandlungen auf einer von E. Solvay einberufenen Zusammenkunft (30. Oktober bis 3. November 1911), mit einem Anhange über die Entwicklung der Quantentheorie vom Herbst 1911 bis Sommer 1913*. Halle a.S.: Knapp.

Evans, Evan J. 1913. "The Spectra of Helium and Hydrogen." *Nature* 92: 5.

Fedak, William A., and Jeffrey J. Prentis. 2002. "Quantum Jumps and Classical Harmonics." *American Journal of Physics* 70: 332–344.

Fellows, Fred H. 1985. *J. H. Van Vleck: The Early Life and Work of a Mathematical Physicist*. PhD diss., University of Minnesota.

Feynman, Richard P., Robert B. Leighton, and Matthew Sands. 1964. *The Feynman Lectures on Physics*. 3 Vols. Reading, MA: Addison-Wesley.

Fölsing, Albrecht. 1993. *Albert Einstein: Eine Biographie*. Frankfurt am Main: Suhrkamp. Page references to abridged English translation (by Ewald Osers): *Albert Einstein: A Biography*. New York: Viking, 1997.

Föppl, August. 1897–1910. *Vorlesungen über Technische Mechanik*. Vol. 1–6. Leipzig: Teubner.

Forman, Paul. 1970. "Alfred Landé and the Anomalous Zeeman Effect, 1919–1921." *Historical Studies in the Physical Sciences* 2: 153–261.

Foster, John Stuart, and Mary Laura Chalk. 1929. "Relative Intensities of Stark Components in Hydrogen." *Proceedings of the Royal Society of London A* 123: 108–118.

Fowler, Alfred. 1912. "Observations of the Principal and Other Series in the Spectrum of Hydrogen." *Monthly Notices of the Royal Astronomical Society* 73: 62–71.

———. 1913a. "The Spectra of Helium and Hydrogen." *Nature* 92: 95–96.

———. 1913b. "Comment on [Bohr 1913f]." *Nature* 92: 232–233.

Franck, James, and Gustav Hertz. 1913. "Messung der Ionisierungsspannung in Verschiedenen Gasen." *Deutsche Physikalische Gesellschaft. Verhandlungen* 15: 34–44.

———. 1914. "Über die Erregung der Quecksilberresonanzlinie 253, 6 $\mu\mu$ durch Elektronenstösse." *Deutsche Physikalische Gesellschaft. Verhandlungen* 16: 512–517.

———. 1916. "Über Kinetik von Elektronen und Ionen in Gasen." *Physikalische Zeitschrift* 17: 409–416, 430–440.

———. 1919. "Die Bestätigung der Bohrschen Atomtheorie im optischen Spektrum durch Untersuchungen der unelastischen Zusammenstöße langsamer Elektronen mit Gasmolekülen." *Physikalische Zeitschrift* 20: 132–143.

Franck, James, and Paul Knipping. 1919. "Die Ionisierunsspannungen des Heliums." *Physikalische Zeitschrift* 20: 481–488.

Franck, James, and Fritz Reiche. 1920. "Über Helium und Parhelium." *Zeitschrift für Physik* 1: 154–160.

Frank, Philipp. 1947. *Einstein: His Life and Times*. New York: Alfred Knopf.

Franklin, Allan. 1986. *The Neglect of Experiment*. Cambridge: Cambridge University Press.

———. 2013. "Millikan's Measurement of Planck's Constant." *The European Physical Journal H* 38: 573–594.

Friedman, Robert Marc. 2001. *The Politics of Excellence: Behind the Nobel Prize in Science*. New York: Henry Holt.

Fues, Erwin. 1927. "Störungsrechnung." Chap. 4 in *Handbuch der Physik*. Vol. 5. *Grundlagen der Mechanik. Mechanik der Punkte und starren Körper*, edited by Richard Grammel, pp. 131–177. Julius Springer.

Galison, Peter. 1981. "Kuhn and the Quantum Controversy." *British Journal for the Philosophy of Science* 32: 71–85.

——. 1987. *How Experiments End*. Chicago: University of Chicago Press.

——. 2008. "The Assassin of Relativity." In Galison, Holton, and Schweber (2008, Ch. 13, pp. 185–204).

Galison, Peter, Gerald Holton, and Silvan S. Schweber, eds. 2008. *Einstein for the 21st Century: His Legacy in Science, Art, and Modern Culture*. Princeton: Princeton University Press.

Gearhart, Clayton A. 1990. "Einstein before 1905: The Early Papers on Statistical Mechanics." *American Journal of Physics* 58: 460–480.

——. 2002. "Planck, the Quantum, and the Historians." *Physics in Perspective* 4: 170–215.

——. 2010. " 'Astonishing Successes' and 'Bitter Disappointment': The Specific Heat of Hydrogen in Quantum Theory." *Archive for History of Exact Sciences* 64: 113–202.

——. 2014. "The Franck–Hertz Experiments, 1911–1914. Experimentalists in Search of a Theory." *Physics in Perspective* 16: 293–343.

——. 2017. "James Franck, the Ionization Potential of Helium, and the Experimental Discovery of Metastable States." *Studies In History and Philosophy of Modern Physics* 60: 95–109.

Geiger, Hans. 1910. "The Scattering of the α-Particles by Matter." *Proceedings of the Royal Society of London A* 83: 492–504.

Geiger, Hans, and Ernest Marsden. 1909. "On a Diffuse Reflection of the α-Particles." *Proceedings of the Royal Society of London A* 82: 495–500.

——. 1913. "The Laws of Deflexion of α-Particles Through Large Angles." *Philosophical Magazine* 25: 604–623. Reprinted in Birks (1962, pp. 205–227).

Gibbs, Josiah Willard. 1902. *Elementary Principles in Statistical Mechanics*. New York: Scribner.

Giulini, Domenico, and Norbert Straumann. 2006. "Einstein's Impact on the Physics of the Twentieth Century." *Studies in History and Philosophy of Modern Physics* 37: 115–173.

Glitzer, Karl. 1917. "Spektroskopischer Vergleich zwischen den Theorien des starren und des deformierbaren Elektrons." *Annalen der Physik* 52: 608–630.

Goldberg, Stanley. 1976. "Max Planck's Philosophy of Nature and His Elaboration of the Special Theory of Relativity." *Historical Studies in the Physical Sciences* 7: 125–160.

Goldstein, Herbert. 1950. *Classical Mechanics*. Cambridge, MA: Addison-Wesley.

Goldstein, Herbert, Charles P. Poole, and John L. Safko. 2002. *Classical Mechanics*. 3rd ed. San Francisco: Addison-Wesley.

Gooday, Graeme, and Daniel Jon Mitchel. 2013. "Rethinking 'Classical Physics." Chap. 24 in *The Oxford Handbook of the History of Physics*, edited by Jed Z. Buchwald and Robert Fox, pp. 721–764. Oxford: Oxford University Press.

Gordin, Michael D. (2020). *Einstein in Bohemia*. Princeton: Princeton University Press.

Grotrian, Walter. 1928. *Graphische Darstellung der Spektren von Atomen und Ionen mit ein, zwei und drei Valenzelektronen*. Berlin, Heidelberg: Springer-Verlag.

Gutzwiller, Martin C. 1990. *Chaos in Classical and Quantum Mechanics*. New York: Springer.

Haas, Arthur Erich. 1910a. "Über die elektrodynamische Bedeutung des Planckschen Gesetzes und über eine neue Bestimmung des elektrischen Elementarquantums und der Dimensionen des Wasserstoffatoms." *Sitzungberichte der Kaiserlichen Akademie der Wissenschaften* (Vienna), *Abteilung IIa* 119:119–144. Reprinted in Hermann (1965c).

——. 1910b. "Der Zusammenhang des Planckschen elementaren Wirkungsquantums mit den Grundgrössen der Elektronentheorie." *Jahrbuch der Radioaktivität und Elektronik* 7: 261–268. Reprinted in Hermann (1965c).

——. 1910c. "Über eine neue theoretische Methode zur Bestimmung des elektrischen Elementarquantums und des Halbmessers des Wasserstoffatoms." *Physikalische Zeitschrift* 11: 537–538.

Hamel, Georg. 1912. *Elementare Mechanik*. Leipzig and Berlin: Teubner.

Hartley, W. N. 1883. "On Homologous Spectra." *Journal of the Chemical Society. Transactions* 43: 390–400.

Heber, O., R. Golser, D. Berkovits, Y. Toker, M. Eritt, M. L. Rappaport, and D. Zajfman. 2006. "Lifetimes of the negative molecular hydrogen ions: H_2^-, D_2^-, and HD^-." *Physical Review A* 73: 060501(1–4).

Heilbron, John L. 1964. *A History of Atomic Structure from the Discovery of the Electron to the Beginning of Quantum Mechanics*. PhD diss., University of California, Berkeley.

——. 1966. "The Work of H. G. J. Moseley." *Isis* 57: 336–364. Reprinted in facsimile in Heilbron (1981, pp. 230–258).

——. 1968. "The Scattering of α and β Particles and Rutherford's atom." *Archive for History of Exact Sciences* 4: 247–307. Reprinted in facsimile in Heilbron (1981, pp. 85–145).

——. 1974. *H. G. J. Moseley. The Life and Letters of an English Physicist, 1887–1915*. Berkeley, Los Angeles, London: University of California Press.

——. 1977. "Lectures on the History of Atomic Physics, 1900–1922." In *History of Twentieth-Century Physics*, edited by Charles Weiner, pp. 40–108. New York: Academic Press. Reprinted in facsimile in Heilbron (1981, pp. 14–82).

——. 1981. *Historical Studies in the Theory of Atomic Structure*. New York: Arno Press.

——. 1986. *The Dilemmas of an Upright Man: Max Planck as a Spokesman for German Science*. Berkeley, Los Angeles, London: University of California Press.

——. 2013. "Nascent Science. The Scientific and Psychological Background to Bohr's trilogy." In Aaserud and Heilbron (2013, pp. 105–200).

Heilbron, John L., and Thomas S. Kuhn. 1969. "The Genesis of the Bohr Atom." *Historical Studies in the Physical Sciences* 1: 211–290. Reprinted in facsimile in Heilbron (1981, pp. 149–228).

Heisenberg, Werner. 1922. "Zur Quantentheorie der Linienstruktur und der anomalen Zeeman-effekte." *Zeitschrift für Physik* 8: 273–297.

——. 1925. "Über die quantentheoretische Umdeutung kinematischer und mechanischer Beziehungen." *Zeitschrift für Phyik* 33: 879–893. English translation in Van der Waerden (1967, pp. 261–276).

Hentschel, Klaus. 2012. "Walther Ritz's Theoretical Work in Spectroscopy. Focussing on Series Formulas." In *Le destin douloureux de Walther Ritz (1878–1909): Physicien théoricien de génie*, edited by Jean-Claude Pont, pp. 129–156. Archives de l'état du Valais.

Hermann, Armin. 1965a. "Die Entdeckung des Stark-Effektes." In Hermann (1965b, Pp. 7–16).

——. 1965b. *Dokumente der Naturwissenschaft. Abteilung Physik*. Bd. 6. *Johannes Stark. Paul S. Epstein. Der Stark Effekt*. Stuttgart: Ernst Battenberg Verlag.

——. ed. 1965c. *Dokumente der Naturwissenschaften. Abteilung Physik*. Bd. 10. *Arthur Erich Haas. Der erste Quantensatz für das Atom*. Stuttgart: Ernst Battenberg Verlag.

——. 1969. *Frühgeschichte der Quantentheorie (1899–1913)*. Mosbach in Baden: Physik Verlag. Translation: Hermann (1971).

——. 1971. *The Genesis of Quantum Theory (1899–1913)*. Cambridge, MA: The MIT Press. Translation of Hermann (1969).

Hertz, Heinrich. 1888a. "Über die Ausbreitungsgeschwindigkeit der elektromagnetischen Wirkungen." *Königlich Preussische Akademie der Wissenschaften* (Berlin). *Sitzungsberichte*: 197–210. Reprinted in *Annalen der Physik* 34 (1888): 551–569 and in *Gesammelte Werke von Heinrich Hertz*, Vol. 2, pp. 115–132. Leipzig: Barth, 1891.

——. 1888b. "Die Kräfte elektrischer Schwingungen, behandelt nach der Maxwellschen Theorie." *Annalen der Physik* 36: 1–22. Reprinted in *Gesammelte Werke von Heinrich Hertz*, Vol. 2, pp. 147–170. Leipzig: Barth, 1891.

Herzberg, Gerhard. 1945. *Atomic Spectra and Atomic Structure.* New York: Dover.

Hettner, Gerhard. 1922. "Die Bedeutung von Rubens' Arbeiten für die Plancksche Strahlungs-formel." *Die Naturwissenschaften* 10: 1036.

Hirosige, Tetu. 1971. "The Van den Broek Hypothesis." *Japanese Studies in the History of Science* 10: 143–162.

Hoffmann, Banesh. 1972. *Albert Einstein: Creator and Rebel.* New York: Viking Press.

Hoffmann, Dieter. 2001. "On the Experimental Context of Planck's Foundation of Quantum Theory." *Centaurus* 43: 240–259.

——. ed. 2008. *Max Planck: Annalen Papers.* Weinheim: Wiley.

Horton, Frank, and Ann Catherine Davies. 1919. "An Experimental Determination of the Ionization Potential for Electrons in Helium." *Proceedings of the Royal Society of London A* A95: 408–429.

Hoyer, Ulrich. 1973. "Über die Rolle der Stabilitätsbetrachtungen in der Entwicklung der Bohrschen Atomtheorie." *Archive for History of Exact Sciences* 10: 177–206.

Ishiwara, Jun. 1915. "Die universelle Bedeutung des Wirkungsquantums." *Proceedings of the Tokyo Mathematico-Physical Society* 8: 106–116. Annotated translation, "The Universal Meaning of the Quantum of Action" (by Karla Pelogia and Carlos Alexandre) in *The European Journal of Physics H* 42(2017):523–536.

Jacobi, Carl Gustav Jacob. 1884. *Vorlesungen über Dynamik.* Berlin: G. Reimer.

Jähnert, Martin. 2016. *Practicing the Correspondence Principle in the Old Quantum Theory: A Transformation Through Application.* PhD diss., Technische Universität, Berlin.

Jammer, Max. 1966. *The Conceptual Development of Quantum Mechanics.* New York: McGraw-Hill.

Janssen, Michel. 2003. "The Trouton Experiment, $E = mc^2$, and a Slice of Minkowski Space-Time." In *Revisiting the Foundations of Relativistic Physics,* edited by Abhay Ashtekar, Robert S. Cohen, Don Howard, Jürgen Renn, Sahotra Sarkar, and Abner Shimony, pp. 27–54. Dordrecht: Kluwer.

——. 2019. "Arches and Scaffolds: Bridging Continuity and Discontinuity in Theory Change." In *Beyond the Meme. Development and Structure in Cultural Evolution,* edited by Alan C. Love and William C. Wimsatt, pp. 95–199. Minneapolis: University of Minnesota Press.

Janssen, Michel, and Christoph Lehner, eds. 2014. *The Cambridge Companion to Einstein.* Cambridge: Cambridge University Press.

Janssen, Michel, and Matthew Mecklenburg. 2007. "From Classical to Relativistic Mechanics: Electromagnetic Models of the Electron." In *Interactions: Mathematics, Physics and Philosophy, 1860–1930,* edited by V. F. Hendricks, K. F. Jørgensen, J. Lützen, and S. A. Pedersen, pp. 65–134. Berlin: Springer.

Janssen, Michel, and Jürgen Renn. 2015. "Arch and scaffold: How Einstein found his field equations." *Physics Today* (November): 30–36.

Janssen, Michel, and John Stachel. 2004. *The Optics and Electrodynamics of Moving Bodies.* Preprint 265. Berlin: Max Planck Institute for the History of Science.

Jeans, James Hopwood. 1901. "The Mechanism of Radiation." *Philosophical Magazine* 2: 421–455.

——. 1905a. "The Dynamical Theory of Gases and of Radiation." *Nature* 72: 101–102.

——. 1905b. "On the Partition of Energy between Matter and Aether." *Philosophical Magazine* 10: 91–98.

——. 1905c. "On the Laws of Radiation." *Proceedings of the Royal Society of London A* 76: 545–552.

——. 1908. "Zur Strahlungstheorie." *Physikalische Zeitschrift* 9: 853–855.

——. 1909. "Temperature-radiation and the Partition of Energy in Continuous Media." *Philosophical Magazine* 17: 229–254.

Jeans, James Hopwood. 1910. "On Non-Newtonian Mechanical Systems and Planck's Theory of Radiation." *Philosophical Magazine* 20: 943–954.

———. 1914a. "Discussion on Radiation." In *Report of the Eighty-third Meeting of the British Association for the Advancement of Science, Birmingham 1913, September 10–17*, 376–386. London: John Murray.

———. 1914b. *Report on Radiation and the Quantum Theory.* London: The Physical Society of London ("The Electrician" Printing & Publishing Co).

Jeffreys, Harold. 1924. "On Certain Approximate Solutions of Linear Differential Equations of the Second Order." *Proceedings of the London Mathematical Society* 23: 428–436.

Joas, Christian, and Christoph Lehner. 2009. "The Classical Roots of Wave Mechanics: Schrödinger's Transformations of the Optical-mechanical Analogy." *Studies in History and Philosophy of Modern Physics* 40: 338–351.

Joas, Christian, Christoph Lehner, and Jürgen Renn, eds. 2008. *HQ-1: Conference on the History of Quantum Physics.* Preprint 350. Berlin: Max Planck Institute for the History of Science.

Jordi Taltavull, Marta. 2017. *Transformation of Optical Knowledge from 1870 to 1925: Optical Dispersion between Classical and Quantum Physics.* PhD diss., Humboldt University, Berlin.

Jungnickel, Christa, and Russell McCormmach. 1986. *Intellectual Mastery of Nature.* 2 Vols. Chicago: University of Chicago Press.

Kangro, Hans. 1970. *Vorgeschichte des Planckschen Strahlungsgesetzes. Messungen und Theorien der spektralen Energieverteilung bis zur Begründung der Quantenhypothese.* Wiesbaden: Franz Steiner Verlag. English translation: Kangro (1976).

———. ed. 1972. *Planck's Original Papers in Quantum Physics.* London: Taylor & Francis. Translations: Dirk ter Haar and Stephen G. Brush.

———. 1976. *Early History of Planck's Radiation Law.* London/New York: Taylor & Francis/Crane, Russak & Company. Translation (with revisions and supplements) of Kangro (1970). Translator: R. E. W. Maddison.

Kao, Molly. 2015. "Unification and the Quantum Hypothesis in 1900–1913." *Philosophy of Science* 82: 1200–1210.

Kaufmann, Walter. 1906. "Über die Konstitution des Elektrons." *Annalen der Physik* 20: 487–553.

Kayser, Heinrich Gustav Johannes, and Carl Runge. 1890. "Über die Spectren der Alkalien." *Annalen der Physik* 277: 302–320.

Keller, Alex. 1983. *The Infancy of Atomic Physics: Hercules in his Cradle.* Oxford: Clarendon.

Kelvin, Lord (William Thomson). 1901. "Nineteenth Century Clouds Over the Dynamical Theory Of Heat and Light." *Philosophical Magazine* 2: 1–40. Page reference to reprint as 'Appendix B' (pp. 486–527) in Lord Kelvin, *The Baltimore Lectures on Molecular Dynamics and the Wave Theory of Light.* London: Clay, 1904.

Kemble, Edwin Crawford. 1921. "The Probable Normal State of the Helium Atom." *Philosophical Magazine* 42: 123–133.

Kennedy, Robert E. 2012. *A Student's Guide to Einstein's Major Papers.* Oxford: Oxford University Press.

Kent, Norton. 1914. "Five Lithium Lines and Their Magnetic Separation." *Astrophysical Journal* 40: 337–355.

Kirchhoff, Gustav. 1859a. "Über die Fraunhofer'schen Linien." *Monatsberichte der Königlichen Preussischen Akademie der Wissenschaften* (Berlin): 662–665. Reprinted in Kirchhoff (1898, pp. 3–5).

———. 1859b. "Über den Zusammenhang zwischen Emission und Absorption von Licht und Wärme." *Monatsberichte der Königlichen Preussischen Akademie der Wissenschaften* (Berlin): 783–787. Page references to reprint in Kirchhoff (1898, pp. 6–10).

———. 1860. "Über das Verhältnis zwischen dem Emissionsvermögen und dem Absorptionsvermögen der Körper für Wärme und Licht." *Annalen der Physik* 109: 275–301. Reprinted in Kirchhoff (1898, pp. 11–36). English translation: "On the Relation between the Radiating and Absorbing Powers of Different Bodies for Light and Heat." *Philosophical Magazine* 20 (1860): 1–21.

———. 1897. *Vorlesungen über Mechanik.* 4th ed. Leipzig: Teubner.

———. 1898. *Abhandlungen über Emission und Absorption.* Edited by Max Planck. Leipzig: Verlag von Wilhelm Engelmann.

Kirsten, Christa, and Hans-Günther Körber. 1975. *Physiker über Physiker.* Vol. 1. *Wahlvorschläge zur Aufname von Physikern in die Berliner Akademie 1870 bis 1929 von Hermann v. Helmholtz bis Erwin Schrödinger.* Berlin: Akademie Verlag.

———. 1979. *Physiker über Physiker.* Vol. 2. *Antrittsreden Erwiderungen bei der Aufnahme von Physikern in die Berliner Akademie Gedächtsnisreden 1870 bis 1929.* Berlin: Akademie Verlag.

Klein, Felix, and Arnold Sommerfeld. 1897–1910. *Über die Theorie des Kreisels. Heft 1–4.* Leipzig: Teubner.

Klein, Martin J. 1959. "Ehrenfest's Contributions to the Development of Quantum Statistics." *Proceedings of the Koninklijke Nederlandse Akademie van Wetenschappen. Series B. Physical Sciences* 62: 41–50, 51–62.

———. 1962. "Max Planck and the Beginning of the Quantum Theory." *Archive for History of Exact Sciences* 1: 459–479.

———. 1963a. "Planck, Entropy, and Quanta, 1901–1906." *The Natural Philosopher* 1: 83–108.

———. 1963b. "Einstein's First Paper on Quanta." *The Natural Philosopher* 2: 59–86.

———. 1964. "Einstein and the Wave–Particle duality." *The Natural Philosopher* 3: 3–49.

———. 1965. "Einstein, Specific Heats, and the Early Quantum Theory." *Science* 148: 173–180.

———. ed. 1967. *Letters on Wave Mechanics.* New York: Philosophical Library.

———. 1970a. *Paul Ehrenfest.* Vol. 1. *The Making of a Theoretical Physicist.* Amsterdam: North-Holland.

———. 1970b. "Maxwell, his Demon, and the Second Law of Thermodynamics." *American Scientist* 58: 84–97.

———. 1974. "Einstein, Boltzmann, and the Mechanical World View." In *XIVth International Congress of the History of Science. Tokyo & Kyoto, Japan, 19–27 August 1974. Texts of Symposia (Proceedings,* no. 1). Science Council of Japan.

———. 1977. "The Beginnings of Quantum Theory." In *Proceedings of the International School of Physics "Enrico Fermi" 1972: History of Twentieth Century Physics,* edited by C. Weiner, pp. 1–39. New York and London: Academic Press.

Klein, Martin J., Abner Shimony, and Trevor Pinch. 1979. "Paradigm lost? A review symposium." *Isis* 4: 429–440.

Kossel, Walter. 1914. "Bemerkung zur Absorption homogener Röntgenstrahlen. II." *Deutsche Physikalische Gesellschaft. Verhandlungen* 16: 953–963.

Kox, Anne J. 1993. "Einstein and Lorentz: More Than Just Good Colleagues." In Beller, Cohen, and Renn (1993, pp. 43–58).

———. 1997. "The discovery of the electron: II. The Zeeman Effect." *European Journal of Physics* 18: 139–144.

———. 2013a. "Hendrik Antoon Lorentz's Struggle With Quantum Theory." *Archive for History of Exact Sciences* 67: 149–170.

———. 2013b. "The Discovery of the Stark Effect and its Early Theoretical Explanations." *Annalen der Physik* 525 (5): A63-A66.

———. 2014. "Einstein on Statistical Physics." In Janssen and Lehner (2014, pp. 103–116).

Kragh, Helge. 1977. "Chemical aspects of Bohr's 1913 theory." *Journal of Chemical Education* 54: 208–210.

——. 1999. *Quantum Generations. A History of Physics in the Twentieth Century.* Princeton: Princeton University Press.

——. 2003. "Magic number: A Partial History of the Fine-Structure Constant." *Archive for History of Exact Sciences* 57: 395–431.

——. 2012. *Niels Bohr and the Quantum Atom. The Bohr model of Atomic Structure, 1913–1925.* Oxford: Oxford University Press.

Kramers, Hendrik A. 1919. "Intensities of Spectral Lines. On the Application of the Quantum Theory to the Problem of the Relative Intensities of the Components of the Fine Structure and of the Stark Effect of The Lines of the Hydrogen Spectrum." *Det Kongelige Danske Videnskabernes Selskab. Skrifter. Naturvidenskabelig og Matematisk Afdeling* 8 (3.3): 285–386. Reprinted in Kramers (1956, pp. 3–108).

——. 1923. "Über das Modell des Heliumatoms." *Zeitschrift für Phyik* 13: 312–341.

——. 1924a. "The Law of Dispersion and Bohr's Theory of Spectra." *Nature* 113: 673–676. Reprinted in Van der Waerden (1967, pp. 177–180).

——. 1924b. "The Quantum Theory of Dispersion." *Nature* 114: 310–311. Reprinted in Van der Waerden (1967, pp. 199–201).

——. 1926. "Wellenmechanik und halbzahlige Quantisierung." *Zeitschrift für Physik* 39: 828–840.

——. 1956. *Collected Scientific Papers.* Edited by H. B. G. Casimir. Amsterdam: North Holland.

Kramers, Hendrik A., and Werner Heisenberg. 1925. "Über die Streuung von Strahlung durch Atome." *Zeitschrift für Physik* 31: 681–707. English translation in Van der Waerden (1967, pp. 223–252).

Kries, Johannes von. 1886. *Die Principien der Wahrscheinlichkeitsrechnug.* Freiburg: Akademische Verlagsbuchhandlung.

Kuhn, Thomas S. 1962. *The Structure of Scientific Revolutions.* Chicago: University of Chicago Press. 4th (50th anniversary) ed.: 2012.

——. 1978. *Black-Body Theory and the Quantum Discontinuity, 1894–1912.* Oxford: Oxford University Press. Reprinted, including Kuhn (1984) as a new afterword, as Kuhn (1987).

——. 1984. "Revisiting Planck." *Historical Studies in the Physical Sciences* 14: 231–252. Reprinted in Kuhn (1987, pp. 349–370).

——. 1987. *Black-Body Theory and the Quantum Discontinuity, 1894–1912.* 2nd ed. Chicago: University of Chicago Press. Reprint of Kuhn (1978) including Kuhn (1984) as a new afterword.

Kuhn, Thomas S., John L. Heilbron, Paul Forman, and Lini Allen. 1967. *Sources for History of Quantum Physics. An Inventory and Report.* Philadelphia: The American Philosophical Society.

Lakatos, Imre. 1970. "Falsification and the Methodology of Scientific Research Programmes." In *Criticism and the Growth of Knowledge,* edited by Imre Lakatos and Alan Musgrave, pp. 91–196. Cambridge: Cambridge University Press.

Landé, Alfred. 1919a. "Das Serienspektrum des Heliums." *Physikalische Zeitschrift* 20: 228–234.

——. 1919b. "Adiabatenmethode zur Quantelung gestörter Elektronensysteme." *Deutsche Physikalische Gesellschaft. Verhandlungen* 21: 578–584.

——. 1919c. "Eine Quantenregel für die räumliche Orientierung von Elektronenringen." *Deutsche Physikalische Gesellschaft. Verhandlungen* 21: 585–588.

——. 1920. "Störungstheorie des Heliums." *Physikalische Zeitschrift* 21: 114–122.

——. 1921a. "Über den anomalen Zeemaneffekt. I." *Zeitschrift für Physik* 5: 231–241.

——. 1921b. "Über den anomalen Zeemaneffekt. II." *Zeitschrift für Physik* 7: 398–405.

——. 1922. "Zur Theorie der anomalen Zeeman- und magneto-mechanischen Effekte." *Zeitschrift für Physik* 11: 353–363.

——. 1923a. "Termstruktur und Zeemaneffekt der Multipletts." *Zeitschrift für Physik* 15: 189–205.

——. 1923b. "Termstruktur und Zeemaneffekt der Multipletts. Zweite Mitteilung." *Zeitschrift für Physik* 19: 112–123.

Langevin, Paul. 1905. "Magnétisme et théorie des électrons." *Annales de chimie et de physique* 5: 70–127.

Langevin, Paul, and Maurice de Broglie, eds. 1912. *La théorie du rayonnement et les quanta. Rapports et discussions de la réunion tenue à Bruxelles, du 30 octobre au 3 novembre 1911, sous les auspices de M. E. Solvay.* Paris: Gauthier-Villars.

Langley, Samuel Pierpont. 1881. "On a Thermal Balance." *The Chemical News and Journal of Physical Science* 43: 6.

Langmuir, Irving. 1912. "The Dissociation of Hydrogen into Atoms." *Journal of the American Chemical Society* 34: 860–877.

Larmor, Joseph. 1897. "On the theory of the Magnetic Influence on Spectra; and on the Radiation of Moving Ions." *Philosophical Magazine* 44: 503–512.

Laue, Max. 1911. *Das Relativitätsprinzip.* Braunschweig: Vieweg.

Lazaroff-Puck, Cameron. 2015. "Gearing up for Lagrangian dynamics. The Flywheel Analogy in Maxwell's 1865 Paper on Electrodynamics." *Archive for History of Exact Sciences* 69: 455–490.

Lehner, Christoph. 2014. "Einstein's Realism and His Critique of Quantum Mechanics." In Janssen and Lehner (2014, pp. 306–353).

Lenard, Philipp. 1900. "Erzeugung von Kathodenstrahlen durch ultraviolettes Licht." *Annalen der Physik* 2: 359–375.

——. 1902. "Ueber die lichtelektrische Wirkung." *Annalen der Physik* 8: 149–198.

——. 1903. "Über die Beobachtung lagsamer Kathodenstrahlen mit Hilfe der Phosphoreszenz und über Sekundärentstehung von Kathodenstrahlen." *Annalen der Physik* 12: 449–490.

Leone, Matteo, Alessandro Paoletti, and Nadia Robotti. 2004. "A Simultaneous Discovery: The Case of Johannes Stark and Antonino Lo Surdo." *Physics in Perspective* 6: 271–294.

Livceing, G. D., and J. Dewar. 1879. "On the Spectra of Sodium and Potassium." *Proceedings of the Royal Society of London A* 29: 398–402.

Lo Surdo, Antonino. 1913. "Sul fenomeno analogo a quello di Zeeman nel campo elettrico." *Rendiconti della Reale Accademia dei Lincei* 22: 664–666. Reprinted in German translation: "Über das elektrische Analogon des Zeeman-Phänomens." *Physikalische Zeitschrift* 15 (1914): 122.

Lodge, Sir Oliver. 1914. "Continuity (Presidential Address)." In *Report of the Eighty-third Meeting of the British Association for the Advancement of Science, Birmingham 1913, September 10–17,* 2–42. London: John Murray.

Loettgers, Andrea. 2003. "Samuel Pierpont Langley and His Contributions to the Empirical Basis of Blackbody Radiation." *Physics in Perspective* 5: 262–280.

Lombardi, Anna Maria. 2003. "The Bolometer and the Spectro-bolometer as Steps towards the Black-body Spectrum." *Nuncius* 18: 825–840.

Lorentz, Hendrik Antoon. 1892. "La théorie électromagnétique de Maxwell et son application aux corps mouvants." *Archives Néerlandaises des Sciences Exactes et Naturelles* 25: 363–552. Reprint: Leyden: Brill, 1892. Also reprinted in Lorentz (1934–9, Vol. 2, pp. 164–343).

——. 1895. *Versuch einer Theorie der electrischen und optischen Erscheinungen in bewegten Körpern.* Leyden: Brill. Reprinted in Lorentz (1934–9, Vol. 5, pp. 1–138).

Lorentz, Hendrik Antoon. 1897. "Ueber den Einfluss magnetischer Kräfte auf die Emission des Lichtes." *Annalen der Physik* 63: 278–284. French translation: Lorentz 1898.

——. 1898. "Influence du champ magnétique sur l'émission lumineuse." *L'éclairage électrique: Revue hebdomadaire d'électricité* 14: 435–438. French translation of Lorentz 1897. Reprinted in Lorentz (1934–9, Vol. 3, pp. 40–46).

——. 1903. "On the Emission and Absorption by Metals of Rays of Heat of Great Wavelength." *Proceedings of the Koninklijke Nederlandse Akademie van Wetenschappen. Series B. Physical sciences* 5: 666–685. Reprinted in Lorentz (1934–9, Vol. 3, pp. 155–175).

——. 1904. "Electromagnetische verschijnselen in een stelsel dat zich met willekeurige snelheid, kleiner dan die van het licht, beweegt." *Koninklijke Akademie van Wetenschappen te Amsterdam. Wis- en Natuurkundige Afdeeling. Verslagen van de Gewone Vergaderingen* 12: 986–1009. Translation: "Electromagnetic Phenomena in a System Moving With Any Velocity Smaller Than That of Light." *Koninklijke Akademie Van Wetenschappen te Amsterdam. Section of Sciences. Proceedings* 6: 809–831. Translation reprinted in Lorentz (1934–9, Vol. 5, pp. 172–197) and (without the final section) in Einstein et al. (1952, pp. 11–34).

——. 1905. "The Motion of Electrons in Metallic Bodies." *Proceedings of the Koninklijke Nederlandse Akademie van Wetenschappen. Series B. Physical sciences*: 438–453, 585–593, 684–691.

——. 1908a. *Le partage de l'énergie entre la matière pondérable et l'éther.* Rome: R. Accademia dei Lincei. Lecture, Rome, April 8, 1908 at the 4th Congresso Internazionale dei Matematici, April 6–11, 1908. Reprinted in slightly revised version as Lorentz (1908c) and Lorentz (1909a). Reprinted with additional note as Lorentz (1909b).

——. 1908b. "Zur Strahlungstheorie." *Physikalische Zeitschrift* 9: 562–563. Reprinted in Lorentz (1934–9, Vol. 7, pp. 344–346).

——. 1908c. "Le partage de l'énergie entre la matière pondérable et l'éther." *Il Nuovo Cimento* 16: 5–34. Slightly revised reprint of Lorentz (1908a).

——. 1909a. "Le partage de l'énergie entre la matière pondérable et l'éther." *Revue générale des sciences pures et appliquées,* no. 20: 14–26. Slightly revised reprint of Lorentz (1908a).

——. 1909b. "Le partage de l'énergie entre la matière pondérable et l'éther." In *Atti del IV Congresso Internazionale dei Matematici (Roma, 6–11 Aprile 1908).* 3 Vols, 1: 145–165. Rome: R. Accademia dei Linicei. Reprint of Lorentz (1908a) with additional note. Page references to reprint in Lorentz (1934–9, Vol. 7, pp. 317–343).

——. 1916. *Les théories statistiques en thermodynamique: conférences faites au Collège de France en novembre 1912.* Leipzig: Teubner.

——. 1934–1939. *Collected Papers.* Edited by Pieter Zeeman and Adriaan D. Fokker. The Hague: Nijhoff.

——. 2008. *The Scientific Correspondence of H. A. Lorentz.* Vol. 1. Edited by Anne J. Kox. New York: Springer.

——. 2018. *The Scientific Correspondence of H. A. Lorentz.* Vol. 2. *The Dutch Correspondents.* Edited by Anne J. Kox. New York: Springer.

Lorentz, Hendrik Antoon, Albert Einstein, and Hermann Minkowski. 1913. *Das Relativitätsprinzip. Eine Sammlung von Abhandlungen.* Leipzig: Teubner.

Lorentz, Hendrik Antoon, Albert Einstein, Hermann Minkowski, and Hermann Weyl. 1922. *Das Relativitätsprinzip. Eine Sammlung von Abhandlungen.* 4th ed. Leipzig: Teubner. Translation: Einstein et al. (1952).

Loschmidt, Joseph. 1876. "Über den Zustand des Wärmegleichgewichtes eines Systems von Körpern mit Rücksicht auf die Schwerkraft." *Kaiserliche Akademie der Wissenschaften* (Vienna). *Mathematisch-Naturwissenschaftliche Classe. Zweite Abteilung. Sitzungsberichte* 73: 128–142.

Lummer, Otto, and Eugen Jahnke. 1900. "Über die Spectralgleichung des schwarzen Körpers und des blanken Platins." *Annalen der Physik* 308: 283–297.

Lummer, Otto, and Ernst Pringsheim. 1899a. "Die Vertheilung der Energie im Spectrum des schwarzen Körpers." *Deutsche Physikalische Gesellschaft. Verhandlungen* 1: 23–41.

——. 1899b. "1. Die Vertheilung der Energie im Spectrum des schwarzen Körpers und des blanken Platins; 2. Temperaturbestimmung fester glühender Körper." *Deutsche Physikalische Gesellschaft. Verhandlungen* 1: 215–235.

——. 1900. "Über die Strahlung des schwarzen Körpers für lange Wellen." *Deutsche Physikalische Gesellschaft. Verhandlungen* 2: 163–180.

——. 1908. "Über die Jeans-Lorentzsche Strahlungsformel." *Physikalische Zeitschrift* 9: 449–450.

Lyman, Theodore. 1922. "The Spectrum of Helium in the Extreme Ultraviolet." *Nature* 110: 278–279.

Martinez, Alberto A. (2004). "Ritz, Einstein, and the Emission Hypothesis." *Physics in Perspective* 6: 4–28.

Marx, Erich, and Karl Lichtenecker. 1913. "Experimentelle Untersuchung des Einflusses der Unterteilung der Belichtungszeit auf die Elektronenabgabe in Elster und Geitelschen Kaliumhydrürzellen bei sehr schwacher Lichtenergie." *Annalen der Physik* 41: 124–160.

Maury, Antonia C., and Edward C. Pickering. 1897. "Spectra of Bright Stars Photographed with the 11-inch Draper Telescope as part of the Henry Draper Memorial." *Annals of the Astronomical Observatory of Harvard College* 28: 1–128.

Maxwell, James Clerk. 1860. "Illustrations of the Dynamical Theory of Gases. Pt I. On the Motions and Collisions of Perfectly Elastic Spheres. Pt. II. On the Process of Diffusion of Two or More Kinds of Moving Particles Among One Another." *Philosophical Magazine.* 19: 19-32 (I); 20: 21-37 (II).

——. 1861–1862. "On Physical Lines of Force, Pts. I–IV." *Philosophical Magazine.* 21 (1861): 161–175 (I), 281–291, 338–345 (II); 23 (1862): 12–24 (III), 85–95 (IV). Reprinted in *The Scientific Papers of James Clerk Maxwell,* edited by W. D. Niven, Vol. 1, pp. 451–513. Cambridge: Cambridge University Press, 1890.

——. 1871. *Theory of Heat.* London, New York: Longmans, Green & Co. Page references to a reprint (New York: Dover, 2001), edited by Peter Pesic, of the 9th edition (1888).

——. 1873. "Molecules." *Nature* 9 (September): 437–441. "Lecture delivered before the British Association at Bradford." Page references to reprint in Maxwell (1952, Vol. 2, pp. 361–377).

——. 1875. "On the Dynamical Evidence of the Molecular Constitution of Bodies." *Nature* 11 (March): 357–359. Page references to reprint in Maxwell (1952, Vol. 2, pp. 418–438).

——. 1952. *Scientific Papers of James Clerk Maxwell.* Edited by W. D. Niven. Dover. Two volumes bound as one. Originally published (in two volumes) in 1890 by Cambridge University Press.

McCormmach, Russell. 1966. "The Atomic Theory of John William Nicholson." *Archive for the History of the Exact Sciences* 3: 160–184.

——. 1967. "Henri Poincare and the Quantum Theory." *Isis* 58: 37–55.

——. 1970. "H. A. Lorentz and the Electromagnetic View of Nature." *Isis* 61: 459–497.

Mehra, Jagdish, and Helmut Rechenberg. 1982a. *The Historical Development of Quantum Theory.* Vol. 1. *The Quantum Theory of Planck, Einstein, Bohr and Sommerfeld: Its Foundations and the Rise of Its Difficulties.* New York, Heidelberg, Berlin: Springer.

——. 1982b. *The Historical Development of Quantum Theory.* Vol. 2. *The Discovery of Quantum Mechanics 1925.* New York, Heidelberg, Berlin: Springer.

——. 1982c. *The Historical Development of Quantum Theory.* Vol. 3. *The Formulation of Matrix Mechanics and Its Modifications 1925–1926.* New York, Heidelberg, Berlin: Springer.

Mehra, Jagdish, and Helmut Rechenberg. 1982d. *The Historical Development of Quantum Theory.* Vol. 4. Part 1. *The Fundamental Equations of Quantum Mechanics. 1925–1926.* Part 2. *The Reception of the New Quantum Mechanics. 1925–1926.* New York, Heidelberg, Berlin: Springer.

———. 1987. *The Historical Development of Quantum Theory.* Vol. 5. *Erwin Schrödinger and the Rise of Wave Mechanics.* Part 1. *Schrödinger in Vienna and Zurich 1887–1925.* Part 2. *The Creation of Wave Mechanics. Early Response and Applications 1925–1926.* New York, Heidelberg, Berlin: Springer.

———. 2000–2001. *The Historical Development of Quantum Theory.* Vol. 6. *The Completion of Quantum Mechanics 1926–1941.* Part 1. *The Probability Interpretation and the Statistical Transformation Theory, the Physical Interpretation, and the Empirical and Mathematical Foundations of Quantum Mechanics 1926–1932.* Part 2. *The Conceptual Completion of the Extension of Quantum Mechanics 1932–1941. Epilogue: Aspects of the Further Development of Quantum Theory 1942–1999. Subject Index: Volumes 1 to 6.* New York, Heidelberg, Berlin: Springer.

Meitner, Lise. 1925. "Die γ-Strahlung der Actiniumreihe und der Nachweis, dass die γ-Strahlen erst nach Erfolgtem Atomzerfall emittiert werden." *Zeitschrift für Physik* 34: 807–818.

Michelson, Vladimir Alexandrovich. 1887. "Essai théorique sur la distribution de l'énergie dans les spectres des solides." *Journal de Physique Théorique et Appliquée* 6: 467–479.

———. 1888. "Theoretical Essay on the Distribution of Energy in the Spectra of Solids." *Philosophical Magazine* 25: 425–435.

Midwinter, Charles, and Michel Janssen. 2013. "Kuhn Losses Regained: Van Vleck From Spectra to Susceptibilities." In Badino and Navarro (2013, pp. 137–205).

Miller, Arthur I. 1981. *Albert Einstein's Special Theory of Relativity. Emergence (1905) and Early Interpretation (1905–1911).* Reading, MA: Addison- Wesley. Reprinted in 1998 (Springer, New York). Page references are to this reprint.

Millikan, Robert A. 1916a. "Einstein's Photoelectric Equation and Contact Electromotive Force." *Physical Review* 7: 18–32.

———. 1916b. "A Direct Photoelectric Determination of Planck's 'h'." *Physical Review* 7: 355–388.

———. 1917. *The Electron: Its Isolation and Measurement and the Determination of Some of its Properties.* Chicago: University of Chicago Press.

———. 1950. *The Autobiography of Robert A. Millikan.* New York: Prentice-Hall.

Minkowski, Hermann. 1909. "Raum und Zeit." *Physikalische Zeitschrift* 20: 104–111. English translation in Einstein et al. (1952, pp. 75–91).

Moseley, Henry Gwyn Jeffreys. 1913. "The High-frequency Spectra of the Elements." *Philosophical Magazine* 26: 1024–1034.

———. 1914. "The High-Frequency Spectra of the Elements. Part II." *Philosophical Magazine* 27: 703–713.

Mosharrafa, A. M. 1922. "On the Stark Effect for Strong Electric Fields." *Philosophical Magazine* 44: 371–373.

Nagaoka, Hantaro. 1904. "Kinetics of a System of Particles Illustrating the Line and the Band Spectrum and the Phenomena of Radioactivity." *Philosophical Magazine* 8: 445–455.

Nagel, Bengt. 1982. "The Discussion Concerning the Nobel Prize for Max Planck." In *Science, Technology and Society in the Time of Alfred Nobel,* edited by Carl Gustaf Bernhard, Elisabeth Crawford, and Per Sörbom, pp. 361–376. New York: Pergamon.

Nakane, Michiyo. 2015. "The Origins of Action-Angle Variables and Bohr's Introduction of Them in a 1918 Paper." In Aaserud and Kragh (2015, 290–309).

Natanson, Władyslaw. 1911. "Über die statistische Theorie der Strahlung." *Physikalische Zeitschrift* 12: 659–666.

Nath, Biman B. 2013. *The Story of Helium and the Birth of Astrophysics.* New York: Springer.

Nauenberg, Michael. 2016. "Max Planck and the Birth of the Quantum Hypothesis." *American Journal of Physics* 84: 709–720.

Navarro, Jaume. 2012. *A History of the Electron: J. J. and G. P. Thomson.* Cambridge: Cambridge University Press.

Navarro, Luis, and Enric Pérez. 2004. "Paul Ehrenfest on the Necessity of Quanta (1911): Discontinuity, Quantization, Corpuscularity, and Adiabatic Invariance." *Archive for the History of the Exact Sciences* 58: 97–141.

——. 2006. "Paul Ehrenfest: The Genesis of the Adiabatic Hypothesis, 1911–1914." *Archive for the History of the Exact Sciences* 60: 209–267.

Needell, Allan A. 1980. *Irreversibility and the Failure of Classical Dynamics: Max Planck's Work on the Quantum Theory, 1900–1915.* PhD diss., Yale University.

——. 1988. "Introduction." In *The Theory of Heat Radiation,* by Max Planck, pp. xi–xlv. New York: American Institute of Physics.

Nernst, Walther. 1906. "Über die Berechnung chemischer Gleichgewichte aus thermischen Messungen." *Königliche Gesellschaft der Wissenschaften zu Göttingen. Mathematisch-physikalische Klasse, Nachrichten*: 1–40.

——. 1910. "Untersuchungen über die spezifische Wärme bei tiefen Temperaturen. II." *Königlich Preussische Akademie der Wissenschaften* (Berlin). *Sitzungsberichte*: 262–282.

——. 1911a. "Über neuere Probleme der Wärmetheorie." *Königlich Preussische Akademie der Wissenschaften* (Berlin). *Sitzungsberichte*: 65–90.

——. 1911b. "Zur Theorie der spezifischen Wärme und über die Anwendung der Lehre von den Energiequanten auf physikalisch-chemische Fragen überhaupt." *Zeitschrift für Elektrochemie und angewandte physikalische Chemie* 17: 265–275.

——. 1911c. "Untersuchungen über die spezifische Wärme bei tiefen Temperaturen. III." *Königlich Preussische Akademie der Wissenschaften* (Berlin). *Sitzungsberichte*: 306–315.

——. 1913. *Thermodynamics and Chemistry.* New Haven, CT: Yale University Press.

Nernst, Walther, Fritz Koref, and Frederick A. Lindemann. 1910. "Untersuchungen über die spezifische Wärme bei tiefen Temperaturen. I." *Königlich Preussische Akademie der Wissenschaften* (Berlin). *Sitzungsberichte*: 247–261.

Nernst, Walther, and Frederick A. Lindemann. 1911. "Spezifische Wärme und Quantentheorie." *Zeitschrift für Elektrochemie und angewandte physikalische Chemie* 17: 817–827.

Nicholson, John William. 1911a. "On the Number of Electrons Concerned in Metallic Conduction." *Philosophical Magazine* 22: 245–266.

——. 1911b. "A Structural Theory of the Chemical Elements." *Philosophical Magazine* 22: 864–889.

——. 1911c. "The Spectrum of Nebulium." *Monthly Notices of the Royal Astronomical Society* 72: 49–64.

——. 1911d. "The Constitution of the Solar Corona, I: Protofluorine." *Monthly Notices of the Royal Astronomical Society* 72: 139–150.

——. 1912. "The Constitution of the Solar Corona. II." *Monthly Notices of the Royal Astronomical Society* 72: 677–692.

——. 1914a. "The Constitution of Nebulae." *Monthly Notices of the Royal Astronomical Society* 74: 486–506.

——. 1914b. "The High-frequency Spectra of the Elements and the Structure of the Atom." *Philosophical Magazine* 27: 90–1003.

——. 1914c. "Atomic Structure and the Spectrum of Helium." *Philosophical Magazine* 28: 90–103.

Nordheim, Lothar, and Erwin Fues. 1927. "Die Hamilton-Jacobische Theorie der Dynamik." Chap. 3 in *Handbuch der Physik*. Vol. 5. *Grundlagen der Mechanik. Mechanik der Punkte und starren Körper*, edited by Richard Grammel, pp. 91–130. Julius Springer.

Norton, John D. 1993. "The Determination of Theory by Evidence: The Case for Quantum Discontinuity, 1900–1915." *Synthese* 97: 1–31.

———. 2006. "Atoms, Entropy, Quanta: Einstein's Miraculous Argument of 1905." *Studies in History and Philosophy of Modern Physics* 37: 71–100.

———. 2008. "Einstein's Miraculous Argument of 1905: The Thermodynamic Grounding of Light Quanta." In Joas, Lehner, and Renn (2008, Vol. 1, pp. 63–78).

Nye, Mary Jo. 1972. *Molecular Reality: A Perspective on the Scientific Work of Jean Perrin*. London: Macdonald.

Okun, Lev B. 1989. "The Concept of Mass." *Physics Today* 42 (June): 31–36.

Ornstein, Leonard Salomon. 1910. "Enige opmerkingen over de mechanische grondslagen der warmteleer." *Koninklijke Akademie van Wetenschappen te Amsterdam. Wis- en Natuurkundige Afdeeling. Verslagen van de Gewone Vergaderingen* 19: 809–823.

Pais, Abraham. 1982. '*Subtle is the Lord ...*' *The Science and the Life of Albert Einstein*. Oxford: Oxford University Press.

———. 1986. *Inward Bound. Of Matter and Forces in the Physical World*. Oxford: Clarendon.

Paschen, Friedrich. 1896–1897. "Über Gesetzmässigkeiten in den Spectren fester Körper. 2 Pts." *Annalen der Physik*. 58: 455–492 (I); 60: 662–723 (II).

———. 1899a. "Über die Vertheilung der Energie im Spektrum des schwarzen Körpers bei niederen Temperaturen (I)." *Königlich Preussische Akademie der Wissenschaften* (Berlin). *Sitzungsberichte*: 405–420.

———. 1899b. "Über die Vertheilung der Energie im Spektrum des schwarzen Körpers bei höheren Temperaturen (II)." *Königlich Preussische Akademie der Wissenschaften* (Berlin). *Sitzungsberichte*: 959–976.

———. 1901. "Ueber das Strahlungsgesetz des schwarzen Körpers." *Annalen der Physik* 4: 277–298.

———. 1916. "Bohrs Heliumlinien." *Annalen der Physik* 50: 901–940.

Paschen, Friedrich, and Ernst Back. 1912. "Normale und anomale Zeemaneffekte." *Annalen der Physik* 39: 897–932.

———. 1913. "Normale und anomale Zeemaneffekte. Nachtrag." *Annalen der Physik* 40: 960–970.

Pauli, Wolfgang. 1923. "Über die Gesetzmässigkeiten des anomalen Zeemaneffektes." *Zeitschrift für Physik* 16: 155–164.

———. 1924. "Zur Frage der Zuordnung der Komplexstrukturterme in starken und schwachen äusseren Feldern." *Zeitschrift für Physik* 20: 371–387.

———. 1926. "Quantentheorie." *Handbuch der Physik* 23: 1–278.

———. 1946. "Remarks on the History of the Exclusion Principle." *Science* 103: 213–215.

———. 1979. *Wissenschaftlicher Briefwechsel mit Bohr, Einstein, Heisenberg u.a. BandI: 1919–1929/ Scientific Correspondence with Bohr, Einstein, Heisenberg, a.o. Vol. I: 1919–1929*. Edited by Armin Hermann, Karl von Meyenn, and Victor F. Weisskopf. New York, Heidelberg, Berlin: Springer.

Pérez, Enric. 2009. "Ehrenfest's Adiabatic Theory and the Old Quantum Theory, 1916–1918." *Archive for the History of the Exact Sciences* 63: 81–125.

Perrin, Jean. 1913. *Les Atomes*. Paris: Librairie Félix Alcan. English translation by D. Ll. Hammick, *Atoms* (London: Constable, 1916). Reprint of second English edition of 1923: Woodbride, CT: Ox Bow Press, 1990.

Pickering, Edward C. 1896. "Stars Having Peculiar Spectra. New Variable Stars in Crux and Sygnus." *Astrophysical Journal* 4: 369–370.

——. 1897. "The Spectrum of ζ Puppis." *Astrophysical Journal* 5: 92–94.

Plancherel, Michel. 1913. "Beweis der Unmöglichkeit ergodischer mechanischer Systeme." *Annalen der Physik* 42: 1061–1063.

Planck, Max. 1879. *Über den zweiten Hauptsatz der mechanischen Wärmetheorie*. Munich: Ackermann. Dissertation. Reprinted in Planck (1958, Vol. 1, pp. 1–61).

——. 1882. "Verdampfen, Schmelzen und Sublimieren." *Annalen der Physik* 15: 446–475. Reprinted in facsimile (without the final footnote) in Planck (1958, Vol. 1, pp. 134–163) and in Hoffmann (2008, pp. 185–214).

——. 1891. "Allgemeines zur neueren Entwicklung der Wärmetheorie." *Verhandlungen der Gesellschaft Deutscher Naturforscher und Ärzte*: 56–61. Page references to reprint in Planck (1958, Vol. 1, pp. 372–381).

——. 1895. "Absorption und Emission elektrischer Wellen durch Resonanz." *Königlich Preussische Akademie der Wissenschaften* (Berlin). *Sitzungsberichte*: 289–301. Reprinted in *Annalen der Physik* 57 (1896): 1–14, reprinted in facsimile in Planck (1958, Vol. 1, pp. 445–558) and in Hoffmann (2008, pp. 420–433).

——. 1896. "Über elektrische Schwingungen, welche durch Resonanz erregt und durch Strahlung gedämpft werden." *Königlich Preussische Akademie der Wissenschaften* (Berlin). *Sitzungsberichte*: 151–170. Reprinted in *Annalen der Physik* 60 (1897): 577–599, reprinted in facsimile in Planck (1958, Vol. 1, pp. 466–488) and in Hoffmann (2008, pp. 434–456).

——. 1897–1899. "Über irreversible Strahlungsvorgänge. 1.–5. Mittheilung." *Königlich Preussische Akademie der Wissenschaften* (Berlin). *Sitzungsberichte*: 57–68, 715–717, 1122–1145 (1897), 449–476 (1898), 440–480 (1899). Page references to reprint in Planck (1958, Vol. 1, 493–504 (I), 505–507 (II), 508–531 (III), 532–559 (IV), 560–600 (V)).

——. 1900a. "Über irreversible Strahlungsvorgänge." *Annalen der Physik* 1: 69–122. Reprinted in facsimile in Planck (1958, Vol. 1, pp. 614–667) and in Hoffmann (2008, pp. 461–514).

——. 1900b. "Entropie und Temperatur strahlender Wärme." *Annalen der Physik* 1: 719–737. Reprinted in facsimile in Planck (1958, Vol. 1, 668–686) and in Hoffmann (2008, pp. 515–533).

——. 1900c. "Über eine Verbesserung der Wien'schen Spectralgleichung." *Deutsche Physikalische Gesellschaft. Verhandlungen* 2: 202–204. Reprinted in Planck (1958, Vol. 1, 687–689). Page references to the English translation in Kangro (1972, pp. 35–45).

——. 1900d. "Zur Theorie des Gesetzes der Energieverteilung im Normalspektrum." *Deutsche Physikalische Gesellschaft. Verhandlungen* 2: 237–245. Reprinted in Planck (1958, Vol. 1, pp. 698–706). Page references to the English translation in Kangro (1972, pp. 35–45).

——. 1901a. "Über das Gesetz der Energieverteilung im Normalspektrum." *Annalen der Physik* 4: 553–563. Reprinted in facsimile in Planck (1958, Vol. 1, pp. 717–727) and in Hoffmann (2008, pp. 537–547). Page references to (a slightly abbreviated) English translation Shamos (1959, pp. 305–313).

——. 1901b. "Über die Elementarquanta der Materie und der Elektricität." *Annalen der Physik* 4 (1901): 564–566, reprinted in facsimile in Planck (1958, Vol. 1, pp. 728–730) and in Hoffmann (2008, pp. 548–550).

——. 1901c. "Über irreversible Strahlungsvorgänge. Nachtrag." *Königlich Preussische Akademie der Wissenschaften* (Berlin). *Sitzungsberichte*: 544–555. Reprinted (in a slightly different version) in *Annalen der Physik* 6 (1901): 818–831, reprinted in facsimile in Planck (1958, Vol. 1, pp. 744–757) and in Hoffmann (2008, pp. 551–564).

Planck, Max. 1901d. "Über die Verteilung der Energie zwischen Aether und Materie." *Archives Néerlandaises des Sciences Exactes et Naturelles.* Vol. 6. *Livre jubilaire offert à la Société Hollandaise des Sciences à Haarlem par les amis de J. Bosscha*: 55–66. Page references to reprint in *Annalen der Physik* 9 (1902): 629–642, reprinted in facsimile in Planck (1958, Vol. 1, pp. 731–743) and in Hoffmann (2008, pp. 586–598).

——. 1906a. *Vorlesungen über die Theorie der Wärmestrahlung.* Leipzig: Barth. Reprinted in Planck (1988, 241–470).

——. 1906b. "Die Kaufmannschen Messungen der Ablenkbarkeit der β-Strahlen in ihrer Bedeutung für die Dynamik der Elektronen." *Deutsche Physikalische Gesellschaft. Verhandlungen* 8: 418–432. Reprinted in *Physikalische Zeitschrift* 7 (1906): 753–759 (760–761: discussion); and in Planck (1958, Vol. 2, pp. 121–135).

——. 1910. "Zur Theorie der Wärmestrahlung." *Annalen der Physik* 31: 758–768. Reprinted in facsimile in Planck (1958, Vol. 2, pp. 237–247) and in Hoffmann (2008, pp. 599–609).

——. 1911a. "Eine neue Strahlungshypothese." *Deutsche Physikalische Gesellschaft. Verhandlungen* 13: 138–148. Reprinted in Planck (1958, Vol. 2, 249–259).

——. 1911b. "Zur Hypothese der Quantenemission." *Königlich Preussische Akademie der Wissenschaften* (Berlin). *Sitzungsberichte*: 723–731. Reprinted in Planck (1958, Vol. 2, pp. 260–268).

——. 1912. "Über die Begründung des Gesetzes der schwarzen Strahlung." *Annalen der Physik* 37: 642–656. Reprinted in facsimile in Planck (1958, Vol. 2, pp. 287–301) and in Hoffmann (2008, pp. 610–624).

——. 1913. *Vorlesungen über die Theorie der Wärmestrahlung.* 2nd ed. (1st ed.: 1906). Leipzig: Barth. Page references to English translation: *The Theory of Heat Radiation* (New York: Dover, 1991). Reprinted in Planck (1988, 1–239).

——. 1914. *Dynamische und statistische Gesetzmässigkeit.* Leipzig: Barth. Reprinted in Planck (1958, Vol. 3, pp. 77–90).

——. 1916. "Die Physikalische Struktur des Phasenraumes." *Annalen der Physik* 50: 385–418. Reprinted in facsimile in Planck (1958, Vol. 2, pp. 386–419) and in Hoffmann (2008, pp. 654–687).

——. 1920. *Die Entstehung und bisherige Entwicklung der Quantentheorie.* Leipzig: Barth. Nobel lecture held in Stockholm, June 2, 1920. Reprinted in Planck (1958, Vol. 3, pp. 121–136). Page references to English translation ("The Origin and Development of the Quantum Theory"): Oxford: Clarendon Press, 1922.

——. 1928. "Hendrik Antoon Lorentz." *Die Naturwissenschaften* 16: 549–555. Page references to reprint in Planck (1958, Vol. 3, pp. 343–349).

——. 1943. "Zur Geschichte der Auffindung des physikalischen Wirkungsquantum." *Die Naturwissenschaften* 31: 153–159. Page references to reprint in Planck (1958, Vol. 3, pp. 255–267).

——. 1949. *Wissenschaftliche Selbstbiographie.* Leipzig: Barth. Page references to reprint in Planck (1958, Vol. 3, pp. 374–401). English translation (by Frank Gaynor): *Scientific Autobiography and Other Papers* (New York: Philosophical Library, 1949).

——. 1958. *Physikalische Abhandlungen und Vorträge.* 3 Vols. Braunschweig: Vieweg.

——. 1988. *The Theory of Heat Radiation.* New York: American Institute of Physics. With an introduction by Allan A. Needell.

Poincaré, Henri. 1890. "Sur les équations de la dynamique et le problème des trois corps." *Acta Mathematica* 13: 1–270.

——. 1892–1899. *Les Méthodes Nouvelles de la Méchanique Céleste.* 3 Vols. Paris: Gauthier-Villars. Vol. 1 (1892), Vol. 2 (1893), Vol. 3 (1899).

——. 1911. "Sur la théorie des quanta." *Comptes Rendus de l'Académie des Science* 153: 1103–1108.

——. 1912. "Sur la théorie des quanta." *Journal de Physique Théorique et Appliquée* 2: 5–34.

Preston, Thomas. 1899. "General Law of the Phenomena of Magnetic Perturbations of Spectral Lines." *Nature* 59: 248.

Pyenson, Lewis. 1985. *The Young Einstein. The Advent of Relativity.* Bristol and Boston: Adam Hilger.

Rayleigh, Lord (John Willam Strutt). 1900a. "The Law of Partition of Kinetic Energy." *Philosophical Magazine* 49: 98–118. Page reference to reprint in Rayleigh (1903, Vol. 4, pp. 433–451).

——. 1900b. "Remarks Upon the Law of Complete Radiation." *Philosophical Magazine* 49: 539–540.

——. 1903. *Scientific Papers.* Cambridge: Cambridge University Press.

——. 1905a. "The Dynamical Theory of Gases and Radiation." *Nature* 72: 54–55.

——. 1905b. "The Constant of Radiation as Calculated from Molecular Data." *Nature* 72: 243–244.

Regener, Erich. 1909. "Über Zählung der α-Teilchen durch die Szintillation und über die Größe des elektrischen Elementarquantums." *Königlich Preussische Akademie der Wissenschaften* (Berlin). *Sitzungsberichte*: 948–965.

Renn, Jürgen. 2000. "Einstein's Controversy with Drude and the Origin of Statistical Mechanics: A New Glimpse from the 'Love Letters'." In *Einstein, the Formative Years, 1879–1909,* edited by Don Howard and John Stachel, pp. 107–158. Boston: Birkhäuser.

——. 2005. "Einstein's Invention of Brownian Motion." *Annalen der Physik* 14 Supplement: 23–37.

Renn, Jürgen, and Robert Rynasiewicz. 2014. "Einstein's Copernican Revolution." In Janssen and Lehner (2014, pp. 38–71).

Ritz, Walter. 1903. "Zur Theorie der Serienspektren." *Annalen der Physik* 317: 264–310.

——. 1908a. "Magnetische Atomfelder und Serienspectren." *Annalen der Physik* 35: 660–696.

——. 1908b. "On a New Law of Series Spectra." *Astrophysical Journal* 28: 237–243.

Robertson, Peter. 1979. *The Early Years. The Niels Bohr Institute, 1921–1930.* Copenhagen: Akademisk Forlag.

Robotti, Nadia. 1983. "The Spectrum of ζ Puppis and the Historical Evolution of Empirical Data." *Historical Studies in the Physical Sciences* 14: 123–146.

Rosenfeld, Léon. 1936. "La première phase de l'évolution de la théorie des quanta." *Osiris* 2: 149–196.

——. 1963. "Introduction." In Bohr (1963).

Rosenthal, Artur. 1913. "Beweis der Unmöglichkeit ergodischer Gassysteme." *Annalen der Physik* 42: 796–806.

Rubens, Heinrich. 1899. "Ueber die Reststrahlen des Flussspathes." *Annalen der Physik* 69: 576–588.

Rubens, Heinrich, and Ferdinand Kurlbaum. 1900. "Über die Emission langwelliger Wärmestrahlen durch den schwarzen Körper bei verschiedenen Temperaturen." *Königlich Preussische Akademie der Wissenschaften* (Berlin). *Sitzungsberichte*: 929–941.

——. 1901. "Anwendung der Methode der Reststrahlen zur Prüfung des Strahlungsgesetzes." *Annalen der Physik* 4: 649–666.

Rubinowicz, Adalbert. 1918a. "Bohrsche Frequenzbedingung und Erhaltung des Impulsmomentes, I. Teil." *Physikalische Zeitschrift* 19: 441–445.

——. 1918b. "Bohrsche Frequenzbedingung und Erhaltung des Impulsmomentes, II. Teil." *Physikalische Zeitschrift* 19: 465–474.

Runge, Carl. 1902. "Über den Zeemaneffekt der Serienlinien." *Physikalische Zeitschrift* 3: 441–442.

——. 1907. "Über die Zerlegung der Spektrallinien im magnetischen Felde." *Physikalische Zeitschrift* 8: 232–237.

Runge, Carl, and Friedrich Paschen. 1896. "On the Spectrum of Cléveite Gas." *Astrophysical Journal* 3: 4–28.

——. 1900. "Studium des Zeeman-effektes im Quecksilberspektrum." *Physikalische Zeitschrift* 1: 480–481.

Rutherford, Ernest. 1911. "The Scattering of α and β Particles by Matter and the Structure of the Atom." *Philosophical Magazine* 21: 669–688. Reprinted in Birks (1962, pp. 182–204) and in Ter Haar (1967, pp. 108–131).

Rutherford, Ernest, and Hans Geiger. 1908. "The Charge and Nature of the α-Particle." *Royal Society of London. Proceedings A* 81: 162–173.

Rydberg, Janne Robert. 1890. "Recherche sur la constitution des spectres d'émission des éléments chimiques." *Kongliga Svenska Vetenskaps-Academiens Handlingar* 23: 1–155.

——. 1896. "The New Elements of Cléveite Gas." *Astrophysical Journal* 4: 91–96.

——. 1897. "The New Series in the Spectrum of Hydrogen." *Astrophysical Journal* 6: 233–238.

Rynasiewicz, Robert. 2015. "The (?) Correspondence Principle." In Aaserud and Kragh (2015, pp. 175–199).

Rynasiewicz, Robert, and Jürgen Renn. 2006. "The Turning Point for Einstein's Annus Mirabilis." *Studies in History and Philosophy of Modern Physics* 37: 5–35.

Scerri, Eric. 2007. *The Periodic Table. Its Story and Its Significance.* Oxford: Oxford University Press.

Schott, George Adolphus. 1907. "On the Electron Theory of Matter and on Radiation." *Philosophical Magazine* 13: 189–213.

Schrödinger, Erwin. 1926a. "Quantisierung als Eigenwertproblem (Zweite Mitteilung)." *Annalen der Physik* 79: 489–527. Translation in Schrödinger (1982, pp. 13–40).

——. 1926b. "Quantisierung als Eigenwertproblem. Dritte Mitteilung: Störungstheorie, mit Anwendung auf den Starkeffekt der Balmerlinien." *Annalen der Physik* 80: 437–490. Translation in Schrödinger (1982, pp. 62–101).

——. 1982. *Collected Papers on Wave Mechanics.* Providence, RI: American Mathematical Society Chelsea Publishing. Third (augmented) English edition.

Schulmann, Robert. 1993. "Einstein at the Patent Office: Exile, Salvation, or Tactical Retreat." In Beller, Cohen, and Renn (1993, pp. 17–24).

——. 2014. "Einstein's Politics." In Janssen and Lehner (2014, pp. 421–454).

Schwarzschild, Karl. 1916. "Zur Quantenhypothese." *Königlich Preussische Akademie der Wissenschaften* (Berlin). *Sitzungsberichte*: 548–568.

Segrè, Emilio. 1980. *From X-Rays to Quarks. Modern Physicists and Their Discoveries.* San Francisco: W. H. Freeman. Reprint: New York: Dover, 2007.

Serwer, Daniel. 1977. "*Unmechanischer Zwang*: Pauli, Heisenberg, and the Rejection of the Mechanical Atom, 1923–1925." *Historical Studies in the Physical Sciences* 8: 189–256.

Seth, Suman. 2010. *Crafting the Quantum. Arnold Sommerfeld and the Practice of Theory, 1890–1926.* Cambridge, MA: The MIT Press.

Shamos, Morris H., ed. 1959. *Great Experiments in Physics. Firsthand Accounts from Galileo to Einstein.* New York: Dover.

Sigmund, Karl. 2017. *Exact Thinking in Demented Times. The Vienna Circle and the Epic Quest for the Foundations of Science.* New York: Basic Books.

Sime, Ruth Lewin. 1996. *Lise Meitner. A Life in Physics.* Berkeley: University of California Press.

Sobel, Dava. 2016. *The Glass Universe: How the Ladies of the Harvard Observatory Took the Measure of the Stars.* New York: Penguin.

Sommerfeld, Arnold. 1904a. "Zur Elektronentheorie. I. Allgemeine Untersuchung des Feldes eines beliebig bewegten Elektrons." *Königliche Gesellschaft der Wissenschaften zu Göttingen. Mathematisch-physikalische Klasse. Nachrichten*: 99–130. Reprinted in Sommerfeld (1968, Vol. 2, pp. 39–70).

———. 1904b. "Zur Elektronentheorie. II. Grundlage für eine allgemeine Dynamik des Elektrons." *Königliche Gesellschaft der Wissenschaften zu Göttingen. Mathematisch-physikalische Klasse. Nachrichten*: 363–439. Reprinted in Sommerfeld (1968, Vol. 2, pp. 71–147).

———. 1905. "Zur Elektronentheorie. III. Ueber Lichtgeschwindigkeits- und Ueberlichtgeschwindigkeits-Elektronen." *Königliche Gesellschaft der Wissenschaften zu Göttingen. Mathematisch-physikalische Klasse. Nachrichten*: 201–235. Reprinted in Sommerfeld (1968, Vol. 2, pp. 148–182).

———. 1909. "Über die Verteilung der Intensität bei der Emission der Röntgenstrahlen." *Physikalische Zeitschrift* 10: 969–976. Reprinted in Sommerfeld (1968, Vol. 4, pp. 369–376).

———. 1910a. "Zur Relativitätstheorie I. Vierdimensionale Vektoralgebra." *Annalen der Physik* 32: 749–776. Reprinted in Sommerfeld (1968, Vol. 2, pp. 217–257).

———. 1910b. "Zur Relativitätstheorie II. Vierdimensionale Vektoranalysis." *Annalen der Physik* 33: 649–689. Reprinted in Sommerfeld (1968, Vol. 2, pp. 217–257).

———. 1911a. "Das Plancksche Wirkungsquantum und seine allgemeine Bedeutung für die Molekularphysik." *Physikalische Zeitschrift* 12: 1057–1069. Reprinted in Sommerfeld (1968, Vol. 3, pp. 1–19).

———. 1911b. "Über die Struktur der γ-Strahlen." *Königlich Bayerische Akademie der Wissenschaften zu München. Mathematisch-physikalische Klasse. Sitzungsberichte*: 1–60. Reprinted in Sommerfeld (1968, Vol. 4, pp. 377–436).

———. 1912. "Sur l'application de la théorie de l'élément d'action aux phénomènes moléculaires non périodiques." In Langevin and de Broglie (1912, pp. 313–392).

———. 1913. "Der Zeemaneffekt eines anisotrop gebundenen Elektrons und die Beobachtungen von Paschen-Back." *Annalen der Physik* 40: 748–774. Reprinted in Sommerfeld (1968, Vol. 3, pp. 20–46).

———. 1914. "Zur Voigtschen Theorie des Zeeman-Effektes." *Königliche Gesellschaft der Wissenschaften zu Göttingen. Mathematisch-physikalische Klasse. Nachrichten*: 207–229. Reprinted in Sommerfeld (1968, Vol. 3, pp. 47–69).

———. 1915a. "Zur Theorie der Balmerschen Serie." *Königlich Bayerische Akademie der Wissenschaften zu München. Mathematisch-physikalische Klasse. Sitzungsberichte*: 425–458. Reprinted in Sommerfeld (2013). English translation: Sommerfeld (2014a).

———. 1915b. "Die Feinstruktur der Wasserstoff- und der Wasserstoff-ähnlichen Linien." *Königlich Bayerische Akademie der Wissenschaften zu München. Mathematisch-physikalische Klasse. Sitzungsberichte*: 459–500. Reprinted in Sommerfeld (2013). English translation: Sommerfeld (2014b).

———. 1916a. "Zur Quantentheorie der Spektrallinien. Pts. I and II." *Annalen der Physik* 51: 1–94. Reprinted in Sommerfeld (1968, Vol. 3, pp. 172–265).

———. 1916b. "Zur Quantentheorie der Spektrallinien. Pt. III." *Annalen der Physik* 51: 125–167. Reprinted in Sommerfeld (1968, Vol. 3, pp. 266–308).

———. 1916c. "Zur Quantentheorie der Spektrallinien. Ergänzungen und Erweiterungen." *Königlich Bayerische Akademie der Wissenschaften zu München. Mathematisch-physikalische Klasse. Sitzungsberichte*: 131–182. Page references to reprint in Sommerfeld (1968, Vol. 3, pp. 326–377).

Sommerfeld, Arnold. 1916d. "Zur Theorie des Zeemaneffektes der Wasserstofflinien mit einem Anhang über den Starkeffekt." *Physikalische Zeitschrift* 17: 491–507. Reprinted in Sommerfeld (1968, Vol. 3, pp. 309–325).

——. 1916e. "Die Quantentheorie der Spektrallinien und die letzte Arbeit von Karl Schwarzschild." *Die Umschau* 20: 941–946. Reprinted in Sommerfeld (1968, Vol. 4, pp. 507–512).

——. 1918a. "Die Drudesche Dispersionstheorie vom Standpunkte des Bohrschen Modelles und die Konstitution von H_2, O_2, and N_2." *Annalen der Physik* 53: 497–550. Reprinted in Sommerfeld (1968, Vol. 3, pp. 378–431).

——. 1918b. "Atombau und Röntgenspektren, I Teil." *Physikalische Zeitschrift* 19: 297–307.

——. 1919. *Atombau und Spektrallinien.* 1st ed. Braunschweig: Vieweg.

——. 1920. "Allgemeine spektroskopische Gesetze, insbesondere ein magnetooptischer Zerlegungssatz." *Annalen der Physik* 63: 221–263.

——. 1921. *Atombau und Spektrallinien.* 2nd ed. Braunschweig: Vieweg.

——. 1922a. *Atombau und Spektrallinien.* 3rd ed. Braunschweig: Vieweg.

——. 1922b. "Quantentheoretische Umdeutung der Voigt'schen Theorie des anomalen Zeeman-Effektes vom D-Linientypus." *Zeitschrift für Physik* 8: 257–272. Reprinted in Sommerfeld (1968, Vol. 3, pp. 609–624).

——. 1923. *Atomic Structure and Spectral Lines.* London: Methuen. Translation (by Henry L. Brose) of Sommerfeld (1922a).

——. 1924. *Atombau und Spektrallinien.* 4th ed. Braunschweig: Vieweg.

——. 1968. *Gesammelte Schriften.* 4 Vols. Edited by Fritz Sauter. Braunschweig: Vieweg.

——. 2000. *Wissenschaftlicher Briefwechsel. Band 1: 1892–1918.* Edited by Michael Eckert and Karl Märker. Berlin, Diepholz, München: Deutsches Museum. Verlag für die Geschichte der Naturwissenschaften und der Technik.

——. 2004. *Wissenschaftlicher Briefwechsel. Band 2: 1919–1951.* Edited by Michael Eckert and Karl Märker. Berlin, Diepholz, München: Deutsches Museum. Verlag für die Geschichte der Naturwissenschaften und der Technik.

——. 2013. *Die Bohr–Sommerfeldsche Atomtheorie. Sommerfelds Erweiterung des Bohrschen Atommodells 1915/16 kommentiert von Michael Eckert.* Berlin: Springer. With an introduction by Eckert (2013a).

——. 2014a. "On the theory of the Balmer series." *The European Physical Journal H* 9: 157–177. English translation of Sommerfeld (1915a).

——. 2014b. "The Fine Structure of Hydrogen and Hydrogen-like lines." *The European Physical Journal H* 39: 179–204. English translation of Sommerfeld (1915b).

Speziali, Pierre, ed. 1979. *Albert Einstein. Michele Besso. Correspondance 1903–1955.* Paris: Hermann.

Stachel, John, ed. 2005. *Einstein's Miraculous Year. Five Papers that Changed the Face of Physics.* Princeton: Princeton University Press. Centenary Edition. Originally published in 1998.

Staley, Richard. 2005. "On the Co-Creation of Classical and Modern Physics." *Isis* 96: 530–558. This paper is incorporated into Chs. 9–10 of Staley (2008).

——. 2008. *Einstein's Generation. The Origins of the Relativity Revolution.* Chicago: University of Chicago Press. Chs. 9–10 incorporate Staley (2005).

Stark, Johannes. 1911. *Prinzipien der Atomdynamik.* Vol. 2. *Die elementare Strahlung.* Leipzig: Hirzel.

——. 1913a. "Beobachtungen über den Effekt des elektrischen Feldes auf Spektrallinien." *Königlich Preussische Akademie der Wissenschaften* (Berlin). *Sitzungsberichte*: 932–946. Reprinted (with minor corrections) as Stark (1914a).

——. 1913b. "Observation of the separation of spectral lines by an electric field." *Nature* 92: 401.

———. 1914. "Beobachtungen über den Effekt des elektrischen Feldes auf Spektrallinien. I. Quereffekt." *Annalen der Physik* 43: 965–982. Reprint (with minor corrections) of Stark (1913a).

———. 1915. "Beobachtungen über den Effekt des elektrischen Feldes auf Spektrallinien. V. Feinzerlegung der Wasserstoffserie; VI. Polarisierung und Verstärkung einer Serie." *Annalen der Physik* 48: 193–209 (V), 210–235 (VI).

Stark, Johannes, and Heinrich Kirschbaum. 1914. "Beobachtungen über den Effekt des elektrischen Feldes auf Spektrallinien. III. Abhängigkeit von der Feldstärke; IV. Linienarten, Verbreiterung." *Annalen der Physik* 43: 991–1016 (III); 1017–1047 (IV).

Stark, Johannes, and Georg Wendt. 1914. "Beobachtungen über den Effekt des elektrischen Feldes auf Spektrallinien. II. Längseffekt." *Annalen der Physik* 43: 983–990.

Stefan, Josef. 1879. "Über die Beziehungen zwischen der Wärmestrahlung und der Temperatur." *Kaiserliche Akademie der Wissenschaften* (Vienna). *Mathematisch-Naturwissenschaftliche Classe. Zweite Abteilung. Sitzungsberichte* 79: 391–428.

Stehle, Philip. 1994. *Order, Chaos, Order. The Transition from Classical to Quantum Physics*. Oxford: Oxford University Press.

Stone, A. Douglas. 2005. "Einstein's Unknown Insight and the Problem of Quantizing Chaos." *Physics Today* 58 (August): 37–43.

———. 2008. "Einstein's Unknown Contribution to Quantum Theory." In Galison, Holton, and Schweber (2008, Ch. 18, pp. 270–286).

———. 2013. *Einstein and the Quantum: The Quest of the Valiant Swabian*. Princeton: Princeton University Press.

Stuewer, Roger H. 1975. *The Compton Effect: Turning Point in Physics*. Canton, MA: Science History Publications.

———. 1983. "The Nuclear Electron Hypothesis." In *Otto Hahn and the Rise of Nuclear Physics*, edited by William R. Shea, pp. 19–67. Dordrecht: Reidel.

———. 2014. "The Experimental Challenge of Light Quanta." In Janssen and Lehner (2014, pp. 143–166).

Taylor, T. S. 1909. "On the Retardation of Alpha Rays by Metals and Gases." *Philosophical Magazine* 18: 604–619.

Ter Haar, Dirk, ed. 1967. *The Old Quantum Theory*. Oxford: Pergamon.

Thiesen, Max. 1900. "Über das Gesetz der schwarzen Strahlung." *Deutsche Physikalische Gesellschaft. Verhandlungen* 2: 116–121.

Thomson, Joseph John. 1900–1901. "Indications relatives à la constitution de la matière fournies par les recherches récentes sur le passage de l'électricité à travers les gaz." In *Rapport présentés au Congrès International de Physique réuni a Paris en 1900*. 3: 138–151. Paris.

———. 1903. "The Magnetic Properties of Systems of Corpuscles Describing Circular Orbits." *Philosophical Magazine* 6: 673–693.

———. 1904. "On the Structure of the Atom." *Philosophical Magazine* 7: 237–265.

———. 1906. "On the Number of Corpuscles in an Atom." *Philosophical Magazine* 11: 769–781.

———. 1912a. "Ionization by Moving Electrified Particles." *Philosophical Magazine* 23: 449–457.

———. 1912b. "Further Experiments on Positive Rays." *Philosophical Magazine* 24: 209–253.

Thomson, William. 1851. "On the Dynamical Theory of Heat; With Numerical Results Deduced from Mr. Joule's Equivalent of a Thermal Unit and M. Regnault's Observations on Steam." *Transactions of the Royal Society of Edinburgh* 20: 261–288.

Thorndike Greenspan, Nancy. 2005. *The End of the Certain World. The Life and Science of Max Born. The Nobel Physicist Who Ignited the Quantum Revolution*. New York: Basic Books.

Uffink, Jos. 2006. "Insuperable Difficulties: Einstein's Statistical Road to Molecular Physics." *Studies in History and Philosophy of Modern Physics* 37: 36–70.

Uhlenbeck, George Eugene, and Samuel Goudsmit. 1925. "Ersetzung der Hypothese vom unmechanischen Zwang durch eine Forderung bezüglich des inneren Verhaltens jedes einzelnen Elektrons." *Die Naturwissenschaften* 13: 953–954.

Van den Broek, Antonius Johannes. 1913. "Die Radioelemente, das periodische System und die Konstitution der Atome." *Physikalische Zeitschrift* 14: 32–41.

Van der Waerden, Bartel Leendert, ed. 1967. Sources of Quantum Mechanics. Amsterdam: North-Holland. Reprint: New York: Dover, 1968.

Van Leeuwen, Hendrika Johanna. 1919. *Vraagstukken uit de electronentheorie van het magnetisme.* PhD diss., Rijks-Universiteit Leiden.

——. 1921. "Problèmes de la théorie électronique du magnétisme." *Journal de Physique et le Radium* 2: 361–377.

Van Vleck, John H. 1922a. "The Dilemma of the Helium Atom." *Physical Review* 19: 419–420.

——. 1922b. "The Normal Helium Atom and Its Relation to the Quantum Theory." *Philosophical Magazine* 44: 842–869.

——. 1924a. "The Absorption of Radiation by Multiply Periodic Orbits, and Its Relation to the Correspondence Principle and the Rayleigh–Jeans law. Part I. Some Extensions of the Correspondence Principle." *Physical Review* 24: 330–346. Reprinted in Van der Waerden (1968, pp. 203–222).

——. 1924b. "The Absorption of Radiation by Multiply Periodic Orbits, and Its Relation to the Correspondence Principle and the Rayleigh–Jeans law. Part II. Calculation of Absorption by Multiply Periodic Orbits." *Physical Review* 24: 347–365.

——. 1926. *Quantum Principles and Line Spectra. Bulletin of the National Research Council* 10, Pt. 4. Washington, D. C.: National Research Council.

——. 1932. *The Theory of Electric and Magnetic Susceptibilities.* Oxford: Oxford University Press.

——. 1974. "Acceptance speech." In *Koninklijke Nederlandse Akademie van Wetenschappen. Bijzondere bijeenkomst der afdeling natuurkunde … 28 september 1974 … voor de plechtige uitreiking van de Lorentz- medaille aan Prof. Dr. J. H. Van Vleck.*

Voigt, Woldemar. 1901. "Über das elektrische Analogon des Zeemaneffektes." *Annalen der Physik* 4: 197–208.

——. 1907. "Betrachtungen über die komplizierteren Formen des Zeemaneffektes." *Annalen der Physik* 24: 193–224.

——. 1913a. "Die anormalen Zeemaneffekte der Spektrallinien vom D-typus." *Annalen der Physik* 42: 210–230.

——. 1913b. "Weiteres zum Ausbau der Koppelungstheorie der Zeemaneffekte." *Annalen der Physik* 41: 403–440.

Von Neumann, John. 1927. "Wahrscheinlichkeitstheoretischer Aufbau der Quantenmechanik." *Königliche Gesellschaft der Wissenschaften zu Göttingen. Mathematisch-physikalische Klasse. Nachrichten*: 245–272.

Warburg, Emil. 1913. "Bemerkung zu der Aufspaltung der Spektrallinien im elektrischen Feld." *Deutsche Physikalische Gesellschaft. Verhandlungen* 15: 1259–1266.

Weber, Heinrich Friedrich. 1872. "Die spezifische Wärme des Kohlenstoffs." *Annalen der Physik and Chemie* 147: 311–319.

——. 1875. "Die spezifischen Wärmen der Elemente Kohlenstoff, Bor und Silicium." *Annalen der Physik und Chemie* 4: 367–423, 553–582.

Wentzel, Gregor. 1923. "Zur Rydbergschen Termformel und über einen von ihr abweichenden Serientypus." *Zeitschrift für Physik* 19: 53–66.

———. 1926. "Eine Verallgemeinerung der Quantenbedingungen für die Zwecke der Wellen-mechanik." *Zeitschrift für Physik* 38: 518–529.

Wheaton, Bruce R. 1983. *The Tiger and the Shark. Empirical Roots of Wave–Particle Dualism.* Cambridge: Cambridge University Press.

Whiddington, Richard. 1911. "The Production of Characteristic Röntgen Radiation." *Proceedings of the Royal Society of London. Series A* 85: 323–323.

Wien, Wilhelm. 1893. "Eine neue Beziehung der Strahlung schwarzer Körper zum zweiten Hauptsatz der Wärmetheorie." *Königlich Preussische Akademie der Wissenschaften* (Berlin). *Sitzungsberichte*: 55–62.

———. 1894. "Temperatur und Entropie der Strahlung." *Annalen der Physik* 52: 132–165.

———. 1896. "Über die Energievertheilung im Emissionsspectrum eines schwarzen Körpers." *Annalen der Physik* 58: 662–669.

———. 1900a. "Les Lois théoriques der rayonnement." In *Rapport présentés au Congrès International de Physique réuni a Paris en 1900*. 2: 23–40. Paris: Gauthier-Villars.

———. 1900b. "Über die Möglichkeit einer elektromagnetischen Begründung der Mechanik." *Archives Néerlandais des Sciences Exactes et Naturelles* 2: 96–107. Reprinted in *Annalen der Physik* 5(1901): 501–513.

———. 1909. "Theorie der Strahlung." In *Encyklopädie der Mathematischen Wissenschaften mit Einschluss ihrer Anwendungen*. Vol. 5, Pt. 3, edited by Arnold Sommerfeld, pp. 282–357. Leipzig: Teubner.

———. 1930. *Wilhelm Wien:Aus dem Leben und Werken eines Physikers*. Edited by Karl Wien. Leipzig: Barth.

Wilson, Harold A. 1912. "The Theory of Spectral Series." *Philosophical Magazine* 23: 660–663.

Wilson, William. 1915. "The Quantum Theory of Radiation and Line Spectra." *Philosophical Magazine* 29: 795–802.

———. 1916. "The Quantum of Action." *Philosophical Magazine* 31: 156–162.

———. 1956. "John William Nicholson. 1881–1955." *Biographical Memoirs of Fellows of the Royal Society* 2: 209–214.

Yagi, Eri. 1964. "On Nagaoka's Saturnian Atomic Model." *Japanese Studies in the History of Science* 3: 29–47.

Yourgrau, Wolfgang, and Stanley Mandelstam. 1979. *Variational Principles in Dynamics and Quantum Theory*. 3rd rev. ed. New York: Dover.

Zahn, C. T., and A. A. Spees. 1938. "A Critical Analysis of the Classical Experiments on the Variation of Electron Mass." *Physical Review* 53: 511–521.

Zeeman, Pieter. 1896. "Over den invloed eener magnetisatie op den aard van het door een stof uitgezonden licht." *Koninklijke Akademie van Wetenschappen te Amsterdam. Wis- en Natuurkundige Afdeeling. Verslagen van de Gewone Vergaderingen* 5: 181–185. English translation: "On the Influence of Magnetism on the Nature of the Light Emitted by a Substance." *Philosophical Magazine* 43 (1897) 226–239.

Zermelo, Ernst. 1896a. "Über einen Satz der Dynamik und die mechanische Wärmetheorie." *Annalen der Physik* 57: 485–494.

———. 1896b. "Über mechanische Erklärungen irreversibler Vorgänge. Eine Antwort auf Hrn. Boltzmann's 'Entgegnung'." *Annalen der Physik* 59: 793–801.

Index